11th Biennial Autonomous VTOL Technical Meeting 2025

Phoenix, Arizona, USA
4-6 February 2025

ISBN: 979-8-3313-2545-9

Printed from e-media with permission by:

Curran Associates, Inc.
57 Morehouse Lane
Red Hook, NY 12571

Some format issues inherent in the e-media version may also appear in this print version.

Vertical Flight Society
2700 Prosperity Ave, Suite 275
Fairfax, VA 22031
USA

Phone: (703) 684-6777
Fax: (703) 739-9279

www.vtol.org

Additional copies of this publication are available from:

Curran Associates, Inc.
57 Morehouse Lane
Red Hook, NY 12571 USA
Phone: 845-758-0400
Fax: 845-758-2633
Email: curran@proceedings.com
Web: www.proceedings.com

TABLE OF CONTENTS

Transformative Vertical Flight 2025

11th Biennial Autonomous VTOL Technical Meeting &
12th Annual Electric VTOL Symposium
Phoenix, AZ, USA | Feb. 4–6, 2025

Vertical Flight Society Arizona Chapter

Vertical Flight Society

Thank you to our sponsors!

Diamond Sponsor

Honeywell

Platinum Sponsor

BAE SYSTEMS

Gold Sponsors

wisk

'TORAY'
Innovation by Chemistry

DASSAULT SYSTEMES

BOEING

PIASECKI AIRCRAFT CORPORATION

Unither BIOELECTRONICS

DAEDALEAN
APPLIED INTELLIGENCE

BLOOMY

Joby

SKYDRIVE

NEAR EARTH AUTONOMY

Silver Sponsors

ARIZONA
COMMERCE AUTHORITY

EvoLito
The Power of Electric Flight

PHX
Phoenix Sky Harbor
International Airport

PIVOTAL

STELL

Advertisers

GOAERO

SMG
CONSULTING

Monday, February 3, 2025: Registration Desk Hours 2:00 pm–5:00 pm (South Ballroom Foyer)	
8:00 am–5:00 pm: Short Course on Electric VTOL Design, Dr. James Wang (Room 161–162)	
Tuesday, February 4, 2025: Registration Desk Hours 7:00 am–5:00 pm (South Ballroom Foyer)	
Plenary Sessions in Room 164-166	
Welcome Keynotes	
8:00–8:20	Welcomes: **Colton Marchessault** (Boeing), VFS Arizona Chapter President, and **Angelo Collins**, VFS Executive Director
8:20–8:50	**Dave Shilliday**, VP/GM of UAM/UAS, Honeywell Aerospace Technologies (sponsor)
8:50–9:10	**Kate Gallego**, Mayor, City of Phoenix
9:10–9:30	**Chris Speights**, Chief Engineer, Vertical Lift, The Boeing Company (sponsor)
9:30–10:15 am Refreshment Break (Exhibit Hall: South Ballroom	
Government Keynotes: Chaired by Angelo Collins, VFS Executive Director	
10:15–10:40	**Derek Gowanlock**, Research Flight Test Engineer, National Research Council Canada
10:40–11:10	**Dr. Michael Patterson**, ConOps and Analysis Lead, AAM Mission Integration Office, NASA
11:10–11:40	**LTC Shawn Naigle, PhD, US Army**, Dep Asst Dir, Design, Simulation & Experimentation, US Army DEVCOM AvMC TDD
11:40–12:00	GoAERO Update: **Gwen Lighter**, CEO, GoAERO
12:00–1:30 pm Lunch (Exhibit Hall: South Ballroom)	
Electric VTOL Leaders: Chaired by Elan Head, Senior Editor, The Air Current	
1:30–1:55	**Tom Anderson**, COO, Archer Aviation
1:55–2:15	**Kristen Costello**, Regulatory Affairs Lead, Beta Technologies
2:15–2:35	**Ryan Naru**, Aviation Policy & Regulatory Affairs Lead, Joby Aviation (sponsor)
2:35–2:55	**Ken Karklin**, CEO, Pivotal (sponsor)
2:55–3:15	**Mikaël Cardinal**, VP, E-Aviation, Organ Delivery Systems, Unither Bioelectronics (sponsor)
3:15–3:45 pm Refreshment Break (Exhibit Hall: South Ballroom)	
Autonomous VTOL Leaders: Chaired by Ajay Sehgal, Senior Technical Fellow, KBR	
3:45–4:05	**Dr. Yemaya Bordain**, CCO & President of Americas, Daedalean AI (sponsor)
4:05–4:25	**Lyle Chamberlain**, CTO, Near Earth Autonomy (sponsor)
4:25–4:45	**Erick Corona**, Director of Airspace Operational Integration, Wisk Aero (sponsor)
4:45–4:55	Wrap-up: **Ariel Louie**, Co-Technical Director, VFS Arizona Chapter
5:00–6:30 pm Reception (Exhibit Hall: South Ballroom)	

Visit our exhibitors in the South Ballroom

Thank you to the VFS Arizona Chapter for their essential support of this event:

- *Technical Chairs: **Ariel Louie** and **Brett Sjostrom** (Boeing)*
- *Chapter President: **Colton Marchesseault** (Boeing)*

Vertical Flight Society Arizona Chapter www.vfsaz.org

Wednesday, February 5, 2025: Registration Desk Hours 7:00 am–5:00pm (South Ballroom Foyer)

Autonomous VTOL Technical Meeting (Room 161–162)

Time	Session
8:00–9:30	**Titans of Autonomy: *Invited Presentations*** *Session Chair: Colton Marchesseault, The Boeing Company*
8:00–8:30	**Collision-Resilient Drones – High-Fidelity Simulations and Precision Control Studies:** Prof. Wenlong Zhang, Arizona State University
8:30–9:00	**Flight Dynamics, Control and Autonomy for Army Vertical Lift Systems:** Dr. Jeffery Lusardi, US Army
9:00–9:30	**Concept of Operations for Airspace Integration of AAM**: Nate Isbell, SkyGrid

Electric VTOL Symposium (Rooms 164–166)

Time	Session
8:00–9:30	**Session 1. eVTOL Progress** *Chair: Ben Goldstein, AAM Report, Aviation Week*
8:00–8:20	Kenji Nakagawa, Dir, AirMobility Business Development, SkyDrive *
8:30–8:50	Tiago Gomes, Head of Aeronautics, Eve Air Mobility
8:50–9:10	Dr. Simon Briceno, CCO, Jaunt Air Mobility
9:10–9:30	Pulkit Agrawal, Lead of Regulatory Affairs UAS/UAM, and Dan Newman, CTO, Honeywell Aerospace *

9:30–10:00 am Refreshment Break (Exhibit Hall: South Ballroom)

Time	Autonomous VTOL Technical Meeting
10:00-12:00	**Vehicle Design I** *Session Chair: Prof. Wenlong Zhang, Arizona State Univ.*
10:00–10:30	**Enhancing Yaw Control Authority and Resilience for eVTOL Aircraft through Rotor Canting Configuration:** Prof. James Wang, Nanyang Technological Univ. Singapore
10:30–11:00	**Application of an Inverse Airfoil Design for Improved Proprotor Performance:** Elvin Puchuela, Georgia Institute of Technology
11:00–11:30	**Building Enhanced Resilience into Aviation**: Larry Young, NASA Ames
11:30-12:00	

Time	Electric VTOL Symposium
10:00-12:00	**Session 2-3. AAM Perspectives** *Chair: Tom Perkowski, Founding Principal, Apex View*
10:00–10:20	Tom Perkowski, Founding Principal, Apex View
10:20–10:40	Willi Tacke, CEO, eFlight Journal
10:40–11:00	Sergio Cecutta, Partner, SMG Consulting
11:00–11:20	Elan Head, Senior Editor, The Air Current
11:20-11:40	James Wang, Director of eVTOL Research & Innovation Centre, Nanyang Technical University, Singapore
11:40-12:00	Discussion

12:00–1:00 pm Lunch (Exhibit Hall: South Ballroom)

Time	Autonomous VTOL Technical Meeting
1:00–3:00	**Autonomy and AI** *Session Chair: Gary Klein, The Boeing Company*
1:00–1:30	**VTOL UAS Auto-recovery Using A Tested Long-Term Motion Prediction Method to Define the Deck Environment (*Invited Presentation*)**: Prof. Stefano Brizzolara (VA Tech/Fincantieri Marinette Marine)
1:30–2:00	**Minimum Waypoint Spacing for VTOL Rotorcraft Trajectory Guidance using B-Spline Midpoint Refinement:** Bryan Chu, Arizona State University
2:00–2:30	**Application of Machine Learning to Maximize Power Density of Propulsion Motors for VTOL Applications:** Tadashi Sawata, Collins Aerospace
2:30–3:00	**Canadian Vertical Lift Autonomy Demonstration — High Level Concepts for the Development of Autonomy (*Invited Presentation*)**: Kris Ellis, National Research Council Canada

Time	Electric VTOL Symposium
1:00–2:00	**Session 4. Hybrid-Electric VTOL** *Chair: Peter Schmidt, COO, Transcend Air*
1:00-1:20	Eric Bartsch, CEO, VerdeGo Aero
1:20-1:40	Ray McFall, Emerging & Disruptive Capabilities Lead, Pratt & Whitney
1:40-2:00	Darek Zatorski, Principal Engineer, GE Aerospace
2:00–3:00	**Session 5. Hydrogen-Electric VTOL** *Chair: Mike Hirschberg, VFS Director of Strategy*
2:00-2:20	John Scott, PM Hydrogen Propulsion Technology, Piasecki Aircraft Corp. *
2:20-2:40	Gad Shaanan, CEO, Unmanned Aerospace
2:40-3:00	Ian Randall, CEO, MagLev Aero

3:00–3:30 pm Refreshment Break (Exhibit Hall: South Ballroom)

Time	Autonomous VTOL Technical Meeting
3:30–5:00	**Simulation & Testing I** *Session Chair: Dr. Kendra Befort, The Boeing Company*
3:30–4:00	**Exploring Transition Profiles for the RAVEN Multi-Tiltrotor Aircraft:** Tristan Huang, Georgia Institute of Technology
4:00–4:30	**Aerodynamic Design and Analysis of An Autonomous eVTOL Using Computational Fluid Dynamics:** Prof. Chunhua Sheng, University of Toledo
4:30–5:00	**Airfoil Table Generation Comparison Utilizing XFOIL and UNS2D Flow Solvers:** Gianmarco Sahragard-Monfared, NASA Ames
5:00–5:30	**Sensitivity Analysis and Aerodynamic Modeling of eVTOL Low-speed Predicted Handling Qualities:** Prof. Giuseppe Quaranta, Politecnico di Milano

Time	Electric VTOL Symposium
3:30–4:15	**Session 6. Autonomous VTOL Panel** *Chair: Elan Head, Senior Editor, The Air Current*
	• Dr. Yemaya Bordain, CCO & President of Americas, Daedalean AI * • Dr. Hector Xu, CEO, Rotor Technologies • Howard Kass, VP, Government and Legal Affairs, Skyryse
4:15-5:15	**Session 7. Arizona AAM Stakeholders Panel** *Chair: Taylor Alberstadt, Honeywell **
	• Angela Aspito, Managing Dir, Southwest Mission Acceleration Center • Sarah J. Demory, COO/Assistant Aviation Director, City of Phoenix * • Prof. Tony Waas, Director, School for Engineering of Matter, Transport and Energy, Arizona State University • Marisa Walker, Executive VP, Infrastructure, AZ Commerce Authority

Thursday, February 6, 2025: Registration Desk Hours 7:00 am–2:00 pm (South Ballroom Foyer)

Autonomous VTOL Technical Meeting (Rooms 161–162)

8:00–9:30	**Vehicle Design II** *Session Chair: Ariel Louie, VFS AZ Co-Technical Director*
8:00–8:30	**A Future for Rotorcraft that Engages with New Application Domains:** Larry Young, NASA Ames
8:30–9:00	**Novel Control of a Propeller Driven Rotor as a Swashplateless System:** Robert Brown, US Navy
9:00–9:30	**Improvements to the Da Vinci Aerial Screw as a VTOL Thrusting Alternative:** Ron Berlin, University of Maryland

Electric VTOL Symposium (Rooms 164–166)

8:00–9:30	**Session 8. Manufacturing & Development** *Chair: Mike Hirschberg, VFS Director of Strategy*
8:00-8:25	Matt Farides, VP Business Development, Evolito *
8:25-8:50	DeWayne Howell, Application Development and Expert Service Field Mgr, Toray Advanced Composites *
8:50-9:10	Dr. Behnam Afsharpoya, Sr Industry Process Consultant, Dassault Systèmes *
9:10-9:30	Peter Blume, CEO, Bloomy Controls *

9:30–10:00 am Refreshment Break (Exhibit Hall: South Ballroom)

10:00 12:00	**Electric Propulsion** *Session Chair: Colton Marchesseault, The Boeing Company*
10:00–10:30	**Considerations for both Silicon and Wide-Bandgap Semiconductors in Electric Aircraft:** Bill Dillard, Microchip
10:30–11:00	**Assessment of Retrofitting an Existing Helicopter with Electric Propulsion Technologies:** Dounya Jarrad, Georgia Institute of Technology
11:00–11:30	**Testing Battery Sensitivities of Electric Aircraft Systems Using a Battery Surrogate:** Grant Gothing, Bloomy Controls, Inc.
11:30-12:00	**Silicon Batteries with High Energy and Power Density for Electric Mobility:** Ionel Stefan, Amprius Technologies, Inc.

10:00-11:00	**Session 9-10. Regulatory Pathways for AAM** *Chair: David Dunning, GAMA*
10:00-10:05	David Dunning, Director of Global Innovation & Policy, General Aviation Manufacturers Association
10:05-10:30	Kristen Costello, Regulatory Affairs Lead, Beta Technologies
10:30-10:55	Ryan Naru, Aviation Policy & Regulatory Affairs Lead, Joby Aviation *
10:55-11:20	Tom Gunnarson, Lead of Regulatory Affairs, Wisk Aero*
11:20-11:45	Robert Bassey, Airports Design & Construction Engineer, Office of Airports (AAS-110), FAA
11:45-12:00	Recap: David Dunning, GAMA, and Mike Hirschberg, Vertical Flight Society

12:00–1:00 pm Lunch (Exhibit Hall: South Ballroom)

1:00–3:00	**Simulation & Testing II** *Session Chair: Dr. Kendra Befort, The Boeing Company*
1:00–1:30	**CFD Analysis of Quadrotor Biplane Tailsitter Hover-to-Cruise Transition:** Paulo Arias, University of Maryland
1:30–2:00	**Aeroelastic Simulation of Complex VTOL Configurations:** Connor Schultz and Dr. Thomas Hays, Answer Engineering
2:00–2:30	**Emerging CFD Simulation Technologies for eVTOL Aerodynamics, Fluid Management, and Powertrain Optimization:** Jun Li, AVL Mobility Technologies, Inc.
2:30–3:00	**Agricultural Application using Autonomous UAS:** Daniel Wachspress, Continuum Dynamics Inc.

1:00–2:00	**Session 11. Commercial Air Transport eVTOL** *Chair: Andy Mearns, Multicopter Aerospace Consulting*
1:00-1:20	Furqan, CEO, Sora Aviation
1:20-1:40	Peter Schmidt, COO, Transcend Air
1:40-2:00	Andy Mearns, Multicopter Aerospace Consulting

2:00–3:00	**Session 12. Logistics eVTOL Development Panel** *Chair: Jonny Gilchrist, FalconWorks, BAE Systems*

- Jonny Gilchrist, Head of US Business Development, FalconWorks, BAE Systems *
- Mark Butkiewicz, VP, Applied Technology Sector, SURVICE Engineering
- James O'Neill, Deputy CTO, Malloy Aeronautics

3:00–3:30 pm Refreshment Break (Foyer: Room 161–166)

3:30–5:00	**Enabling Systems & Human SI** *Session Chair: Ariel Louie, VFS AZ Co-Technical Director*
3:30–4:00	**The Potential Role of Digital Twins for UAS/Ship Interface Artificial Intelligence (AI) Systems:** Prof. Stefano Brizzolara, VA Tech/Fincantieri Marinette Marine
4:00–4:30	**Real-Time Control Power Margin and Gust Estimation for Distributed Electric Propulsion (DEP) Aircraft:** Robert McKillip, Continuum Dynamics Inc.
4:30–5:00	**Novel Human Factors Challenges, Unexpected Events, and Safety Hazards Associated with Operation of Battery-powered eVTOL Aircraft:** Kendall Carmody, Florida Institute of Technology

3:00–5:00	**Session 13. Cargo & Military Applications** *Chair: Dan Newman, CTO, Honeywell AAM*
3:30-3:50	Lauren Wolfe, SPRINT Performance Lead, Aurora Flight Sciences
3:50-4:10	Andrew Clare, CEO, Elroy Air
4:10-4:30	James Dorris, CEO, Odys Aviation
4:30-4:50	Discussion
4:50-5:00	Wrap-Up: Mike Hirschberg

* = sponsor

Enhancing Yaw Control Authority and Resilience for eVTOL Aircraft through Rotor Canting Configuration

Archit Krishna Kamath	Thanaraj T.	Xiaoqiang Lu	James Wang	Mir Feroskhan
Research Fellow	Research Fellow	Research Fellow	Professor	Assistant Professor

School of Mechanical and Aerospace Engineering, Nanyang Technological University, Singapore

ABSTRACT

Maximizing yaw control authority is a critical challenge in the design of electric vertical takeoff and landing (eVTOL) aircraft, as it directly affects stability and maneuverability during hover, transition, and cruise. With distributed electric propulsion at the core of modern eVTOL designs, innovative strategies such as rotor canting offer new avenues to enhance yaw control authority. This study investigates how lateral and longitudinal rotor cant angles can be leveraged to improve yaw control authority in an eight-rotor, lift-plus-cruise eVTOL configuration. By systematically comparing multiple rotor spin configurations and performing sensitivity analyses with incremental cant angles, we establish the relationship between cant geometry and yaw control authority. Our results suggest that orthogonal canting yields superior yaw control authority by effectively maximizing moment generation, thereby enhancing eVTOL resilience in challenging flight conditions.

INTRODUCTION

Electric Vertical Take-off and Landing (eVTOL) aircraft have garnered significant attention in recent years as a promising solution for on-demand urban air mobility (UAM). By leveraging distributed electric propulsion systems, these vehicles offer efficient, quiet, and sustainable transportation for passengers, cargo, and medical evacuation, effectively bypassing ground congestion in growing metropolitan areas (Ref. 1). Beyond improving accessibility, eVTOL platforms are poised to reduce emissions and operational costs compared to conventional helicopters or small fixed-wing aircraft, fueling rapid advancements and diverse design approaches (Ref. 2).

Among the many eVTOL concepts under development, several broad categories have emerged, including tilt-rotor, tilt-wing, lift-plus-cruise, and vectored-thrust configurations (Ref. 3). Each arrangement addresses the key challenges of vertical take-off and landing, forward cruise efficiency, and robust flight control across the entire flight envelope. For instance, tilt-rotor and tilt-wing eVTOLs permit dynamic reorientation of rotors, enabling efficient hover and cruise with the same propulsors. By contrast, lift-plus-cruise concepts employ separate propulsors for hover and forward-flight regimes independently (Ref. 4). Across these variations, yaw control remains pivotal for ensuring flight stability and maneuverability, especially as aircraft become larger, carry heavier payloads, or encounter gusty atmospheric conditions.

Beyond aerodynamic performance, resilient design has also become paramount in eVTOL development, particularly because of the potential severity of system faults such as motor failures. Historical incidents, including the 1970s New York Airways helicopter crash (Ref. 5), highlight the need for dependable urban air transport. More recently, a well-known eVTOL prototype accident further underscored concerns about eVTOL resilience (Ref. 6). Current regulations require that aircraft remain controllable even with one motor inoperative (Ref. 7), while the probability of a single failure leading to a catastrophic event must be kept extremely low (Ref. 8). This underscores the importance of robust fault-tolerant strategies, which can be implemented through passive approaches, maintaining functionality under component failures (Refs. 9, 10), or active strategies that diagnose and reconfigure controls during flight (Refs. 11–13). At the design stage, hardware redundancies improve reliability but add weight, reducing endurance. To address this trade-off, researchers have explored motor position optimization (Refs. 14, 15) and hoverability analyses (Refs. 16, 17) to ensure stable control after a failure. In tilt-enabled designs, the redistribution of loads following a motor failure becomes especially crucial for maintaining roll, pitch, and yaw authority.

In conventional multi-rotor eVTOLs, yaw authority often relies on counter-rotating propellers. To initiate yaw to the right, for example, thrust on counterclockwise (CCW) propellers is increased and thrust on clockwise (CW) propellers is decreased, generating a net yaw moment.

In contrast, vectored-thrust eVTOLs, such as tilt-rotor or tilt-wing designs, can adjust rotor orientation to produce horizontal thrust components for yaw control. However, in many lift-plus-cruise concepts, rotors remain in a fixed orientation, limiting the ability to tilt for additional yaw authority. In such cases, canting the motors—angling them laterally or longitudinally relative to the vertical axis—adds supplemental thrust vectors that can significantly enhance yaw control without relying exclusively on differential rotor speeds.

Several eVTOL prototypes already leverage motor canting.

Presented at the Vertical Flight Society's 11th Biennial Autonomous Technical Meeting, Phoenix, AZ, USA, Feb. 4–6, 2025.

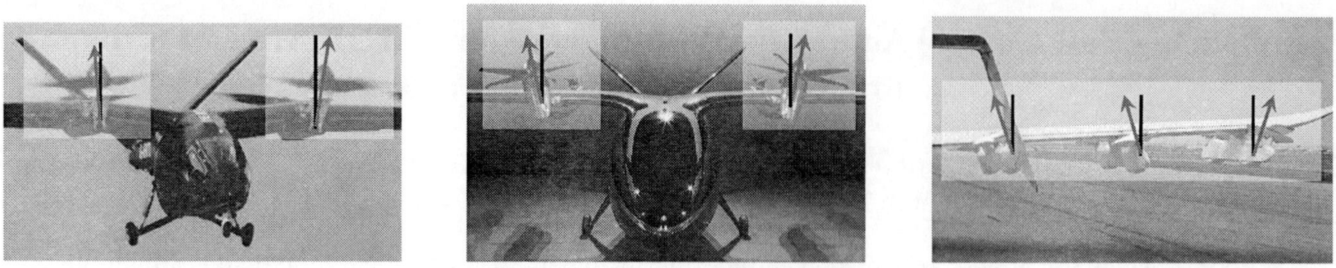

(a) Vertical Aerospace (b) Archer Aviation (c) Wisk Cora

Figure 1: Canted angles for motors on existing eVTOL designs for (a) Vertical Aerospace, (b) Archer Aviation and (c) Wisk Cora. The black line depicts vertical while red arrow depicts canted direction.

For example, Figure 1a shows a Vertical Aerospace demonstrator (an eight-rotor lift-plus-cruise design) with rotors slightly canted outward relative to the fuselage, generating horizontal thrust components that aid yaw control. Similarly, Archer Aviation's Midnight eVTOL, Figure 1b, a twelve-rotor lift-plus-cruise design,employs uniform outward canting, while Wisk Cora Figure 1c, another twelve-rotor design, uses a non-uniform canting strategy featuring inward-canted inner motors and outward-canted middle and outer motors.

Recent research has also examined the influence of rotor canting on yaw performance. One study investigated differential torsional canting in a two-passenger quadrotor, reporting improved control sensitivity about the yaw axis as confirmed by handling qualities assessments (Ref. 18). Another study explored torsional and flapwise canting on a quadrotor and its impact on control authority (Ref. 19).

These examples underscore the wide range of canting solutions employed in the industry, reflecting various performance considerations yet sharing the goal of achieving responsive yaw control. Recognizing this diversity, the present work focuses on systematically analyzing how different motor canting configurations influence yaw control authority. Specifically, the study first isolates lateral canting to assess its direct impact on yaw authority, then examines longitudinal canting individually, and finally explores dual-axis canting to determine how lateral and longitudinal angles interact. By establishing the relationship between cant geometry and yaw performance, this study provides insights for improving yaw control, complementing ongoing eVTOL resilience efforts.

An eight-motor planar eVTOL serves as the baseline for this investigation, offering a representative platform for quantifying how cant angles, spin directions, and thrust allocations influence yaw dynamics. By doing so, this study bridges the gap between conventional yaw control approaches and emerging requirements for enhanced eVTOL resilience.

MATHEMATICAL MODELING

Consider an eVTOL aircraft with n lift motors R_i for $i = 1,2,\ldots,n$, each located at coordinates (x_i, y_i) in a body-fixed system where the x-axis is forward, the y-axis is to the right, and the z-axis is downward (NED convention). The center of gravity (CG) is placed at the origin. Figure 2 illustrates an example with eight lift motors in this coordinate system. To maintain a steady hover, the net forces and net moments, collectively represented by the wrench vector $\mathbf{F} = [T, u_\phi, u_\theta, u_\psi]$, must all vanish, equivalently, the norm of $\mathbf{F}$ should approach zero. Here, T represents the total thrust generated by all lift motors, u_ϕ represents the rolling moment about the x-axis, u_θ represents the pitching moment about the y-axis, and u_ψ represents the yawing moment about the z-axis.

Each lift motor may be given two cant angles: α_i, which tilts the rotor forward about the y-axis, where $\alpha_i \in [\alpha_{\min}, \alpha_{\max}]$, and β_i, which tilts it to the right about the x-axis, where $\beta_i \in [\beta_{\min}, \beta_{\max}]$. In the absence of tilts, the thrust vector of each rotor is aligned with the positive z-axis (downward). A positive α_i introduces a forward component to the thrust, while a positive β_i introduces a rightward component. Because these horizontal components contribute to yaw authority, the new wrench vector can be expressed as $\mathbf{F} = \left[T,\, u_\phi,\, u_\theta,\, u_{\psi_d} + u_{\psi_x} + u_{\psi_y}\right]$, where u_{ψ_x} and u_{ψ_y} represent

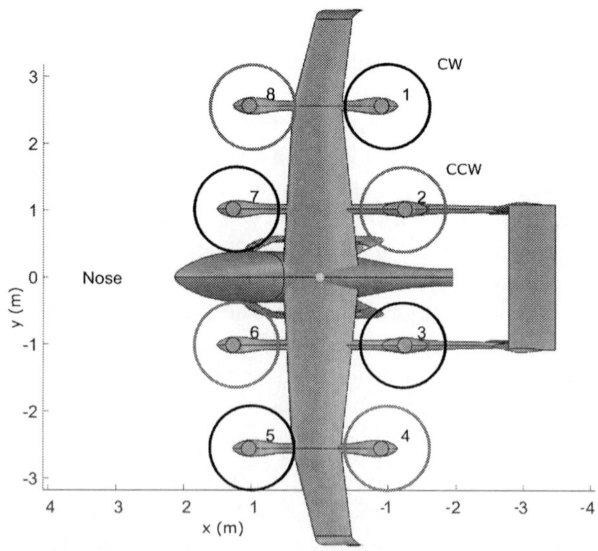

Figure 2: Eight rotor lift-plus-cruise eVTOL configuration. Clockwise and counterclockwise spinning motors are denoted by black and red circles, respectively. Body frame of eVTOL is defined using North-East-Down convention.

the additional yaw-producing components in the x–y plane.

Figure 3 illustrates how the rotor thrust T_i (initially along z_1) is rotated about the y_1-axis by α_i and also about the x_2-axis by β_i. Denoting the intermediate thrust vectors by $T_i^{(1)}$, $T_i^{(2)}$, and $T_i^{(3)}$ (all having magnitude T_i), the resulting rotation matrix applied to the original thrust vector $[0,\ 0,\ T_i]^\mathsf{T}$ yields:

$$\begin{bmatrix} T_{i_x} \\ T_{i_y} \\ T_{i_z} \end{bmatrix} = \begin{bmatrix} \cos\alpha_i & -\sin\alpha_i\sin\beta_i & \sin\alpha_i\cos\beta_i \\ 0 & \cos\beta_i & \sin\beta_i \\ -\sin\alpha_i & -\cos\alpha_i\sin\beta_i & \cos\alpha_i\cos\beta_i \end{bmatrix} \begin{bmatrix} 0 \\ 0 \\ T_i \end{bmatrix}$$
$$= \begin{bmatrix} T_i\sin\alpha_i\cos\beta_i \\ T_i\sin\beta_i \\ T_i\cos\alpha_i\cos\beta_i \end{bmatrix}, \tag{1}$$

where the original thrust along z is redistributed into three components $T_{i_x}, T_{i_y}, T_{i_z}$. The generalized thrust and moment balance equations for n dual-axis canted rotors can then be summarized as:

$$T = \sum_{i=1}^{n} T_i\cos\alpha_i\cos\beta_i - mg = 0, \tag{2}$$

$$u_\phi = \sum_{i=1}^{n} T_i\cos\alpha_i\cos\beta_i|y_i|\,\mathrm{sign}(y_i) = 0, \tag{3}$$

$$u_\theta = \sum_{i=1}^{n} T_i\cos\alpha_i\cos\beta_i|x_i|\,\mathrm{sign}(x_i) = 0, \tag{4}$$

$$u_\psi = \sum_{i=1}^{n}\Big[(-1)^i K_i T_i\cos\alpha_i\cos\beta_i + T_i\sin\alpha_i\cos\beta_i|y_i|\,\mathrm{sign}(y_i) + T_i\sin\beta_i|x_i|\,\mathrm{sign}(x_i)\Big] = 0. \tag{5}$$

where T_i is the thrust magnitude of rotor i, K_i is the reaction torque constant that distinguishes clockwise and counterclockwise spinning rotors, embedded within the alternating term $(-1)^i$. The function $\mathrm{sign}(\cdot)$ accounts for the geometric placement of each rotor relative to the center of gravity (CG). As each motor tilts forward or sideways, its thrust vector contributes differently to the overall roll, pitch, and yaw moments. By solving these equations, one can identify the thrust allocations and rotor tilt angles required to maintain a stable hover condition while exploiting the additional yaw control authority from rotor canting.

METHODOLOGY

To evaluate yaw control authority in an eVTOL aircraft, this study employs a comprehensive framework to analyze rotor thrusts, cant angles, and their contributions to yaw moment generation, as presented in Figure 4. The yaw components are systematically calculated for lateral, longitudinal, and combined canting configurations, providing insights into their respective effects on yaw dynamics.

Input Parameters: The analysis begins by defining motor thrusts $[T_1, T_2, \ldots, T_n]$, the center of gravity (CG) location at $(0,0)$, motor positions (x_i, y_i), and thrust constants $K = \mathrm{diag}(k_1, k_2, \ldots, k_n)$. Rotor cant angles are represented as α and β, where the former is lateral cant about the y-axis and the latter is longitudinal cant about the x-axis.

Yaw Moment Calculation: The total net yaw moment u_ψ arises due to three primary components : (1) the *reaction torque* $u_{\psi d}$, (2) the *lateral cant* $u_{\psi x}$, and (3) the *longitudinal cant* $u_{\psi y}$. Their respective expressions are as such:

$$u_{\psi d} = \sum_{i=1}^{n}(-1)^i K_i T_i\cos\alpha_i\cos\beta_i, \tag{6a}$$

$$u_{\psi x} = \sum_{i=1}^{n}\big(T_i\sin\alpha_i\cos\beta_i\,|y_i|\,\mathrm{sign}(y_i)\big), \tag{6b}$$

$$u_{\psi y} = \sum_{i=1}^{n}\big(T_i\sin\beta_i\,|x_i|\,\mathrm{sign}(x_i)\big). \tag{6c}$$

In Equation (6a), $u_{\psi d}$ captures the yaw moment generated by the rotors' spin directions, where $(-1)^i$ encodes whether rotor i is clockwise or counterclockwise, and K_i represents the rotor's thrust-to-torque constant. In Equation (6b) $u_{\psi x}$ is computed from each rotor's sideward cant angle β_i and its offset along x_i. In Equation (6c), $u_{\psi y}$ is computed using each rotor's longitudinal cant angle α_i and its offset along y_i. By summing these three components, the total net yaw moment follows as: $u_\psi = u_{\psi d} + u_{\psi x} + u_{\psi y}$.

Constraints: The analysis adheres to physical feasibility constraints, ensuring thrust balance and non-negativity. The total thrust generated by the motors must counteract the gravitational force, as represented in Equation (5).

Canting Configurations: For the computation of yaw moment, three canting configurations are considered. In the case of lateral canting only, the yaw moment consists of the reaction moment ($u_{\psi d}$), which arises from rotor thrust imbalance, and the yaw moment due to lateral cant ($u_{\psi x}$), resulting in a total yaw moment of $u_\psi = u_{\psi d} + u_{\psi x}$. For longitudinal canting only, the yaw moment is composed of the reaction moment ($u_{\psi d}$) and the yaw moment due to longitudinal cant ($u_{\psi y}$), giving $u_\psi = u_{\psi d} + u_{\psi y}$. Finally, for the combined canting configuration, all contributions are included: the reaction moment ($u_{\psi d}$), the yaw moment due to lateral cant ($u_{\psi x}$), and the yaw moment due to longitudinal cant ($u_{\psi y}$), resulting in a total yaw moment of $u_\psi = u_{\psi d} + u_{\psi x} + u_{\psi y}$.

Rotor Spin Convention

Two main conventions are introduced in this study to define the direction of each rotor's spins. These conventions standardize the analysis and provide an intuitive way to visualize how the rotor spin contributes to yaw control authority.

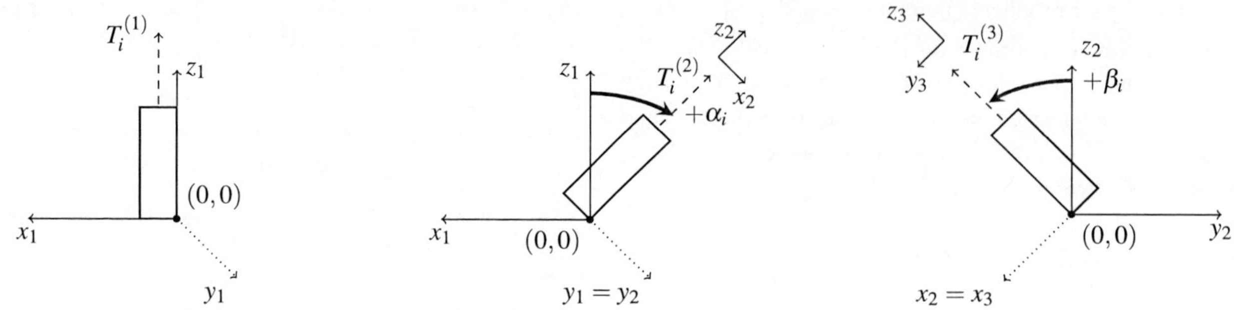

(a) No cant (initial thrust) (b) Cant about the y-axis (c) Cant about the x-axis

Figure 3: Progression from (a) no cant to (b) single-axis cant and (c) dual-axis cant. In (a), the rotor thrust $T_i^{(1)}$ is aligned with z_1. In (b), the rotor is tilted forward by α_i around the y_1-axis, yielding new axes (x_2, z_2) and a reoriented thrust $T_i^{(2)}$. In (c), the rotor is further tilted by β_i around the x_2-axis, resulting in a second new frame (x_3, y_3, z_3) and a final thrust vector $T_i^{(3)}$.

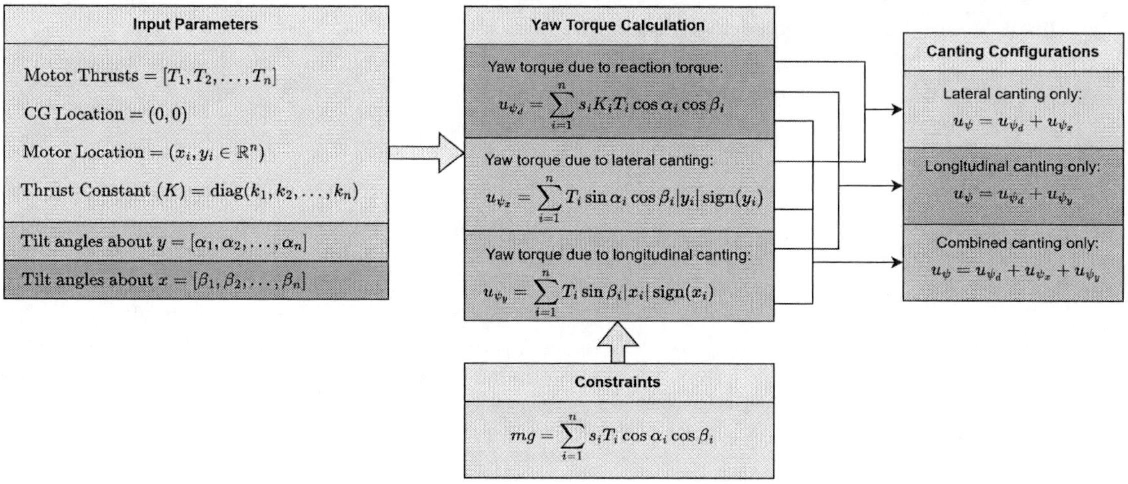

Figure 4: Flowchart illustrating the yaw moment calculation framework for lateral, longitudinal, and combined canting configurations with input parameters and constraints.

Rotor Spin Alternating Pattern : This pattern governs the sequence in which adjacent rotors spin clockwise (CW) or counterclockwise (CCW), are denoted red and blue, respectively. In Type 1, the spin direction changes after every rotor, yielding a strictly alternating pattern (CW, CCW, CW, CCW, etc.), as shown in Figure 5a. In Type 2, the direction changes after every two rotors, resulting in two consecutive CW rotors followed by two consecutive CCW rotors (Figure 5b).

Front Rotor Spin Direction This convention sets whether the front pair of rotors spin inward (IN) or outward (OUT). Under IN, shown in Figure 5c, the front rotors spin inward, directing airflow toward the aircraft's centerline. Under OUT, shown in Figure 5d, the front rotors spin outward, pushing airflow away from the fuselage centerline.

Yaw Moment Computation Approach

The yaw moment computation begins by defining the eVTOL's key physical parameters, its mass, gravitational acceleration, hover thrust requirement, and each rotor's maximum thrust and thrust-to-reaction torque characteristics. The rotors are positioned so that the eVTOL's nose aligns with the x-axis in the body-fixed frame, as shown in Figure 2, and each rotor is assigned a spin direction according to the chosen configuration. A set of cant angles α_i, β_i is also specified for evaluation.

A critical component of this framework is the thrust allocation mechanism, which distributes the total vertical thrust required for hover among the rotors while ensuring that no rotor exceeds its physical (motor) constraints. To achieve a positive yaw moment (rightward turn), priority is given to CCW rotors for providing vertical thrust. If the CCW rotors alone are insufficient to meet the hover thrust requirement, their thrust is set to the allowable maximum, and the remaining thrust is distributed among the CW rotors. This allocation strategy ensures that the total hover thrust is achieved while maintaining rotor operations within their capacity limits.

Once thrusts are allocated, the model computes each rotor's *reaction torque* (accounting for spin direction, thrust level) as well as *horizontal thrust components* introduced by the rotor cant. Summing these contributions yields the net yaw moment about the z-axis, with rightward yaw direction being positive. By systematically varying the rotor spin configurations and cant angles, this methodology offers comprehensive insight into how best to leverage cant angles and cant direction for robust yaw control in an eVTOL.

RESULTS AND DISCUSSION

The results are presented to illustrate how rotor cants affect yaw-moment generation. The investigation first focuses on lateral canting, where rotors are tilted sideways and outward. This is followed by an analysis of longitudinal canting, where rotors are tilted forward or backward relative to the aircraft's centerline. Lastly, the study evaluates combined lateral and longitudinal cants to determine how their interaction can maximize the overall yaw moment.

Lateral Cant

In the lateral canting study, each rotor was tilted sideways/outwards over a range of cant angles β, incrementally increased from 0 to 45 degrees. These cants modify each rotor's thrust vector, thereby generating yaw moments from both reaction torque (due to spin direction) and the lateral-cant contribution. Figure 6 depicts how the rotor cants alter the thrust directions in both front and top views. As illustrated in Figure 7, the total yaw moment in this scenario combines two major components: the reaction-based yaw moment and the yaw moment from lateral cant, while the yaw moment from longitudinal cant is zero.

Figure 8 expands on these findings by comparing different rotor spin patterns (Type 1 and Type 2) and front-rotor orientations (IN and OUT). Each configuration uniquely affects the balance between reaction-based yaw moment and the lateral-cant contribution. Spin arrangements that align these two effects tend to produce stronger positive yaw, whereas those with opposing interactions reduce the net yaw. Consequently, careful selection of rotor spin conventions is essential to maximizing the combined yaw moment for effective yaw control.

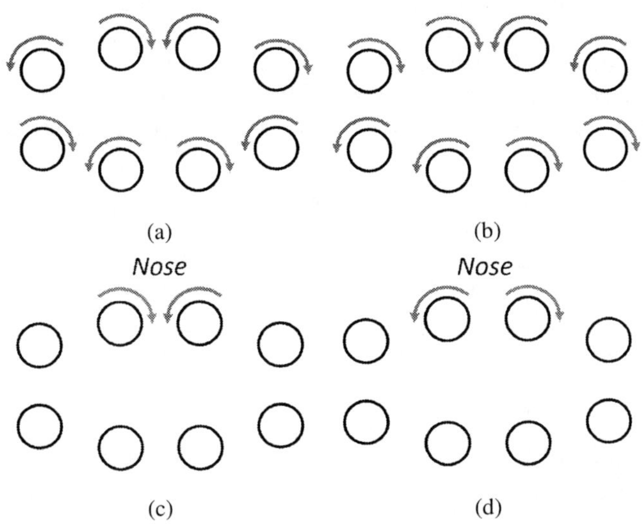

Figure 5: Rotor Spin Conventions: (a) Type 1 alternates the spin direction every one rotor while, (b) Type 2 alternates every two rotors, (c) [IN] configuration with the front rotors spinning inward relative to the nose, and (d) [OUT] configuration with the front rotors spinning outward.

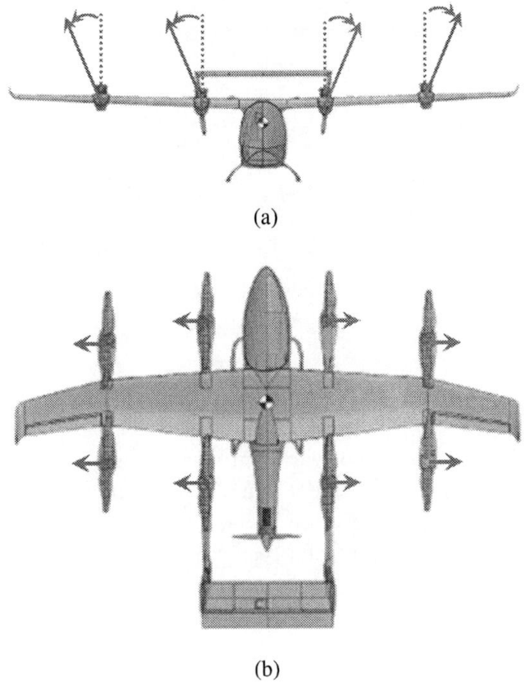

Figure 6: Thrust vector from the laterally canted motors illustrated in (a) front view and (b) top view.

The Type 2 configuration generally produces a larger total yaw moment than Type 1, as seen in Figure 8. In Type 2, grouping pairs of same-spin rotors effectively turns the eVTOL into a "giant quadrotor" with four rotor arms, each composed of two same-spin rotors. This arrangement better aligns each rotor's contribution, yielding a stronger yaw overall. By contrast, Type 1 uses a strictly alternating spin pattern, so certain CCW rotors may oppose the rightward yaw moment generated by other CCW rotors in adjacent quadrants, reducing the net yaw.

Under the OUT spin convention, the net rightward yaw moment can become negative. Initially, the reaction-based yaw moment tends to rotate the aircraft to the right. However, the lateral-cant contribution exerts an opposing leftward yaw moment. As the cant angle grows, this opposing leftward yaw moment overwhelms the moment generated from reaction torque and ultimately reverses the direction of the net yaw moment. Consequently, the OUT configuration becomes less favorable for yaw control at higher cant angles.

Interestingly, the same yaw moment magnitude observed in Type 2 can be achieved with Type 1 by opposing the lateral cant angle, demonstrating the flexibility of the Type 1 configuration. The illustration of the canting can be seen in Figure 9. Here, the plot in Figure 10, it can be seen that the right yaw

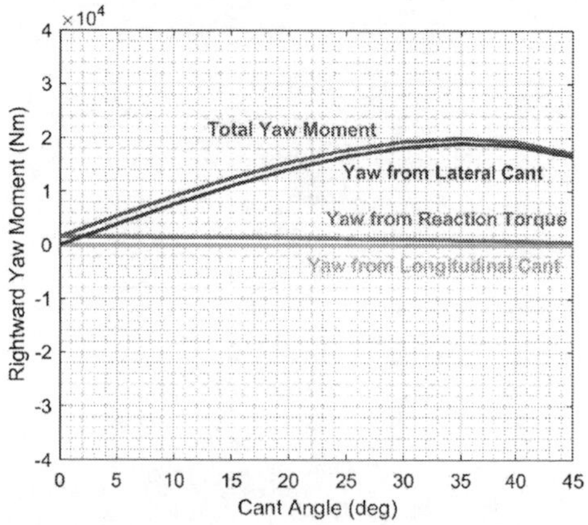

Figure 7: Contribution of yaw components for the lateral cant scenario. Total yaw moement (blue) is the sum of torque from lateral cant (purple), reaction torque (orange), and negligible torque from longitudinal cant (yellow).

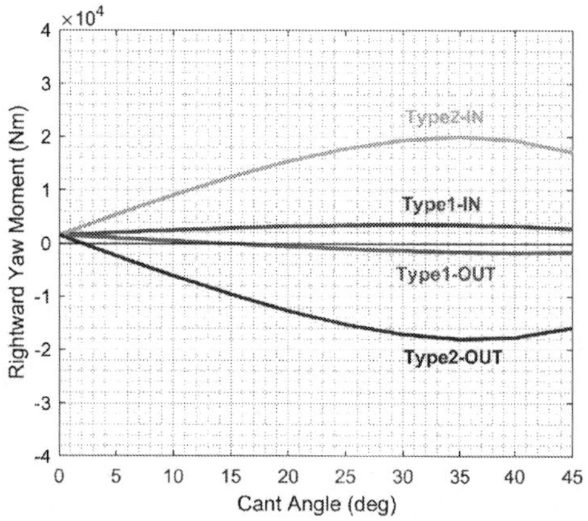

Figure 8: Right yaw moment for different lateral cant angles in Type 1 and Type 2 spin alternating patterns with IN and OUT spin configurations.

moment generated by opposing the lateral cant matches the same that of Type 2-IN spin configuration.

Quantifying Fault-Tolerance for Type 1 and Type 2 Spin Configurations

To further evaluate the impact of spin configuration on fault tolerance, we performed thrust-redistribution optimizations for single-rotor failure modes. These optimizations enforce zero net roll, pitch, yaw, and altitude changes, thereby maintaining stable flight. The objective is to minimize total power consumption, and no upper bound is imposed on individual rotor thrust.

Tables 1 and 2 present the percentage increase in rotor power for each rotor when a single rotor fails under Type 1 and Type 2 spin configurations, respectively. Each failure case is indicated by "Ri," signifying that rotor i has failed (thrust = 0). The table entries show how the remaining seven rotors must increase their power to keep the aircraft's attitude and altitude unchanged.

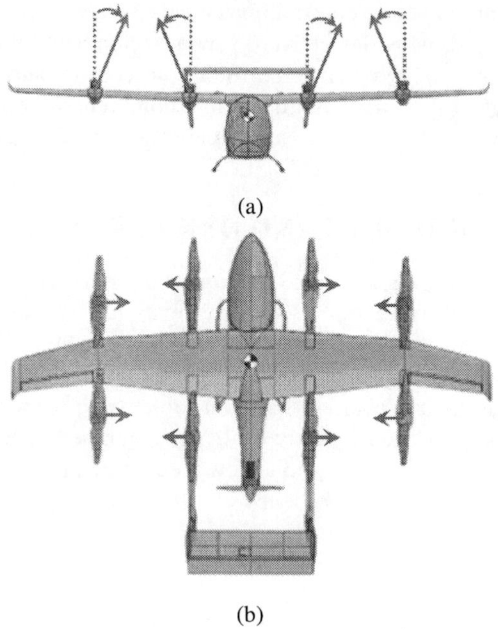

Figure 9: Thrust vectors from oppositely facing laterally canted motors, shown in (a) front view and (b) top view.

These tables illustrate the extent to which each rotor's power demand must increase to compensate for a failed rotor. In Type 1 (Table 1), no single rotor's power approaches excessively high values, owing to the aircraft's design as two overlapping "quadrotors." Meanwhile, in Type 2 (Table 2), a single rotor (often the closest same-spin partner) can see a more dramatic power increase, particularly when it must compensate simultaneously for roll, pitch, and yaw imbalances caused by the failed rotor.

The findings emphasize the superior fault-tolerance characteristics of Type 1, which benefits from a design resembling two overlapping "quadrotors." When a single rotor fails, the optimization results in no individual rotor in Type 1 exceeding approximately 2.44 times its nominal power. This is achieved

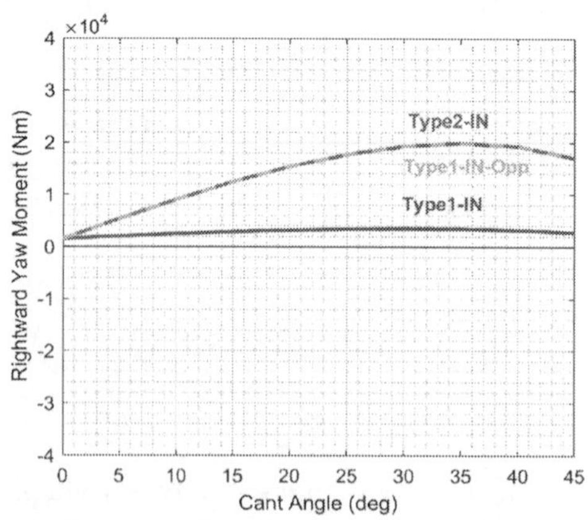

Figure 10: Right yaw moment for different cant angles in Type 1-IN configuration with opposing cant. This comparison is compared with Type 1-IN and Type 2-In configuration.

Table 1: Final Power as a Ratio of Nominal Power for Type 1 Spin Configuration under Single-Rotor Failure

Failure Mode	R1	R2	R3	R4	R5	R6	R7	R8
None	1.0	1.0	1.0	1.0	1.0	1.0	1.0	1.0
R1	0.0	1.9	1.9	1.2	0.5	0.6	**2.4**	0.5
R2	**2.1**	0.0	0.5	1.7	0.9	0.8	0.8	2.2
R3	1.7	0.5	0.0	**2.1**	2.2	0.8	0.8	0.9
R4	1.2	1.9	1.9	0.0	0.5	**2.4**	0.6	0.5
R5	0.6	0.6	**2.2**	0.6	0.0	1.5	2.0	1.4
R6	0.8	0.8	0.7	**2.2**	1.8	0.0	0.7	1.7
R7	**2.2**	0.7	0.8	0.8	1.7	0.7	0.0	1.8
R8	0.6	**2.2**	0.6	0.6	1.4	2.0	1.5	0.0

Table 2: Final Power as a Ratio of Nominal Power for Type 2 Spin Configuration under Single-Rotor Failure

Failure Mode	R1	R2	R3	R4	R5	R6	R7	R8
None	1.0	1.0	1.0	1.0	1.0	1.0	1.0	1.0
R1	0.0	**2.9**	0.9	1.4	0.7	1.2	1.0	0.7
R2	**2.5**	0.0	0.6	1.0	1.3	1.0	1.3	1.2
R3	1.0	0.6	0.0	**2.5**	1.2	1.3	1.0	1.3
R4	1.4	0.9	**2.9**	0.0	0.7	1.0	1.2	0.7
R5	0.7	1.2	1.1	0.7	0.0	**3.0**	1.0	1.4
R6	1.4	0.9	1.5	0.9	2.5	0.0	0.8	0.9
R7	0.9	1.5	0.9	1.4	0.9	0.8	0.0	**2.5**
R8	0.7	1.1	1.2	0.7	1.4	1.0	**3.0**	0.0

because the additional load is distributed more evenly. The nearest same-spin rotor only compensates for roll and yaw, while the nearest opposing rotor handles roll and pitch. This balanced redistribution reduces the strain on any single rotor.

In contrast, Type 2's design, which pairs same-spin rotors more closely and effectively forms a single large "quadrotor" with four arms, demonstrates less favorable fault-tolerance performance. In this configuration, the nearest same-spin rotor must not only offset the lost thrust but also compensate for roll, pitch, and yaw. This results in a significantly higher power demand of up to 2.97 times the rotor's nominal value, placing a greater burden on individual components.

From the entries in Tables 1 and 2, it is evident that the Type 1 configuration consumes a lower maximum power percentage compared to the Type 2 configuration. It is important to note, however, that the current calculations indicate a maximum power of 2.44 times the nominal value for the Type 1 configuration, which may not be practically desirable. This value is derived based on a predefined rotor diameter and rotor placement position. Adjusting these parameters can reduce the maximum power percentage for the Type 1 aircraft. Nonetheless, the overall conclusion that the Type 1 configuration is more power-efficient than the Type 2 configuration remains valid.

Longitudinal Cant

In the longitudinal cant configuration, each rotor is angled forward or backward relative to the eVTOL's centerline, as illustrated in Figure 11. Unlike the lateral-cant scenario, longitudinal cant places each rotor farther from the center of gravity (CG) along the aircraft's lengthwise axis. This greater lever arm significantly increases the potential yaw moment. As a result, forward or backward rotor cants can generate higher yaw torques, making them highly effective for yaw control.

Figure 12 presents the resulting yaw-moment trends for various rotor spin patterns under longitudinal cant. A key observation is that the Type 2 alternate-spin configuration attains a substantially higher rightward yaw moment than Type 1. Because Type 2 keeps pairs of same-spin rotors grouped, their forward-backward cants reinforce one another's yaw effect, leading to the larger net yaw moment observed.

Orthogonal Cant

The orthogonal cant configuration combines both lateral and longitudinal cants so that each rotor's tilt is oriented orthogonally to the lever arm from eVTOL's center of gravity. By maximizing this lever arm, orthogonal canting achieves notably higher yaw moment generation. Figure 13 illustrates the eVTOL in this configuration from the top, side, and front views, highlighting the thrust vectors produced by each rotor for lateral-only, longitudinal-only and orthogonal cant configurations.

Figure 14 compares the yaw moments across different spin configurations, including both Type 1 and Type 2 alternating spin patterns, as well as IN and OUT spin configurations. Among these, the Type 1-IN configuration provides the highest magnitude of rightward yaw moment. This is because the

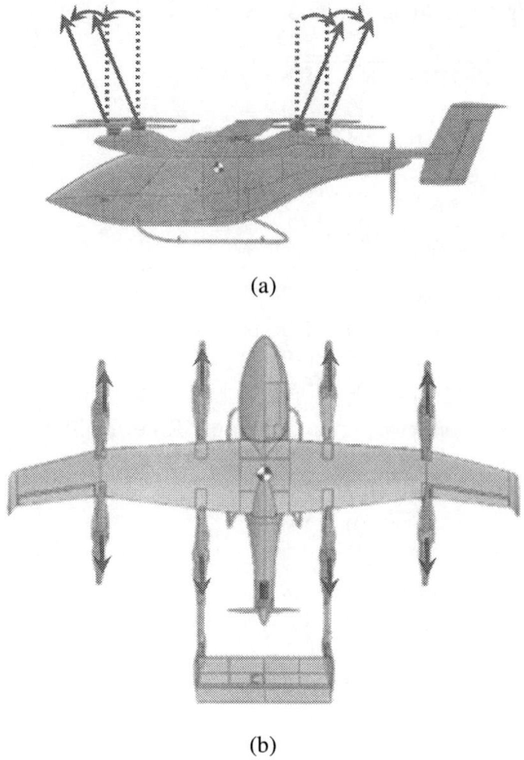

Figure 11: Thrust vectors from oppositely facing laterally canted motors, shown in (a) front view and (b) top view.

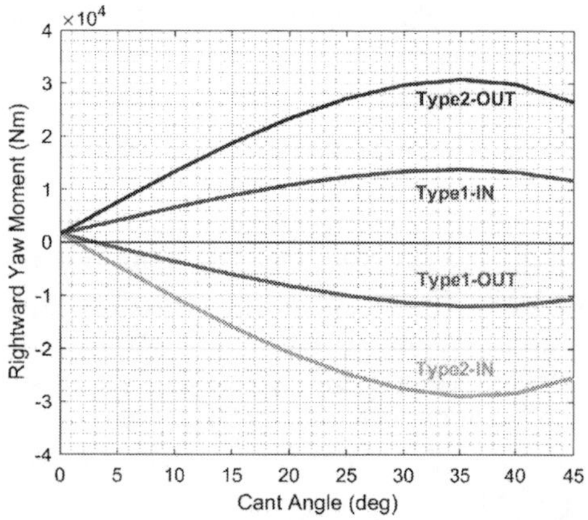

Figure 12: Rightward yaw moment for different longitudinal cant angles in Type1 and Type2 spin alternating patterns with IN and OUT spin configurations.

rotors that contribute to rightward yaw are canted in directions that further enhance the reaction torque. In contrast, the Type 2 alternating pattern produces lower yaw moment magnitudes, as the canting directions are less effective in aligning with the reaction torque to generate rightward yaw.

The results, shown in Figure 15, further compare the yaw moment generated by orthogonal cant with those from lateral canting (Type 1-IN-Opp), longitudinal canting (Type 1-IN), and lateral canting with the opposing Type 1-IN-Opp configuration. The orthogonal cant configuration produces

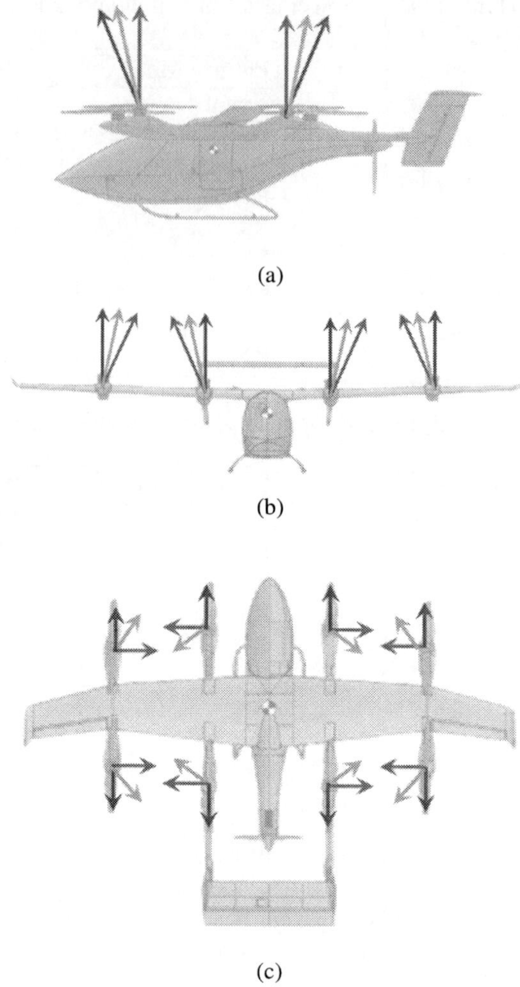

Figure 13: Thrust vectors for lateral-only, longitudinal-only and orthogonal cant configurations, shown in (a) side view, (b) front view and (c) top view.

Figure 14: Right yaw moment for different orthogonal cant angles in Type1 and Type2 spin alternating patterns with IN and OUT spin configurations.

yaw moments significantly higher than those achieved with either lateral or longitudinal cant. This superiority is primarily attributed to the maximization of the lever arm distance

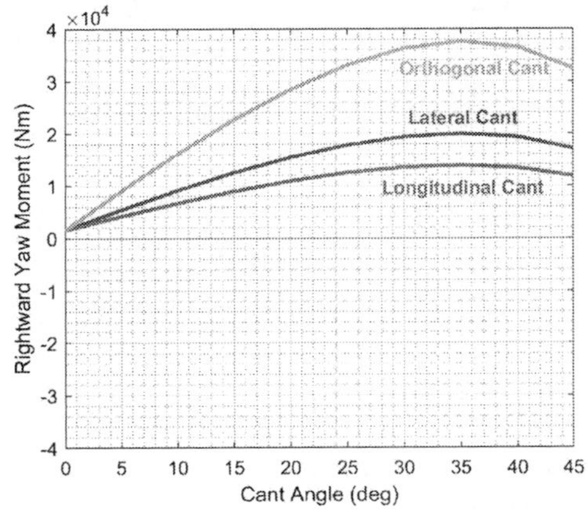

Figure 15: Rightward yaw moment for lateral, longitudinal, and orthogonal canting configurations, where lateral canting scenario specifically refers to opposing lateral canting.

achieved by aligning the canting direction orthogonally to the lever arm. This alignment allows for optimal moment generation, exceeding the capabilities of both lateral and longitudinal canting configurations.

After analyzing the yaw moments generated by various canting configurations, the yaw acceleration ($\dot{r}$) can be calculated based on the eVTOL's moment of inertia about the yaw axis. For the eVTOL in question, the yaw moment of inertia (I_{zz}) is estimated to be $20,000\,\mathrm{kg\cdot m^2}$. Using this value, the yaw acceleration can be determined using the relation $\dot{r} = \frac{u_\psi}{I_{zz}}$, where u_ψ is the net yaw moment generated. The yaw acceleration is directly proportional to u_ψ and inversely proportional to I_{zz}. The resultant rightward yaw acceleration for three different cant angle configurations are presented in Figure 16, which compares lateral (opposing cant), longitudinal, and orthogonal canting configurations.

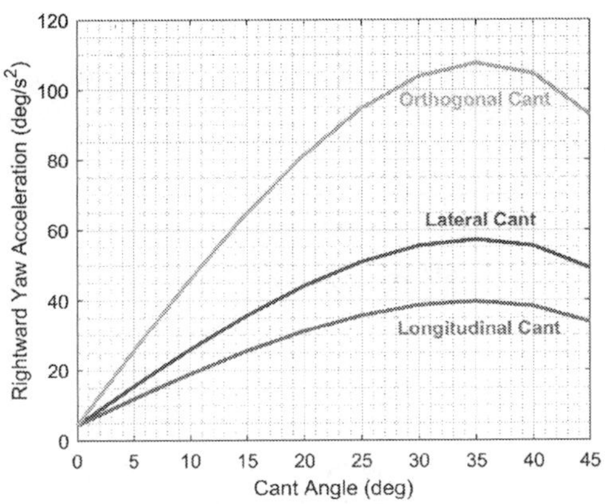

Figure 16: Rightward yaw acceleration for lateral, longitudinal, and orthogonal canting configurations. Note that the lateral canting scenario specifically refers to opposing lateral canting.

To demonstrate the impact of canting on yaw control, consider a scenario with a cant angle of 10 deg and a sustained rightward yaw maneuver lasting 2 seconds. The resulting yaw rate varies significantly based on the canting configuration. With lateral opposing canting, the eVTOL achieves a yaw rate of 40 deg/s. Longitudinal canting improves the yaw rate to 50 deg/s. Notably, orthogonal canting delivers the highest yaw rate, reaching 110 deg/s. These findings highlight the superior effectiveness of orthogonal canting in enhancing yaw performance and control authority.

Impact of Moment of Inertia on Yaw Acceleration

The study further investigates the impact of varying the yaw moment of inertia (I_{zz}) on yaw acceleration. The moment of inertia was shifted between $15,000\,\mathrm{kg\cdot m^2}$ and $25,000\,\mathrm{kg\cdot m^2}$, compared to the baseline value of $20,000\,\mathrm{kg\cdot m^2}$. Since yaw acceleration ($\dot{r}$) is inversely proportional to the moment of inertia, an increase in I_{zz} results in a decrease in yaw acceleration, while a decrease in I_{zz} leads to a corresponding increase in yaw acceleration.

This effect is clearly demonstrated at a 10-degree cant angle, as shown in Figure 17. At the baseline moment of inertia ($20,000\,\mathrm{kg\cdot m^2}$), the yaw acceleration is $25.94\,\mathrm{deg/s^2}$. When the moment of inertia decreases to $15,000\,\mathrm{kg\cdot m^2}$, the yaw acceleration increases significantly to $34.59\,\mathrm{deg/s^2}$. Conversely, an increase in the moment of inertia to $25,000\,\mathrm{kg\cdot m^2}$ results in a reduced yaw acceleration, illustrating the direct relationship between the moment of inertia and yaw dynamics.

Variation of Yaw Acceleration with Thrust Capacity

The maximum thrust per rotor in the considered eVTOL is estimated to be 5750 N, a value chosen to ensure the eVTOL can provide sufficient thrust (1.9g) in case of a single rotor failure. To evaluate the sensitivity of yaw acceleration to rotor performance, the maximum thrust capacity was reduced by 10% and 20%, effectively lowering the thrust available for allocation in all canting configurations.

As shown in Figure 18, reducing the rotor capacity significantly decreases the maximum achievable yaw acceleration.

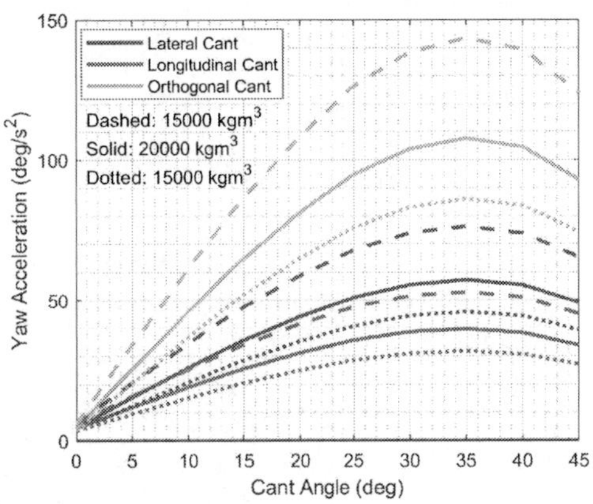

Figure 17: Rightward yaw acceleration for lateral, longitudinal, and orthogonal canting configurations for different moments of inertia.

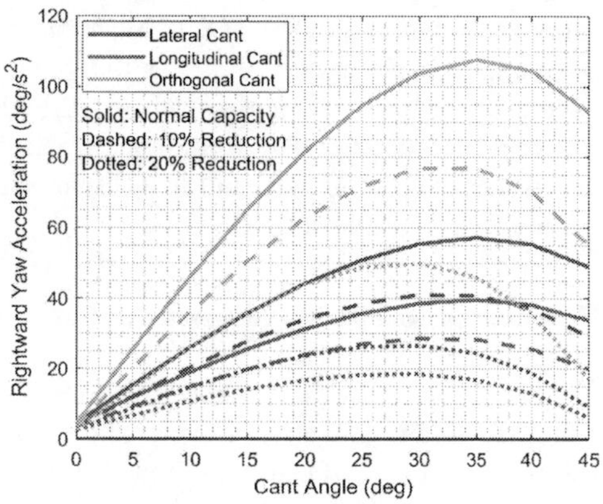

Figure 18: Rightward yaw acceleration for lateral, longitudinal, and orthogonal canting configurations for different maximum rotor thrust capacity.

For a 20% reduction in rotor capacity, the yaw acceleration is reduced by nearly 54%, underscoring the sensitivity of yaw control authority to rotor performance. The dashed and dotted lines in the figure represent the 10% and 20% reductions, respectively, highlighting the proportional impact on yaw dynamics across lateral (opposing cant), longitudinal, and orthogonal canting configurations.

Achieving Maximum Steady Yaw Rate

Once the eVTOL has accelerated to a specific yaw rate, it must continuously generate enough yaw moment to overcome the aerodynamic drag that opposes its rotation. This opposing yaw moment arises primarily from the fuselage, inner booms, outer booms, and vertical fins, all of which experience tangential airflow as the vehicle yaws.

To quantify this effect, each relevant vertical surface is discretized into small strips, and the local drag force on each strip is calculated based on its tangential velocity, the strip's area, and an appropriate drag coefficient. Specifically, for an eVTOL yawing with an angular rate ω, the local tangential velocity at a strip located at distance r_i from the yaw axis is ωr_i. The drag force on that strip is then given by

$$F_{\text{drag},i} = \frac{1}{2}\rho\,(\omega \cdot r_i)^2 C_{D,i} A_i, \tag{7}$$

where ρ is the air density, $C_{D,i}$ is the drag coefficient for the strip, and A_i is its area.

Multiplying each strip's drag force by its moment arm r_i (the distance from the yaw axis) and summing over all strips yields the total aerodynamic yaw moment:

$$u_{\psi,\text{drag}} = \sum_i \left(F_{\text{drag},i} \cdot r_i\right), \tag{8}$$

For the arbitrary 2500 kg eVTOL considered in this work, sustaining a steady yaw rate of 60 deg/s necessitates overcoming an aerodynamic yaw moment of approximately 911 Nm, 58% of which originates from the vertical fins alone. As shown in Figure 15, the yaw moment available from the rotors at 0 deg cant angle in each of the examined configurations (lateral-opposing cant, longitudinal cant and orthogonal cant), the aircraft can provide sufficient yaw moment to maintain 60 deg/s steady yaw rate.

CONCLUSION

This study shows that yaw control authority in an eVTOL depends strongly on the geometric configuration of its rotors, each rotor's thrust capacity, and the thrust-to-reaction torque relationship. Although rotor canting provides a significant influence on yaw authority, the ultimate performance emerges from the interplay of these factors.

The present eVTOL demonstrates superior yaw control authority when using an opposing lateral cant, with fault tolerance effectively achieved by the Type 1-IN rotor spin configuration. Longitudinal canting further augments the benefits of lateral canting but raises the risk of propeller interference with the boom structure when inward cant is employed. Orthogonal canting maximizes yaw moment generation by aligning the cant direction orthogonally to the lever arm, though it requires inward cant of the inboard rotors—necessitating additional clearance. Notably, even with zero rotor cant, this eVTOL is capable of sustaining a steady yaw rate of 60 deg/s under the examined conditions.

One important limitation here is that maintaining hover demands more than 50% of the maximum thrust the rotors can provide, thereby constraining the available thrust margin for yaw generation and affecting the overall relationship between yaw moment and cant angle. If hovering required significantly less than 50% of each rotor's maximum thrust, the aircraft could rely on fewer rotors to sustain lift, leaving more thrust capacity to induce large yaw moments.

This finding underscores that yaw control authority is highly sensitive to both the thrust capacity of the rotors and the thrust allocation strategy. Future studies should therefore examine a broader range of hover thrust requirements to identify optimal canting configurations and fully assess the operational trade-offs.

ACKNOWLEDGMENTS

This research is supported by A*STAR under its RIE2025 Manufacturing Trade and Connectivity (MTC) Industry Alignment Fund - Pre- Positioning (IAF-PP), with Award No. M23L5a0002. Sincere gratitude is extended to Dr. Andrew Barnes and Mr. Andrian Pratama for their invaluable assistance and support provided during this study.

REFERENCES

1. Goyal, R., Reiche, C., Fernando, C., and Cohen, A., "Advanced Air Mobility: Demand Analysis and Market Potential of the Airport Shuttle and Air Taxi Markets," *Sustainability*, Vol. 13, No. 13, Jul. 2021, p. 7421.

2. Lim, R., Deahl, E., and Ng, J., "Advanced Air Mobility (AAM): Implications for Transport Service Providers," *Alton Aviation Consultancy*, Jan. 9, 2024. [Accessed: Jan. 17, 2025].

3. Ugwueze, O., Statheros, T., Bromfield, M. A., and Horri, N., "Trends in eVTOL aircraft development: the concepts, enablers, and challenges," *AIAA Scitech 2023 Forum*, 2023, p. 2096.

4. Ducard, G. J. J., and Allenspach, M., "Review of designs and flight control techniques of hybrid and convertible VTOL UAVs," *Aerospace Science and Technology*, Vol. 118, Nov. 2021, p. 107035.

5. National Transportation Safety Board, "Aircraft Accident Report - New York Airways, Inc., Sikorsky S61-L N618PA, Newark, New Jersey," Sep. 27, 1979, [Report]. Accessed Apr. 30, 2024.

6. Scerri, A., "Lessons learned from Joby's 2022 eVTOL accident," *Vertical Mag*, Mar. 24, 2024. [Online]. Available: https://verticalmag.com/opinions/lessons-learned-from-jobys-2022-evtol-accident/. Accessed Apr. 30, 2024.

7. ASTM International, "ASTM F3298-19, Standard Specification for Design, Construction, and Verification of Lightweight Unmanned Aircraft Systems (UAS)," Mar. 2019. [Online]. Available: https://www.astm.org/Standards/F3298.htm. Accessed Apr. 30, 2024.

8. EASA, "Special Condition for small-category VTOL aircraft," Special Condition for VTOL and Means of Compliance, July 2, 2019. [Online]. Available: https://www.easa.europa.eu/en/document-library/product-certification-consultations/special-condition-vtol. Accessed Apr. 30, 2024.

9. Yogi, S. C., Kamath, A. K., Singh, P., and Behera, L., "Trajectory Tracking Control of a Quadrotor UAV using an Auto-tuning Robust Sliding Mode Controller," *2022 International Conference on Unmanned Aircraft Systems (ICUAS)*, IEEE, 2022, pp. 345–353.

10. Kamath, A. K., Yogi, S. C., Behera, L., Nair, R. R., and Nahavandi, S., "Back-stepping Nonlinear Integral Sliding Mode Control of a Vision Augmented 3 DoF Quadrotor in the presence of System Uncertainties and Bounded Disturbances," *2022 IEEE 17th International Conference on Control & Automation (ICCA)*, IEEE, 2022, pp. 541–546.

11. Thanaraj, T., Low, K. H., and Ng, B. F., "Actuator fault detection and isolation on multi-rotor UAV using extreme learning neuro-fuzzy systems," *ISA Transactions*, 2023.

12. Thanaraj, T., Ng, B. F., and Low, K. H., "Active Fault Tolerant Control on Multi-rotor UAVs using ELM-based Neuro-Fuzzy System," *AIAA AVIATION 2022 Forum*, American Institute of Aeronautics and Astronautics, Jun. 2022.

13. Fourlas, G. K., and Karras, G. C., "A survey on fault diagnosis and fault-tolerant control methods for unmanned aerial vehicles," *Machines*, Vol. 9, No. 9, 2021, p. 197.

14. Quan, Q., Cui, G., and Du, G.-X., "Controllable probability and optimization of multicopters," *Aerospace Science and Technology*, Vol. 119, 2021, p. 107162.

15. Thanaraj, T., Govind, S., Roy, A., Ng, B. F., and Low, K. H., "A reliability framework for safe octorotor UAV flight operations," *2023 International Conference on Unmanned Aircraft Systems (ICUAS)*, IEEE, 2023, pp. 1013–1020.

16. Mochida, S., Matsuda, R., Ibuki, T., and Sampei, M., "A geometric method of hoverability analysis for multirotor UAVs with upward-oriented rotors," *IEEE Transactions on Robotics*, Vol. 37, No. 5, 2021, pp. 1765–1779.

17. Mochida, S., Onuki, R., Kawagoe, T., Ito, T., Ibuki, T., Funada, R., and Sampei, M., "Hoverability Analysis and Development of a Quadrotor Only with Clockwise Rotors," *2022 IEEE/RSJ International Conference on Intelligent Robots and Systems (IROS)*, IEEE, 2022, pp. 7558–7564.

18. K. Atci, T. Jusko, A. Štrbac, and F. Guner, "Impact of differential torsional rotor cant on the flight characteristics of a passenger-grade quadrotor," *CEAS Aeronautical Journal*, pp. 1–16, 2024, Springer.

19. R. Niemiec and F. Gandhi, "Effect of rotor cant on trim and autonomous flight dynamics of a quadcopter," in *Proc. AHS International 74th Annual Forum and Technology Display*, Phoenix, AZ, 2018.

Application of an Inverse Airfoil design for Improved Proprotor Performance

Elvin J. Puchuela
Undergraduate Student
Georgia Institute of Technology
Atlanta, GA, USA

Lakshmi N. Sankar
Regents Professor and Sikorsky Professor
Georgia Institute of Technology
Atlanta, GA, USA

ABSTRACT

Significant progress has been made to improve the aeroacoustic characteristics of an existing UAV/ drone (RQ-7B). A four-bladed pusher propeller that replaces the existing two-bladed propeller was designed using public domain analysis. A four-bladed rotor, operating at a reduced RPM and with an acoustic footprint substantially lower than the baseline two-bladed rotor, was chosen for its comparable propulsive efficiency. Further improvements in rotor performance have been achieved by using advanced low drag airfoils specifically designed for this rotor, with lower drag coefficients over a broad range of lift coefficients.

NOTATION

c	Local chord
C_p	Pressure coefficient
$\bar{C_d}$	Average drag coefficient
$\bar{C_l}$	Average lift coefficient
FM	Figure of Merit
F	Tip loss factor
R	Blade radius
r	Radial location
σ	Rotor solidity
ξ	Chordwise square root
x	Chordwise location of a surface point
κ	Induction factor
$\frac{V_\infty}{nD}$	Advance ratio

INTRODUCTION

Over the past several decades, there has been a growing interest in the use of advanced computational tools for the conceptual and preliminary design of propellers, rotors, and proprotors. The conceptual design processes have matured over the past three decades, with conceptual tools capable of designing rotors for conventional rotorcraft and tilt-rotor vehicles, as well as autonomous rotorcraft systems discussed in Refs. 1-7. Research has been conducted in parallel to integrate data from advanced aerodynamic, structural dynamics, and elastic analyzes into these tools. As computational tools matured, it became feasible to design or redesign rotorcraft components using computational fluid dynamics (CFD) tools coupled to comprehensive structural dynamics and aeroacoustics tools discussed in Refs. 8-9. Work has also been done on the development of advanced airfoils with desirable high lift-to-drag ratios, and the incorporation of these airfoils into the tip regions of commercial helicopter rotors, such as the S-76 discussed in Ref. 10.

Presented at the Vertical Flight Society's 11th Biennial Autonomous VTOL Technical Meeting, Phoenix, Arizona, USA, February 4–6, 2025.

RELATED PRIOR WORK

In this study, we revisit the tools developed by the researchers, as discussed in Refs. 8–10, to redesign the pusher propeller for the drone shown in Figure 1. In an earlier study Ref. 11, the present researchers performed an optimization of the rotor planform (variation of chord and twist) and RPM. A combination of low and high-fidelity tools was used. The range of this vehicle is limited by the logistics of radio communication between the aircraft and the monitoring station. Therefore, our earlier study explores the impact of propeller redesign on rotor power consumption, with the resulting impact on endurance, take-off, maximum forward speed, rate of climb, and acoustic footprint. Public domain information available on websites has been used to determine the basic characteristics of the vehicle, including size, weight, performance, and powerplant. The aircraft uses a 38 hp (27 kW) AR741-1101 Wankel engine, operating at a maximum of 7000 RPM. A two-bladed rotor was used in the baseline configuration.

RESEARCH OBJECTIVE

Our earlier propeller design studies in Ref. 11 used the NACA family of airfoils, often used in conceptual studies of propeller and rotor applications. In the present study, we attempt to further improve the performance of the pusher propeller for the Boeing RQ-7B Shadow using custom designed low drag

Figure 1. Boeing AAI RQ-7B Shadow UAV.

airfoils. An inverse design tool, previously developed by the present authors in Refs. 10 has been used.

INVERSE DESIGN METHODOLOGY

The present inverse design approach iteratively modifies the rotor surface $z(x,r)$ where, x, represents the chordwise location of a surface point on the section and, r, represents the radial (or span-wise) location. The objective of the iterative process is to adjust the surface pressure coefficient $C_p(x,r)$, calculated using any convenient flow analysis (CFD or panel method, as applicable), at any point on the surface. This adjustment drives it toward a user-specified target pressure coefficient $C_{p,\text{target}}(x,r)$ by incrementally modifying the ordinate $z(x,r)$ by a small amount $\Delta z(x,r)$ at each iteration.

Since the surface pressure distribution is influenced by the surface ordinate $z(x,r)$, its variation in the chordwise direction, and its curvature (or second derivative), the following heuristic equation is solved:

$$A(\Delta z) + B\frac{(\partial(\Delta z))}{\partial \xi} - C\frac{\partial^2(\Delta z)}{\partial \xi^2} = \Delta C_p(x,r) \qquad (1)$$

In the above equation, ξ is the square root of the nondimensional distance along the chord line measured from the leading edge:

$$\xi = \sqrt{\frac{x}{c}}$$

Where, c, is the local chord, and the three constants A, B, C are user-specified and drive the changes in the surface shape Δz to zero as the computed surface pressure distribution approaches the target values.

In this approach, it is not necessary to specify the target pressure distribution on the entire surface of the rotor section. In regions, the existing surface pressure values can be deemed acceptable, and they can be considered equal to the 'target' values. The differences in the pressure coefficient ΔCp , equal to $C_p(x,r) - C_{p,target}(x,r)$, are set to zero in these regions.

An iterative process is used to gradually change the airfoil geometry to achieve the target pressure values. At each iteration, the surface pressure coefficients $C_p(x,r)$ for the current version of the airfoil shape are calculated using XFOIL, taking into account the effect of the boundary layer on lift and drag. The entire design process, including the calls to XFOIL within the design process, has been automated within a Matlab script. The design process converges quickly, within 1 or 2 seconds, on laptop computer systems.

SAMPLE RESULTS FOR AIRFOIL DESIGN

Figure 2 shows a sample airfoil designed with the inverse design approach, starting with the NACA 0012 airfoil. Figure 3 shows the pressure distribution for NACA 0012 and the target pressure distribution with favorable or near-zero pressure gradients on most of the upper surface, chosen to delay the transition and achieve a drag bucket at a Reynolds number of 3 million at a Mach number of 0.45.

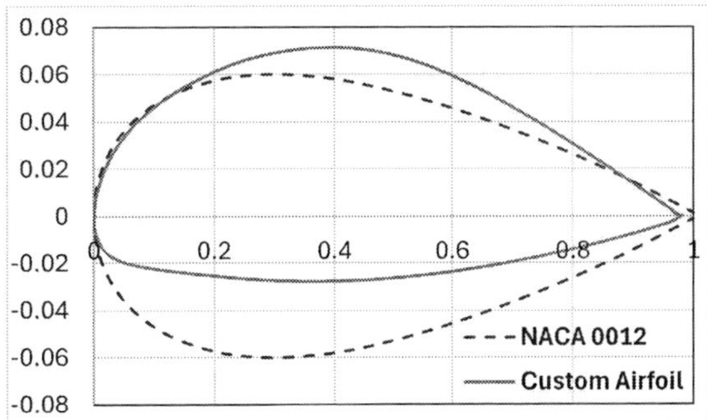

Figure 2. Starting Airfoil and Final Design

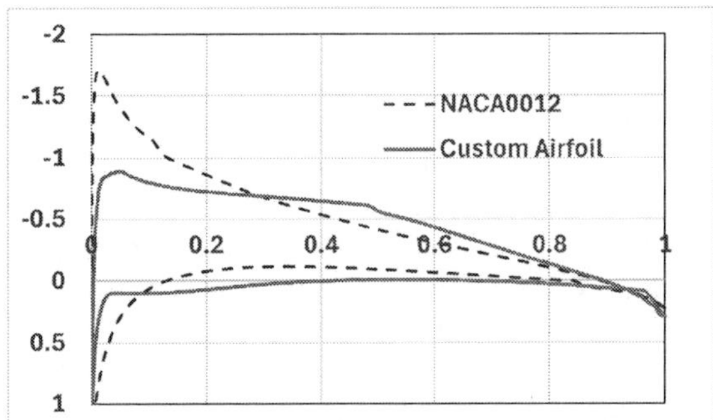

Figure 3. Comparison of Starting and target Pressure Values for a Typical Design, Mach number 0.45, 3 Million Reynold Number, 0.5 Lift Coefficient.

Tables 1 and 2 show the tabulated drag data for the two airfoils computed using XFOIL. The new airfoil is seen to achieve laminar flow on a larger part of the upper surface of the airfoil (up to 47% of the chord) compared to the starting airfoil (11% of the chord) with a nominal lift coefficient of 0.5. Figure 4 also shows these data in a graphical form, and it is observed that the custom airfoil exhibits a drag bucket over narrow range lift coefficients. Furthermore, even at higher angles of attack, where the flow is likely to be turbulent, the drag values for the custom airfoil are seen to be lower than those for the classical NACA 0012 airfoil, because of the reduction in pressure drag. This reduction is due to the thinner turbulent boundary layer over the upper and lower surfaces for the custom airfoil, allowing improved pressure recovery at the trailing edge region.

Classical blade element theory indicates that the thrust coefficient may be approximately expressed as

$$C_\tau \cong \int_0^1 \frac{1}{2}\left(\frac{r}{R}\right)^2 \sigma C_l d\left(\frac{r}{R}\right) \approx \frac{\sigma \overline{C_l}}{6} \tag{2}$$

Here, σ is the solidity of the rotor and $\overline{C_l}$ is the average lift coefficient throughout the blade. The power coefficient may also be approximately written in terms of the thrust coefficient, the induction factor κ that accounts for nonuniform inflow effects, and the average drag coefficient $\overline{C_d}$ as follows:

$$C_P = \kappa \frac{1}{\sqrt{2}} (C_\tau)^{3/2} + \frac{\sigma \overline{C_d}}{8} \tag{3}$$

From the following definition of the figure of merit of a rotor in hover,

$$\mathrm{FM} = \frac{\frac{1}{\sqrt{2}} (C_T)^{3/2}}{C_P} \tag{4}$$

and Figure 4, it is therefore clear that the custom airfoil would allow for an improved figure of merit.

OVERVIEW OF THE PROPELLER DESIGN

The public domain solver XROTOR was first used to explore the design space from the performance (propulsive efficiency) and noise perspectives. The vehicle shown in Figure 1 uses a two-blade pusher propeller with a maximum diameter of 0.7924 meters (31.2 inches), operating at an RPM of 7000 for a fixed power of 30 HP. These values were estimated from vehicle propeller dimensions and operating conditions.

The following design variables were parametrically varied: RPM rotor (2000 through 5000), number of blades (2, 3, 4, 5), and four airfoil families. The available power was assumed to be 80% of the maximum power (38 HP). For each of the combination of design parameters, a new planform was designed using the XROTOR software. A Pareto frontier was generated to identify the rotors with the minimum noise footprint when the vehicle is flying at 1000 feet altitude, with the highest thrust. Appendix A shows some of the most promising combinations.

For a two-bladed rotor operating at 7000 RPM, XROTOR provided a rotor design with a propulsive efficiency of 0.83, with a thrust of 328 Newtons (74 lb) at a cruise speed of 110 knots (56.7 meters per sec). However, XRotor predicts that this rotor had a noise footprint of 73 dB when the propeller operated at an altitude of 1000 feet under International Standard Atmosphere (ISA) conditions , as shown in Figure 5 below.

Table 1. NACA 0012 Airfoil Characteristics at Mach 0.45 and Reynolds 3.00×10^6

Mach: 0.450		**Reynolds:** 3.00×10^6		**Ncrit:** 9.000	
C_L	C_D	C_{Dp}	C_M	Top_Xtr	Bot_Xtr
0.000	0.00545	0.00090	0.000	0.4744	0.4745
0.100	0.00548	0.00094	0.0004	0.3996	0.5495
0.300	0.00591	0.00131	0.0018	0.2433	0.6998
0.400	0.00630	0.00166	0.0029	0.1702	0.7734
0.500	0.00680	0.00215	0.0045	0.1087	0.8412
0.600	0.00737	0.00275	0.0066	0.0691	0.8984
0.700	0.00806	0.00349	0.0095	0.0455	0.9423
0.800	0.00885	0.00434	0.0129	0.0327	0.9736
0.900	0.00975	0.00531	0.0142	0.0253	0.9889
1.100	0.01210	0.00794	0.0145	0.0174	0.9985
1.200	0.01468	0.01097	0.0249	0.0147	1.0000

Table 2. Custom Airfoil Characteristics at Mach 0.45 and Reynolds 3.00×10^6

Mach: 0.450		**Reynolds:** 3.00×10^6		**Ncrit:** 9.000	
C_L	C_D	C_{Dp}	C_M	Top_Xtr	Bot_Xtr
0.000	0.0063	0.00114	−0.0404	0.6661	0.0479
0.100	0.00608	0.00094	−0.0392	0.6472	0.0670
0.200	0.00577	0.00080	−0.0376	0.6247	0.1434
0.300	0.00428	0.00069	−0.0310	0.5844	0.6841
0.400	0.00386	0.00086	−0.0251	0.5272	0.9340
0.500	0.00421	0.00107	−0.0358	0.4712	0.9652
0.600	0.00562	0.00178	−0.0367	0.2776	0.9780
0.700	0.00701	0.00260	−0.0366	0.1345	0.9870
0.800	0.00786	0.00325	−0.0380	0.0824	0.9914
0.900	0.00860	0.00389	−0.0401	0.0540	0.9944
1.000	0.00948	0.00471	−0.0411	0.0331	0.9978

Increasing the number of blades to 4, and lowering the RPM to 5000, yielded an XRotor design with a comparable efficiency of 0.83 and thrust (327 Newtons). However, the noise footprint was considerably lower, as shown in Figure 6 below.

The average lift coefficient was 0.5 for the four-bladed rotor. The custom airfoil exhibits significantly lower profile drag at this lift coefficient, as seen in Figures 4. For this reason, the four bladed rotor at the lower RPM has been considered in this study, through improved performance with custom the low-drag airfoil discussed earlier.

PROPELLER MODELING METHODOLOGY

Various public domain tools are available for modeling and designing propellers. However, many of these tools have some limitations. For example, XROTOR software assumes a parabolic variation of the drag coefficient with the lift coefficient, with the coefficients of the parabola determined from the user-specified input for the minimum drag value and the corresponding lift coefficient. Javaprop, another popular software, includes drag polars for a number of airfoils. However, it was difficult to add additional drag polars to this library.

For these reasons, a propeller analysis based on combined blade element-momentum theory (BEM) was developed as part of this study. Unlike the classical BEM model, this does

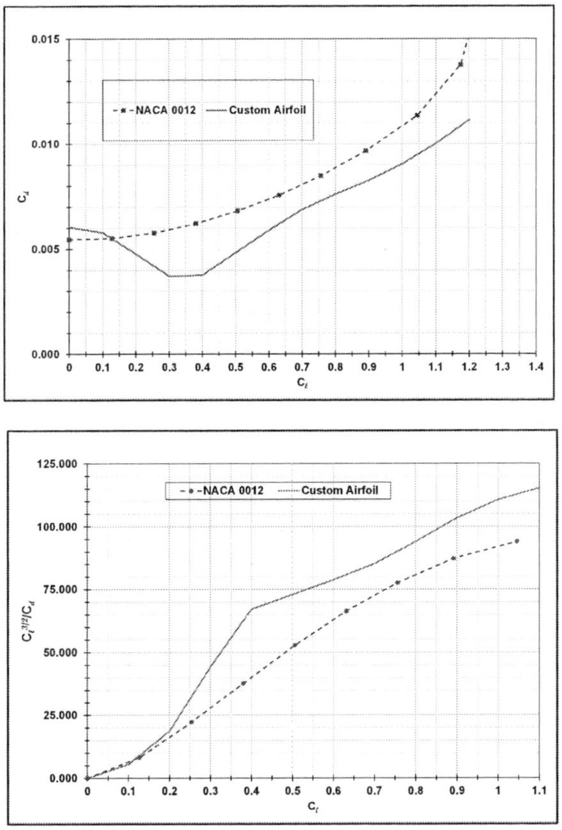

Figure 4. Comparison of Drag Polars for Starting Airfoil and Final design (Reynolds number 3 Million, Mach number 0.45).

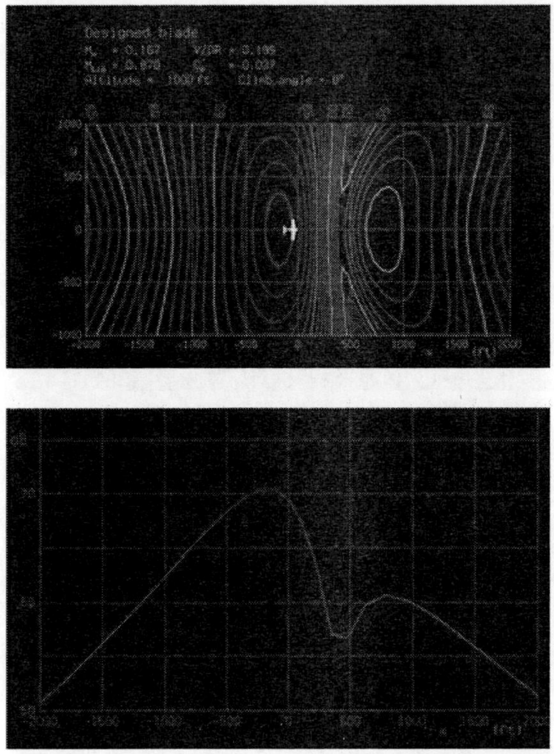

Figure 5. Noise Footprint of the Two-Bladed Rotor at 7000 RPM at an Altitude of 1000 Feet Under ISA Conditions (XROTOR).

not assume a linear variation of lift vs. angle of attack. Swirl losses and tip losses are automatically included, in addition to nonuniform inflow.

The methodology in the following is given in a dimensional form for convenience. However, much of the analysis is done in nondimensional form. All distances and velocities in the formulation shown below have been non-dimensionalized by the tip radius and the tip velocity, respectively. The blade operating parameters, such as RPM and forward speed, and the geometric parameters such as blade diameter, are used only to compute the thrust force and power in dimensional form and allow the propeller to meet the performance requirements of the configuration shown in Figure 1.

Here are the basic steps of the propeller analysis. The blade planform, i.e. the radial variation of chord non-dimensionalized by blade radius R, and twist are imported from Excel spreadsheet. The blade is divided into segments (blade elements). For simplicity, the number of elements and the width of each blade element are based on the location of the supplied blade geometry.

A range of advance ratio values and pitch settings are chosen. The pitch setting was set to zero in the results shown. The advance ratio, J, is defined as $\frac{V_\infty}{nD}$ and is based on the propeller forward speed V_∞ along the shaft axis, the number of rotor revolutions per second 'n', and the diameter of the rotor, D.

For each combination of advance ratio and blade pitch settings, the thrust and power (in their nondimensional form) are computed by summing up the individual contributions from the blade elements.

An iterative process is used to update these quantities in the center of each blade element as follows. The induced inflow velocity v is initialized to zero. The swirl velocity is also initialized to zero at the start of the iteration.

For an annulus of radius r (located in the center of a given blade element) and the corresponding element width 'dr', the generated thrust, dT, may be computed from momentum theory as shown in Figure 7 below:

In this figure, the quantity, F, is the tip loss factor. In this analysis, the tip loss factor has been calculated based on the radial location, the number of blades b, and the total inflow ratio l, equal to (V+v) divided by blade tip speed, as follows:

$$F = \frac{2}{\pi}\arccos\left(e^{-f}\right) \qquad \text{where,} \quad f = \frac{b}{2}\frac{(1-r/R)}{\lambda} \tag{5}$$

The thrust production for each blade segment may be computed from the blade element theory, with the current values of induced velocity v, and the swirl velocity, as shown in figure 8 below.

Equating the thrust of the blade element theory with the corresponding value from the momentum theory, the following updated estimate for the induced velocity may be obtained.

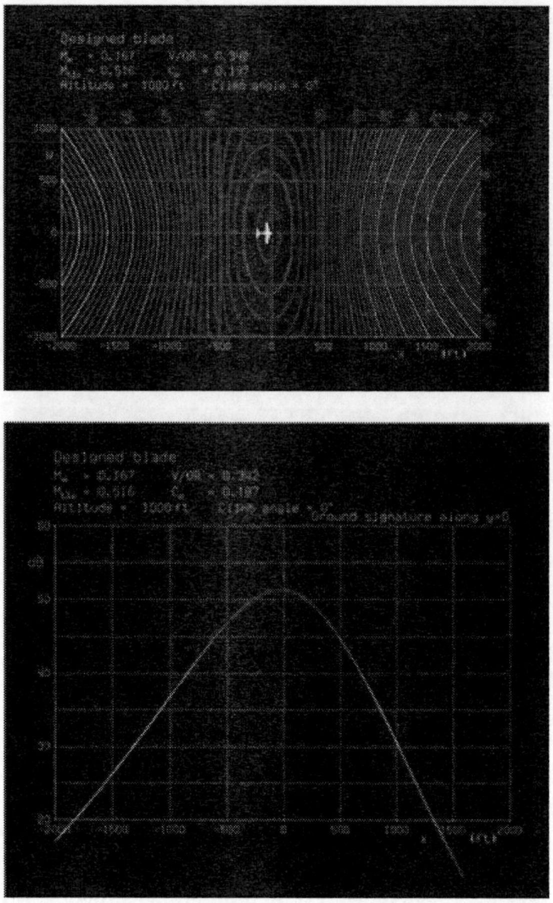

Figure 6. **Noise Footprint of the Four-Bladed Rotor at 4000 RPM at an Altitude of 1000 Feet Under ISA Conditions (XROTOR).**

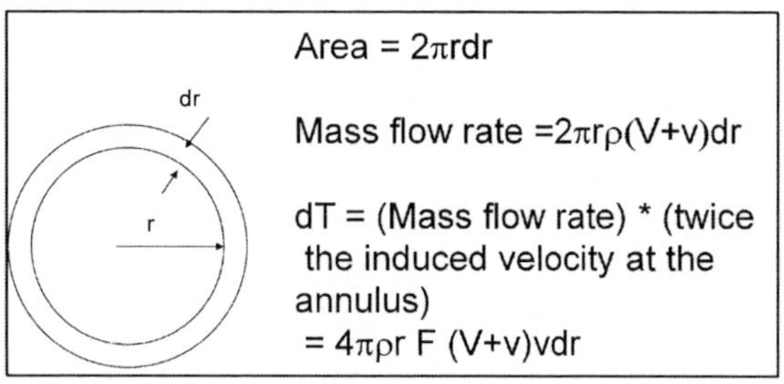

Figure 7. **Thrust production over an Annulus from Momentum Theory.**

$$v = \frac{bc\left[(\Omega r - v_{swirt})^2 + (V+v)^2\right](C_l \cos\phi - C_d \sin\phi)}{8\pi r F(V+v)} \tag{6}$$

where, the right-hand side is computed using the induced velocity and swirl velocity estimates at the center of the blade element from the previous iteration. The lift and drag coefficients are estimated from the incompressible flow drag polars precomputed from XFOIL shown in Figure 4 earlier. The lift coefficient is corrected for compressibility effects using

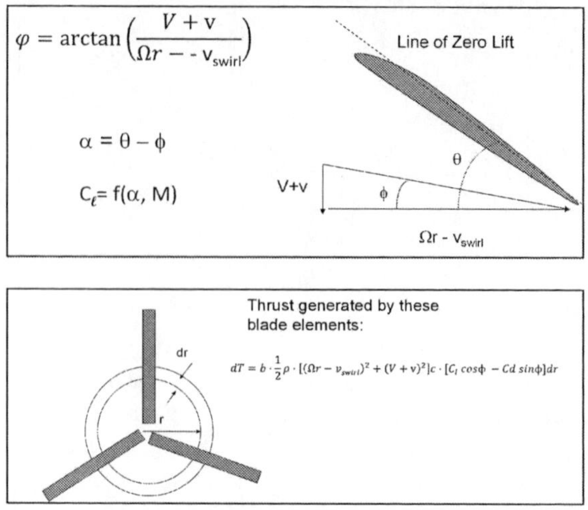

Figure 8. **Thrust production over an Annulus from Blade Element Theory.**

the Prandtl Glauert rule. No compressibility corrections have been applied to the drag coefficients, since the profile drag does not significantly affect the compressibility effects until the onset of transonic flow above the critical Mach number.

The angular momentum associated with the swirl of the flow in the far wake, corresponding the fluid mass crossing the annulus shown in Figure 5 may similarly be expressed as

$$dQ = \rho(V+v)(2\pi r dr) r\,(2v_{swirl}) \tag{7}$$

From blade element theory, this swirl is caused by the torque exerted by the blade elements on the fluid mass, as follows.

$$dQ = \frac{1}{2}\rho r b c\left[(\Omega r - v_{swirl})^2 + (V+v)^2\right](C_l \sin\phi + C_d \cos\phi) \tag{8}$$

Equating these two expressions, the following update for the swirl velocity may be obtained.

$$v_{swirl} = \frac{bc\left[(\Omega r - v_{swirl})^2 + (V+v)^2\right](C_l \sin\phi + C_d \cos\phi)}{8\pi r(V+v)} \tag{9}$$

In equation 9, the right-hand side is computed using the estimates of the induced and swirl velocities from the previous iteration. The tip loss factor shown in equation 5 is also updated at each iteration.

When the iterations have converged, the contributions of the thrust and torque from the individual blade elements may be summed up, and multiplied by the number of blades to compute the total thrust or torque (or power).

The cruise condition of the drone shown in figure 9 corresponds to a forward speed of 110 knots (56.7 meters per second) and an advance ratio of 1.07. It is clear that the custom airfoil rotor has better propulsive efficiency under this operating condition compared to the NACA 0012 airfoil. Even under off-design conditions, the rotor requires less power than the rotor with the classical NACA 0012 airfoil.

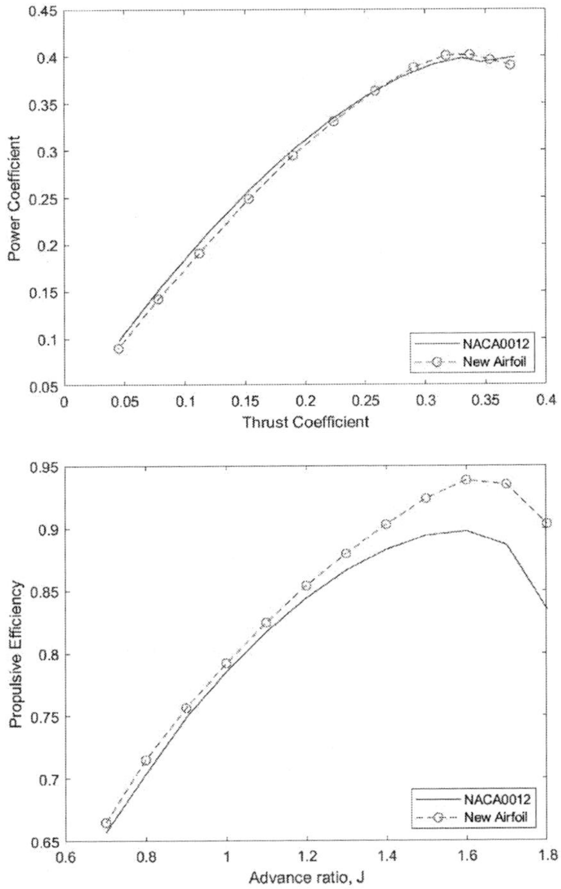

Figure 9. Effect of Custom Airfoils on a Four-Bladed Rotor designed for Optimum Performance at 110 Knot Cruise Conditions.

CONCLUSIONS

Work has been done to improve the aeroacoustics characteristics of an existing UAV/ drone (RQ-7B). A four-bladed pusher propeller that replaces the existing two-bladed propeller was designed using XRotor. A four-bladed rotor operating at a reduced RPM with an acoustic footprint that is substantially lower than the baseline two-bladed rotor, with a comparable propulsive efficiency was chosen.

Further improvements in rotor performance have been achieved through the use of advanced low-drag airfoils, with lower drag coefficients over a broad range of lift coefficients.

Additional work is needed in modeling the baseline rotor and the new rotor with state-of-the-art computational fluid dynamics analyzes with advanced turbulence/transition models to understand the effects of airfoil profile on the flow transition from laminar to turbulent flow. Inflow must be examined to determine whether further improvements are possible through reduction in induced power with a more uniform inflow. This would require modifying the blade twist and taper guided by flow simulations.

Author Contact

Elvin J. Puchuela epuchuela3@gatech.edu
Lakshmi N. Sankar lakshmi.sankar@aerospace.gatech.edu

APPENDIX

RPM	No of Blades	Thrust Newtons	Thrust, lbf	Power, Hf	Efficiency	1/Efficiency	Max Noise 1000 ft Below Flight path
7000	2	328	73.737352	30	0.8318	1.20221207	71
7000	3	329	73.962161	30	0.8338	1.19932838	68
7000	4	328	73.737352	30	0.8318	1.20221207	66
7000	5	327	73.512543	30	0.8287	1.2067093	65
7000	6	326	73.287734	30	0.8252	1.21182744	64
6000	2	328	73.737352	30	0.8326	1.20105693	67
6000	3	331	74.411779	30	0.8395	1.19118523	64
6000	4	332	74.636588	30	0.8405	1.189768	63
6000	5	331	74.411779	30	0.8395	1.19118523	59
6000	6	331	74.411779	30	0.8377	1.19374478	58
5000	2	324	72.838116	30	0.8217	1.21698917	65
5000	3	329	73.962161	30	0.8343	1.19860961	61
5000	4	331	74.411779	30	0.8386	1.19246363	57
5000	5	331	74.411779	30	0.8398	1.1907597	53
5000	6	331	74.411779	30	0.8397	1.19090151	50
4000	2	315	70.814835	30	0.7977	1.25360411	62
4000	3	323	72.613307	30	0.8187	1.22144864	57
4000	4	327	73.512543	30	0.8278	1.20802126	52
4000	5	328	73.737352	30	0.8321	1.20177863	47
4000	6	329	73.962161	30	0.8342	1.1987533	44
3000	2	293	65.869037	30	0.7431	1.3457139	59
3000	3	307	69.016363	30	0.7787	1.2841916	53
3000	4	314	70.590026	30	0.7961	1.2561236	44
3000	5	318	71.489262	30	0.8058	1.24100273	37
3000	6	320	71.93888	30	0.8117	1.23198226	30

REFERENCES

1. Johnson, W.; Yamauchi, G.K.; and Watts, M.E., *NASA Heavy Lift Rotorcraft Systems Investigation*, NASA TP 2005-213467, December 2005.

2. Johnson, W.; Yamauchi, G.K.; and Watts, M.E., "Designs and Technology Requirements for Civil Heavy Lift Rotorcraft," *Proceedings of the American Helicopter Vertical Lift Aircraft Design Conference*, San Francisco, CA, January 18-20, 2006.

3. Johnson, W.; Yeo, H.; and Acree, C., "Performance of Advanced Heavy-Lift, High-Speed Rotorcraft Configurations," *Proceedings of the AHS International Forum on Rotorcraft Multidisciplinary Technology*, Seoul, Korea, October 15-17, 2007.

4. Johnson, W., *NDARC NASA Design and Analysis of Rotorcraft*, NASA/TP–2009-215402.

5. Johnson, W., *NDARC: NASA Design and Analysis of Rotorcraft*, NASA TP 2015-218751, April 2015.

6. Johnson, W.; Withrow-Maser, S.; Young, L.; Malpica, C.; Koning, W.J.F.; Kuang, W.; Fehér, M.; Tuan, A.; Chan, A.; Datta, A.; Chi, C.; Lumba, R.; Escobar, D.; Balaram, J.; Tzanetos, T.; Grip, H., "Mars Science Helicopter Conceptual Design," NASA/TM–2020-220485.

7. Silva, C.; Johnson, W., "Practical Conceptual Design of Quieter Urban VTOL Aircraft," *Presented at the Vertical Flight Society's 77th Annual Forum & Technology Display*, Virtual, May 10-14, 2021.

8. Collins, K.; and Sankar, L.N., "Application of Low and High-Fidelity Simulation Tools to Helicopter Rotor Blade Optimization," *AHS 65th Annual Forum*, May 2009.

9. Collins, K., Bain, K., Rajmohan, N., Sankar, L. N., Egolf, T. A., Janakiram, R. D., Brentner, K., and Lopes, L., "Toward a High-Fidelity Helicopter Rotor Redesign Framework," *AHS 64th Annual Forum*, 2008.

10. Ku, M., Gahlot, A., and Sankar, L. N., "Application of an Inverse Design Methodology to Rotor Tip Design," *Proceedings of the European Rotorcraft Forum*, September 2024.

11. Sankar, L. N., and Tung, C., "A Combined Low and High-Fidelity Approach for an eSTOL Rotor Modeling and Design," *Proceedings of the 2022 VFS Transformative Vertical Flight Technical Meeting*, San Jose, CA, January 24–27, 2022.

Building Enhanced Resilience into Aviation

Larry A. Young[1]
NASA Ames Research Center
Moffett Field, CA 94035

ABSTRACT

The aviation sector is a very complex but generally successful enterprise within the United States. Unfortunately, over multiple decades a series of demonstrable major aviation disruptions have occurred, for a wide-ranging set of reasons. Thus, it is more than timely to consider how to enhance the resilience of the aviation sector. This will entail developing research projects that promote system-of-systems engineering solutions that focus on both vehicle design advances and novel airspace system architectures. Several such potential research initiatives are outlined in this paper.

NOMENCLATURE

A	Rotor disk area, ft^2; $A=\pi R^2$
A_P	Stopped-cycloidal-rotor projected area, ft^2
CTOL	Conventional takeoff and landing
c_R	Mean chord of rotors, ft
c_W	Mean chord of the fixed wing of aircraft, ft
C_P	Rotor, or proprotor, or fan power coefficient, nondim.; $C_P = P/\rho A V_{tip}^3$
C_P^*	Stopped-cycloidal-rotor power coefficient, nondim.; $C_P^* = P_{Total}/\rho A_P V_{blade}^3$
C_T	Rotor, or proprotor, or fan thrust coefficient, nondim.; $C_T = T/\rho A V_{tip}^2$
C_T^*	Stopped-cycloidal-rotor thrust coefficient, nondim.; $C_T^* = T_{Total}/\rho A_P V_{blade}^2$
DOF	Degree-of-freedom, nondim.
i_N	Nacelle tilt, Deg.
i_P	Nacelle pivot, Deg.
L/D	Lift to drag ratio, nondim.
L/D_e	Effective lift to drag ratio for rotorcraft, $L/D_e = WV/P$
P	Vehicle total power,
PAX	Number of passengers, nondim.
RIA	Runway independent aircraft
R	Rotor radius, ft
SOA	State of the art
STOL	Short takeoff and landing
S	(Fixed-) wing planform area, ft^2
V	Cruise velocity, ft/s (m/s)
VTOL	Vertical takeoff and landing
W	Takeoff gross weight of vehicle, lb_f
α	Angle of attack, Deg.
$\Delta D_e/L$	Delta effective drag to lift ratio, nondim.

INTRODUCTION

The aviation sector is a very complex but generally successful enterprise within the United States. It spans a spectrum of stakeholders that range from: the traveling public, to aircraft researchers and developers, to airport operators, to airlines and other aircraft operators, to the public/government

[1] Presented at the 11th Biennial Autonomous VTOL Technical Meeting, Phoenix, AZ, Feb 4-6, 2025.

regulators and officials. Overall, the aviation sector is critical to the smooth operation of society. After a series of demonstrable major disruptions to aviation over multiple decades, it is appropriate to consider how to enhance the overall resilience of the aviation sector. This will entail developing research programs that promote system-of-systems engineering solutions that focus on vehicle design advances and novel airspace system architectures. Several such potential research initiatives will be outlined in this paper.

BACKGROUND

Since the Oil Crisis of the 1970's, aviation has seen several major disruptions. These disruptions include the 2001 9/11 terrorist attacks, the 2008 recession, the 2020-2021 coronavirus pandemic, and emerging impacts from climate change. It is appropriate to consider both from a vehicle and an airspace system/networks perspective how to build additional 'resilience' into the aviation sector.

To accomplish this goal, it is necessary to consider engineering solutions to make the aviation sector more resilient to anticipated and potential future crises. To help define those solutions it is important to consider the following questions:

1. How can the aviation fleet be quickly 'right sized' and/or re-configured to meet the challenges of rapid changes in the economic operational environment?
2. How can aviation be refined to improve economic opportunities for urban and rural/regional communities in the US?
3. How can aviation respond to changes stemming from environmental challenges, including climate change?
4. How can aviation evolve to respond to the ever-increasing challenges of disaster relief and emergency response?

PRELIMINARY DISCUSSION OF PROBLEM

The following are some foundational principles that can aid in addressing the above questions:

I. Vertical lift (and short takeoff and landing (STOL)) aerial vehicles, particularly those having hybrid-electric propulsion, have the potential to expand into and fill several subsectors of the aviation sector, Ref. 1.

II. Autonomous systems and robotics technology also potentially have a profound influence on aviation.

III. Aviation can and must expand to new markets and to new applications to meet critical society needs. This, in turn, will economically sustain aviation during difficult times.

IV. Modular and distributed vehicles and/or subsystems could radically change aircraft design and operations.

V. Wholly new types of amphibious and multimodal vehicles and networks also have an opportunity to transform urban and regional aviation markets.

VI. A broad spectrum of vehicle sizes and capabilities will be required for building aviation resilience (e.g., disaster relief and emergency response, Ref. 2).

These foundational principles and their implications for addressing the question of building enhanced resilience into the aviation system will be discussed. Figures 1-2 are some examples of how runway independent aircraft (VTOL, VSTOL, and STOL aircraft) will be a vital component of that proposed enhanced resilience.

Figure 1. (VTOL, STOL, and V/STOL) Runway Independent Aircraft (RIA) to build enhanced aviation resiliency (Ref. 1, 3-4)

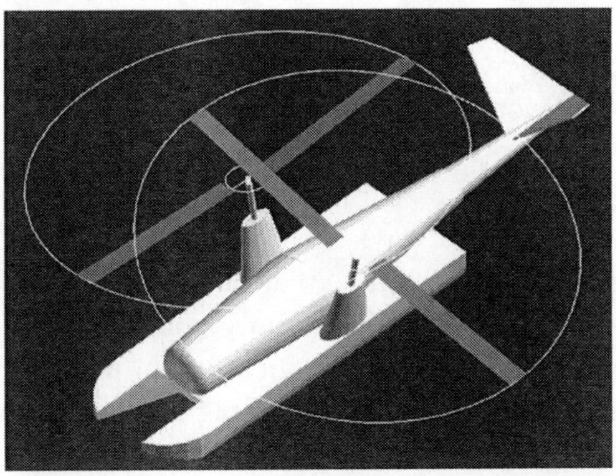

Figure 2. An emerging possible role for amphibious rotorcraft/aircraft (Ref. 5)

Figure 3 illustrates a possible new generation of STOL or V/STOL aircraft to maximize mission flexibility by being able to operate from vertiports, (short or long runway) airports, and even operate amphibiously. It emphasizes the use of distributed, and heterogenous rotors, and tilts the wings and rotor nacelles independently to enable this V/STOL flexibility.

Figure 3. Hybrid tiltwing/tiltrotor aircraft with heterogeneous rotors and propellers for flexible V/STOL operation

The remaining portion of this paper will be divided into two parts. The first part of the paper will consider *system-of-systems architecture issues* related to developing new airspace and aerial transport systems for the future. The second part of the paper will examine *vehicle design considerations* for new classes of vehicles that might respond to making the aviation system of the future more resilient. Some novel or speculative/exotic (fixed- and rotary-wing) aircraft configurations are introduced. Finally, there is an appendix at the end of the paper that summarizes some initial design requirements and vehicle sizing considerations for potential amphibious V/STOL aircraft that might be responsive to adapting to future climate change impacts as well as overall trying to enable sustainable, green aviation for regional commercial rotorcraft.

SYSTEM-OF-SYSTEMS AIRSPACE/AERIAL TRANSPORT ARCHITECTURE OPPORTUNITIES

Airspace resilience must be provided for: (1) short-term or transient unexpected events, (2) moderate-duration single-occurrence acute, or periodic chronic occurrences, that may or may not be anticipable and (3) long-term, progressive trends related to technological, environmental, or socioeconomic changes. Engineering solutions for resilience for airspace systems that fall within each of these three categories of problems will present unique challenges. These solutions will be engineered into existing or future systems.

Airspace Flexibility to Respond

An example of airspace flexibility is the ability to accommodate self-organizing emergency networks using vertical lift aerial vehicles. One early study into this type of emergency VTOL networks is Ref. 2. This issue of repurposing or commandeering urban air mobility vehicles during emergencies was also touched upon in Ref. 6; this is analogous to purposing public transit buses to support hurricane evacuations. Accordingly, the development of novel aerial transportation networks should seek to balance out commercial transport interests and potential public service interest (utility) of such potential dual-purpose vehicles. Reference 7 discusses this opportunity for dual-purpose aerial vehicles in the context of 'delivery drones' being pressed into public service disaster relief and emergency response (DRER) efforts, i.e., instead of delivering online shopping purchases, if an emergency occurs, the same drones could be delivering small packages of food aid or first-aid medical supplies.

Optimizing for Multimodal Mobility Systems

A few rotorcraft aerial transport network concepts have been, and are being, explored at NASA Ames. These include 'Skimmer,' 'New Nomads,' 'Pogo,' and 'Unity' networks. Each aerial transportation network concept seeks to introduce some novel mission capability to urban and regional aerial transport of people and goods.

'Skimmer' networks are networks for amphibious VTOL aerial vehicles (e.g., Fig. 2). Refer to Ref. 5 for more details. As interest in urban aerial mobility grows, it is worthwhile to consider whether amphibious vertical takeoff and landing vehicles can play an important role in providing such mobility.

'New Nomads' networks are about integrating ground mobility with aerial mobility with transport of not just passengers but work- and living-space 'habitats' modules, Ref. 8 (Fig. 4). Such habitat modules could be transported by autonomous tugs on roadways and/or transported by 'skycrane' like VTOL platforms. This would potentially address not only traffic congestion but other critical urban planning challenges, such as providing for adequate cost-effective housing and reduced office-space infrastructure to promote economic growth.

Figure 4. New Nomad short-range multirotor configuration (Ref. 8)

'Pogo' networks are a hybrid-electric VTOL/VSTOL vehicles analog to bus and rail networks (Fig. 5). For example, a linear network of vertiport charging stations for hybrid VTOL/VSTOL vehicles could potentially be overlayed with respect to existing rail stations and freeway rest stops. This would balance vehicle per-charge range with a modest number of large rail-station-like vertiports/charging stations. Developing regional CTOL electric-propulsion vehicles is worthy of research but, likely, such longer-range CTOL vehicles will be hybrid-electric in nature. This will reduce their carbon-footprint by some modest amount but will not have the payoff of all-electric vehicles. The only way to have longer-range all-electric vehicles is to string together pogo-like hops (a series of short-haul flights) with moderately frequent recharging and/or swapping of vehicles.

Figure 5. Notional Pogo Network Vehicle (AI-generated image using Microsoft Copilot); railway-like regional air mobility

Figure 6 is another possible notional vehicle tilting ducted fan configuration (see Refs. 9-10) that might be responsive to the Pogo network concept. Unlike the more conventional ducted fan aerial vehicle portrayed in the Fig. 5 artwork, the Fig. 6 tilting ducted fan vehicle employs oval ducts and embedded tandem fans/rotors to provide a simple flight control approach that is partly inspired by quadrotor drone flight control in hover and low speed, helicopter-mode, edgewise forward flight.

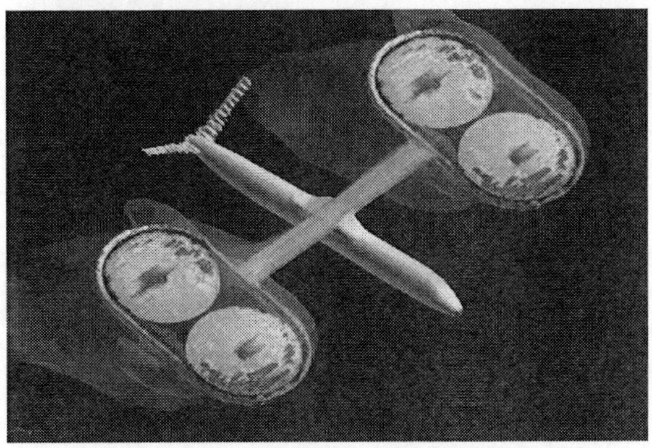

Figure 6. Another tilting ducted fan aerial vehicle (depicting early transition/conversion)

Several possible hybrid-electric tiltwing, tilt-duct, tiltrotor vehicles (e.g., Refs. 1, 9-10) could potentially support regional air mobility with eVTOL or eSTOL requirements. Mid-fidelity CFD predictions of another notional, potential Pogo network vehicle (based off the 'Flex Type-A' concept noted in Ref. 1) is shown in Fig. 7a-d.

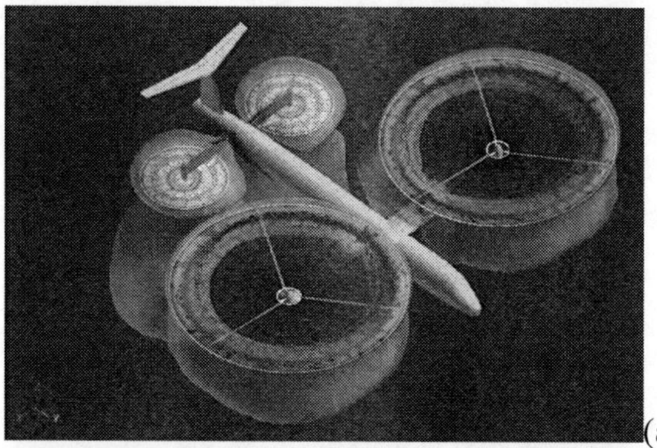
(a)

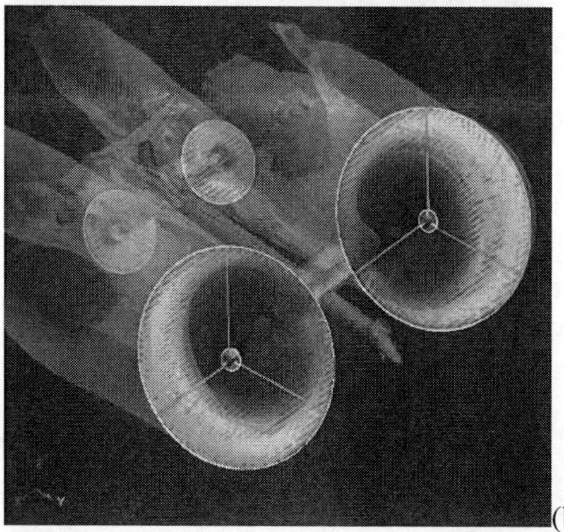
(b)

(c)

(d)

Figure 7. CFD of a Pogo network vehicle concept: (a) hover, (b) transition or conversion, (c) early cruise, and (d) late cruise

'Unity' networks are aerial vehicle networks that link local or rural communities with urban centers (Fig. 8). Such networks, through a combination of CTOL, VSTOL, and VTOL electric-propulsion vehicles encompassing moderate-sized passenger-carrying or cargo-carrying vehicles and small aerial robots, would increase regional economic strength as well as improve connectivity with nearby urban centers. An example of the criticality of this need is emergency hospital services for rural communities that no longer have local hospitals.

Figure 8. Notional Unity Network Vehicle (AI-generated image using Microsoft Copilot)

Mid-fidelity CFD predictions of a Unity network vehicle is shown in Fig. 9a-b; predictions performed with the RotCFD software tool, Ref. 23. Multirotor vehicle configurations, with or without ducts surrounding the rotors, have become quite popular with respect to the urban air mobility (UAM) developer community. UAM vehicles with modest design adjustments might make good candidate vehicles for Unity networks.

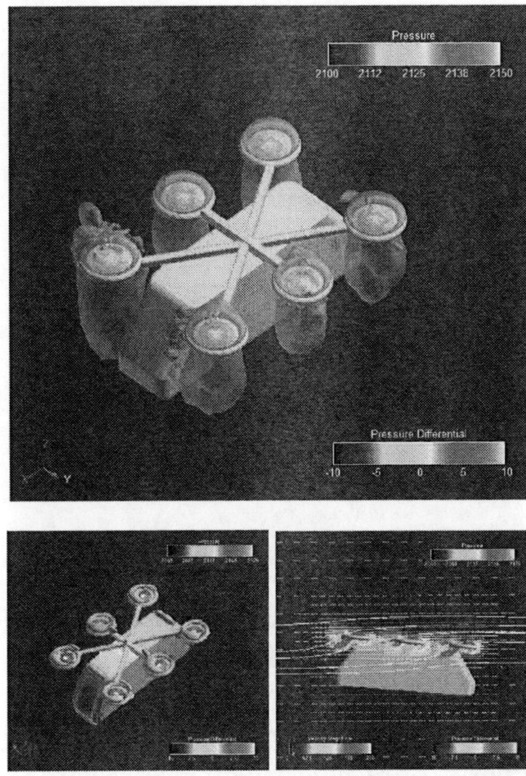

(a)

(b)

Figure 9. Mid-fidelity CFD (isosurfaces of velocity magnitude, body surface pressures, and rotor disk differential pressures) of a Unity network vehicle concept: (a) hover and (b) forward flight

Sustainable/green aviation will only be fully realizable if both the vehicle design and the system-of-systems network architecture are considered in a combined holistic manner. Figure 10 illustrates a notional integrated set of Pogo (regional) and Unity (local) networks along the US west coast. This continues the analogy of railway networks in the conceptualization and study of the Pogo and Unity networks.

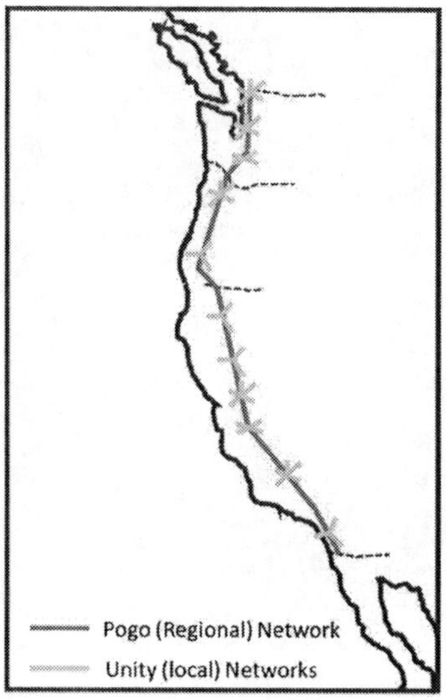

Figure 10. Unity and Pogo Networks integrated together in a notional west coast system

Vertiports (Large and Small) Everywhere

Vertiports for large and modest-sized aircraft for passenger/cargo-carrying are currently being discussed in detail by researchers. Smaller neighborhood vertiports, automated 'sentinel stations,' and even kiosk-type vertiports have not been fully considered. Some examples of limited work in this area are Refs. 7 and 11. Future research needs to focus more on this area of investigation.

This issue regarding the potential introduction of many vertiports – both large passenger-carrying vehicles (Fig. 11) and smaller 'drones' – is an important consideration in attempting to build in enhance resilience into the aviation system of the future.

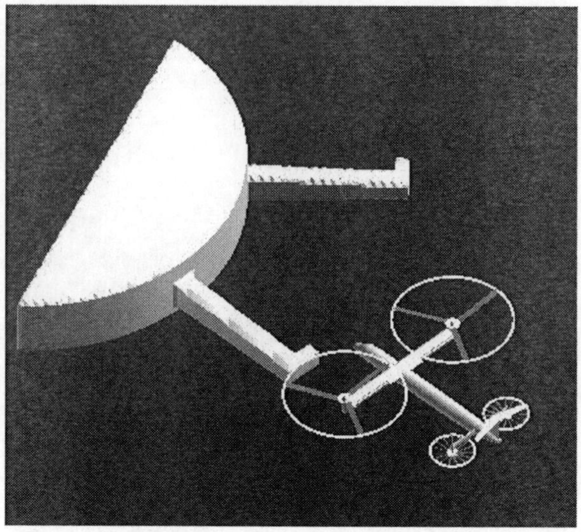

Figure 11. Large vertiports for passenger-carrying rotorcraft at airports

Merging/Evolving UTM and UAM and Radically Transforming Air Traffic Control

This paper suggests several research opportunities that, if pursued, will necessitate a radical transformation of current UAV Traffic Management (UTM), urban air mobility (UAM), and air traffic control (ATC) concepts or architectures. Some near-term research opportunities that are primarily focused on the intersections of vehicle and airspace research are outlined below:

1. Taking a vehicle sizing tool like NDARC (Ref. 12) and exporting out input files for NASA airspace analysis tools like FACET (Ref. 13), etc.

2. Taking ANOPP (Ref. 14) acoustic results and developing unique low-noise flight trajectories for novel aircraft configurations that could be fed into airspace tools.

3. Using airspace tools to what-if novel flight trajectories that then could be used to specify mission/design requirements for

sizing tools like NDARC or flight control handling-quality tools like FlightCODE (See Ref. 15).

4. Taking results from FlightCODE to define emergency maneuvers and/or contingency flight paths that then could be factored or incorporated into airspace simulations.

Vehicle-centric and airspace-centric analysis tools currently require a great deal of manual effort to exchange/share results with each other – especially for novel aircraft and/or rotorcraft. To improve and/or automate the process for information exchange between analysis tools could be a critical modeling and simulation need for the future. The above list of analysis tools and the opportunities for exchange or sharing of results is only a partial summary of the overall analysis needs. Relevant parsing tools may also need to be developed for efficient transfer of data between the various analysis tools.

Aerial Robotic Ecosystems

It is possible in the far future that aerial (or hybrid aerial/ground/water) robots will come to number almost as many as birds in the sky thereby resulting in large aerial robotic ecosystems, aka 'mech life.' Clearly centralized control of such aerial robotic ecosystems will not be feasible. Further, this aerial robotic ecosystem is going to have to safely coexist with more conventional aerial vehicles and conventional airspace management systems and strategies.

This additional issue regarding the potential introduction of whole 'aerial robotic ecosystems' that must nonintrusive interact with and or cooperate with passenger-carry aircraft is yet another example of a huge potential leap in aviation system complexity that might occur in the mid- to far-term future. One example of such an 'aerial robotic ecosystem' is the forest replanting aerial robots of Ref. 16. The nascent ideas of Ref. 16 have now become mainstream with the introduction of aerial robots performing actual forestry services, Ref. 17.

VEHICLE DESIGN OPPORTUNITIES

There is clearly a role for advanced aerial vehicle design in building in enhanced resilience into our aviation system. A few ideas for addressing such enhanced resilience via aircraft design (and the associated system-of-systems architectures employing such vehicles) will now be presented. It is important to note though that the presented ideas are to inspiration other researchers and developers to propose their own ideas.

Expanding Vertical Lift Aerial Vehicles into all Aviation Sectors

Work throughout the mid-1990's to the late-2000's showed the potential for civil tiltrotor aircraft to meet short haul and regional commercial aircraft markets (e.g., Refs. 3-4). The potential was also studied and demonstrated for tiltrotor aircraft to satisfy humanitarian relief and disaster relief and emergency response missions (e.g., Ref. 2). These studies somewhat fell to the sidelines, but the results are still just as valid today as much as they were in the past.

Enabling Modular and/or Distributed Aircraft

Advanced structural design and manufacturing concepts, particularly for composite structures, coupled with advanced flight computer and controls and possible propulsion electrification (all- or hybrid-electric), could enable more flexible airframe/vehicle configurations.

Monohull airframes could use swappable 'plugs' in the fuselage to reduce cabin size and passenger loads, while still being aerodynamically tailored, respectively, to be low drag in both configurations. These swappable plugs should be enabled such that very little retrofitting/remanufacturing is required; ideally this could be accomplished almost on-demand. Twin hull CTOL aircraft have in the past been examined before by NASA. But this was primarily from a fuel efficiency perspective and not from a 'resilience' perspective. Twin hull tiltrotor aircraft have also been briefly studied, Ref. 1. Swappable wing extensions should also be studied in the context of resilience engineering.

Monohull airframes could be replaced by shorter fineness-ratios multiple hulls that could be mated at multiple attach points on the primary wing. These multiple hulls could be staggered spanwise (with the respect to the primary wing) or even staggered vertically in a closely packed geometric arrangement. Hulls could be removed or added to increase cabin space and passenger loads on-demand. Two twin hull tiltrotor configurations were initially examined in Ref. 1. Figure 12 illustrates yet another twin hull tiltrotor conceptual design approach.[2]

Figure 12. One possible twin hull tiltrotor configuration (artwork AI-generated by Microsoft Copilot)

One way of 'right-sizing' aircraft is to form large aircraft from multiple small aircraft using various 'parasite' aircraft ensembles; for example, see Ref. 24. Parasite aircraft formations can be formed with individual aircraft with swept or un-swept wings or, alternatively, small aircraft with oblique or scissor (pivoting) wings. Such parasite aircraft ensembles could be applied to both fixed-wing (CTOL) and rotary-wing (VTOL) vehicles. Closely spaced rotors or proprotors can also potentially be accommodated in parasite aircraft ensembles by using inter-meshing rotors.

Refer to Fig. 13 for a modular high-speed rotorcraft with tilting rotors. Previous work has studied modular rotorcraft (multirotor configurations with non-tilting rotors in edgewise flight). This, however, is the first presentation of a modular rotorcraft with tilting rotors. Each individual rotorcraft

[2] This figure includes one of a few AI-generated images in this paper. AI images proved to be useful for early conceptualization or ideation but did suffer from image artifacts that ideally require manual artwork editing to improve the images. Such artwork editing was not performed for this paper to show the reader some of the artwork artifacts/defects that might occur with the current generation of software, including extraneous rotor/propellers being included in the artwork as well as extraneous wings, tail surfaces, and landing gear. On the other hand, some of the graphic art details (backgrounds, cabin interiors, etc.) were compelling from a first-order perspective.

element must be attached at the other rotorcraft (tiltrotor aircraft) wingtips with some level of structural/mechanical fixation and rigidity. The key limitation of this type of modular tilting rotor configuration is that the individual rotorcraft elements are limited to linear arrays in forming the aggregate or ensemble whole versus the two-dimensional matrix aggregates that can be formed with modular multirotor configurations.

(a)

(b)

Figure 13. Modular high-speed rotorcraft – three elements, in an array, comprising the rotorcraft (in hover configuration): (a) planform view and (b) isometric view

Figure 14 is a five-element modular rotorcraft array. Each rotorcraft element has a pair of side-by-side tractor, or a pusher, type tilting rotors. This is very evocative of World War II wingtip-mounted 'parasite flier' aircraft concepts, though adapted to rotorcraft or tiltrotor aircraft versus conventional aircraft.

(a)

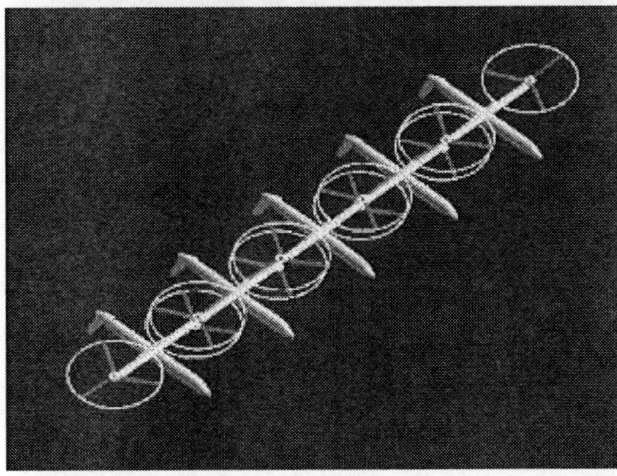
(b)

Figure 14. Modular high-speed rotorcraft – five elements, in an array, comprising the rotorcraft (in hover configuration): (a) planform view and (b) isometric view

These modular tilting rotor configurations could be assembled on the factory floor as a semi-permanent configuration or, even more ambitiously, assembled on the flight line to meet the needs of each individual flight.

Figure 15 presents some initial mid-fidelity CFD predictions for the three-element modular high-speed rotorcraft illustrated in Fig. 13 for hover, transition, and cruise conditions. Wing and fuselage surface pressures are shown; additionally, isosurfaces of velocity magnitude are also shown to highlight the rotor wakes in hover, transition, and cruise.

(a)

(b)

(c)

Figure 15. Modular high-speed rotorcraft – three elements, in an array, comprising the rotorcraft: (a) hover, (b) transition, and (c) cruise

Figure 16a-d presents planform views of the three-element tilting modular rotorcraft configuration for an angle of attack sweep. The wing upper surface pressures are shown as well as rotor wake velocity magnitude isosurfaces in cruise.

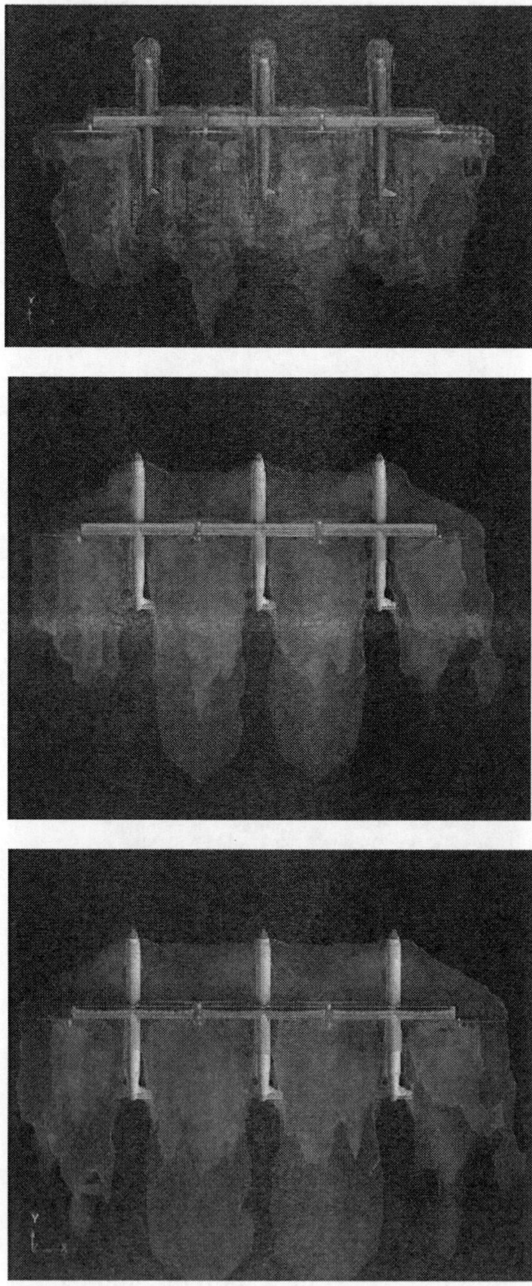

(a)

(b)

(c)

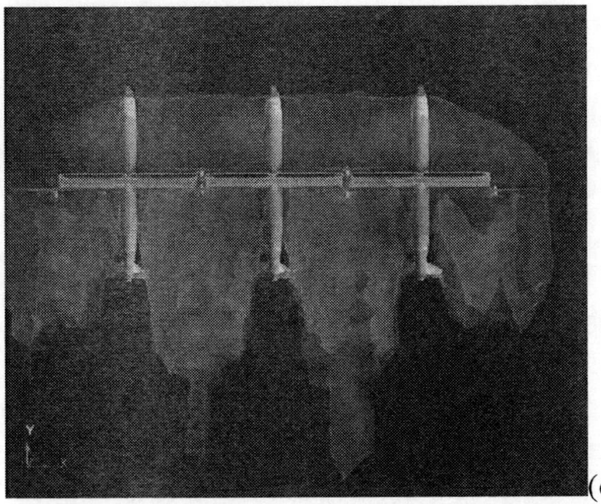

(d)

Figure 16. Modular tilting rotor (three-element) aircraft configuration initial mid-fidelity CFD predictions for an angle-of-attack sweep in cruise: (a) AOA=0 Deg., (b) AOA =2 Deg., (c) AOA=4 Deg., (d) AOA=6 Deg.

Figure 17a-c presents an illustrative set of mid-fidelity CFD predictions that considers '…the sum of its parts' question[3] for modular rotorcraft, in this case modular tilting rotor aircraft in linear arrays/arrangements, e.g., Figs. 13-14. The anticipated increase in lift efficiency because of the aggregate increase in wingspan is clearly seen in Fig. 17a. However, there does appear to be an increase in net parasite drag due to the complexity of the geometry at the wing tips of some of the vehicle elements; this complexity results from aerodynamic interference effects stemming from some wing tips effectively having two nacelles (one for a tractor proprotor and one for a pusher proprotor) mounted (in proximity) to them. Future work will have to examine how to reduce net parasite drag for any modular tilting rotor (aggregate/ensemble) configuration. This increase in parasite drag with the three element vehicle configuration is why its lift-to-drag ratio is less than that of the one element (tractor) configuration.

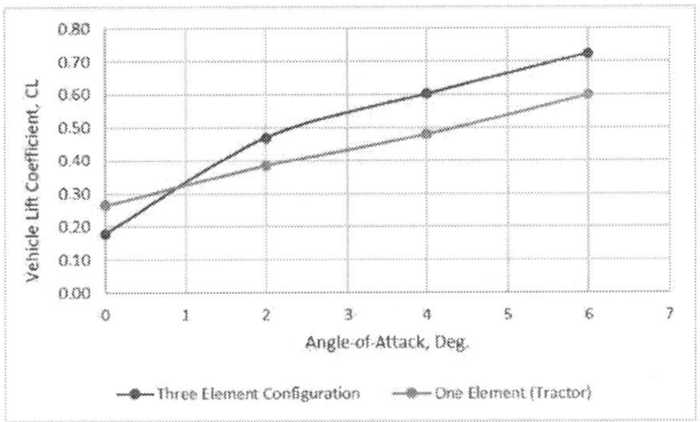

(a)

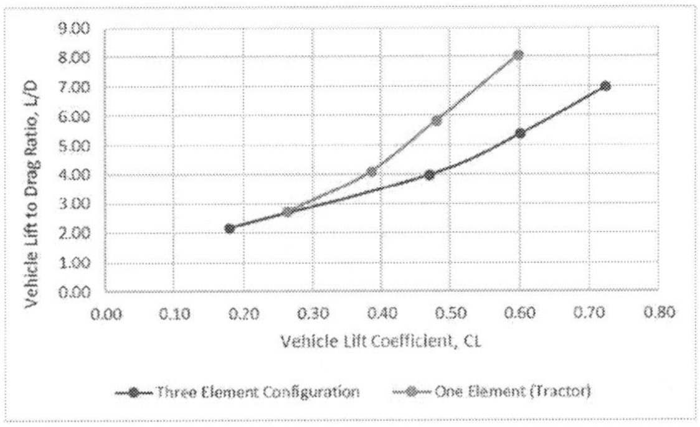

(b)

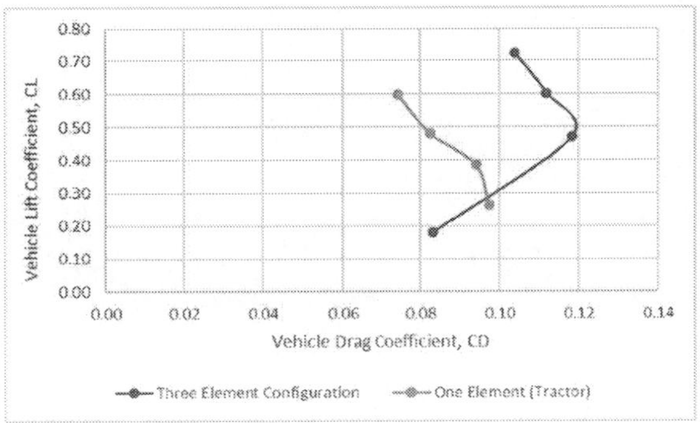

(c)

Figure 17. Mid-fidelity CFD predictions of aggregate modular tilting rotor aircraft: (a) CL versus AOA, (b) L/D versus C_L and (c) C_L versus C_D

[3] The complete quote is attributed to Aristotle: "The whole is greater than the sum of its parts."

Exotic Modular Conventional Takeoff and Landing (CTOL) Aircraft

In the late nineties, exotic nonlinear lifting surface CTOL aircraft were being considered to address future aviation, Ref. 18. Figure 18 is concept artwork to introduce a novel 'skytrain'[4] vehicle that could use ring-wings or elliptical/oval-wings as compact lifting surfaces. Ring-wings and elliptical/oval-wings do not completely eliminate trailed tip vortices but, perhaps, reduce their strength. Accordingly, it is possible that optimized ensembles of ring or oval wings might reduce the net magnitude of induced velocities from one set of elliptical/oval-wings on another subsequent downstream set of wings.

Figure 18. 'Skytrain" CTOL ring-wing vehicles (AI-generated image by Microsoft Copilot)

Figure 19 is an isometric view of a CAD/CFD model for the skytrain CTOL concept. Instead of using ring wings, a pair of tandem oval wings is employed for each modular 'car' of the skytrain ensemble or aggregate. It is initially assumed in this early modeling that some amount of kinematic and/or aeroelastic degrees-of-freedom (one DOF as shown, i.e., a lateral back and forth 'wagging') is allowed between 'cars' but this is purely notional at this point. The lead car has a singular cabin like element added to its front for streamlining and reduction of drag. The trailing car has a 'caboose' like tail added to it, again for streaming and reduction of drag. The 'joints' between cars is again purely notional (again to be evocative of train-like inter-car connections); clearly the geometry/transition of these 'joints' between cars would be critical to keep overall skytrain drag to a minimum. The use of ring wings, or elliptical wings, or oval wings (as in this CAD/CFD model) is intended to reduce trailed vorticity (and induced angular velocities) to a minimum to minimize wing-to-wing aerodynamic interference effects. The oval wings being shown in Fig. 19 (and the ring wings in Fig. 18) are of a constant chord all along the perimeter or circumference of the wings. This is not likely optimal. It is more likely, and something appropriate for future study, to consider tapering the oval or elliptical wings in the vicinity of the more circular arcs of the wings (i.e., at their 'tips'). The whole oval wing versus elliptical wing – and intermediate wing design space – is a very interesting topic for future study. Finally, the fineness ratio of the oval/elliptical wings should be explored; one extreme is a ring wing and the other extreme is a purely planar biplane with edge-plates configuration). Finally, each 'car' has

[4] The notional skytrain CTOL concept presented is not referring to the Vancouver, BC, 'SkyTrain' metro railway, or the Phoenix airport 'Sky Train' metro, or even the C-47A "Skytrain" WWII paratrooper carrier aircraft (which was also used to tow gliders in combat).

its own twin pair of ducted fan propulsors, or turbofan engines, nested in between the leading oval wing of the car.

The initial work in this paper, that of the aerodynamics of oval wings, somewhat indirectly follows previous work in Refs. 9-10 that studied oval ducts for tilting ducted fan VTOL aircraft.

The purpose of discussing such an exotic, speculative aircraft configuration (as was done for the modular tilting rotor vehicles of Figs. 13-14) is as an attempt to promote 'out-of-the-box' thinking and not necessarily an advocacy for one or more vehicle concept. But more than anything, both vehicle concepts seek to address in principle the assertion of this paper that resiliency in the future aviation system will be tied to the overarching theme of rightsizing aircraft for markets demands through use of some level of vehicle modularity.

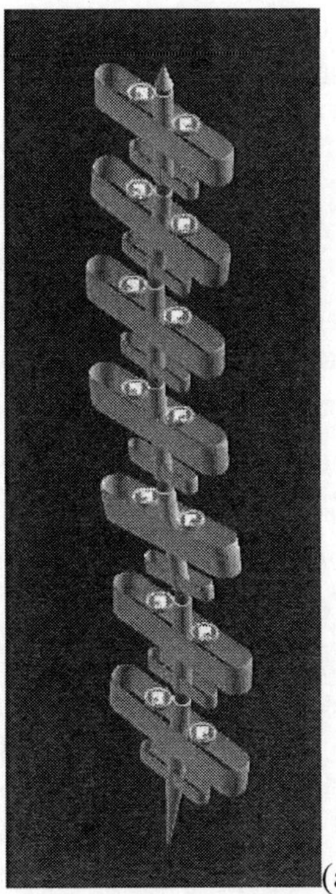
(a)

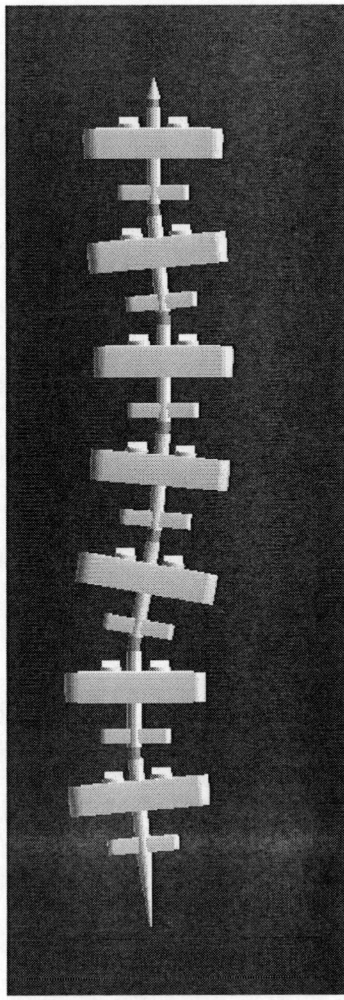
(b)

Figure 19. Skytrain (modular CTOL vehicle concept): (a) isometric view and (b) planform view

Figure 20a-b presents some mid-fidelity CFD results for the skytrain notional configuration shown in Fig. 19 for a one and two car configuration at zero angle-of-attack. A three car configuration is shown in Fig. 21a.

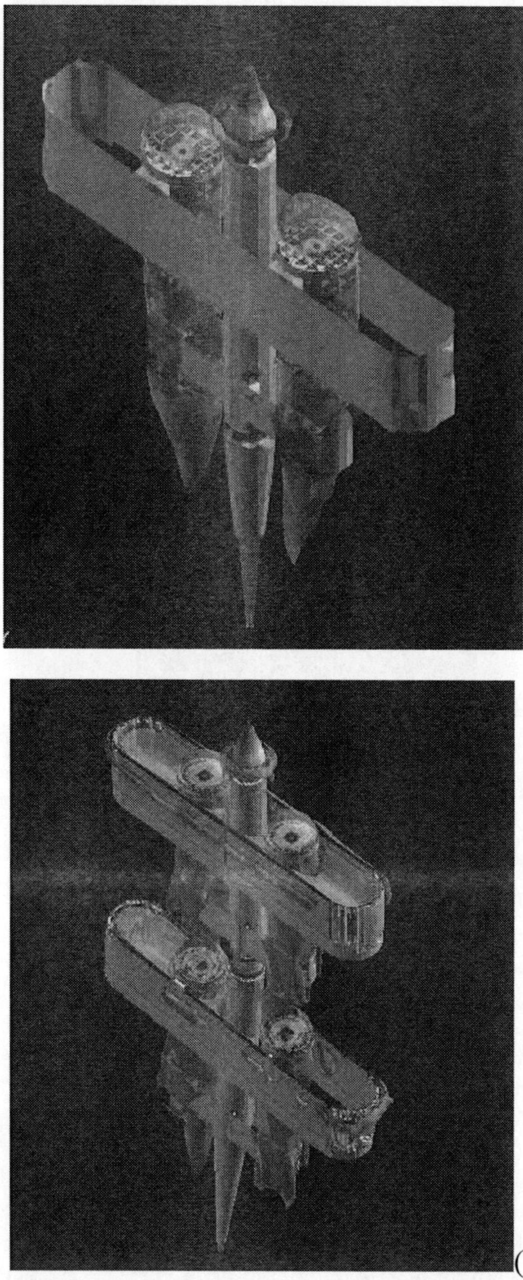

(a)

(b)

Figure 20. Mid-fidelity CFD (isosurfaces of velocity magnitude, body surface pressures, and rotor disk differential pressures) of a skytrain vehicle concept at AOA=0 Deg.: (a) one 'car' and (b) two car

Advances in magnetic levitation (maglev) rail systems are also making it possible to consider the use of maglev for aircraft takeoff and landing, e.g., Ref. 19. Use of maglev for skytrain vehicles might make for an interesting complementary or parallel study. If maglev does have some role to play as to aerial vehicle surface locomotion (taxiing and takeoff), then an even more speculative line of investigation might be to consider whether a skytrain configuration might also play a dual role in a multi-modal transportation system. The skytrain could be used in place of railway surface transport by elevated trains but instead by a skytrain on the ground being propelled by maglev until it gets to the airport where it could takeoff and fly on wings and turbofan engines.

Figure 21a-d presents some limited mid-fidelity CFD results for a three-car skytrain vehicle concept configuration. (CFD models with the mid-fidelity RotCFD software tool had difficulty with four or more skytrain cars; hopefully this won't pose a problem for future investigations.)

Figure 21a-d presents wing and fuselage surface pressures as well as isosurfaces of flow field velocity magnitude (allowing visualization of turbo-fan exhaust/wakes as well as the accelerating flow above the wing leading edges). The increasing leading-edge upper suction pressures (deeper blue hue with increased suction pressure) with angle of attack can be clearly seen in Fig. 21a-d. From such predictions the aerodynamic interference effects of one set of wings on other (downstream) wings can be assessed.

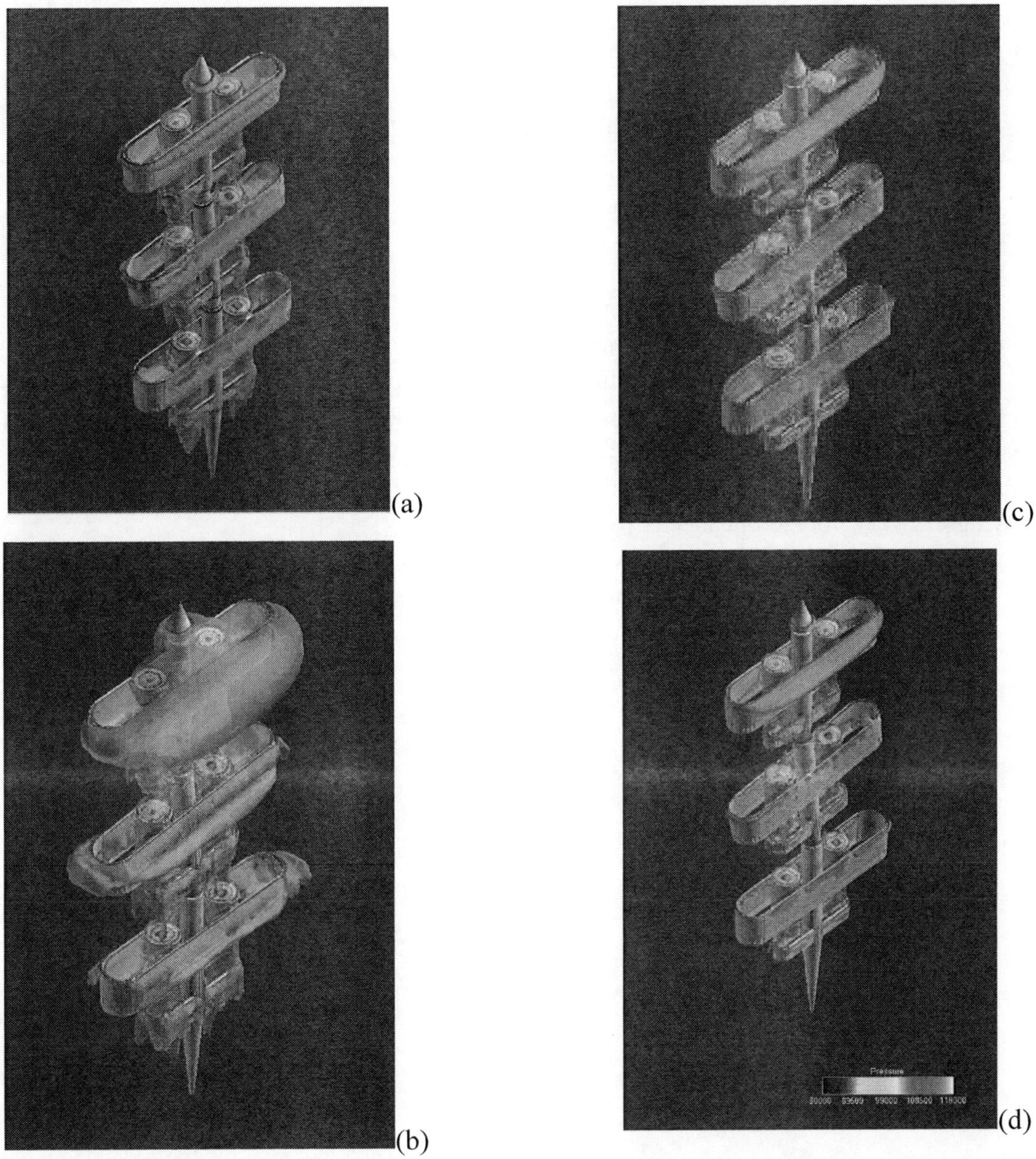

Figure 21. Mid-fidelity CFD (isosurfaces of velocity magnitude, body surface pressures, and rotor disk differential pressures) of a top/planform isometric view of skytrain three-car vehicle concept: (a) AOA=0 Deg., (b) AOA=2, (c) AOA=4 Deg., (d) AOA=6 Deg.

Figure 22a-c presents an illustrative set of initial predictions that considers '…the sum of its parts' question[5] for modular skytrain eCTOL or CTOL concept, e.g., Fig. 19. It is clear from the initial mid-fidelity CFD results that the use of the oval wings did not reduce the magnitude of the induced velocity aerodynamic interference effects of forward wings influencing the aft, downstream, wing as much as might be hoped. The lift curve slope of the three-car configuration is only about sixty percent of the one-car baseline. Part of the problem might lie in the gridding fidelity of the automated grid body fitting employed during this modeling; the gridding along the circumference of the 'oval' section of the wings is very coarse and needs to be further improved/refined. This is clearly an area of future aerodynamic investigation to make this exotic aircraft concept feasible.

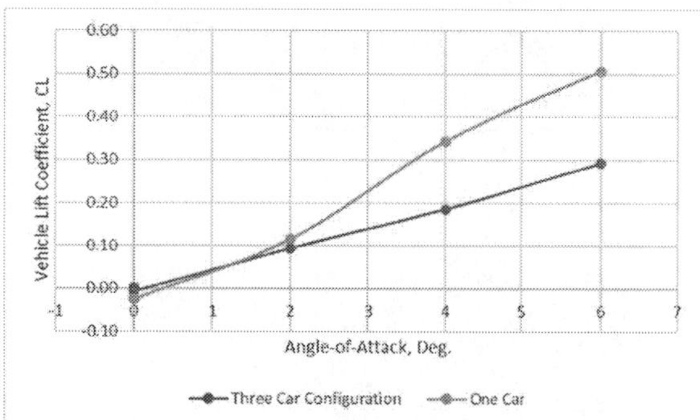

(a)

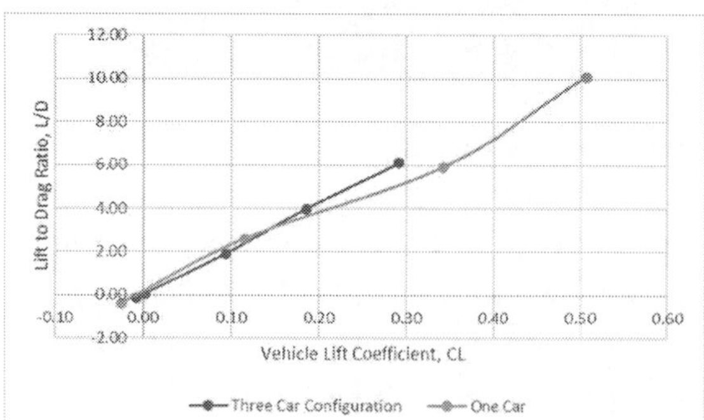

(b)

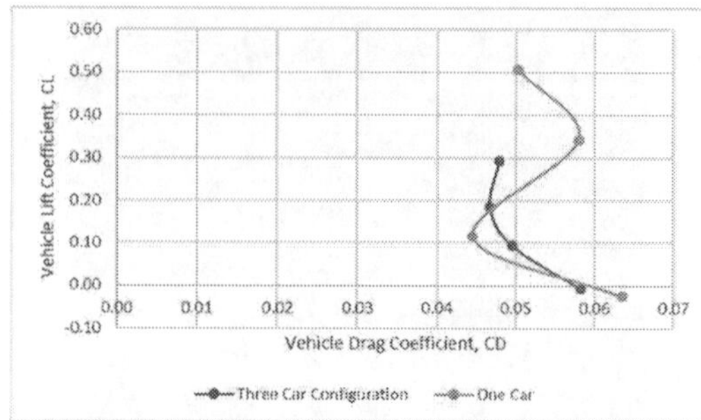

(c)

Figure 22. Mid-fidelity CFD predictions of aggregate modular skytrain concept: (a) CL versus AOA, (b) L/D versus CL and (c) CL versus CD

Providing 'Plug and Play' Propulsion Systems and Airframe Structures

Modularity of aircraft – especially with respect to propulsion system elements to quickly go from one mode of operation such as VTOL flights, to STOL flights, to CTOL flights – could be the key factor in the realization of sustainable aviation for rotorcraft and other runway independent aircraft.

With respect to propulsion system modularity, a turboshaft-driven electric generator 'pallet' could be swapped out with a large battery pack to quickly go from a hybrid-electric to an all-electric aircraft. If battery packs could be swiftly removed or installed on-the-ground between flights then a single V/STOL hybrid-electric runway independent aircraft could efficiently, sustainably fly on-demand VTOL, STOL, or CTOL flight profiles and, thereby, optimize its takeoff weight and overall flight performance efficiency for each individual flight.

[5] The complete quote is attributed to Aristotle: "The whole is greater than the sum of its parts."

Further, from an overall vehicle design reconfigurability perspective, it should be feasible (and aspirational from an engineering perspective) to quickly exchange propulsion systems on a given airframe without substantial modifications: exchange a turbofan with pusher or tractor advanced propfans or even more radical to exchange these propulsion systems with rotary-wing side-by-side rotors (or tiltrotor proprotors) and gearbox/turboshaft engines. Or, alternatively, it should be theoretically feasible (and aspirational from an engineering perspective) to quickly exchange a CTOL 'cruise optimized' wing with a VTOL 'jump/takeoff optimized' wing for side-by-side or tiltrotor type rotorcraft configurations. See Ref. 20 for one example of a study investigating such design reconfigurability, or rather 'conversion' capability.

Transitioning from Urban Air Mobility to Regional Aerial Transportation Systems

The utility of urban air mobility would be greatly increased if the technology were extended to regional aerial systems, i.e., flights between neighboring cities. This might, in turn, increasingly justify examining hybrid-electric propulsion versus all-electric propulsion. It may, as well, require examining how to better incorporate multimodality mobility into the overall transportation system (e.g., networks comprised of amphibious vertical takeoff and landing aerial vehicles, Ref. 5).

Recent advancements within the urban air mobility sector have begun the transition from urban to regional markets, albeit with still relatively small, short-range, and slow aircraft. But incremental progress is still progress. Reference 21 recently announced the flight of a remotely piloted hybrid-electric VTOL aircraft (using hydrogen-based fuel cells) to over 500 miles (or over 800 km); this aircraft was a near-production nominally passenger-carrying UAM/eVTOL vehicle.

From the perspective of trying to preserve a sustained healthy research and development community, continuing to build the urban air mobility market while at the same time beginning to explore the regional market is a good thing. This paper asserts that large commercial passenger-carrying (PAX>50) hybrid-electric vehicles that can fly at least tiltrotor SOA speeds (>250knots or >130 m/s) and modest ranges (>500 miles or >800 km) should ultimately be the rotorcraft research community's goal for the future.

Embracing 'Electric Aircraft' and not just Electric Propulsion

NASA over decades has invested a substantial amount of research into active flow control and active aeroservoelastic actuators to enhance the aerodynamic performance and efficiency of aircraft as well as to flight dynamics and structural/vibratory load alleviation. However, active flow control techniques place significant demands on electric power – above and beyond that for providing for all- or hybrid-electric propulsion. Active rotor control servo-actuators also demand a considerable amount of electric power as well. The complexity of active flow control and aeroservoelastic actuators and the above noted added power demand have kept these well studied performance-enhancing systems from being implemented to-date on production aircraft.

Reference 6 made some of the first suggestions to the aerospace research community as to embracing the 'electric rotorcraft' concept instead of focusing solely on rotorcraft with electric propulsion. As noted in Ref. 6, if very large amounts of electric power storage or production is already going to be installed onboard an aircraft for enabling electric propulsion, then it is a relatively small conceptual leap to consider implementing active flow control and/or active surfaces/actuators that are also electrically powered. In that case, an 'electric rotorcraft' could theoretically become so much more than an aircraft with electric propulsion, however singularly important that is. Instead, the introduction of aircraft with large electric energy storage might enable introduction of a suite of long-researched technologies to promote increased aircraft efficiency and expanded operational envelopes.

Expanding Autonomous System Technology into all Aircraft Life-Cycle Phases

In a workforce context, people are costly. Economic downturns result in significant layoffs. Expanding automation and autonomous system technology into all life-cycle phases of aircraft would potentially moderate, over time, large swings in aviation workforce. However, such expansion of automation in aviation must emphasize safety and stability.

Most urban air mobility, aka eVTOL, developers are currently pursuing human piloted aircraft so as get FAA certified as quickly as possible. Only a few developers are currently seriously pursuing passenger-carrying autonomous aerial vehicles. Reference 6 devoted a portion of its discussion as to the system analysis and economics of passenger-carrying autonomous aerial vehicles for metropolitan aerial transportation systems. This preliminary analysis laid the groundwork for a more in-depth consideration of this research question in the future. In the context of building in enhanced resilience in aviation, the potential use of autonomous aerial vehicles for commercial passenger transport will require extensive community assessment over the next couple of decades.

Responding to Wholly New Aviation Markets

A considerable amount of innovation is currently underway within the rotorcraft community. Urban air mobility vehicles, aka eVTOL, are well on their way to certified production. Work also continues in a steady sense on 'delivery drones.' Several other transformative classes of vehicles and new mission are also being explored. The future of aviation needs to consider not just legacy/state-of-the-art aircraft and rotorcraft but must remain sufficiently adaptable to respond to new aerial vehicles and missions. Among some of these future aerial vehicles and missions are as follows: (a) HEROS, or novel crewed aerial vehicles and robotic aerial platforms for rapid response disaster relief and emergency response missions; (b) Good Stewards, or robotic aerial platforms for environmental monitoring and biodiversity or wildlife conservation; (c) "How to save the world, one life (and UAV) at a time!", or use of autonomous aerial vehicles and aerial robots for disaster relief and emergency response; (d) the development of 'mech life' to monitor, conserve, and take remedial actions to sustain the billions of humans on the planet and the whole of its flora and fauna to achieve a sustainable world.

Such an expanded aviation system would be challenging if only one or two new application domains were emerging, but there is potentially a whole gamut of novel aerial vehicles and aerospace operational models and business cases being considered for the mid- to far-term future. Such potential increased complexity of the aviation system argues for improved system analysis and systems engineering tools.

FUTURE WORK

This paper is merely an introductory recommendation that the aerospace engineering community try partially redirecting some of its research efforts towards trying to enhance the resiliency of the aviation sector. As to the specifics of what that might entail, a short summary of concepts – ranging from vehicle-centric to air transportation system centric – is presented; to kick-start such rotorcraft research community discussion, a set of concepts to build in such resilience is also presented.

Ultimately future work in this area will also be performed, in parallel, to the complementary Climate-adjusted Air Transportation Systems (CATS) project (Ref. 22). CATS is funded by the NASA Convergent Aeronautics Solutions (CAS) project. CATS seeks to focus on building enhanced resilience into Aviation specifically in response to anticipated climate changes foreseen in the 2050 and beyond timeframe.

The appendix of this paper will touch upon some of the more aircraft design related aspects of the CATS project. This includes the sizing of a V/STOL vehicle (notional vehicle shown in Fig. 3) responding to aircraft design requirements that reflect anticipated climate impact effects.

CONCLUDING REMARKS

This work seeks to present a bold challenge for the aviation community: define new system-of-systems architectures to greatly enhance aviation resilience to adequately respond to a manifold set of potential disrupters, including the influence of climate change, economic downturns, energy crises, pandemics, and disasters of all types, etc. It is the broad assertion of this paper that vertical lift aerial vehicles will be a crucial factor in building future aviation resilience.

There is an opportunity to learn from the lessons from several decades of aviation disruptions. The future is largely unknown, but it seems reasonable to anticipate that future aviation disruptions will occur. It is time to take steps now to develop more resilient aerial vehicle design space and airspace systems.

The proposed research theme of building enhanced aviation resilience is highly relevant to the needs of the Nation, as well as a powerful unifying theme to focus and advance critical technologies for future vehicle design and airspace system development. This 'resilience' research theme can readily encompass some of the current NASA research portfolio and, therefore, represents not so much a radical change in direction but an expansive evolution.

In addition to the theme of building enhanced resilience in the aviation system, this work continues considering important questions related to sustainable, or 'green,' aviation. Because of the power intensive aspects of rotorcraft over conventional fixed-wing aircraft it is a challenging problem to consider how to implement 'sustainable rotorcraft.' It is the assertion of this study and similar earlier work, that it will be through a combination of both novel vehicle designs, new runway independent aircraft (which includes rotorcraft, VTOL, and STOL aircraft), compatible networks (of airports and vertiports), and new aviation operational business models (or use cases).

Author contact: Larry Young,
larry.a.young@nasa.gov

ACKNOWLEDGMENTS

This work is for future generations of aerospace engineering innovators. In part, this work was inspired by the early career researchers and interns within the NASA Ames Research Center Aeromechanics Office. The author would also like to acknowledge the valuable extended discussions with the CATS project team over the past several months: Parimal Kopardekar (ARC-A), Nicole Fichera (GRC), Eric Brubaker (LaRC), Jennifer Hinkel (GRC), Heidi Cozby (GRC), Anne L. Miller (Intern, LaRC), Abigail Glenn-Chase (ARC-AT), and Marcus Johnson (ARC-A).

REFERENCES

1. Young, L.A., "What is a Tiltrotor? A Fundamental Reexamination of the Tiltrotor Aircraft Design Space," AHS International Technical Meeting on Aeromechanics Design for Transformative Vertical Flight, San Francisco, CA, January 16-18, 2018.
2. Gibson, T.; Jagielski, M.; Barber, J., Chung. W., and Young, L., "Employment of Conceptual Civil Tiltrotor Aircraft in Disaster Response," 68th Annual Forum of the AHS International, Fort Worth, TX, May 1-3, 2012.
3. Young. L.A., et al, "A Study of Civil Tiltrotor Aircraft in NextGen Airspace," AIAA-2010-9106, 10th AIAA Aviation Technology. Integration, and Operations (ATIO) Conference, Fort Worth, TX, Sept. 13-15, 2010.
4. Young, L.A., et al, "Civil Tiltrotor Aircraft Operations," 11th AIAA Aviation Technology, Integration, and Operations (ATIO) Conference, Virginia Beach, VA, Sept. 20-22, 2011.
5. Young, L.A., "Urban Aerial Mobility Networks using Amphibious Vertical Takeoff and Landing Vehicles," 9th Biennial Autonomous VTOL Technical Meeting & 8th Annual Electric VTOL Symposium, Jan. 26–28, 2021.
6. Alonso, J.J., Arneson, H.M., Melton, J.E., Vegh, J.M., Walker, C., and Young, L.A., "System-of-Systems Considerations in the Notional Development of a Metropolitan Aerial Transportation System: Implications as to the Identification of Enabling Technologies and Reference Designs for Extreme Short Haul VTOL Vehicles with Electric Propulsion," NASA TM 2017-218356, September 2017.
7. Young, L.A., "Smart Precise Rotorcraft InTerconnected Emergency Services (SPRITES)," AIAA Science and Technology Forum and Exposition (SciTech 2018), Kissimmee, Florida, January 8-12, 2018.
8. Young, L.A., "A Transportation and Habitation System Architecture enabled by Multi-Modality Mobility Platforms," NASA TM (Soon to be published).
9. Young. L.A., "Novel Tilting Ducted-Fan Aerial Vehicle Configurations," 10th Autonomous VTOL Technical Meeting and 10th Annual Electric VTOL Symposium, Mesa, AZ, January 24-26, 2023.
10. Young, L.A., "Novel Conceptual Designs for Stopped-Rotor Aerial Vehicles and Other High-Speed Rotorcraft," 6th Decennial VFS Aeromechanics Specialists' Conference, Santa Clara, CA, February 6-8, 2024.
11. Young, L.A., "A Multi-Modality Mobility Concept for a Small Package Delivery

UAV," 7th AHS Technical Meeting on VTOL Unmanned Aircraft Systems, Mesa, AZ, January 24-26, 2017.
12. Johnson, W., "NDARC: NASA Design and Analysis of Rotorcraft Theory," NASA/TP–20220000355/Vol 1.
13. Beard, S., "Future ATM Concept Evaluation Tool (FACET)", June 21, 2017; https://aviationsystems.arc.nasa.gov/research/modeling/facet.shtml; last accessed December 31, 2024.
14. Lopes, L.V. and Burley, C.L., "ANOPP2 User's Manual – Version 1.2," NASA/TM–2016–219342, October 2016.
15. Malpica, C., Suh, P., Silva, C., "Flight Dynamics Conceptual Design Exploration of Multirotor eVTOL," Presented at the Vertical Flight Society's 80th Annual Forum and Technology Display, Montreal, Quebec, Canada, May 7–9, 2024.
16. Sandquist, A., Toclanny, A., Ahern, S., "Design of an Uninhabited Aerial Vehicle for Assistance in Reforestation," student intern paper, Aeromechanics Office, NASA Ames Research Center, August 8, 2017.
17. Squires, A., "The National Forest System, Tree Planting Projects, Wildfire;" https://spectrum.ieee.org/top-robotics-stories-2024?utm_source=roboticsnews&utm_medium=email&utm_campaign=roboticsnews-12-31-24&utm_content=httpsspectrumieeeorgtoproboticsstories2024&mkt_tok=NzU2LUdQSC04OTkAAAGXvUcbXEuLKG-vvxdRoNZY2htuljLWNAlt41-dtjPh9F98fmROrs7s9DRiIr70zbFKRbR8_ab3_avyoMQS61pX2AaypaRuJ2h1Hs8KJPmKUXHh; last accessed December 31, 2024.
18. Kroo, I., McMasters, J., and Smith, S.C., "Highly Nonplanar Lifting Systems," *Transportation Beyond 2000: Technologies Needed for Engineering Design*, September 26-28, 1995.
19. Rohacs, D. and Rohacs, J., "Magnetic Levitation Assisted Take-Off and Landing (feasibility study – GABRIEL concept)," Progress in Aerospace Sciences, Vol. 85, August 2016, Pgs. 33-50; https://www.sciencedirect.com/science/article/abs/pii/S0376042116300070; last accessed November 29, 2024.
20. Kottapalli, A.P. and Harris, F.D., "Converting a C-130 Hercules into a Compound Helicopter: A Conceptual Design Study," American Helicopter Society Aeromechanics Specialists' Conference, San Francisco, CA, January 20-22, 2010.
21. Reim, G., "AFWERX Says Agility Prime eVTOLs Lack Range, Shifts To Hybrid," Aviation Week, posted on December 16, 2024; https://aviationweek.com/defense/aircraft-propulsion/afwerx-says-agility-prime-evtols-lack-range-shifts-hybrid?utm_medium=email&utm_source=rasa_io&utm_campaign=newsletter; last accessed December 19, 2024.
22. Kopardekar, P.H., Johnson, M.A., Young, L.A., Cozby, H.A., Fichera, N.A., Hinkel, J.M., Miller, A.L., and Brubaker, E.R., "A Survey of Climate Impacts on Aviation and Considerations for Future Mitigations," AIAA Aviation Forum, Las Vegas, NV, July 21-25, 2025; soon to be published.
23. Rajagopalan, G., Thistle, J., Polzin, W., "The Potential of GPU Computing for Design in RotCFD," AHS Technical Meeting on Aeromechanics Design for Vertical Lift, Holiday Inn at Fisherman's Wharf, San Francisco, CA, January 16-18, 2018
24. Wolverton, M., Project Tip-Tow, Aviation History, January 2012 ; reposted at https://www.historynet.com/project-tip-tow/; last accessed January 15, 2025.

APPENDIX – AIRCRAFT DESIGN PERSPECTIVE TO THE CLIMATE-ADJUSTED AIR TRANSPORTATION SYSTEMS PROJECT

The Climate-adjusted Air Transportation System (CATS) project is a recently initiated NASA Convergent Aeronautics Solutions (CAS) project 'Discovery' effort; see Ref. 12. CATS is envisioned as filling an important 'gap' in the current NASA

Aeronautics Research Mission Directorate (ARMD) research portfolio. Though ARMD and many other organizations are striving mightily to try reach carbon-neutral goals for the aviation sector by 2050, it is also becoming apparent that technological steps towards climate adjustment, or rather adaptation, should also be pursued by the aviation research community.

CATS is still in the very earliest of stages, but a substantial amount of literature review, problem scope definition, and high-level planning discussion has already occurred. The overall CATS effort will help lay the groundwork to debate and define a balance between climate mitigation and adaptation (for enhanced resilience).

A key element of the early efforts of the CATS effort is to do the tool development necessary to conduct robust data mining and data translation from NASA and other organizations climate model predictions to develop the inputs required for NASA ARMD airspace and aircraft analysis tools. One major point of consideration, it that it is currently unclear whether sea levels will rise in the 2050 or later time frame of a sufficient magnitude to require consideration of aviation system options such as building more airports on artificial islands and/or the use of amphibious aircraft as necessary steps to climate adaptation.

From an airspace operations and airport network perspective, a new type of airport network system (as distinguished from the current hub and spoke type networks) may need to be developed. Alternate airport networks might need to be proposed such as: decentralized or distributed networks; incorporating small airports and vertiports and VSTOL (aka Runway Independent Aircraft) into large airport networks; amphibious 'airfields' and amphibious aircraft integrated airport networks; and/or other multimodal network solutions. Such novel airport network concepts should be explored to increase airspace transportation system resilience in the 2050 and beyond time frame that adapts the airspace system to projected climate disruptions.

New types of aircraft specifically tailored for climate-adaptation will likely be required in the 2050 and beyond timeframe to successful adapt to climate changes that might still occur despite ongoing research and development efforts into sustainable aviation by NASA, other governmental research organizations, and Industry. NASA has a long history of studying 'runway independent aircraft' (RIA) – i.e., VTOL (vertical takeoff and landing) and STOL (short takeoff and landing) aircraft of various types. But such studies haven't directly considered the implication of RIA in the context of climate mitigation and climate adaption. Hybrid electric regional CTOL (conventional takeoff and landing) aircraft are currently being studied by NASA ARMD, but only a very small amount of research is being currently studied for hybrid-electric large, regional VTOL and/or STOL aircraft.

(a)

(b)

Figure A1. Amphibious runway-independent-aircraft (tiltrotors) conceptual artwork (AI-generated images by Microsoft Copilot): (a) pontoons and (b) seaplane-like hulls

General requirements for an aircraft design element to the CATS project could potentially for an exploratory design study of a V/STOL (vertical and short takeoff and landing) aircraft are outlined below:

1. A required minimum range of 310 miles (500km) on hybrid electric propulsion; a cruise speed of 300knots and 25kft is to be assumed;
2. Aircraft should carry 53 passengers and crew total (assumed per passenger/crew weight of 250lbf);
3. Aircraft should have a minimum of 25% of hover power and 25% of helicopter-mode edgewise forward flight power provided by electrical energy; remaining propulsion energy can be provided by combustion-based energy sources (turboshaft engines); all airplane-mode cruise power can be provided by turboshaft engines; this minimum target might be adjusted through initial studies and discussion;
4. The aircraft should be capable of occasionally landing/taking off vertically, taking off and landing on very short runway lengths (in addition to conventional runways), and taking off and landing occasionally on littoral waterways. From a life cycle mission perspective, an aircraft should have the following attributes: (a) 10% of all life-cycle flights should be vertical takeoff and landing flights that include 2 minute hover, (b) 20% of flights that require short takeoff and landing from very short runway lengths of 500ft, and (c) 50% of flights from large airports with conventional runways of 11kft and (d) 20% of flights requiring amphibious, but otherwise conventional, takeoff and landing for littoral waterways (e.g., Boeing 314 'Clipper' like takeoff and landings).
5. The aircraft should be capable of VTOL, STOL, or CTOL takeoff and landings for high and hot conditions anticipated in the 2050 time frame (5k ISA+30°C, or an increase of 10°C from 'LCTR2' design conditions, https://rotorcraft.arc.nasa.gov/Public

ations/files/AHSAcreeFinal_Tiltrotor_1039_12696.pdf); last accessed November 30, 2024.

6. The amphibious takeoff and landings can use lengths on water of 2000ft in sea state of 3 ((1 ft 8 in to 4 ft 1 in), "Slight") and winds of 30 +/- 15mph;

Some initial mid-fidelity CFD and rotorcraft sizing analysis for the notional vehicle shown in Fig. A2a-c was performed for four different missions phases noted above in requirement #4.

(a)

(b)

(c)

Figure A2. Amphibious CATS Flyer: (a) VTOL hover, (b) STOL takeoff, and (c) cruise

Table A1 summarizes the results of a simple rotorcraft sizing analysis for the VTOL mission, which is the most challenging mission from a gross weight perspective. This simple rotorcraft sizing analysis was performed using the methodology outlined in Ref. 16. The sizing presented in Table A1 assumes that modular battery packs – and electric motors in parallel in two input combiner gearboxes – and jet-fuel-based turboshaft engines are used for this version of a CATS amphibious flyer. Promising advances in hydrogen-based fuel cells might present an alternate or augmentation to jet-fuel-based turboshaft engines but that will have to be studied more closely in the future.

Table A1. Resulting sizing weight table: VTOL mission profile

Proprotor (larger rotors) Disk Loading	973.2	N/m^2
Propeller (smaller rotors) Disk Loading	1621.9	N/m^2
Proprotor (larger rotors) Radii	4.3	m
Propeller (smaller rotors) Radii	2.3	m
Number of Proprotor Blades	4.0	Nondim.
Number of Propeller Blades	4.0	Nondim.
Proprotor (larger rotors) solidity	0.12	Nondim.
Propeller (smaller rotors) solidity	0.20	Nondim.
Propeller Tip Speeds	238.2	m/s
Nominal (Mean) Airplane-Mode Cruise Wing Loading	4984.6	N/m^2
Wing Span	14.2	m
Number of Electric Motors per Rotor	1.0	Nondim.
Aircraft Total Hover Power	7386.1	kW
Advance Ratio	0.21	Nondim.
Nominal Total Helicopter-Mode Forward Flight Power	5659.9	kW
Ratio of lift carried by the aggregate of tiltwing like propellers to total lift (in hover)	0.5	Nondim.
Vehicle Effective Lift over Drag in Airplane-Mode Cruise	8.0	Nondim.
Nominal (Mean) Airplane-Mode Cruise Power	5059.4	kW
Energy from Battery over Total Mission Energy	0.09	Nondim.
Prescribed Fraction of Power Delivered by Electric Motors in Hover versus Turboshaft Engines	0.25	Nondim.
Prescribed Fraction of Power Delivered by Electric Motors in Helicopter-Mode Forward Flight versus Turboshaft Engines	0.25	Nondim.
Prescribed Fraction of Power Delivered by Electric Motors in Airplane-Mode Cruise versus Turboshaft Engines	0.0	Nondim.
Payload (PAX+Crew)	6022.7	kg
Total Weight of Rotors	1459.7	kg
Fuselage Weight	2665.1	kg
Wing Weight	3472.5	kg
Total Turboshaft Engines and Drive Train Weight	3214.4	kg
Total Fuel Weight	1399.0	kg
Total Battery Weight	2756.0	kg
Total Electric Motor Weight	926.0	kg
Total Fixed Equipment Weight	2279.0	kg
Total TOGW =	21578.9	kg

The third requirement from the above list of mission/design requirements, that only 25% of hover power and 25% of helicopter-mode edgewise forward flight be provided by electrical power, seems to be a rather modest 'sustainable aviation' goal. It is apparent, though, from initial sensitivity study results that increasing these percentages results in substantial weight growth in the vehicle design. This is hardly surprising for rotorcraft. Still, part of the future discussion must be the question of how much growth in vehicle weight is acceptable to meet attendant electric-propulsion goals. This is not just an open issue for rotorcraft but all aircraft that might be designed to meet sustainable aviation goals. The approach of this study is to chip away at the problem, to make slow advancements, to seek to meet modest goals now rather than later.

One way to compensate for the power intensive nature of VTOL is to make the battery packs onboard aircraft modular so that: (a) if battery technology upgrades become available than new, better batteries can be easily swapped out with older batteries, and, most importantly, (b) the aircraft could ideally be reconfigured between VTOL, STOL, and CTOL modes of operation while on the ground, in between flights, in parallel with aircraft refueling with jet-fuel (or hydrogen if fuel-cells are implemented). If the case can be made that such a mode of operation (reconfiguring between flights with modular propulsion elements or batteries) can be performed and is responsive to an economically viable business model and (airport and vertiport) network architecture model than sustainable aviation through (partial) electric propulsion might be viable for rotorcraft.

It is this possibility of using modular batteries that are installed during VTOL, partially installed during STOL, and completely removed for CTOL flights that is considered in Fig. A3. Figure A3 estimates the increased aircraft range as batteries are removed and flight profiles are adjusted for all the major modes of operation suggested in the Fig. A2 amphibious V/STOL (hybrid tiltwing/tiltrotor) aerial vehicle.

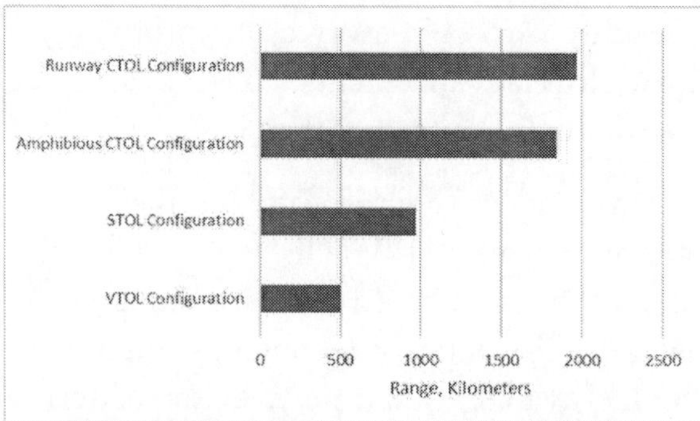

Figure A3. Range for other mission profiles (STOL, CTOL on runway, and CTOL amphibious takeoff) if modular battery packs were removed, if not needed, for a particular flight

Whether climate change impacts will reach to the level that flooding and sea level rise will significantly US airports to the point that artificial islands for airports and the use of amphibious aircraft for commercial travel will be required to adapt is still a question subject to additional research, modeling, and simulation. What does seem more likely, though, is that there are low-lying island nations in the Pacific Ocean – nations which are also highly dependent on tourism – where amphibious commercial transport aircraft might be a well needed lifeline.

VTOL UAS Auto-recovery Using A Tested Long-Term Motion Prediction Method to define the Deck Environment

Professor (H) Bernard Ferrier, FRINA, FRAeS
Head, Dynamic Interface Lab.
Syntek Technologies, Inc.
Fairfax, Virginia, USA

Commander B. Watson, RN
Navy Dev-MAvn T&E, Chief
Naval Test Advisor, Royal Navy
Portsmouth, UK

Professor M. R. Belmont
School of Engineering
University of Exeter
Exeter, Devon UK

Professor J. T. Christmas
Department of Computer Science
University of Exeter
Exeter, Devon UK

ABSTRACT

Launch, recovery, and deck handling operational performance on smaller ship platforms like Corvettes, Frigates and Destroyers are qualified as the most challenging tasks in the UAS ship-deployment of a VTOL Uncrewed Air System (UAS). One of the main hurdles is the random nature of seaway-created deck motions coupled with ship structure disturbed air wake patterns. The MoD has supported a range of work aimed at bringing Quiescent Period Prediction (QPP) technology to fruition. QPP firstly requires Wave Profiling RADAR to measure the sea wave system out to approximately 2km in the region around a vessel. Secondly these measurements are employed in a wave propagation model to predict the actual wave forces acting on a vessel. Using the wave predictions as inputs to a vessel model makes possible to predict the actual (deterministic as opposed to statistical) motions of a vessel. Wave systems naturally alternate groups of large waves with smaller waves, this property, combined with the predictive ability, allows to identify the quietest (most quiescent) periods in which to conduct wave limited naval operations. Naval mission planners in the Royal Navy, and elsewhere in the World, appreciate the need to maintain rapid, but safe, deck tempo. The fundamental concept is to measure remote sea surface profiles to predict the future wave forces acting upon a vessel. The objective is to expand ship operating deck limits to approximately Sea State 6+. The deck definitions generally empirically measured by using standard rating scales, are replaced by instrumented devices reporting the status of the deck prior to touch-down. In this paper, a thorough discussion describing the QPP deck measuring devices designed to replace piloted cueing is provided. Theory, previous simulation studies and current at-sea testing along with data results, are also discussed. To conclude, the interface of the deck measuring device into the next version of the UK UAS system, is provided.

The results of the RADAR trial indicated that the RADAR data was reliable, with the RADAR images matching the physical map. The two-dimensional surface plot showed both the RADAR blocking fence along with an additional target. An additional observation concerning the operation over the deck whilst the ship is experiencing a quiescent ship motion period. The coupled secondary effect documents minimized air wake confusion. This is owing to fewer ship structure excursions into and out of the air flow. To better define deck airflow around the ship the integration of a Doppler LIDAR instrumented federate is proposed. This is meant to predict the future vessel air wake and look for quiescent periods in this paralleling the vessel motion QPP technique.

NOMENCLATURE

DI	Dynamic Interface Study
DOF	Degree-of-freedom
EI	Energy Index (calculated by LPD)
HLA	High Level Architecture
LPD	Landing Period Designator
NATOPS	Naval Air Training and Operating Procedures Standardization
RAO	Response Amplitude Operator
RN	Royal Navy (UK)
RNAS	Royal Navy Air Station (UK)
SHOL	Ship-Helicopter Operational Limit
SMP	Ship Motion Program
SMS	Ship Motion Simulation (NAV routines)
TCS	Tactical Control Station
UAV/S	Uncrewed Aerial Vehicle/System
USV	Uncrewed Surface Vessel/System
WOD	Wind-over-deck

1. INTRODUCTION

Uncrewed Air Systems (UAS) are seen by many to be force multipliers and the capabilities they present, once integrated into a digital battlespace will, in time, change the face of maritime warfare. The area of Ship Air Integration is a familiar domain for crewed aircraft but without the pilot for almost instantaneous decisions and workload assessments the operation and integration of UAS on board is in its infancy. To improve mission effectiveness and deck activity tempo, the RN is investing in is the development of Quiescent Period Prediction. This is an application that has utility across all Uncrewed Systems and one that can not only improve safety but increase capability.

Awareness of the ship's position and movement in the ocean is key to air operations. To be able to predict this and target the right time to deploy and recover UAS/USV vehicles, will give Command greater utility and flexibility (figure 1).

Figure 1 – Early UAS – Ship Launch and Recovery

The operational challenge is to define in a timely manner periods of low deck motion activity (quiescence). This is particularly critical when involving other robotic vehicles as found in the integrated Expeditionary Strike Group plan **[1]**,

MoD/University of Exeter's Quiescent Period Prediction (QPP) purpose of is to forecast (more than a minute(s) depending on the ship and recovery platform) acceptable deck motion periods for **safe and predictable** launch and recovery of piloted and remotely piloted air and sea vehicles. With a view of conducting automated UAS launch and recovery, the specific objectives are: (i) to both give the operator guidance on the optimum ship's heading and speed to initiate launch/recovery vehicular activity and (ii) to predict periods of limited flight deck motion (for safest recovery) over timescales of typically 60 to 120 seconds ahead. The goal is to expand ship UAS/USV operating deck limits to approximately Sea State +6.

From the measured sea surface, a short-term deterministic wave model can be constructed allowing the wave system to be propagated to the ship's location. The ship's response to the wave spectrum is calculated to determine the level of quiescence at the time of encounter. The results are mapped dynamically on speed-polar visual graphics providing the ship's best course to steer to ensure encountering the quiescent zone in a timely manner. This procedure defines the ship motion aspect of the deck environment. A user advice system displays the predicted quiescent periods as the suggested best periods in which to execute an operation.

Such quiescent periods occur very frequently because as is well known groups of large waves typically alternative with groups of small waves. The latter being significantly less than the mean. So clearly one of the means to improve the safety of high sea state recoveries and to safely extend these further is to predict when such quiescent periods of ship motion will occur and their duration.

This article describes the MoD/Royal Navy sponsored development of the UK's QPP System with discussions on test procedures, timely seaway mapping and ship motion characteristics specifically to complete a motion sensitive task like UAS launch and recovery. Operating over deck whilst the ship is experiencing a quiescent period has the secondary effect of minimized air wake confusion. This is owing to fewer ship structure excursions into and out of the air flow. A dynamic diagram is shown that is used to both create the "best course to steer" and the best time to execute an operation.

2. QUIESCENT PERIOD PREDICTION (QPP)

Quiescent Period Prediction (QPP) is a class of technologies that use remote sea surface profile measurements to predict the future wave forces acting on a vessel in space and time. Such technologies have prediction time horizons measured in minutes and offer predictions of the actual motions of the vessel (figure 2). The purpose of this type of technology is to identify so called "Quiescent Periods" in the next few minutes. Knowing by how long a low motion period will begin and how long that period will last, makes it possible to safely execute a motion sensitive operation. These launch and recovery event may be possible even when the mean prevailing conditions exceed the traditionally considered safe deck limits.

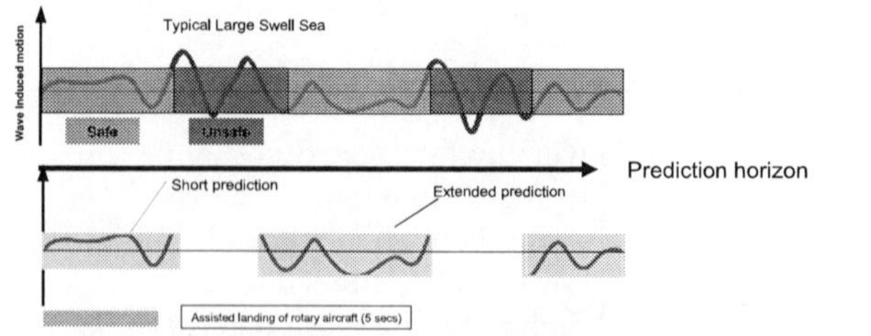

Figure 2 – Relationship between Wave induced motion and prediction.

A QPP system achieves this by using a remote wave sensor system to measure the sea surface several hundred meters in around the ship **[2]**. From the measured sea surface, a short-term deterministic wave model is constructed allowing the wave system to be propagated to the ship's location **[3]**. The ship's response to the wave spectrum is calculated to determine the level of quiescence at the time of encounter (figure 2).

The QPP purpose is to provide ship motion conditions and motion predictions with sufficient time to launch, recover and complete other motion sensitive tasks regardless of the seaway. Based on lessons learned from the use of a pre-prototype QPP on trials in the North Atlantic and Irish Sea, it is conceivable, that once integrated, shipboard communities could operate in much higher sea conditions recovering the air/sea vehicles with elevated confidence and reliability. The caveat is that SHOL can be constructed meeting basic deck limit x vehicle interface criteria **[4]**. As stated the system is designed to safely expand ship operating and deck recovery envelopes. The computational flow diagram corresponding to QPP system components is displayed in figure 3.

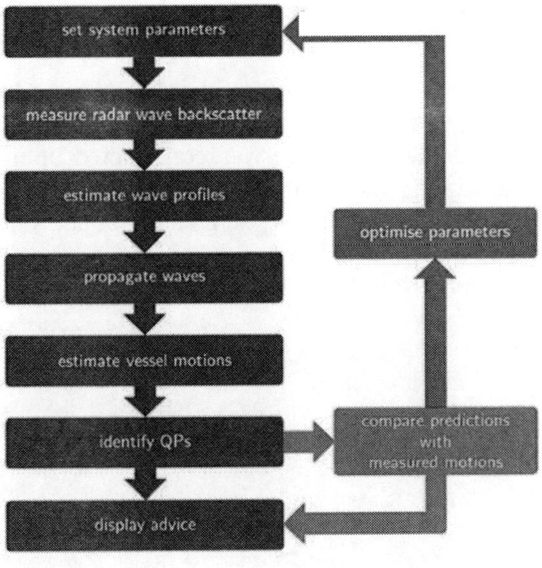

Figure 3 – QPP Computational Flow Diagram

A key part of the system is the wave sensor device, which can be composed of Light Detection and Ranging (LIDAR) or Radio Detection and Ranging (RADAR). A LIDAR system will require a mechanism to scan a laser beam over the sea surface, whereas a radar system can cover a larger area with a single radar pulse. The performance of a candidate sensor will need to be evaluated in a trial, and if proven, the data collected can then be used to validate a simulation of a complete QPP system.

The QPP software package is designed to be also capable of being integrated into a full motion simulator (in the crewed case) and a TCS (in the UAV case). This is a means to estimate system effectiveness as a function of simulated ship motion, visual environment, and synthetic operational systems, and to compare the results to related recorded at-sea data **[4]**. Such a simulator will also help ship operators gain confidence in the new technology.

A QPP system would typically also include a Landing Period Designator (LPD) sub-system to provide an emergency last minute wave-off capability.

2.1 QPP PROCESS

QPP requires first using wave profiling radar to measure the sea wave system out to approximately 2km in the region around a vessel. Second these measurements are employed in a wave propagation model to predict the actual wave forces acting on a vessel. Using the wave predictions as inputs to a vessel model makes it possible to predict the actual (deterministic as opposed to statistical) motions of a vessel.

2.1.1 SIGNAL PROCESSING

The method of measurement uses a typical shipboard wave RADAR and fixed time or a shipboard LIDAR. The RADAR methodology is summarized in Figure 4.

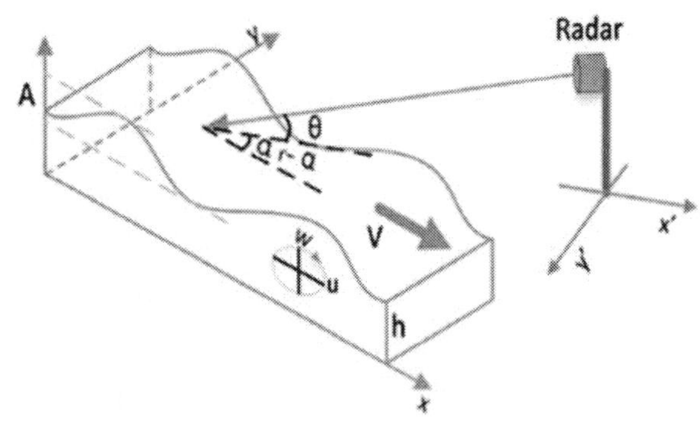

Figure 4 – Radar Wave Measurement

Traditionally marine radar is used sea surface currents, and directional wave power density spectra. This only delivers statistical wave information. The novel aspect of RADAR use in QPP is to measure phase resolved directional wave spectra allowing actual wave shape to be determined. As described in reference [5], ship born wave radars typically operate at acute angled incidence to achieve the desired mapping range of coverage. This introduces gaps in the data due to so called wave shadowing as illustrated in Fig 5b. The development of methods to directly invert marine radar images that can cope with such errors to return accurate sea surface elevations is the subject of the much current research.

The wave data is then re-mapped onto a uniformly distributed equivalent. Typically, the backscatter image using algorithmic filters resembles real-world wave forms as shown in figure 5 **[6]**.

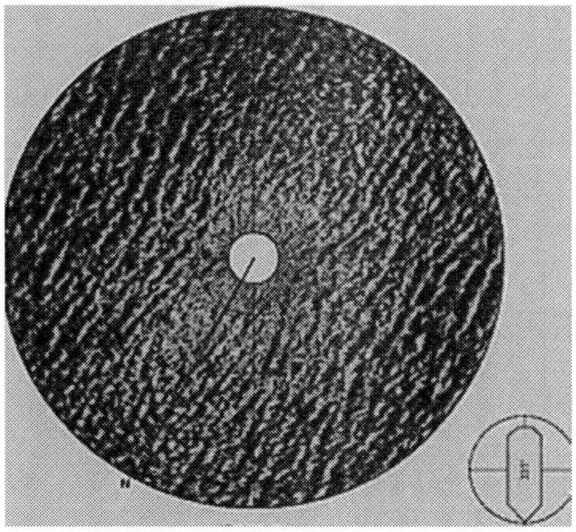

Figure 5: Backscatter Image

2.1.2 THE DETAILS OF WAVE SHADOWING

Wave shadowing shown in Fig 5b is ~~a~~ acknowledged as the major source of error in the wave measurement process. Here large nearby waves, or waves with specific slope values, can block radar returns from more distant sections of the sea surface. Methods have been developed to address such errors, but they require identification of the boundaries of the regions where wave shadowing occurs.

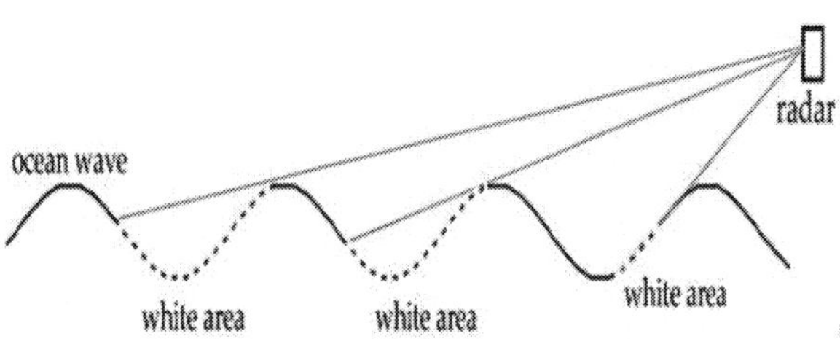

Fig. 5b - Radar Wave Masking (or shadowing)

The standard approach to estimating wave profiles, **[7]** from RADAR data involves using a three-dimensional Fourier Transform (there are two wavenumber axes and one angular frequency axis). This method involves using the data from many separate RADAR scans. The wave profile evolves in shape as it propagates, hence wave shadowing is present at different places in different RADAR scans, thus by using data from many scans there is a degree of compensation of the wave shadow effects.

The effect of this phenomenon can be described as the modulation of the smooth wave profile by a sequence of pulses. This means the direction wave spectrum, $W_s(k,\varphi)$, of sea data that is significantly shadowed, is given by the convolution of the spectrum of the pulse waveform, $W_p(k,\varphi)$, with that of the true wave spectrum, $W(k,\varphi)$. Here, k is the wavenumber and φ the wave direction. Given that $W_s(k,\varphi)$ is used for prediction purposes, shadowing very significantly affects prediction accuracy. Consequently, various techniques have been developed to attempt to compensate for this effect **[8]**.

Typically the short wavelength limit is used in such wave profiling RADAR work which neglects interference and diffraction effects. Diffraction in particular introduces significant errors in the wave profile estimation process. To obtain experimental data on diffraction effects under highly controlled conditions, where the geometries involved are well defined, land-based experiments were conducted. In these a radar blocking fence was set up and the consequent radar returns which included the anticipated shadow region caused by the fence. The results showed that purely geometric modelling of the radar shadowing by the barrier is not adequate to describe the findings, indicating significant diffraction effects were present.

QPP radar trials were held at RNAS Culdrose, Prendannak Airfield in Cornwall (UK). A test rig composed of scaffolding was constructed upon which the QPP radar kit was installed (figure 6). The test rig featured a radar-blocking screen to reduce emissions towards the Control Tower. Cabling was strung from the test rig to the Predannak Control Tower.

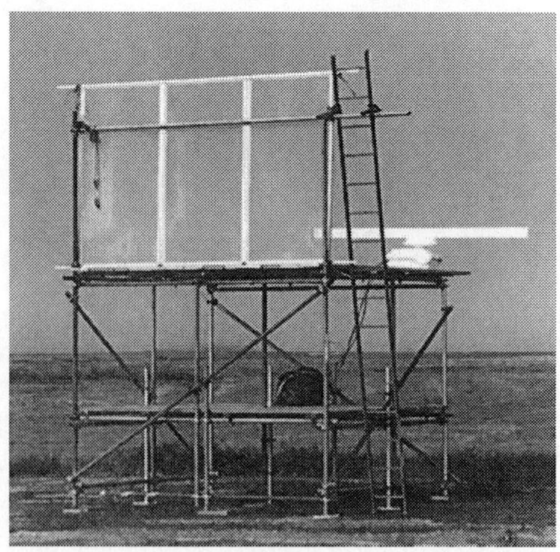

Figure 6 – QPP Radar Test Rig

The radar kit is a vertically polarised, 8 ft, X-band radar system which has been procured for measuring the sea's surface shape around a ship in real time. The purpose of the current test is to demonstrate functionality of the ensemble of equipment with the wave defining software. To demonstrate the effects of wave shadowing, a radar reflecting fence was constructed at 100 metres from the radar installation. The second scenario was repositioning the fence another 100 metres. The third scenario was recorded without a fence. Wave shadowing could alter the wave profiling. The Univ of Exeter software is designed to account for the missing wave force contributions. Trials were conducted displaying the reflection from the Airfield with visual results (figure 7).

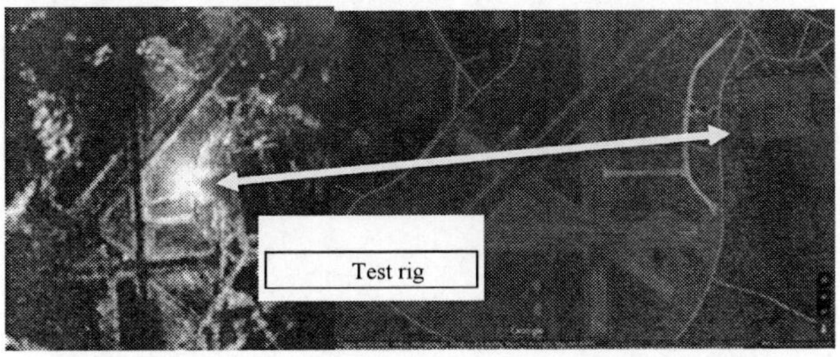

Figure 7 – Inverse Radar Image(left) of Airfield (right)

2.1.3. BACKSCATTER CONVERSION

Whether X-band or S-band radar systems are used, the concept is to measure and then from such "backscatter" data estimate the wave structures. Detailed explanation on the computation of wave forms is given in reference **[9]**. The definition furnishes wave radial velocity, amplitude, direction, and spatial speed. With wave definitions various computational methodologies may be applied, including both frequency domain and deterministic analyses.

In irregular wave fronts, the radar backscatter may miss wave frequency, period, and entire waves hidden behind a closer larger wave **[10]**. As might be anticipated typically, the backscatter image resembles real-world wave forms.

2.1.4 WAVE PREDICTOR

This part of the QPP process takes the backscatter image and converting them into measured wave components (figure 7).

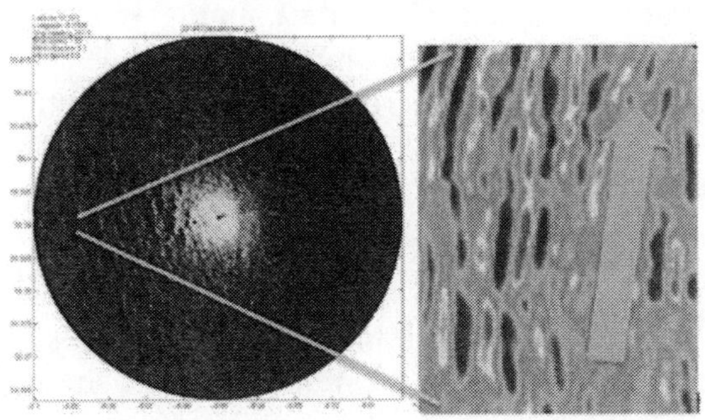

Figure 7 –Backscatter converted to Wave image

The measured waves contain most of the energies causing the ship to move. These waves are generalized into a distribution of wave frequencies at some remote location and propagates those components to the point of encounter by the ship. The ship is moving with forward speed and so the wave surface is varying both temporally and spatially. The theoretical model used to describe the propagation of the wave field is based on classical wave theory.

2.1.5 VESSEL PREDICTOR

Once the measured wave elevations are known and the wave frequency model adjusted for the point of the ship encounter, a virtual motion of the ocean ship model is computed **[11]**.

As shown in figure 8, the seaway frequency distribution is crossed with the ship's transfer functions (response amplitude operator) producing a distribution of harmonic components. These a summed over a short period of time to create 6 degrees of freedom time histories. The deck limits of the motion sensitive task are compared to determine if a particular heading is appropriate.

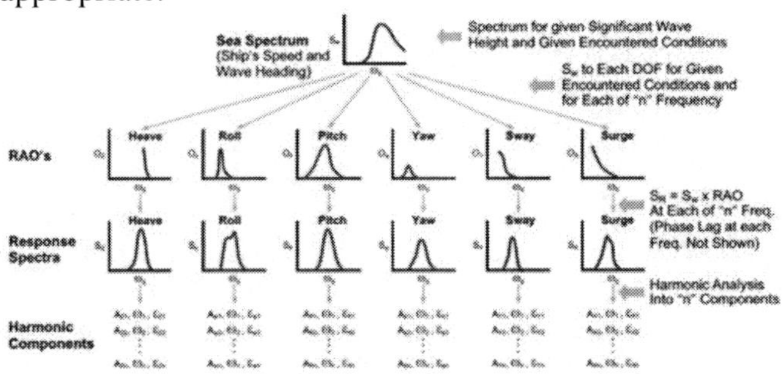

Figure 8 – Conversion of Sea to ship motion

The ship response spectrum is created as the product of the RAO and the driving sea spectrum over the entire range of frequencies. In QPP the conversion to phase-preserved real-time ship responses is given by the Fourier transform in Equation 1:

$$F_i(j\omega,\theta) = W_{enc}(j\omega,\theta,V,\phi)\, H_i(j\omega,\theta) \qquad (1)$$

where:

$F_i(j\omega,\theta)$: Fourier transform « i^{th}» mode

$H_i(j\omega,\theta)$: RAO (transfer function) for « i^{th}»

$W_{enc}(j\omega,\theta,V,\phi)$: Encountered Wave spectrum

RAO corrected ship heading, ϕ and

Ship's speed, V

A typical time history equation is given by Equation 2:

$$A_Z = \sum_{n=1}^{k} \left(A_{Z_n} \cos\left(w_n - e_{Z_n} \right) \right) \quad (2)$$

Where:

A_Z: DOF amplitude

ω: a circular frequency

e: phase angle

2.2 APPLICATION PROGRAM

The long horizon forecast program objective is to produce a design for a sea going QPP system based around an intelligent decision-making wrapper (the AI shell) that exploits data from existing proven hardware and can learn from its previous activity. It applies verified short horizon prediction (Landing Period Designator or LPD) indices in mapping operationally acceptable seaway regions.

2.2.1. AIR VEHICLE ON DECK DEFINITION

To recap why the RN is sponsoring the development of the QPP. The purpose of QPP is to forecast (more than a minute(s) depending on the ship and recovery platform) acceptable deck motion periods for **safe and predictable** launch and recovery of air and sea vehicles. Using the wave predictions as inputs to a vessel model makes it possible to predict the actual (deterministic as opposed to statistical) motions of a vessel. This means, there is enough advanced notice of the onset of a quiescent event to setup and recover, for example, a UAS vehicle. This impacts directly or indirectly mission effectiveness and deck tempo.

Motion of an aircraft on the flight deck is calculated in terms of ship motion as a function of the aircraft model. The aircraft model is considered an extension of the ship. The model is defined by aircraft landing gear footprint; on-deck location and orientation.

With QPP on-board, the air vehicle's launch and recovery events occur as it has prior to the development of QPP. The main difference is the availability in advance of known timed conditions acceptable for air operations.

The aircraft experiences ship transferred forces and moments that create rectilinear and angular accelerations on the air vehicle. The accelerations can be numerically integrated to determine the position and attitude of the helicopter relative to the ship as function of time, for various ship motions **[12].** A deck limit is indicated when an air vehicle motion event occurs, such as an on-deck slide.

Turnover incidents are static or dynamic in character. Static turnover is the same as on shore. The resolved weight vector migrates beyond either the friction forces causing the aircraft to displace or the reaction forces causing the aircraft to turnover. Figure 9 is a simplified description of dynamic turnover caused by the rotor disk (uneven loading of the rotor) and/or by ship motion. These undesirable events describe a QPP forecasted out-of-limit scenario which would cause the ship to avoid

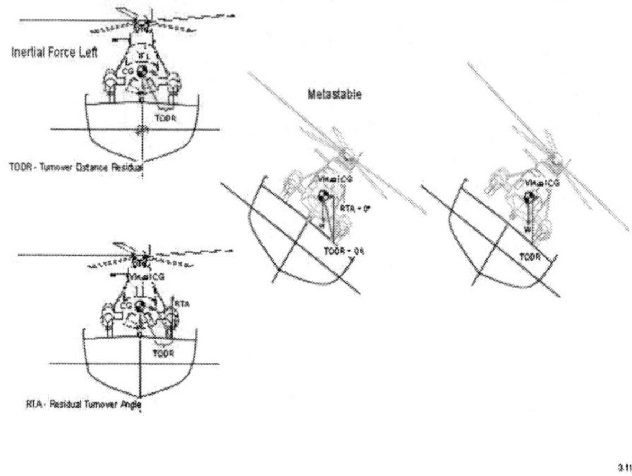

Figure 9 – On-deck Dynamic Turnover

2.2.2 SHORT HORIZON MOTION PREDICTION

An important addition to the QPP system is the Landing Period Designator (LPD). This determines and displays real-time information about current and near- term future motion (1/4 cycle) of any vessel as a function of airframe operational limits. As it does not attempt long horizon time predictions, only rely on actual vessel motions, it can be used to provide an emergency last minute waive-off capability for QPP. Having the ability to measure real-time and near-future deck motions eliminates the need for excessively conservative roll and pitch deck limits **[13].** These limits are expressed as the ship's Energy Index, which is an empirical derivation representing ship motion as a function of aircraft limits to help guide air vehicles to safe deck landings regardless of the actual seaway.

The Energy Index formulation identifies ship's quiescent periods. These periods are interpreted in terms of aircraft and deck limit definitions. This index provides information on the motion that the ship could exhibit in the near-term future using real-time motion measurements of angular displacements and velocities and linear velocities and accelerations. It capitalizes on the rate at which a vessel can displace owing to natural hydrodynamic forces.

The energy index equation is based on an empirical derivation. It measures lateral and vertical velocities and accelerations as well as roll and pitch angular displacements and velocities weighted by dynamic coefficients. The equation is the least squares of the various parameters and terms representing real-time ship - aircraft interface motion.

Eq.(3)

$$EI=\sqrt{[a_1\phi^2 + a_2\phi'^2 + a_3\theta^2 + a_4\theta'^2 + a_5z'^2 + a_6z''^2 + a_7y'^2 + a_8y''^2]}$$

where:

- EI Energy Index value
- ϕ Roll ship angle
- θ Pitch ship angle
- ψ Yaw ship angle
- x Longitudinal displacement
- y Lateral displacement
- z Vertical displacement
- a_1 to a_8 Weighted dynamic coefficients

As indicated in Equation 3, the index contains acceleration, velocity, and displacement terms giving indications of the inertial content of the ship. It is an indication of the amount of motion the ship is likely to travel in the near-term future. All 8 degrees-of-freedom, their rates, and accelerations, are monitored and recorded. This includes the ship's vertical displacement, rate, and acceleration. as shown in Fig. 10.

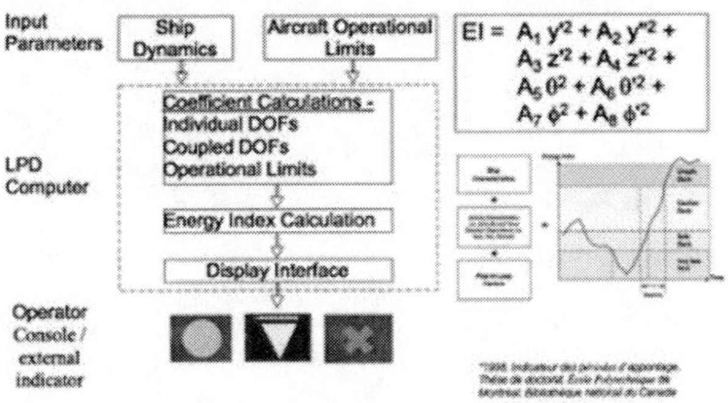

Fig. 10 - Deck Status and Risetime

The time required for the deck motion to rise from minimal motion (or very safe deck) to unacceptable motion is called the rise time. In terms of the EI scale, the rise time is defined as the period that is measured from the end of a green signal to the positive side of the red line. This is given as ($T_3 - T_1$) as shown on Figure 11. The rise time is a thumb print characteristic of the ship's response and remains constant for each ship class.

The very safe deck is a special condition in which there is insufficient energy in the aircraft-ship system to raise the deck out of limit for some defined time-period or rise time, and it is this concept that was used in the simulators to indicate that the deck was safe to initiate landing **[14]**.

Assuming the minimum average rise time for a given vessel remains constant, the aircraft descends within the rise time value and the aircraft is assured a recovery within the deck motion limits. Figure 11 displays graphically beyond static limit operations. The base envelope is taken from 10 knots of ship speed. Any points within the hour-glass structure are conditions within limits and contain no appreciable probability of out-of-limit deck motions. Outside of the structure contains motions which are considered by static reference as out-of-limit **[15]**.

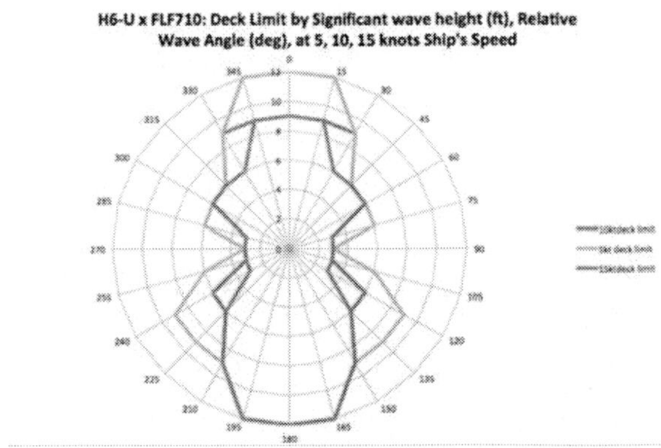

Figure 11: Safe Motion Operations

This is particularly important for robotic vehicles. A schematic representation of how the LPD is used for UAV auto recovery to signal the onset of the descent (and of deck quiescence) is displayed in Figure 12. Throughout the approach and initial hover, LPD is monitoring the deck. It is only at the low hover that the UAV autoland system would accept a Green Deck signal from the LPD. When the EI signals the onset of quiescence and through the uplink sets the air vehicle on its descent to the deck. Short horizon prediction is useful for predictions within 1/4 cycle of ship's motions. QPP is a dimension much longer in predictive horizons but uses the LPD least squares Energy Index to verify QPP's computation at the moment of launch and recovery **[16]**.

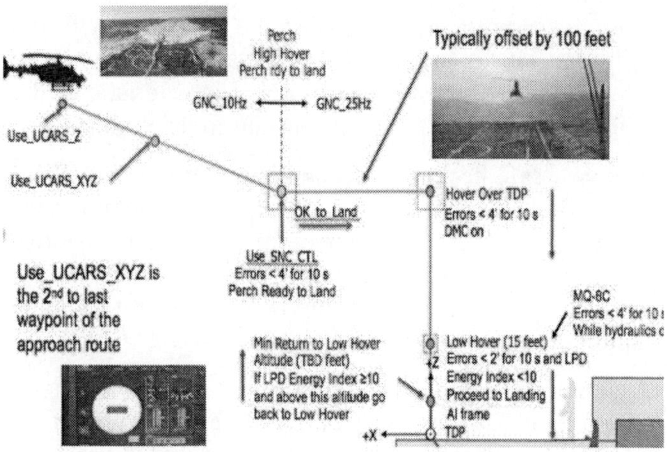

Figure 12: Notional Autoland System

In related trials it was possible to identify every period of low deck activity either by short or long horizon prediction. A simulated scenario is displayed in Figure 13. This compared to Figure 14 displaying actual launch and recovery opportunities.

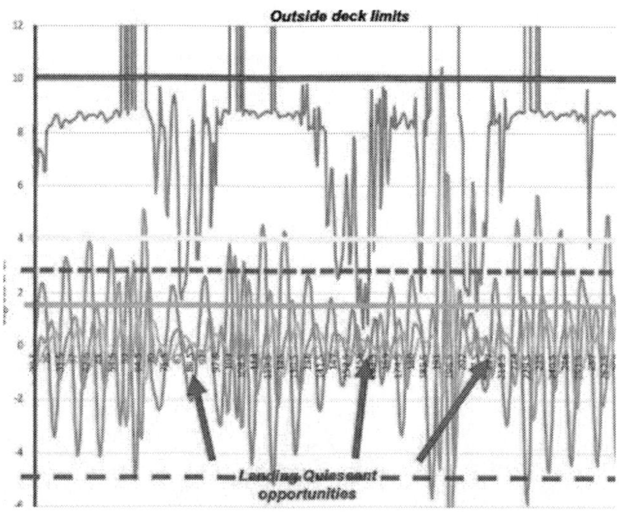

Figure 13: Landing Opportunities, simulated

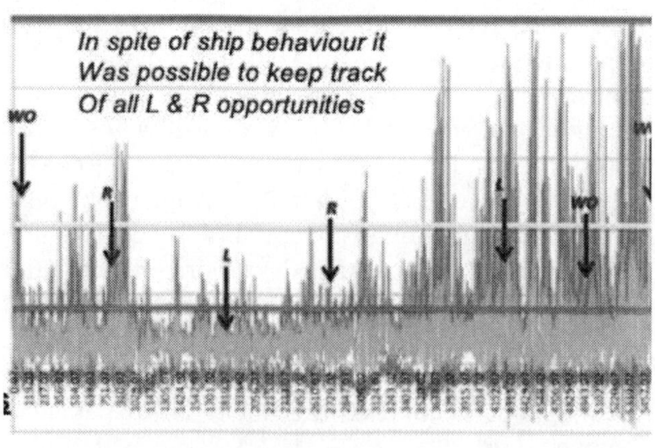

Figure 14: Landing Opportunities, recorded

3.0 QPP BEST COURSE TO STEER

The QPP computation ultimately reduces to Command and Air Crew aides recommending appropriate heading and speed to achieve mission goals. To best describe spatially the measured seas acceptable for flight operations, the energy index is used. It is a short horizon algorithm which is also used as verification that the capture or recovery is occurring within specified limits. As described earlier, one of the computational approaches is to use the QPP algorithms to measure in real-time the seaway around the Mother ship. The converted data are coupled with stochastic ship transfer function information to compute forecasted ship's motions. This is calculated for a given ship's heading and speed. It is estimated the speed-polar will represent the ship's motions against an operational event (e.g. air vehicle recovery) up to 2 km from the initial measurement position.

A speed-polar representation of the seaway to guide ship's command and deck hand activity, appears as a synoptic map created by the lay-over of various measured and computed parameters. Each parameter speed-polar lay-over is application specific computed as a function of the acceptability of ship's conditions to complete a specific motion sensitive task (figure 15). This is based on NAVSEA SEA 851, formally known as the Seakeeeping Department, Active Operator Guidance System's post-processor **[17]**.

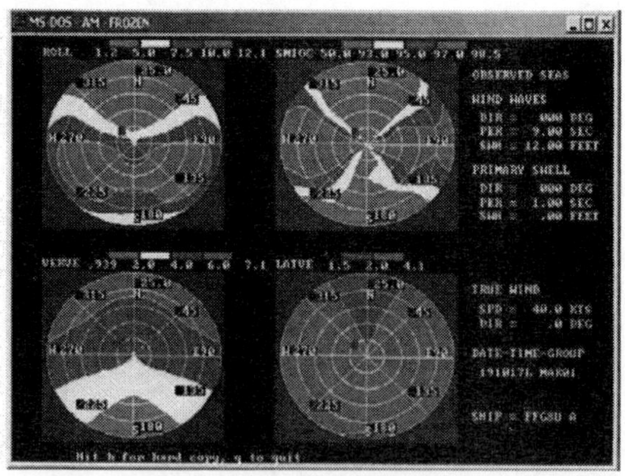

Figure 15 – Parameter Map Speed Polar

A limit is defined by the impact that a certain ship motion condition may impose on the air/ship vehicle's structural integrity or that may result in undesired dynamic response. The sum of these lay overs produces an active operator's guidance map (figure 16). All the various QPP components either exist or require limited developmental modification.

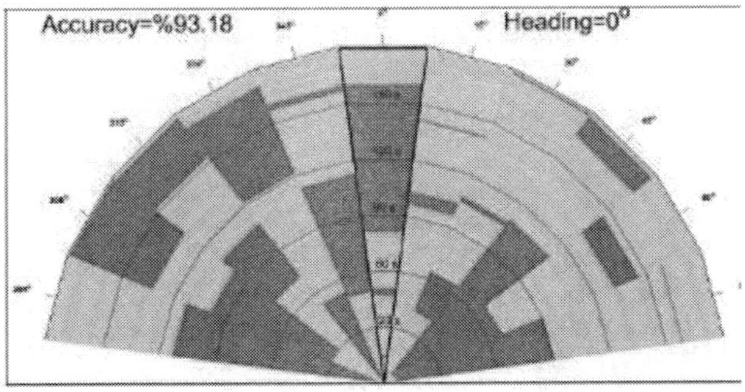

Figure 16 – Best Course to Steer

Multiple solution convergence methodologies may be offered using, for example, real-time long horizon forecasting alone or coupled with seaway real-time mapping providing stochastic (ship transfer function) based distance motion forecasts updated several times per second.

QPP can be applied to all ship motion sensitive activities. Throughout air vehicle approach and initial hover, QPP would be producing operator guidance maps allowing the ship to adopt a heading and speed acceptable for air vehicle recovery. The operator advice system would also predict the occurrence of quiescent periods and thus the best time to execute an operation. It is only during the final stages of recovery that the QPP short-horizon component would be used to verify that the deck was measured in real-time as being in a quiescent period condition.

4.0 CONCLUSIONS

A significant motivation for the development and deployment of VTOL UAS, is their exploitation in situations or conditions in which human exposure would by highly risky. Awareness and prediction of the ship's position and movement in the ocean is key to air operations by targeting the right time to deploy and recover. MoD and the University of Exeter's Quiescent Period Prediction (QPP) objective is to provide ship motion conditions and motion predictions with sufficient forecasted time (over a minute) to launch, recover and complete other motion sensitive tasks regardless of the seaway. The class of technology represented by the QPP approach is being researched by numerous highly competent academic groups. The fundamental concept is to measure remote sea surface profiles to predict the future wave forces acting upon a vessel. QPP's approach functions in all-weather conditions. The application of the QPP process has the potential of delivering a paradigm shift in the way vehicles and platforms approach and depart from a ship. The significant step will be the assembly of a Toughbook-based prototype in 2024. This will be followed by rigorous sea trials further defining the QPP system capabilities.

In a related but future QPP study activity, it was observed that certain ship deck condition tendencies were apparently related. Operating over deck whilst the ship is experiencing a quiescent ship motion period normally has the secondary effect of minimized air wake confusion. This is owing to fewer ship structure excursions into and out of the air flow. To better define deck airflow around the ship it is proposed to integrate a Doppler LIDAR instrumented federate. The purpose of this is to predict the future vessel air wake and look for quiescent periods in this paralleling the vessel motion QPP technique. Preliminary test results from a proof-of-concept trial, are briefly discussed. The ability to fully define a ship's motion and air wake condition as a function of a recovery device greatly enhances the overall system operational capability whilst fundamentally reducing the inherent risks associated with platform manipulation in high sea conditions.

ACKNOWLEDGEMENTS

The authors wish to pay tribute to the technical competence and outstanding professionalism of retired Dr. John Duncan former head of SBA, his staff and associated team members. This report is dedicated to the memory of NAWCAD's late Chief Engineer Robert Ernst, and to the late Dr. Peter J F O'Reilly and Dean Carico, fathers of the Dynamic Interface discipline.

REFERENCES

1) GALDORISI, G. and ROWLEY, J. (2022). Engineering Unmanned Surface Vehicles Into an Integrated Unmanned Solution. *Naval Engineers Journal*. Vol 134, No. 1. Alexandria.
2) BELMONT M.R., CHRISTMAS J., DANNENBERG J., DUNCAN J., DUNCAN J.M and FERRIER B. An Examination of the Feasibility of Linear Deterministic Sea Wave Prediction in Multi-Directional Seas Using Wave Profiling Radar: Theory, Simulation and Sea Trials. *Journal of Atmospheric and Oceanic Technology*, Vol. 31, **No. 7**, pp1601-1614, 2014.
3) CROSSLAND, P., FERRIER, B., APPLEBEE, TR. (2004). Providing Decision Making Support to Reduce Operator Workload for Shipboard Air Operations. Royal Institute of Naval Architects. London.
4) HOENCAMP, A (2016). Helicopter- Ship Qualification Testing. Ipskamp Printing. Enschede.
5) STRESSER, M. et al.(2022). On the Interpretation of Coherent Marine Radar Backscatter From Surf Zone Waves. *IEEE Transactions on Geoscience and Remote Sensing*. **Vol 60**. Toms River.

6) Simpson, A., et al. (2019). Wave-by-Wave Forecasting via Assimilation of Marine Radar Data. *Oregon State University Press*. Corvallis.
7) NIETO, BORGE, J.C., RODRIGUEZ, G.R., HESSNER, K., GONZALEZ, P.L. (2004). "Inversion of Marine Radar Images for Surface Wave Analysis". *J. Atmospheric and Oceanic Technology*, ***Vol. 21***, *pp1291-1300.*
8) BELMONT, M.R. (2023). Wave shadowing effects on phase resolved wave spectra estimated from RADAR backscatter: Part 2: The conflation technique and the deconvolution techniques for mitigation of wave shadowing effect. *Ocean Engineering.* ***Vol 267*** *113315-113330 (2023).*
9) *POULTER, E.*M., SMITH, M.J., MCGREGOR, J.A. (2008). S-Band FMCS Radar Measurements of Ocean Surface Dynamics". *Journal of Atmospheric and Ocean Technology.* ***Vol. 10,*** *No 4*
10) CHEN, Z., WANT, Z., ZHAO, C., XIE, F. AND HE, C. (2017), S-Band Doppler Wave Radar System ". *Remote Sensing Journal. MDPI.* ***Vol. 12,*** *No.1.*
11) FERRIER, B., ERNST, R. (2019). Resolving the conflict between ship design and UAV Launch and Recovery deck limits; the development of enhanced dynamic interface study tests. American Helicopter Society Specialist Conference. Mesa
12) APPLEBEE, TR., BAITIS, AE., MEYERS, W.G. (1981) "User's Manual for the Standard Ship Motion Program, SMP" David W. Taylor Naval Ship R&D Centre Report DTNSRDC/SPD-0936-01.
13) BLACKWELL, J AND FEIK, R. A (1988). "A Mathematical Model of the On-Deck Helicopter/Ship Dynamic Interface (U)". Aerodynamics Technical Memorandum 405. Aeronautical Research Laboratory. Melbourne.
14) FERRIER, B. (2008). MoD Quiescent Period Prediction LPD Software Testing Programme. UK080229-1. Bristol.
15) MORRIS, E.L., ZIENKIEWICZ, H.K., BELMONT, M. R (1998). Short term Forecasting of the Sea Surface Shape. *International Journal of Shipbuilding Prog.* **Vol. 45**, *No. 444*. Pp 381-400. Amsterdam.
16) RINA (2022). Ongoing relationship enhances sub rescue capability. *Warship Technology*. March 2015. London
17) FERRIER, B (2013). Deck Motion Characterisation Report. *LPD NSRS/QPP Operations*. UKSEA130219. Bristol.

BIOGRAPHIES

Michael Belmont, BSc, PhD, hold a chair at the University of Exeter UK, where he heads the Marine Dynamics group. He pioneered the new discipline of Deterministic Sea Wave Prediction which allows the short-term estimation of the details of wave driven vessel motion. This makes possible the prediction of quiescent intervals in the motion of naval platforms allowing the safe extension of the range of sea states for the launch and recovery of air, sea surface and sub-surface craft. Recently the same principals have been applied to the prediction of air-wakes over vessels. The goal of his research is the ability to allow the real time deterministic prediction of all the major environmental factors affecting maritime launch and recovery operations.

Jacqueline Christmas is an Associate Professor of Machine Learning in the Computer Science Department at the University of Exeter, UK. After many years working in industry, she obtained a PhD in the field of statistical pattern recognition in 2011. The main focus of her work is the development of machine learning models for remote sensing, with an emphasis on explain ability and measures of uncertainty, especially for potentially infinite sets of real-time observations, such as are collected by long-term, continuously operating sensors. Her particular interest is in the use of inexpensive, off-the-shelf sensors, such as video, radar and still cameras, to remotely sense difficult dynamic environments in real time. Since 2012 this work has been directed towards the measurement and prediction of sea waves, based on physical oceanographic theory.

Bernard Ferrier, PhD, FRINA, FRAeS is a Systems Engineering and Technology Assistant (SETA) in NAVAIR PMA-266 with the primary task of integrating and deployment of Fire Scout MQ-8C and its technologies and defining the operational limitations of the deck environment. Ferrier holds an honourary Professor position at the University of Exeter's College of Engineering. He is the Engineering Head of Syntek Technology's Aircraft-Ship Dynamic Interface Lab in Fairfax, VA. Ferrier led the CL227 UAV dynamic interface program at Bombardier in Montréal. Ferrier was a rotor dynamist and project leader of the dynamic interface program of the marinised AH-64 (Apache) and LHX programs at the McDonnell Douglas Helicopter Company. He received his last doctorate in helicopter/ship interface engineering at the École Polytechnique de Montréal (Québec) Canada.

Commander Brad Watson is the RN's Chief Naval Test Advisor, he is responsible for the delivery of all Fleet Air Arm Test and Evaluation activity, and works within NAVY DEVELOP helping to shape the RN's future Maritime Aviation & Carrier Strike capabilities. He trained as a Sea King Mk7 Airborne Surveillance & Control [SKASaC] Observer and has completed numerous front line, training, staff, and trials appointments over a 25-year career. His early career focused on flying and included deployments to the Middle East continental Europe and the US, this was interspersed with SKASaC training and trials assignments. His career specialisation in test flying began with completion of the RAF's Aerosystems Course and tours at the RNAS Culdrose SKASaC Trial Unit and Boscombe Down followed. Having previously commanded operationally at OF3, working closely with the USMC, his last OF4 Command appointment was as CO 744NAS, the Mission Systems and Weapon Integration Sqn based at Boscombe Down where he also generated a Joint UAS T&E Flight. Aerosystems MSc qualified and having completed Advanced Command and Staff Course, staff assignments have predominantly been within NCHQ, focused on Capability & Acquisition and Ship-Air Integration. Having spent most of his career in Cornwall, Brad now lives in Somerset with his 3 young children.

Minimum Waypoint Spacing for VTOL Rotorcraft Trajectory Guidance using B-Spline Midpoint Refinement

Bryan C. H. Chu
PhD Candidate, GN&C
Arizona State University
Tempe, AZ, USA

Dr. Spring Berman
Associate Professor
Arizona State University
Tempe, AZ, USA

ABSTRACT

Given a cubic piecewise-polynomial spline representation of a Vertical Take-off and Landing (VTOL) rotorcraft guidance trajectory in the monomial basis, we propose a new method for defining the stopping criterion for midpoint refinement of its representation in the B-spline basis. This method incorporates a model of the rotorcraft flight dynamics, consisting of the wind-axes equations of motion for a point-mass particle under the assumptions of constant altitude, constant rotor speed, and negligible side force in the direction mutually perpendicular to the velocity vector and the thrust vector from the main rotor(s). Under such conditions, the rotorcraft's ability to maneuver and change the ground track angle of the velocity vector is limited by its current speed and the available command accelerations along and normal to the velocity vector. The command acceleration limits are a function of available installed engine power, internal acceleration limiters in the vehicle control laws, and maximum allowable body-axis rates and Euler attitudes. Moreover, these acceleration limits constrain the rate of reorientation of the velocity vector. Our method defines the spline control points as the waypoints of the flight plan and iteratively refines these points until their spacing is too close for the vehicle to reorient its velocity vector to align with the bearing to the next waypoint. This stopping criterion permits a dense spacing of waypoints, since the vehicle is not required to fly through successive waypoints or along the legs connecting them. Five motion primitives were used to demonstrate the algorithm. The motion primitives were scaled in size and speed to match ADS-33/MIL-DTL-32742 Mission Task Elements (MTEs) such as the Slalom and Pirouette, as well as future proposed MTEs for VTOL uncrewed autonomous systems (UAS). We evaluated the tracking performance of a simulated single main-rotor tail-rotor helicopter for flight plans based on two of the MTEs, and we examined how closely the wind-axes flight dynamics model matches these high-fidelity simulations for turn maneuvers in different speed regimes. These investigations indicate directions for further development of our method in simulated scenarios and eventual flight tests. The benefit of this method is that it requires only minimal software changes to extend the capability of in-service autopilot and coupled flight director algorithms to fly generic trajectories that can be represented by cubic piecewise-polynomial splines.

NOTATION

Δ = Waypoint (WP) change distance [ft]
V = Wind-axes velocity magnitude [fps]
χ = Wind-axes velocity ground track angle [deg]
n_{xw}, n_{yw}, n_{zw} = Resultant aerodynamic forces along, across, and normal to flight path, normalized by weight [g's]
g = Gravitational acceleration constant [ft/s^2]
$f(x)$ = Functional representation of cubic spline
θ_b = Desired bearing of next WP from previous WP
θ_t = Bearing to point of velocity tangency from previous WP
r_t = Radial distance to point of velocity tangency [ft]

Presented at the Vertical Flight Society's 11th Biennial Autonomous VTOL Technical Meeting, Phoenix, AZ, February 4-6, 2025.

INTRODUCTION

Current flight guidance algorithms for in-service, passenger-carrying VTOL rotorcraft are designed to guide the vehicle along straight-line paths between pairs of sparsely spaced waypoints. While these algorithms are highly effective and robust to disturbances (e.g. wind) during flight, there has long been a need for algorithms that can guide rotorcraft along more complex, curvilinear trajectories. This need is due to the proliferation of advanced automatic trajectory planners that can be used to route autonomous aircraft through densely populated urban areas and re-route them around unexpected obstacles. These trajectory planners typically output trajectories that can be represented by cubic piecewise-polynomial splines, and in some cases require higher-order polynomials.

Most published path-following and trajectory-tracking algorithms for aircraft are designed for fixed-wing bank-to-turn maneuvering and typically require extensive changes to in-service autopilot and guidance software architecture and code.

(Ref. 1) and (Ref. 2) provide an overview of various guidanc strategies for fixed-wing unmanned aerial vehicles (UAVs and quadrotors. (Ref. 3) and (Ref. 4) discuss two of the mos widely-used "modern" (as they were developed many yeai after waypoint-following algorithms for straight-line paths guidance algorithms, which to the authors' knowledge hav yet to be certified on a production VTOL vehicle. These guic ance algorithms are vector-field guidance (VFG) and the nor linear guidance law (NLGL), which is based on proportion navigation missile guidance for following arbitrary trajectc ries. (Ref. 5) uses VFG to command a small unmanned hel copter to follow different planar paths such as a straight line circle, ellipsoid, and cubic spline. NLGLs have received considerable attention since their introduction, with various improvements developed in (Refs. 6–9). While VFG and NLGLs have been shown to significantly improve aircraft tracking performance over legacy straight-line waypoint-following algorithms for trajectories in simulations and experiments with small-scale platforms, these guidance algorithms are not well-suited for certification and implementation on production vehicles. In our prior work (Ref. 10), we proposed a new guidance optimization scheme for spacing waypoints on spline trajectories, the *bounded area minimization algorithm*, that can be implemented with existing waypoint guidance software on VTOL rotorcraft. While the algorithm was effective at reducing cumulative cross-track error when the waypoint spacing was sufficiently large, this benefit was not realized for densely-spaced waypoints.

We develop a flight guidance algorithm that enables VTOL rotorcraft to fly more densely-spaced sequences of waypoints that more closely approximate a given spline trajectory. The proposed algorithm takes advantage of fundamental properties of cubic polynomial splines, one of the most commonly used types of spline for VTOL path following in the literature (Refs. 11–13), to enable legacy waypoint guidance algorithms to fly arbitrary spline trajectories. Specifically, it transforms a cubic spline from its monomial basis form $(1, x, x^2, x^3)$ to the B-spline form, in which the coefficients of the basis functions represent the locations of the spline's control points. The convex hull of the control points forms the *control polygon*, which bounds the entirety of the spline trajectory and is a gross discrete approximation of the spline. Successive *midpoint refinements*, the insertion of additional knots midway between the original spline knots, produces a sequence of control polygons that converge to the spline defined by the original control polygon (Refs. 14–16). After several midpoint refinements, the refined control polygon is visually indistinguishable from the original spline, as illustrated in Figure 1. This is why this methodology is typically used in Computer Aided Geometric Design (CAGD) to approximate splines with straight lines on computer screens.

Given a cubic spline trajectory to be flown by a VTOL rotorcraft, our algorithm defines the control points of the spline as the waypoints of the vehicle's flight plan. To our knowledge, this is the first rotorcraft flight guidance algorithm to define the control polygon of a spline or one of its refinements as the guidance trajectory. This approach requires the specification of a *stopping criterion* for midpoint refinement of the control polygon. The stopping criterion is usually defined in terms of a threshold on the geometric distance between successive control points or the closest geometric distance from the control polygon to the spline. However, if the spline's control points are to be used as waypoints in a vehicle's flight plan, the midpoint refinement stopping criterion should account for the dynamics of the vehicle that follows the flight plan. We propose a stopping criterion that is based on a simplified flight dynamics model consisting of the wind-axes equations of motion for VTOL rotorcraft, which was first introduced in (Ref. 17).

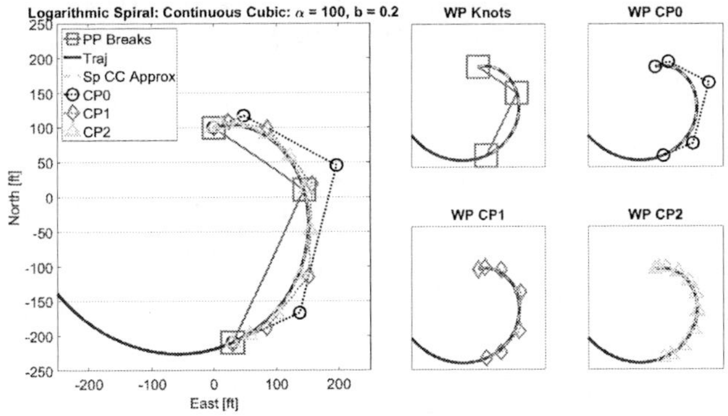

Figure 1. Cubic piecewise-polynomial spline approximation (Sp CC Approx) of a logarithmic spiral trajectory (Traj), with waypoints defined either as knots that connect two piecewise polynomials (PP Breaks), as in our previous work (Ref. 10), or as control points of successive midpoint refinements of the control polygon (CP0, CP1, CP2), as proposed in this paper. The logarithmic spiral can be written in polar coordinates as $r = \alpha \cdot e^{b\theta}$, $\theta \in \mathbb{R}$.

The next three sections discuss details of B-spline theory, the definition of the flight dynamics-based stopping criterion for control polygon midpoint refinement, and a proposed algorithm for computing recommended waypoint sequences along the control polygon based on this stopping criterion. Then, we present the algorithm's recommended waypoint sequences for five motion primitives using the simplified wind-axes flight model at different vehicle speeds and identify conditions under which flight dynamics constraints are violated. Next, we compare trajectories of the wind-axes flight model to those of a full nonlinear VTOL rotorcraft simulation, performed using a MATLAB implementation of the GENHEL model (Refs. 18–20), during right-handed turn maneuvers at different speeds and different maximum resultant aerodynamic forces along and normal to the flight path. Finally, we present cross-track errors from nonlinear vehicle simulations for trajectories based on the ADS-33 Slalom and Emergency Stop Mission Task Elements (MTEs) (Ref. 21). We conclude with a summary of main results of our study and comments on further evaluation and possible extensions of the algorithm.

B-SPLINE THEORY AND MIDPOINT REFINEMENT

Spline Forms and Change of Basis

As discussed in (Ref. 14), any polynomial spline can be converted from its *monomial basis form* to its *B-spline basis form*. We specifically consider piecewise cubic polynomial splines, since they can be used to represent most typical flight maneuvers. A full spline trajectory used for flight planning may consist of up to 150–200 knots with cubic polynomial splines linking successive knots together, but simpler trajectories can also be comprised of as few as 3 knots. A knot (or more precisely, knot point) is a location in 3-dimensional space where two polynomial spline segments are joined together. See Section I.A in (Ref. 10) for a description of the full 3-dimensional representation of a cubic spline segment between two knots. To simplify the derivation of the spline change of basis, here we will consider a single spatial coordinate of the spline, denoted by the cubic polynomial function $f : \mathbb{R} \to \mathbb{R}$, with the understanding that the derivation extends to all three coordinates. Given knots $n-1$ and n with spatial coordinates $f_{n-1} \in \mathbb{R}$ and $f_n \in \mathbb{R}$, respectively, the coordinate f of the nth segment of the piecewise spline, which lies between these two knots, can be represented in the monomial basis as follows:

$$f(x) = a_0 + a_1 x + a_2 x^2 + a_3 x^3, \quad x \in [x_{n-1}, x_n], \qquad (1)$$

where $f(x_{n-1}) = f_{n-1}$ and $f(x_n) = f_n$ are the knot coordinates and $a_i \in \mathbb{R}$, $i = 0,1,2,3$, are constant coefficients. The monomial basis is useful for efficiently evaluating the spline coordinate f at an arbitrary value of $x \in [x_{n-1}, x_n]$.

Other widely-used bases for the spline space include the Lagrange basis, the Bernstein basis (Bézier curves), and the B-spline basis. The B-spline basis is more efficient than the monomial basis for spline construction, since it enables the specification of smoothness requirements on the function f at the knots. In addition, the B-spline basis is the most useful for our application because the coefficients of a B-spline give the locations of the control points which define the spline's control polygon. (Ref. 22) derives the B-spline form by starting with a quadratic spline in the monomial basis, transforming it to the Lagrange basis via the inversion of a Vandermonde matrix, and then using geometric linear combination arguments to arrive at the Bernstein basis for Bézier curves and finally the B-spline form. The B-spline form can be arrived at more directly by use of the Cox-deBoor recursion relationship, but the basis transformation method used in (Ref. 22) highlights the fact that the original spline is unchanged when converting between the monomial and B-spline basis forms. APPENDIX A presents the same transformation equations for a cubic spline that were performed for a quadratic spline in (Ref. 22).

Properties of Spline Control Polygons

The vertices of the spline control polygon comprise the sequence of control points (Ref. 14). The control polygon can be easily computed by converting a spline to its B-spline form and obtaining the control point locations from the B-spline basis coefficients. The control polygon has several useful properties:

- Strong Convex Hull Property: The B-spline curve is contained in the convex hull of its control polygon.
- Variation Diminishing Property: If the B-spline curve is in a plane, no straight line intersects the curve more times than it intersects the curve's control polygon.
- Affine Invariance: If an affine transformation is applied to a B-spline curve, the resulting transformed curve can be constructed from the affine images of the original curve's control points.

The strong convex hull property provides an advantage to approximating a B-spline using midpoint refinement of its control polygon, rather than sampling the spline at its knot locations. A sequence of straight-line paths through adjacent pairs of knots sampled from the spline is not guaranteed to enclose the entire spline and may cut off parts of the spline if the sampling is not dense enough.

MIDPOINT REFINEMENT STOPPING CRITERIA BASED ON FLIGHT DYNAMICS

Although waypoints could theoretically be placed as close together as possible to best approximate the spline in a geometric sense, there are practical limitations in most automatic flight path guidance algorithms that make such oversampling of a spline infeasible. First, there is a limit to the number of waypoints that can be stored in the array buffer that is used to store a complete flight plan sequence in code. Second, since waypoint latitude and longitude coordinates are stored in double-precision format, waypoints placed too closely together may be interpreted as a single waypoint due to round-off error and precision limitations. Finally, a chain of waypoints could be placed so closely together that the removal of some intermediate waypoints might leave the trajectory unaltered, implying that many of the waypoints are redundant.

In addition to the previously mentioned limitations, there are restrictions on waypoint spacing that arise from the constraints on a particular aircraft's flight dynamics. In order to determine a physics-based practical minimum waypoint spacing, we introduce a wind-axes-based simplified rotorcraft model in the next section.

Simplified Wind-Axes Model of Rotorcraft Dynamics

Rotorcraft flight models are notoriously complicated, especially in the low-speed regime where aerodynamic interactions between the rotors and fuselage are highly unpredictable and not amenable to compact integrable equations of motion math models. In order to derive a practical flight dynamics-based stopping criterion for midpoint refinement that could be run in real-time on a guidance, navigation and control flight computer, the model must be relatively simple, while still capturing the salient physical effects on the vehicle dynamics. Given these requirements, we chose an augmented wind-axes model from (Refs. 17,23,24) as the basis for deriving the flight dynamics-based stopping criterion. This model is low-dimensional, derived from first principles, and valid over the entire speed range of VTOL flight, including low-speed, cruise, and high-speed flight conditions. While the model is not as accurate as many higher-fidelity models that include multiple rotor states, Computational Fluid Dynamics (CFD), and inertial cross-couplings, the model's wind-axes formulation, which does not require the vehicle heading to align with the ground track angle, enables its application to the wide variety of flight conditions that are typically encountered by many VTOL aircraft (but not by fixed-wing vehicles). The state-space equations of the wind-axes flight model for 2-dimensional planar flight can be written as:

$$\dot{x} = \begin{bmatrix} \dot{V} \\ \dot{\chi} \end{bmatrix} = \begin{bmatrix} g n_{xw} \\ \pm \frac{g}{V} \sqrt{n_{zw}^2 - 1} \end{bmatrix} \tag{2}$$

where g is the gravitational constant, V is the velocity magnitude, χ is the velocity ground track angle, and (n_{xw}, n_{zw}) represent the resultant weight-normalized forces along and normal to the flight path, respectively. The $\pm$ sign accounts for the fact that a right-hand turn results in increasing χ which corresponds to a positive value of $\dot{\chi}$, whereas a left-hand turn results in decreasing χ and thus a negative value of $\dot{\chi}$. We approximate n_{xw} and n_{zw} as constant step inputs. Step inputs were deemed sufficient for the purpose of determining minimum waypoint spacing to simplify the mathematics, although in practice, a human pilot or autonomous flight computer will typically command ramp inputs or higher-order inputs. A full derivation of the model in Equation (2) and its associated simplifying assumptions are described in APPENDIX B.

In addition to these state-space equations, which describe the evolution of the vehicle's trajectory, (Ref. 17) includes an estimate of the power required to accomplish a given maneuver. In low-speed flight, there are no sources of energy exchange other than installed engine power since there is no excess potential energy from losing altitude or excess kinetic energy from bleeding off speed, and any excess rotor energy is assumed to be negligible due to constant rotor rpm. This power estimate was crucial for evolving the model past a point-mass model that was only applicable to rotorcraft at higher speeds. The power equations from (Ref. 17) are:

$$\begin{aligned} P_R^* &= P_{ip}^* + \frac{1-M}{M} + \frac{1}{4}\frac{f_e}{A} V^{*3} \\ P_{ip}^* &= n_{xw} V^* + \sqrt{(n_{xw})^2 + (n_{zw})^2} \cdot v^* \\ v^* &= \frac{(n_{xw})^2 + (n_{zw})^2}{\sqrt{(P_{ip}^*)^2 + V^{*2} n_{zw}^2}} \end{aligned} \tag{3}$$

with the following definitions of terms:

$$\begin{aligned} P_R &= \text{Power required for steady level flight [hp]} \\ P_{ip} &= \text{Induced + propulsive power above that required} \\ &\quad \text{for steady level flight [hp]} \\ v &= \text{Induced velocity magnitude [fps]} \\ M &= \text{Rotor figure of merit} \\ f_e &= \text{Vehicle equivalent flat-plate drag area [ft}^2\text{]} \\ A &= \text{Rotor disc area [ft}^2\text{]} \\ (\,\cdot\,)^* &= \text{Indicates normalized quantities} \end{aligned} \tag{4}$$

Note that while power required is often a limiting factor in being able to accomplish a maneuver when there are no other sources of energy to exchange, other limiting factors may also need to be accounted for, such as bounds on the allowable attitudes and rates (if the vehicle is piloted) and structural load limitations. Since our waypoint spacing method was found to only have limited applicability to the low-speed flight condition, the power equations from Equation (3) were not used for any of the nonlinear simulations, although this will be an area of future work.

Although Equations (2) and (3) describe a simplified two-state rotorcraft flight dynamics model, it is not straightforward to analytically integrate these state-space equations to solve for the vehicle trajectory as a function of time. Moreover, knowing the exact vehicle position as a function of time, while useful, is not necessary for the solution we are proposing.

Minimum WP Spacing Based on Velocity Vector Bearing

Figure 2 illustrates the most common options for an aircraft's trajectory as the aircraft changes between two legs in its flight plan, e.g., from the leg connecting waypoints $n-1$ and n to the leg connecting waypoints n and $n+1$. In our analysis of minimum waypoint (WP) spacing, we will first assume that the aircraft executes Option 2: Late Change OVERFLY, i.e., overflying its current waypoint before changing to the next waypoint. We consider Option 1: Regular Change at the end of this subsection. We do not address Options 3 and 4 because they are not as commonly exercised and typically are not used to precisely follow a given spline trajectory. These options are often used to reroute a vehicle around pop-up obstacles or to execute a Return to Base maneuver in the event of a lost communication link.

Differently from the OVERFLY option illustrated in Figure 2, we do not assume that the vehicle flies through successive waypoints. Instead, we determine the minimum waypoint spacing, d, based on the location at which the vehicle's velocity vector would have the same direction as the vector from its most recent waypoint to its next one. To illustrate this criterion, Figure 3 displays hypothetical flight paths (solid and dashed brown lines) that start at a waypoint $n-1$. At its current location along the flight path, the aircraft (yellow circle) is at distance r_t and bearing θ_t from waypoint $n-1$. Suppose that θ_b is the desired bearing of the next waypoint, n,

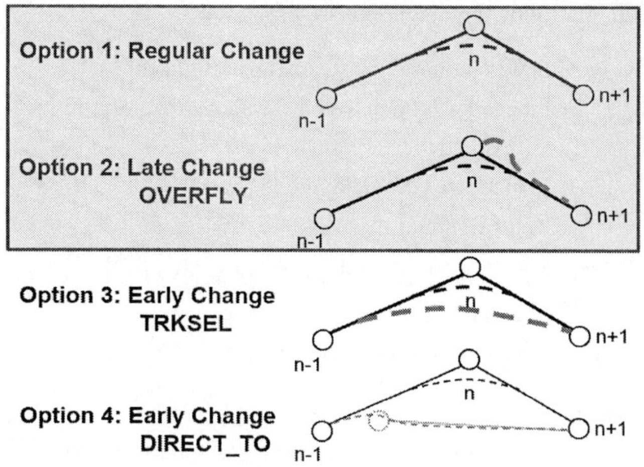

Figure 2. Common Waypoint Change Options (TRKSEL = Track Select)

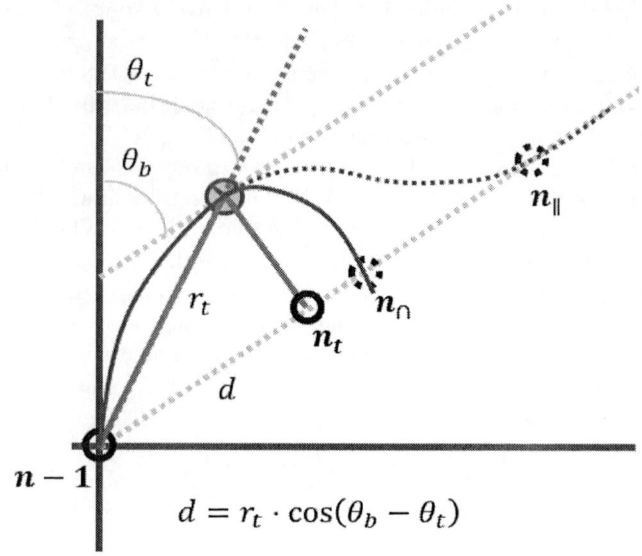

Figure 3. Three Options (n_t, $n_{\cap}$, and $n_{\|}$) for Waypoint n with Desired Bearing θ_b from Waypoint $n-1$

from $n-1$. Three possible options for n with this bearing, n_t, $n_{\cap}$, and $n_{\|}$, are shown along the dashed yellow line extending from waypoint $n-1$, which we will refer to as the *waypoint line*. At the location where the aircraft's velocity vector, which is tangent to its flight path (indicated by the dashed yellow line through the aircraft), has bearing θ_b, the point n_t marks the projection of the aircraft's position onto the waypoint line. The point $n_{\cap}$ marks the intersection of the flight path with the waypoint line in the case where the aircraft's velocity can have arbitrary bearing at this intersection. In the case where the velocity must have bearing θ_b at this intersection, the intersection is marked by the point $n_{\|}$.

From Figure 3, it is evident that the length of the leg (straight-line path) between waypoint $n-1$ and each possible waypoint n is smallest for $n = n_t$ and largest for $n = n_{\|}$. The point $n = n_{\cap}$ could be even closer to waypoint $n-1$ than n_t if the flight path can curve backward toward $n-1$. However, since we approximate the spline trajectory using midpoint refinement of its control polygon, thus defining the sides of the polygon as the legs of the flight plan, we specify that the aircraft's flight path should at least align with these legs; i.e., its velocity vector should converge to the same bearing as each leg, which can be achieved by setting waypoint n to either n_t or $n_{\|}$. If the aircraft has enough space to maneuver, we can also specify that it should converge to zero cross-track error with respect to each leg. Since we aim to define the minimum possible waypoint spacing, we set the waypoint n to n_t, which is closer to waypoint $n-1$ than $n_{\|}$ is. Thus, as illustrated in Figure 3, the minimum waypoint spacing d is defined as:

$$d = r_t \cdot \cos(\theta_b - \theta_t) \tag{5}$$

Note that $d < 0$ for $(\theta_b - \theta_t) \in (\frac{\pi}{2}, \frac{3\pi}{2})$ rad. Since negative waypoint spacings are not possible, d is set to zero when $(\theta_b - \theta_t)$ is in this range (i.e., there is no lower bound on the leg length).

We can determine the minimum waypoint spacing for the Regular Change option, which involves the aircraft changing ground track angle before it reaches its next waypoint, using our analysis for the OVERFLY option. We consider an aircraft in CRUISE flight, under the conditions that ground track angle changes are initiated with bank angle changes and turns are coordinated through either the zeroing of side-slip or zeroing of side-acceleration. The *change distance* Δ is defined as the distance from the next waypoint at which the aircraft switches from the current leg to the next leg. Assuming that the aircraft flies at a constant bank angle during the change and thus inscribes a circle between the two legs, the minimum change distance Δ_{CRUISE} can be computed as follows (see (Ref. 10), Section I.B:)

$$\Delta_{CRUISE} = \frac{V^2}{g \cdot \tan \phi_{max}} \tan \left| \frac{\Delta \psi_L}{2} \right| \tag{6}$$

where $\Delta \psi_L$ is the ground track angle change between successive legs and ϕ_{max} is the maximum allowable bank angle command. This distance can be computed *a priori*, and thus the minimum waypoint spacing d_{RC} for the Regular Change option is given by:

$$d_{RC} = d - \Delta_{CRUISE} \tag{7}$$

where d is defined in Equation (5). Note that the minimum spacing is lower for the Regular Change option than for the OVERFLY option because during Regular Change, the vehicle initiates the ground track angle change before reaching its original target waypoint.

Monotonicity of the Ground Track Angle Function

It is important to confirm that the ground track angle $\chi(t)$ is a well-behaved function that is always either increasing or decreasing with time t, since it is not physically realistic for the aircraft's trajectory to have discontinuities and, as mentioned earlier, we do not consider trajectories that curve backward toward the previous waypoint. Toward this end, we show that $\chi(t)$ is either monotonically increasing or decreasing for fixed values of n_{xw} and n_{zw}. Given an initial velocity magnitude V_o, the state-space model in Equation (2) with the positive value of $\dot{\chi}$ (indicating a right-hand turn) can be integrated analytically to obtain the solution $\chi(t)$:

$$\chi(t) = \frac{1}{n_{xw}} \left(\sqrt{n_{zw}^2 - 1} \right) \cdot \ln \left(\frac{V_o + g n_{xw} t}{V_o} \right), \quad t \geq 0. \tag{8}$$

It can be assumed that $n_{zw} \geq 1$, since aerobatic low-g pushover maneuvers are not being considered, only 2D maneuvers that alter the direction of the velocity vector in the local level plane. For $n_{xw} > 0$ and $t > 0$, we have that $(V_o + g n_{xw} t)/V_o > 1$ and so its natural logarithm is positive, yielding $\chi(t) > 0$. For $n_{xw} < 0$ and $t > 0$, $(V_o + g n_{xw} t)/V_o < 1$ and its natural logarithm is negative, which also yields $\chi(t) > 0$. Thus, irrespective of whether the vehicle is accelerating ($n_{xw} > 0$) or decelerating ($n_{xw} < 0$) during a right-hand turn, $\chi(t)$ is always positive, i.e., monotonically increasing. Using a similar argument, for a negative value of $\dot{\chi}$ (left-hand turn), it can be shown that $\chi(t)$ is always negative, i.e., monotonically decreasing.

Extension to Chains of Waypoints

The proposed criterion for waypoint spacing can be applied to chains of waypoints that form an entire flight plan. Figure 4 shows the path (solid brown line) of an aircraft that flies a chain of three waypoints, labeled $n-1$, n, and $n+1$, in the OVERFLY condition. Waypoint n has been placed at point n_t in Figure 3. As the aircraft approaches waypoint n, the guidance logic will switch to waypoint $n+1$ based on a combination of conditions, such as reaching a certain radial distance from waypoint $n+1$ or crossing the line perpendicular to the leg between waypoints $n-1$ and n that intersects waypoint n. Immediately thereafter, the same criterion can be applied as the vehicle travels from n to $n+1$ to specify that the vehicle aligns its velocity vector in the same direction as the leg between n and $n+1$, which generates the value of d for this leg. Using this criterion, successive waypoints can be similarly defined for the entire spline trajectory. If the aircraft flies the waypoints in the Regular Change condition instead of OVERFLY, the waypoint spacing can be augmented by the minimum change distance defined in Equation (6).

WAYPOINT SPACING ALGORITHM

The proposed algorithm, Algorithm 1, takes as input a set of parameters and specifications that include a cubic piecewise-polynomial spline trajectory, f; the waypoint change option (see Figure 2); and the vehicle's entry speed, V_e, and entry ground track angle, χ_e, defined as the vehicle's speed and ground track angle, respectively, before its first waypoint. The control polygon is computed by converting the piecewise-polynomial spline into its B-spline form and obtaining the locations of the polygon's vertices (the control points) from the

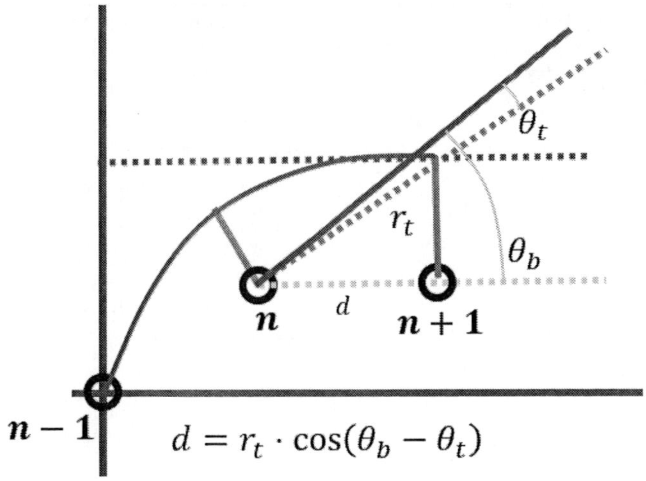

Figure 4. Illustration of Waypoint $n+1$ Placement in a Chain of Three Waypoints

B-spline coefficients. The area A_{kn} enclosed by the spline and the straight-line path through the spline knots, and the area A_{cp} enclosed by the spline and the control polygon, are computed using Green's Theorem, which can be applied to find the area A of a region bounded by a closed curve C (see (Ref. 10), Section II):

$$A = \frac{1}{2} \oint_C (-y dx + x dy) \tag{9}$$

The algorithm performs midpoint refinement of the control polygon to find the polygon that most closely approximates the trajectory, i.e., yields the smallest area A_{cp}, subject to lower bounds on the lengths of its sides, i.e., the legs of the flight plan. Indexing the legs by $l = 1, ..., L$, each leg has length $d_{leg,l}$, and its corresponding lower bound is the minimum waypoint spacing for either the OVERFLY or Regular Change option, denoted by $d_{min,l}$. Midpoint refinements are stopped if $d_{leg,l} < d_{min,l}$ for any leg l. If A_{cp} for the resulting control polygon is less than A_{kn}, then the waypoints are defined as the control points; otherwise, the waypoints are defined as the spline knots. To simplify notation, the subscript l will be dropped from quantities that vary with leg l (e.g., $d_{min,l}$ will be written as d_{min}) when we are referring to an arbitrary leg of the flight plan.

The minimum waypoint spacing d (if explicitly indexed by leg, d_l) in Equation (5) is computed from the position $(x(t_d), y(t_d))$ of the aircraft in an inertial coordinate frame fixed at its previous waypoint (e.g., waypoint $n-1$ in Figure 3), where t_d is the time the aircraft takes to reach distance r_t and bearing θ_t from its previous waypoint. The velocity ground track angle at this time, $\chi(t_d)$, equals θ_b (technically, $\theta_{b,l}$), the desired bearing of the next waypoint from the previous waypoint, and this relation can be used to solve for t_d. Hence, Equation (5) can be written as:

$$\begin{aligned} d &= r_t \cdot \cos(\theta_b - \theta_t) \\ &= \sqrt{x(t_d)^2 + y(t_d)^2} \cdot \cos\left(\theta_b - \tan^{-1}\left(\frac{y(t_d)}{x(t_d)} \right) \right) \end{aligned} \tag{10}$$

Algorithm 1 Waypoint Spacing using Control Polygon Midpoint Refinement with Flight Dynamics-Based Constraints

Require: A cubic piecewise-polynomial spline trajectory, f, constructed using three knots and represented in the monomial basis; waypoint change option (OVERFLY or Regular Change); n_{xw} option (zero or nonzero); n_{zw}; $\Delta\psi_L$; ϕ_{max}; V_e; χ_e; V_f

$A_{kn} \leftarrow$ Area of region enclosed by f and the straight-line path through the knots, from Equation (9)
if ($A_{kn} == 0$) **then** ▷ f is a straight line
  Flight plan waypoints $\leftarrow$ Knot locations
else
  $TOO\_CLOSE \leftarrow 0$ ▷ Boolean variable that indicates a violation of flight dynamics-based constraints on waypoint spacing
  while ($!TOO\_CLOSE$) **do**
    Compute control points $C_0, ..., C_{N_{CP}-1}$, the vertices of the control polygon: on first iteration, via conversion of f to the B-spline basis (see APPENDIX A); on subsequent iterations, via midpoint refinement of the control polygon
    $A_{cp} \leftarrow$ Area of region enclosed by f and the control polygon, from Equation (9)
    for ($l = 1, ..., L$) **do** ▷ The control polygon has L sides (legs of flight plan)
      Compute $\theta_{b,l}$ from the control polygon
      if ($n_{xw} == 0$) **then**
        $t_d, x(t_d), y(t_d) \leftarrow$ Equations (13), (14)
      else
        $n_{xw}, t_d, t_{dec} \leftarrow$ Equations (15), (19), (20)
        $x(t_d), y(t_d) \leftarrow$ Equation (21) if $t_{dec} \geq t_d$; Equation (22) otherwise
      end if
      $d_{min,l} \leftarrow$ Equation (11)
      $d_{leg,l} \leftarrow$ Length of leg l, the distance between control points C_{l-1} and C_l
    end for
    $TOO\_CLOSE \leftarrow 1$ if any $d_{leg,l} < d_{min,l}$; 0 otherwise
  end while
  if ($A_{cp} < A_{kn}$) **then**
    Flight plan waypoints $\leftarrow$ Control points
  else
    Flight plan waypoints $\leftarrow$ Knot locations
  end if
end if
return Flight plan waypoints

Thus, the distance d_{min} is defined as follows for the OVERFLY and Regular Change options:

$$d_{min} = \begin{cases} d \text{ from Eq. (10)}, & \text{OVERFLY} \\ d_{RC} \text{ from Eq. (7) with } d \text{ from Eq. (10)}, & \text{Regular Change} \end{cases} \tag{11}$$

We compute d for two cases: (1) $n_{xw} = 0$, i.e., there is zero along-track acceleration or deceleration; and (2) $n_{xw} \neq 0$. In both cases, there is constant normal acceleration $n_{zw} > 1$ to maintain level flight.

In the case $n_{xw} = 0$, we assume that the waypoints are sufficiently close together such that the aircraft's speed does not change considerably between waypoints, and we set this constant speed to V_e. The state-space model in Equation (2) reduces to:

$$\begin{aligned} V(t) &= V_e = constant \\ \chi(t) &= \frac{g}{V_e}\left(\sqrt{n_{zw}^2 - 1}\right) \cdot t \end{aligned} \tag{12}$$

Setting $\chi(t_d) = \theta_b$ in Equation (12) and solving for t_d yields:

$$t_d = \frac{V_e}{g}\left(\sqrt{n_{zw}^2 - 1}\right)^{-1} \cdot \theta_b \tag{13}$$

Then Equations (12) and (13) can be used to compute the aircraft's position coordinates at time t_d as the following expressions, which are straightforward to integrate:

$$\begin{aligned} x(t_d) &= \int_0^{t_d} V(t) \cdot \cos(\chi(t))dt \\ &= \int_0^{t_d} V_e \cdot \cos\left[\frac{g}{V_e}\left(\sqrt{n_{zw}^2 - 1}\right) \cdot t\right] dt \\ y(t_d) &= \int_0^{t_d} V(t) \cdot \sin(\chi(t))dt \\ &= \int_0^{t_d} V_e \cdot \sin\left[\frac{g}{V_e}\left(\sqrt{n_{zw}^2 - 1}\right) \cdot t\right] dt \end{aligned} \tag{14}$$

In the case $n_{xw} \neq 0$, we assume a constant n_{xw}, approximated as the aircraft's average n_{xw} throughout the maneuver, and compute the aircraft's speed at each waypoint (control point) based on this value of n_{xw}. We define $d_{tot} = \sum_{l=1}^{L} d_{leg,l}$ as the total length of all legs of the flight plan. Let V_j denote the aircraft's speed at control point C_{j-1}, $j = 1, ..., N_{CP}$. V_1 is the aircraft's entry speed, V_e. We assume that the aircraft is decelerating to a desired final speed, V_f, during its maneuver, such that $V_{N_{CP}}$ should be V_f. Typical values of V_f are around 5 knots for a hover target, as most of the effort to achieve hover is applied to bring V_f close to zero. The value of n_{xw} is computed from the constraint that $V_{N_{CP}} = V_f$:

$$n_{xw} = \frac{V_f^2 - V_1^2}{2 \cdot d_{tot}} \tag{15}$$

If $n_{xw} > n_{xw,max}$, n_{xw} is set to $n_{xw,max}$, and a higher value of V_f than desired is accepted. The aircraft's speed at each control point is then given by:

$$V_1 = V_e; \quad V_j = (V_1^2 + 2 \cdot n_{xw} \cdot \sum_{i=1}^{j-1} d_{leg,i})^{1/2}, \quad j = 2, ..., N_{CP} \tag{16}$$

Given the aircraft speed V_j at the control point at the beginning of a leg, the state-space model in Equation (2) can be integrated to obtain:

$$\begin{aligned} V(t) &= V_j + g n_{xw} t \\ \chi(t) &= \int_0^t \frac{g}{V(t)}\left(\sqrt{n_{zw}^2 - 1}\right) dt \end{aligned} \tag{17}$$

Combining the two expressions in Equation (17) and integrating the expression for $\chi(t)$ yields:

$$\chi(t) = \frac{1}{n_{xw}} \left(\sqrt{n_{zw}^2 - 1} \right) \ln \left(\frac{V_j + g n_{xw} t}{V_j} \right) \tag{18}$$

Setting $\chi(t_d) = \theta_b$ in Equation (18) and solving for t_d yields:

$$t_d = \frac{V_j}{g n_{xw}} \left(\exp\left\{ \theta_b \cdot \frac{n_{xw}}{\sqrt{n_{zw}^2 - 1}} \right\} - 1 \right) \tag{19}$$

We can use Equation (17) to solve for the time t_{dec} that the aircraft takes to decelerate to V_{j+1}, the computed speed at the control point at the end of the leg:

$$t_{dec} = \frac{V_{j+1} - V_j}{g n_{xw}} \tag{20}$$

We will first address the case where $t_{dec} \geq t_d$. Equations (18) and (19) can be used to compute the aircraft's coordinates at time t_d:

$$\begin{aligned} x(t_d) &= \int_0^{t_d} V(t) \cdot \left(\cos \left[\frac{\sqrt{n_{zw}^2 - 1}}{n_{xw}} \cdot \ln \left| \frac{V(t)}{V_j} \right| \right] \right) dt \\ y(t_d) &= \int_0^{t_d} V(t) \cdot \left(\sin \left[\frac{\sqrt{n_{zw}^2 - 1}}{n_{xw}} \cdot \ln \left| \frac{V(t)}{V_j} \right| \right] \right) dt \end{aligned} \tag{21}$$

where $V(t)$ is defined in Equation (17). These integrals are not easily solved in closed form, but can be efficiently computed with quadrature. In tests on a 13th Gen Intel Core i7-13700 2.10 GHz processor with 64.0 GB of RAM, it took approximately 0.02 seconds to numerically integrate over 150 instances of Equation (21), without further code optimization. Thus, there are no foreseen barriers to numerically integrating Equation (21) in real-time on a flight control computer.

We next address the case where $t_{dec} < t_d$. In this case, the aircraft performs the maneuver in two phases: (1) turning while decelerating to V_{j+1} from time 0 to t_{dec}, then (2) pure turning at constant V_{j+1} from time t_{dec} to t_d. The coordinates $(x(t_{dec}), y(t_{dec}))$ of the aircraft at time t_{dec} can be computed from Equation (21), with the upper limit of integration t_d replaced by t_{dec} from Equation (20). The aircraft's coordinates at time t_d are then computed by adding $(x(t_{dec}), y(t_{dec}))$ to the expressions in Equation (21), integrated from t_{dec} to t_d:

$$\begin{aligned} x(t_d) &= x(t_{dec}) + \int_{t_{dec}}^{t_d} V_{j+1} \cdot \cos \left[\frac{\sqrt{n_{zw}^2 - 1}}{n_{xw}} \cdot \ln \left| \frac{V_{j+1}}{V_j} \right| \right] dt \\ y(t_d) &= y(t_{dec}) + \int_{t_{dec}}^{t_d} V_{j+1} \cdot \sin \left[\frac{\sqrt{n_{zw}^2 - 1}}{n_{xw}} \cdot \ln \left| \frac{V_{j+1}}{V_j} \right| \right] dt \end{aligned} \tag{22}$$

The ground track angle at time t_d is found by integrating Equation (2) over the two time intervals $[0, t_{dec}]$ (with $V(t)$ defined in Equation (17)) and $[t_{dec}, t_d]$ (with $V(t) = V_{j+1}$):

$$\chi(t_d) = \theta_b = \sqrt{n_{zw}^2 - 1} \left(\frac{1}{n_{xw}} \ln \left| \frac{V_{j+1}}{V_j} \right| + \frac{g}{V_{j+1}} (t_d - t_{dec}) \right) \tag{23}$$

ALGORITHM PERFORMANCE ON MOTION PRIMITIVES

Algorithm 1 was tested on five motion primitives: straight line, sinusoid, semicircle, decaying exponential, and logarithmic spiral. The straight-line primitive was left unscaled, as it was simply a test of the algorithm's ability to reject a case where no control polygon midpoint refinement was required. The sinusoid and semicircle primitives were sized to be applicable to common ADS-33/MIL-DTL-32742 MTEs such as the Slalom and Pirouette. The exponential was sized to match a future proposed MTE, the Emergency Stop, that is tailored specifically for VTOL UAS (Ref. 21). Finally, the logarithmic spiral was sized with an initial radius based on the Pirouette and modeled after a spiral search pattern that is currently in operation on certain cargo/utility VTOL platforms. For all motion primitives, the algorithm was run with the OVERFLY waypoint change and $n_{xw} = 0$ options; for the exponential, the algorithm was additionally run with the $n_{xw} = 0$ option.

The following subsections present the minimum waypoint spacings (i.e., minimum leg lengths), d_{min} from Equation (11), that are generated by Algorithm 1 for each motion primitive (except the straight line), given different aircraft entry speeds V_e. Instead of executing Lines 7-19 based on the Boolean condition in Line 6, these lines were repeated three times to produce an initial control polygon (CP0) and two midpoint refinements (CP1, CP2). This was done to identify any legs of CP0, CP1, and CP2 whose length, d_{leg}, is shorter than the minimum length d_{min} computed for that leg, indicating that this leg violates flight dynamics-based constraints on its length. In addition, the area A_{kn} enclosed by the spline and the baseline (BL) path through the knots and the area A_{cp} enclosed by the spline and each control polygon (CP0, CP1, CP2) are shown for comparison.

Straight Line

The straight-line trajectory is a trivial case, but must be tested to ensure that the algorithm handles it properly. Since the area enclosed by the trajectory and the straight-line path through the waypoints (initially, the knots) is zero, successive midpoint refinements would not reduce this area, so the algorithm terminates without inserting any additional waypoints (control points), as expected.

Sinusoid: Slalom

The sinusoid motion primitive can be used to approximate the ADS-33 Slalom MTE, as shown in Figure 5. It is clear from the figure that the given spline representation of the sinusoid is closely approximated by the control polygon after only one or two midpoint refinements (CP1, CP2). We apply the specification for the Slalom MTE that the aircraft flies at a constant speed throughout the maneuver, which we define as V_e. We compute waypoint spacings for four values of V_e defined in the Slalom MTE standards: 15 knots (desired) and 30 knots (adequate) in Degraded Visual Environments (DVE), and 40

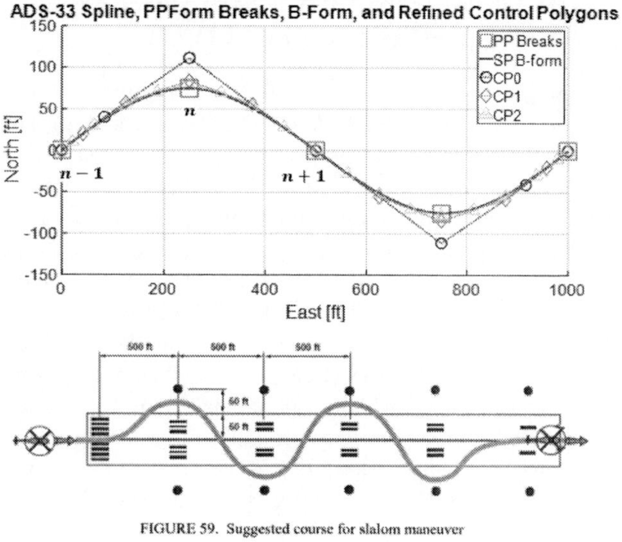

Figure 5. ***Top:*** **Cubic piecewise-polynomial spline approximation of a sinusoid (SP B-form), shown with knots (PP Breaks) and control points of successive midpoint refinements of the control polygon (CP0, CP1, CP2).** ***Bottom:*** **ADS-33 Slalom MTE from (Ref. 25).**

knots (desired) and 60 knots (adequate) in Good Visual Environments (GVE). For CP0, CP1, and CP2, Table 1 displays the values of d_{leg} in bold and the values of d_{min} computed for each entry speed V_e. Values of d_{min} for which $d_{leg} < d_{min}$ are highlighted in red, along with the corresponding entry speed. The table shows that the legs of CP0 all exceed d_{min} for $V_e = 15$ and 30 knots, but two legs are shorter than d_{min} for $V_e = 40$ and 60 knots.

The areas A_{cp} for CP0, CP1, and CP2 are about 90%, 25%, and 7%, respectively, of the area A_{kn}, indicating that the successive midpoint refinements produce flight plans that approximate the spline trajectory more and more closely, thus reducing the tracking error for the maneuver. However, this comes at the cost of incurring violations of flight dynamics-based constraints (henceforth simply referred to as "violations"): the number of legs that are shorter than d_{min} generally increases with successive midpoint refinements and increasing entry speed. No constraints are violated at the lowest entry speed of 15 knots for any of the control polygons.

Semicircle: Pirouette

As shown in Figure 6, the semicircle motion primitive models the ADS-33 Pirouette MTE, a slow, constant-speed circular maneuver with a radius of 100 ft. It represents one of the most challenging low-speed rotorcraft maneuvers and epitomizes "heading independent of ground track," since the nose of the vehicle must always be pointed at the center of the circle as the vehicle moves around the circle's perimeter. Figure 6 shows that the spline representation of the semicircle is very

Table 1. Sinusoid Motion Primitive: Waypoint Spacings for Constant Aircraft Speed

Entry Ground Track Angle: 90 deg

Bounded Area [ft²]		Grd Track Chg [deg]	Separation Distances [ft]										
BL 6123	CP0 (5pts) **4579**	-26,4,44,4	CP	**93**	**180**	**180**	**93**						
			15	23	5	29	5						
			30	91	19	116	19						
			40	161	33	206	33						
			60	[illegible]	75	[illegible]	75						
CP1	CP1 (7pts) **1267**	-26,2,13, 23,13,2	CP	**47**	**91**	**128**	**128**	**91**	**47**				
			15	23	2	14	21	14	2				
			30	91	10	54	84	54	10				
			40	[illegible]	17	96	[illegible]	[illegible]	17				
			60	362	38	217	334	217	38				
CP2	CP2 (11pts) **368**	-26,1,4,6,9, 12,9,6,4,1	CP	**23**	**46**	**67**	**85**	**63**	**63**	**85**	**67**	**46**	**23**
			15	23	1	5	7	10	13	10	7	5	1
			[illegible]	[illegible]	5	19	27	39	51	39	27	19	5
			40	161	9	33	49	70	[illegible]	70	49	33	9
			[illegible]	[illegible]	19	[illegible]	[illegible]	[illegible]	[illegible]	[illegible]	[illegible]	[illegible]	19

RED indicates violation of flight dynamics-based constraints

Table 2. Semicircle Motion Primitive: Waypoint Spacings for Constant Aircraft Speed

Entry Ground Track Angle: 0 deg

Bounded Area [ft²]		Grd Track Chg [deg]	Separation Distances [ft]										
BL 5708	CP0 (5pts) **4292**	0,45,90, 45	CP	**50**	**141**	**141**	**50**						
			05	0	4	3	4						
			10	0	13	9	13						
			15	0	29	20	29						
			30	0	117	79	117						
CP1	CP1 (7pts) **855**	0,27,37, 53,37,27	CP	**25**	**56**	**84**	**84**	**56**	**25**				
			05	0	3	4	4	4	3				
			10	0	10	12	13	12	10				
			15	0	23	27	30	27	23				
			30	0	93	109	119	109	93				
CP2	CP2 (11pts) **102**	0,14,22, 19,22,28,22 ,19,22,14	CP	**13**	**26**	**40**	**49**	**39**	**39**	**49**	**40**	**26**	**13**
			05	0	2	3	2	3	3	3	2	3	3
			10	0	6	5	8	9	11	9	8	9	6
			15	0	14	20	18	20	24	20	18	20	14
			[illegible]	0									

RED indicates violation of flight dynamics-based constraints

closely approximated by the control polygon after two midpoint refinements (CP2). We define the aircraft's speed as V_e and compute waypoint spacings for four speeds defined in the Pirouette MTE standards: 5 knots (desired) and 10 knots (adequate) in DVE, and 15 knots (desired) and 30 knots (adequate) in GVE.

Table 2 displays the same information as Table 1 for the semicircle motion primitive. The table shows that for each control polygon refinement, all legs exceed d_{min} for $V_e = 5$ and 10 knots. For CP0 and CP1, this is also true for $V_e = 15$ knots; however, refining the control polygon to CP2 incurs a violation at this speed (the last leg is shorter than d_{min}). Thus, the semicircle maneuver can likely be performed at speeds greater than 15 knots, depending on how many refinements of the control polygon are applied.

Note that the semicircle motion primitive can be flown using existing guidance algorithms for circular trajectories that are even simpler than control polygon midpoint refinement. A comparison of these methods is a possible area of future work.

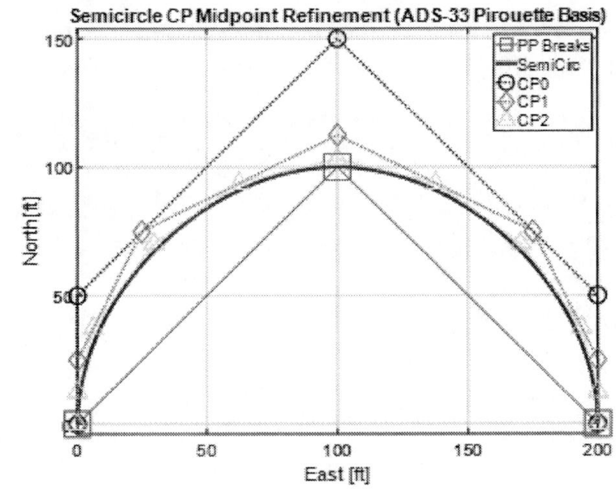

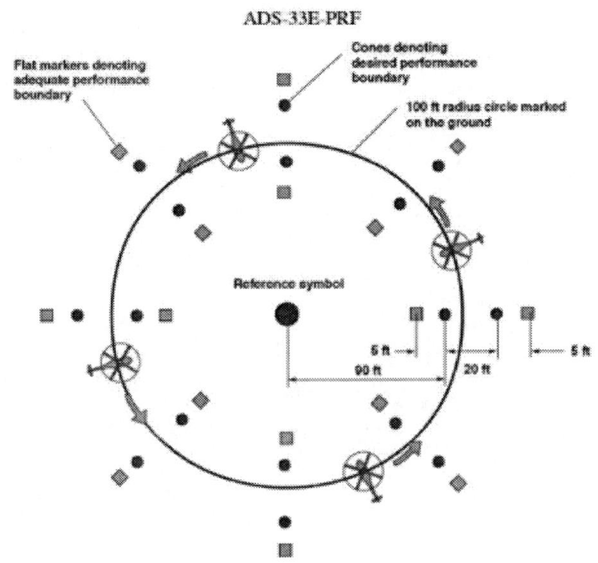

Figure 25. Suggested course for pirouette maneuver

Figure 6. *Top:* **Cubic piecewise-polynomial spline approximation of a semicircle (SemiCirc), shown with knots (PP Breaks) and control points of successive midpoint refinements of the control polygon (CP0, CP1, CP2).** *Bottom:* **ADS-33 Pirouette MTE from (Ref. 25).**

Exponential: Emergency Stop

The Emergency Stop is a newer MTE, first proposed in (Ref. 21), that was designed for VTOL UAS. One of the primary objectives of this maneuver is to *check the ability to accomplish an unexpected stop from a representative autonomous operation speed with limited forward distance travel.* The trajectory of a vehicle during this maneuver can be modeled by the exponential motion primitive. Assuming an aircraft Froude scaling of 1, the desired and adequate values of the pertinent requirements from (Ref. 21) are summarized in Table 3.

The exponential Emergency Stop maneuver used for testing

Table 3. Emergency Stop MTE Performance Standards

Metric	Desired	Adequate
Minimum speed from which deceleration is initiated, V_{decel}	60 kts	50 kts
Maximum allowable forward travel after deceleration is initiated	350 ft	600 ft

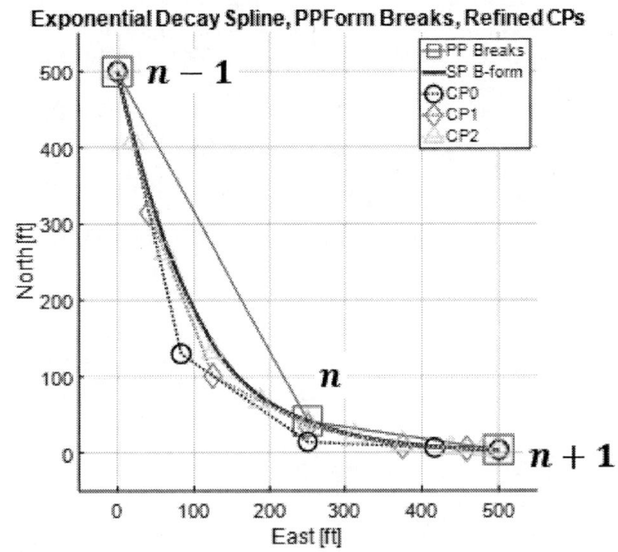

Figure 7. Cubic piecewise-polynomial spline approximation of a decaying exponential (SP B-form), shown with knots (PP Breaks) and control points of successive midpoint refinements of the control polygon (CP0, CP1, CP2).

Algorithm 1 is shown in Figure 7. The aircraft enters from the north while decelerating and turning to the east. The MTE requirements allow for the aircraft to execute the turn in either a coordinated or uncoordinated fashion, as long as it concludes in a steady hover with minimal altitude loss. We first specify that the aircraft flies at a constant speed (i.e., $n_{xw} = 0$), its entry speed V_e, which is set to four values with a maximum of 60 knots, the desired value in Table 3. The aircraft's maximum allowable forward travel was set to 500 ft, which is between the desired and adequate values in the table.

Table 4 displays the same information as the previous tables for the exponential motion primitive. The table shows that for speeds between 40 and 60 knots, each control polygon (CP0, CP1, and CP2) has one or more legs that are shorter than d_{min}, and thus incur a violation. Since these speeds overlap with the range of speeds (50–60 knots) at which the Emergency Stop MTE is meant to be flown, we attempt to eliminate these violations by running Algorithm 1 with $n_{xw} \neq 0$, which modifies the waypoint spacing calculations to account for the vehicle's deceleration during the maneuver.

A comparison of waypoint spacings for the constant-speed (CS; $n_{xw} = 0$) and decelerating (DCL; $n_{xw} \neq 0$) cases for each

Table 4. Exponential Motion Primitive: Waypoint Spacings for Constant Aircraft Speed

Entry Ground Track Angle: 180 deg

Bounded Area [ft²]		Grd Track Chg [deg]	Separation Distances [ft]										
BL 23508	CP0 (5pts) 9347	-13, -43, -32, -1	CP	379	203	167	83						
			15	14	29	26	1						
			30	54	115	102	5						
			40	96	206	181	9						
			60	217	481	408	19						
CP1	CP1 (7pts) 3551	-13, -9, -41, -16, -9, 0	CP	190	230	142	128	83	42				
			15	14	10	28	16	10	0				
			30	54	39	114	64	39	0				
			40	96	70	202	114	70	0				
			60	217	158	454	256	158	0				
CP2	CP2 (11pts) 675	-13, -3, -10, -18, -22, -8, -7, -5, -3, 0	CP	95	152	143	90	69	65	63	63	42	21
			15	14	4	11	18	20	9	8	6	4	0
				54	14	43	70	[illegible]	36	32	23	14	0
			40	96	25	77	125	144	63	56	41	25	0
				[illegible]	57	[illegible]	[illegible]	[illegible]	[illegible]	[illegible]	[illegible]	[illegible]	0

RED indicates violation of flight dynamics-based constraints

Table 5. Exponential Motion Primitive, Control Polygon CP0: Waypoint Spacings for Decreasing Aircraft Speed

Entry Ground Track Angle: 180 deg

Bounded Area [ft²]		n_{xw} Required	Separation Distances [ft]				
BL 23508	CP0 (5pts) 9347	-0.386 [ft/s²]	CP	379	203	167	83
			15 kts CS	14	29	26	1
			15 kts DCL	14	16	8	0
			V_{entry}	15kts	11.1kts	8.2kts	4.7kts
Ground Trk Change [deg]		-1.541 [ft/s²]	30 kts CS	54	115	102	5
			30 kts DCL	54	63	31	0
			V_{entry}	30kts	22.1kts	16.4kts	9.5kts
-13, -43, -32, -1		-2.739 [ft/s²]	40 kts CS	96	[illegible]	[illegible]	0
			40 kts DCL	96	111	55	1
			V_{entry}	40kts	29.5kts	21.9kts	12.6kts
		-6.163 [ft/s²]	60 kts CS	217	481	408	19
			60 kts DCL	217	[illegible]	123	2
			V_{entry}	60kts	44.3kts	32.9kts	19.0kts

RED indicates violation of flight dynamics-based constraints

of the three control polygons is shown in Tables 5, 6, and 7. In the decelerating case, V_f is set to 0 knots, n_{xw} *required* is computed from Equation (15), and the rows labeled V_e in the tables list the aircraft's speed at the start of each leg. For the control polygon CP0, Table 5 shows that accounting for the aircraft's deceleration between control points reduces the minimum distance required to reorient along the spline. Due to this reduction in waypoint spacing, there are no longer any violations at a speed of 40 knots and only one violation at 60 knots, where the magnitude $d_{min,l} - d_{leg,l}$ of the violation is significantly lower than that of the two violations at 60 knots in the constant-speed case. Similarly, for control polygons CP1 and CP2, Tables 6 and 7 show that the decelerating case also reduces the number of violations and their magnitudes relative to the constant-speed case at speeds of 40 and 60 knots. Future versions of the algorithm can account for the particular magnitude $d_{min,l} - d_{leg,l}$ of any violations in determining minimum waypoint spacing, rather than just whether $d_{min,l} - d_{leg,l} > 0$.

Table 6. Exponential Motion Primitive, Control Polygon CP1: Waypoint Spacings for Decreasing Aircraft Speed

Entry Ground Track Angle: 180 deg

Bounded Area [ft²]		n_{xw} Required	Separation Distances [ft]						
BL 23508	CP1 (7pts) 3551	-0.393 [ft/s²]	CP	190	230	142	128	83	42
			15 kts CS	14	10	28	16	10	0
			15 kts DCL	14	6	14	5	2	0
			V_{entry}	15 kts	13.1 kts	10.4 kts	8.4 kts	5.9 kts	3.4 kts
Ground Track Change [deg]		-1.573 [ft/s²]	30 kts CS	54	39	114	64	39	0
			30 kts DCL	54	30	55	20	6	0
			V_{entry}	30 kts	26.3 kts	20.9 kts	16.7 kts	11.7 kts	6.8 kts
-13, -9, -41, -16, -9, 0		-2.796 [ft/s²]	40 kts CS	96	70	[illegible]	114	70	0
			40 kts DCL	96	54	98	35	11	0
			V_{entry}	40kts	35kts	27.8kts	22.3kts	15.7kts	9.1kts
		-6.292 [ft/s²]	60 kts CS	217	158	454	[illegible]	158	0
			60 kts DCL	[illegible]	121	[illegible]	80	24	0
			V_{entry}	60kts	52.5kts	41.8kts	33.4kts	23.5kts	13.6kts

RED indicates violation of flight dynamics-based constraints

Table 7. Exponential Motion Primitive, Control Polygon CP2: Waypoint Spacings for Decreasing Aircraft Speed

Entry Ground Track Angle: 180 deg

Bounded Area [ft²]		n_{xw} Req	Separation Distances [ft]										
BL 23508	CP2 (11pts) 675	15 kts -0.4 [ft/s²]	CP	95	152	143	90	69	66	63	63	42	21
			CS	14	4	11	18	20	9	8	6	4	0
			DCL	14	5	7	9	8	3	2	1	0	0
			V_{entry}	15	14.1	12.5	10.8	9.5	8.4	7.3	5.9	4.2	2.4
Ground Track Change [deg]		30 kts -1.6 [ft/s²]	CS	54	14	43	70	[illegible]	36	32	23	14	0
			DCL	54	13	30	36	33	11	7	4	1	0
			V_{entry}	30	29.2	25	21.5	19	16.9	14.6	11.9	8.4	4.8
-13, -3, -10, -18, -22, -8, -7, -5, -3, 0		40 kts -2.84 [ft/s²]	CS	[illegible]	25	77	[illegible]	[illegible]	63	56	41	25	0
			DCL	96	22	53	64	58	20	13	6	2	0
			V_{entry}	40	37.6	33.3	28.7	25.4	22.5	19.4	15.8	11.2	8.5
		60 kts -6.38 [ft/s²]	CS	217	57	173	281	324	142	125	92	67	0
			DCL	[illegible]	50	120	[illegible]	[illegible]	45	30	14	4	0
			V_{entry}	60	56.3	49.9	43	38	33.7	29.1	23.8	16.8	9.7

RED indicates violation of flight dynamics-based constraints

Logarithmic Spiral: Search Pattern

The logarithmic spiral is a typical search pattern used to cover large distances without having to constantly change the direction of aircraft travel and with only small adjustments in airspeed and bank angle, depending on ambient wind conditions. Starting from the last known location of a target (e.g., a lost person) the aircraft can systematically expand its search radius without the risk of revisiting terrain or missing the target. Figure 8 shows that, as for other motion primitives, the spline representation of the logarithmic spiral is very closely approximated by the control polygon after two midpoint refinements (CP2).

Table 8 displays the same information as the previous tables for the logarithmic spiral. As expected, flight dynamics-based constraints limit the speed at which the spiral can be flown, given the allowable cross-track error metrics. The table shows that violations are incurred for all control polygons at speeds of 30 knots and above, indicating that this waypoint spacing method is not suitable recommended for higher speeds. The results show why the logarithmic spiral is not typically flown with waypoints operationally.

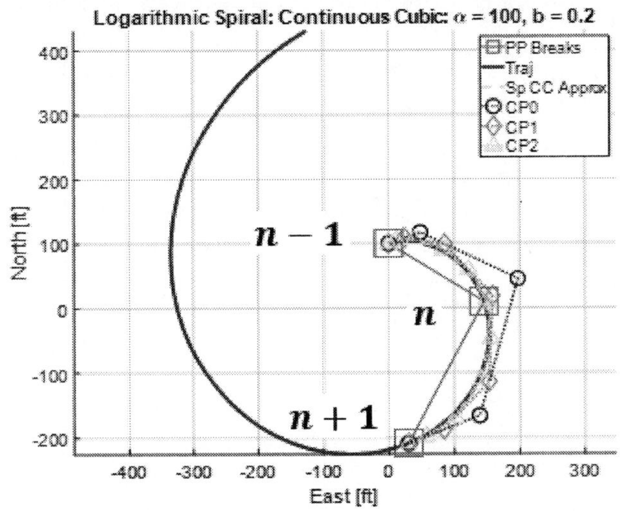

Figure 8. Cubic piecewise-polynomial spline approximation (SP CC Approx) of a logarithmic spiral trajectory (Traj), shown with knots (PP Breaks) and control points of successive midpoint refinements of the control polygon (CP0, CP1, CP2).

Summary of Results for Motion Primitives

As demonstrated in the previous subsections, the proposed algorithm can be used to define a feasible waypoint sequence that follows a complicated trajectory while maintaining legacy guidance algorithms and not requiring an excessive number of waypoints. For the sinusoid, semicircle, and exponential motion primitives, the control polygon refinements generally yield waypoint spacings that do not violate flight dynamics-based constraints at speeds required for the corresponding MTEs (the Slalom, Pirouette, and Emergency Stop). Cases where violations are incurred, and thus the desired MTE would not be flyable with the computed waypoint spacing, can be identified beforehand using our methodology. Because the algorithm was developed based on the simplified wind-axes flight model, which does not include some real-world physical effects, we assessed its utility for waypoint guidance under the more realistic conditions and vehicle dynamics of a high-fidelity nonlinear simulation, as described in the next section.

NONLINEAR SIMULATIONS

The core flight dynamics model used in the nonlinear simulation was GENHEL, a generic nonlinear single main-rotor tail-rotor (MR-TR) helicopter model, with the UH-60A parameters in Table 9. The basis of this nonlinear model and its implementation are discussed in (Refs. 18–20). APPENDIX C describes how the weight-normalized aerodynamic forces (n_{xw}, n_{yw}, n_{zw}) are extracted from this flight model implementation. We first compared the predictions of the simplified wind-axes model and the simulated aircraft trajectories for

Table 8. Logarithmic Spiral Motion Primitive: Waypoint Spacings for Constant Aircraft Speed

Entry Ground Track Angle: 90 deg

Bounded Area [ft²]		Grd Track Chg [deg]	Separation Distances [ft]										
BL 16822	CP0 (5pts) 9617	-68,53, 80,45	CP	116		220			165			51	
			15	28		30			24			29	
			30	112		119			97			117	
			40	199		211			173			207	
			60	[illegible]		[illegible]			[illegible]			[illegible]	
CP1	CP1 (7pts) 2924	-68,26,41, 44,39,28	CP	58		101		133	108		62		26
			15	28		23		28	29		28		24
			30	112		91		114	116		112		96
			[illegible]	[illegible]		[illegible]		[illegible]	[illegible]		[illegible]		[illegible]
			60	448		382		454	453		446		379
CP2	CP2 (11pts) 868	-68,12,23, 21,20,23, 21,20,23, 16	CP	26	63	71	65	80	54	50	46	27	13
			15	26	13	21	20	19	21	20	19	21	16
			[illegible]	[illegible]	51	[illegible]	[illegible]	[illegible]	[illegible]	[illegible]	[illegible]	[illegible]	[illegible]
			40	[illegible]	90	[illegible]	[illegible]	[illegible]	[illegible]	[illegible]	[illegible]	[illegible]	114

RED indicates violation of flight dynamics-based constraints

Table 9. GENHEL UH-60A Example Parameters

Parameter	Variable	Value
Air Density	Atm.rho	1.225 kg/m^3
MR No. Blades	Helo.MR.nb	4
MR Speed	Helo.MR.omega	27 rad/s (258 RPM)
MR Chord	Helo.MR.c	1.73 ft
MR Lift-Curve Slope	Helo.MR.a	5.37 1/rad
MR Flap Hinge Offset	Helo.MR.e	1.25 ft
Primary Mission GW	Helo.Mass	16,260 lb

right-handed turn maneuvers, and then evaluated the simulated aircraft's trajectory-tracking performance for flight plans based on the Slalom and Exponential Stop MTEs. The performance of the simulated aircraft at flying these relatively simple maneuvers can provide insight into flight conditions under which our algorithm can generate feasible waypoint sequences for more complex maneuvers.

Trajectory Comparisons with Wind-Axes Model

The trajectories of the simplified wind-axes model and the nonlinear simulation were compared for two series of right-handed turn maneuvers, flown under the following conditions with different combinations of $(n_{xw,max}, n_{zw,max})$:

- *High-speed cruise:* Entry speed $V_e = 100$ knots, constant bank angle, with ball-centered turn coordination entry conditions
- *Low-speed cruise:* Entry speed $V_e = 20$ knots, changing ground track angle, fixed aircraft nose heading due north (0°)

For these two series of maneuvers, $(n_{xw,max}, n_{zw,max})$ were computed as shown in Tables 10 and 11. In the tables, $\dot{v}_{xc,max}$ and $\dot{v}_{yc,max}$ are the longitudinal and lateral maximum acceleration commands in the local level frame, and a_z is the accelerometer measurement along the z-axis of the body frame.

Table 10. Maximum Acceleration and Bank Angle Commands Converted to $(n_{xw,max}, n_{zw,max})$ For Increasingly Aggressive Maneuvers at High-Speed Cruise (100 kts), Ball-Centered Turn Coordination

$\dot{v}_{xc,max}$ [g's]	ϕ_{max} @$n_{zw,max}$	a_z $(T = W)$	$n_{xw,max}$ $(\dot{v}_{xc,max})$	$n_{zw,max}$ $(1/\cos\phi_{max})$
0.1	10°	1.0	0.1	1.0154
0.2	15°	1.0	0.2	1.0353
0.3	30°	1.0	0.3	1.1547
0.5	45°	1.0	0.5	1.4142
1.0	60°	1.0	1.0	2.0000

Table 11. Maximum Acceleration and Deceleration Commands Converted to $(n_{xw,max}, n_{zw,max})$ For Increasingly Aggressive Maneuvers at Low-Speed Cruise (20 kts); n_{yw} assumed negligible

$\dot{v}_{(x,y)c,max}$ [g's]	MR Thrust Tilt @$n_{zw,max}$	$n_{xw,max}$ $(\dot{v}_{xc,max})$	$n_{zw,max}$ $\sqrt{\dot{v}_{xc}^2 + \dot{v}_{yc}^2 + 1.0}$
0.1	8.0°	0.1	1.0100
0.2	15.8°	0.2	1.0392
0.3	23.0°	0.3	1.0863
0.5	35.3°	0.5	1.2247
1.0	54.7°	1.0	1.7321

Note that in the low-speed cruise case (Table 11), $n_{zw,max}$ does not depend on the maximum bank angle command ϕ_{max}, since the aircraft heading is independent of ground track angle and turns do not have to be coordinated. (Moved text here:) In low-speed flight, the fundamental mechanism of rotorcraft maneuverability is different than in high-speed flight: although the aircraft is capable of making coordinated turns (either ball-centered or zero side-slip), its speed is typically well within bounds imposed by side-slip, structural, or critical wind azimuth limitations, and thus it does not need to perform a coordinated turn to change ground track angle. For example, the most efficient way for an aircraft in low-speed flight to achieve a 180° change in ground track angle is to decelerate, go through hover, and then re-establish its new velocity vector.

Figure 9 compares the planar trajectories of the nonlinear simulation and wind-axes model for the high-speed cruise case with $n_{xw,max} = -0.2$ g's and ϕ_{max} varied from 10° to 60°. The nonlinear simulation 6DOF autopilot was set up with a flight plan that included an OVERFLY waypoint at (0E ft, 500N ft) bearing north and a subsequent waypoint placed immediately behind bearing due south, which forced the autopilot guidance to command and hold a turn at the maximum bank angle to reacquire the next leg. As Figure 9 shows, there is close agreement between the trajectories of the wind-axes model and the nonlinear simulation at the benign bank angles of 10° and 15°, with growing discrepancies as the bank angle increases up to 60° (i.e., as the turn becomes more aggressive). This is in part because it is difficult for the aircraft autopilot to hold a perfectly coordinated turn at high bank angles while maintaining constant deceleration. Thus, a major limitation of any

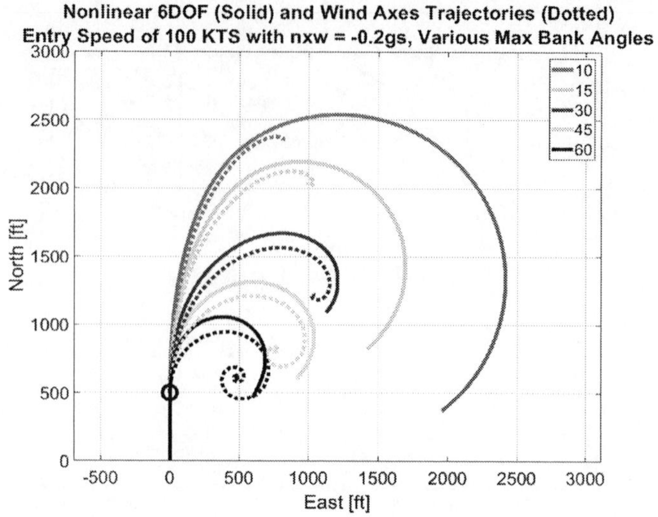

Figure 9. Trajectories of Nonlinear Simulation 6DOF Autopilot (Solid Lines) and Wind-Axes Model (Dotted Lines) for High-Speed Cruise (100 kts) OVERFLY with $n_{xw,max} = -0.2$ g's. Legend shows Maximum Bank Angle Command ϕ_{max} (deg).

automatic flight path guidance algorithm that is based on the simplified wind-axes model will be the ability of the vehicle control laws to maintain commanded attitudes and accelerations so that the model is valid.

Figure 10 shows trajectories of the nonlinear simulation for the low-speed cruise case with $n_{xw,max} = -0.2$ g's and $n_{zw,max} = 1.04$, in a scenario where the vehicle must make a 135° change in ground track angle via a Regular Change maneuver at different change distances Δ from the waypoint. At larger distances Δ of 25-75 ft, the vehicle performs a combined pitch and bank change to acquire the new track. At smaller distances Δ of 5-15 ft, the vehicle decelerates to a hover and backtracks to reacquire the new ground track. Both types of maneuvers are unique to the low-speed regime in that they are both uncoordinated, involving high amounts of side-slip (the nose of the aircraft is pointed due north with excursions $< \pm 1.0°$ during the entire maneuver) and considerable amounts of side-force. It was observed that for the same values of Δ, the vehicle never backtracks for ground track angle changes between -90° and 90° and always backtracks for changes of 180°.

Figure 10 illustrates that the minimum waypoint spacing algorithm must be used with discretion for low-speed flight, due to the types of maneuvers that aircraft can perform at low speeds which are not possible at high speeds. While low-speed uncoordinated turns can be approximated (albeit with considerable error) by the simplified wind-axes model even without the addition of the power equation constraint that we use to determine minimum waypoint spacing, low-speed backtracking maneuvers require a different application and integration of the flight model. In this case, the vehicle simply slows to a hover and resumes the flight plan at the current waypoint, so

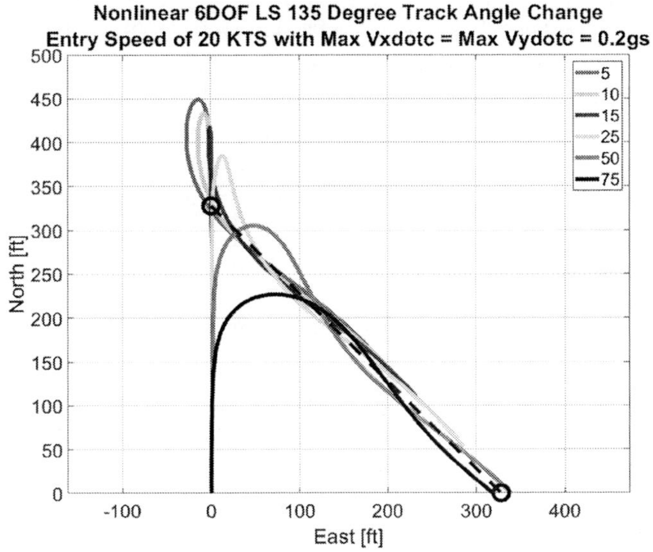

Figure 10. Trajectories of Nonlinear Simulation 6DOF Autopilot for Low-Speed Cruise (20 kts) REGULAR CHANGE, with $n_{xw,max} = -0.2$ g's, $n_{zw,max} = 1.04$ g's. Vehicle Performs 135° Change in Ground Track Angle. Legend shows Change Distance Δ (ft) Before Waypoint Due North of Origin.

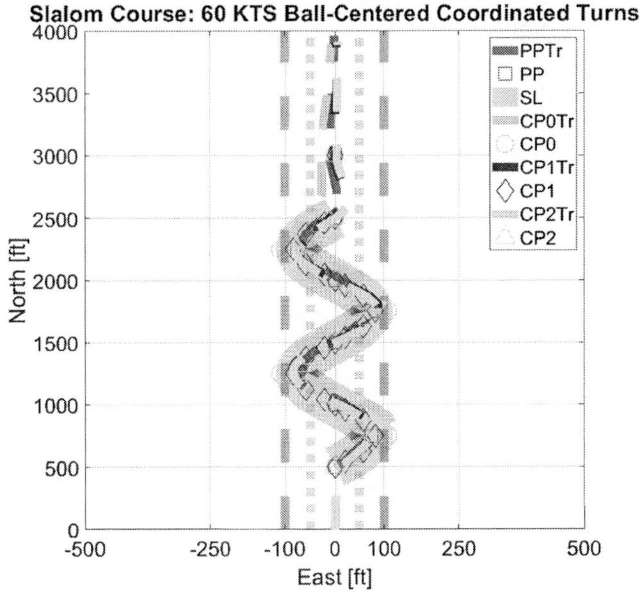

Figure 11. Slalom MTE course (SL) and nonlinear simulation trajectories (PPTr, CP0Tr, CP1Tr, CP2Tr) for flight plans with waypoints defined either as knots (PP) of a cubic piecewise-polynomial spline approximation of the MTE, or as control points of successive midpoint refinements of the control polygon (CP0, CP1, CP2). Vertical dashed lines indicate the maximum and minimum excursion from the centerline required with each turn.

there is no additional waypoint along the projecte flight-path to define. Moreover, the ability of the vehicle to slow to a hover and redirect its velocity vector without turning indicates that the constraint of a minimum distance to the next waypoint is not necessary for velocity vector reorientation in this case. For these reasons, the development of waypoint guidance algorithms that accommodate these types of low-speed maneuvers is a direction for future work.

Performance on Motion Primitives

Since the algorithm was found to be consistently applicable to maneuvers involving high-speed coordinated turns, with the aforementioned limitations on its application to low-speed maneuvers, comparisons of nonlinear simulation trajectories were performed for flight plans based on the Slalom and Emergency Stop MTEs at high entry speeds of 60 knots. The flight plan waypoints were defined either as knots (PP) of the spline approximation to the corresponding sinusoid or exponential motion primitive, or as the control points of successive midpoint refinements of the control polygon (CP0, CP1, CP2). Both MTEs were flown using the Regular Change option.

Figure 11 shows the simulated trajectories for the Slalom MTE, which was flown at constant speed and constant altitude as much as possible, in accordance with the MTE requirements. The change distance Δ was set to 196.8 ft, a constant approximation to the expression for Δ in Equation (6). The figure shows that the trajectories for all flight plans except CP0 meet the MTE requirement that cross-track error not exceed 100 ft from the centerline (some waypoints in CP0 slightly exceed this threshold). Successive midpoint refinements improve the aircraft's tracking error with respect to the spline trajectory, at the cost of adding waypoints in the flight plan. The particular use case will determine the necessary degree of tracking accuracy, and thus the appropriate waypoint sequence.

Table 12 shows the cumulative cross-track error, defined in Appendix A of (Ref. 10), and its maximum value and standard deviation for the CP0, CP1, and CP2 waypoint sequences based on the Slalom MTE. All values are normalized relative to the corresponding values for the PP sequence, which are scaled to 1. The errors were calculated by comparing the simulated vehicle trajectory to a sinusoid with the same wavelength and amplitude as the Slalom MTE. The CP0 sequence results in higher cross-track error values than the PP sequence, which is expected since the initial control polygon is a gross approximation of the spline trajectory. However, the refined control polygons CP1 and CP2 result in decreasing cross-track error relative to the PP sequence, at the cost of increasing numbers of waypoints in the flight plan.

Figure 12 shows the simulated trajectories for the Emergency Stop MTE, in which the aircraft decelerated to $V_f = 0$. The change distance Δ was manually tuned for each waypoint, due to the greater discrepancy between the wind-axes model and the simulated trajectories for this MTE at low speeds than for

Table 12. Slalom MTE Normalized Cross-Track Error (ft) Statistics for Waypoint (WP) Sequences Defined as Spline Knots (PP) or Control Points (CP0, CP1, CP2)

Sequence	Num WPs	Cumulative	Max	StdDev
PP	9	1.00	1.00	1.00
CP0	17	3.08	2.86	3.08
CP1	25	0.75	0.80	0.89
CP2	42	0.59	0.71	0.71

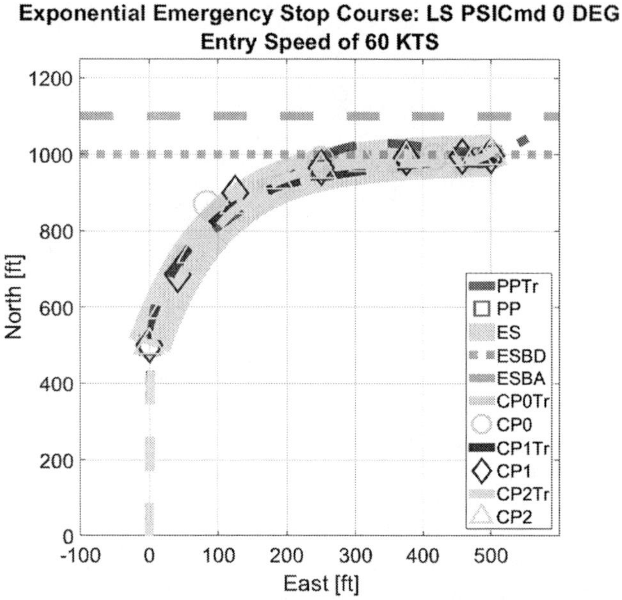

Figure 12. Emergency Stop MTE (ES) and nonlinear simulation trajectories (PPTr, CP0Tr, CP1Tr, CP2Tr) for flight plans with waypoints defined either as knots (PP) of a cubic piecewise-polynomial spline approximation of the MTE, or as control points of successive midpoint refinements of the control polygon (CP0, CP1, CP2). Horizontal dashed lines indicate the boundaries of desired (ESBD) and adequate (ESBA) maximum allowable forward travel.

the Slalom MTE at a constant cruise speed. Note, however, that as long as the aircraft's deceleration to hover during the Emergency Stop does not involve significant values of n_{yw}, the waypoint spacing criteria based on the wind-axes model should still be applicable. Table 13 shows the same cross-track error statistics as Table 12 for the Emergency Stop MTE. The table indicates that, as for the Slalom MTE, successive refinement of the control polygon results in decreasing cross-track error.

CONCLUSIONS

We have shown that control polygon midpoint refinement can be used to generate waypoints for autonomous flight path guidance driven by legacy waypoint guidance algorithms. The critical component of our proposed waypoint spacing algorithm is a robust stopping criterion for midpoint refinement

Table 13. Emergency Stop MTE Normalized Cross-Track Error (ft) Statistics for Waypoint (WP) Sequences Defined as Spline Knots (PP) or Control Points (CP0, CP1, CP2)

Sequence	Num WPs	Cumulative	Max	StdDev
PP	3	1.00	1.00	1.00
CP0	5	0.88	0.75	0.72
CP1	7	0.65	0.41	0.38
CP2	11	0.29	0.31	0.31

based on realistic vehicle flight dynamics. An augmented wind-axes model was used for real-time computation of the stopping criterion. The algorithm was tested on five motion primitives for different vehicle entry speeds to determine conditions under which the computed waypoint sequences do not violate flight dynamics-based constraints. In addition, the tracking performance of high-fidelity nonlinear simulations was assessed for a subset of waypoint sequences, and the wind-axes model was compared to the nonlinear simulations for particular turn maneuvers to identify discrepancies that could affect the applicability of the algorithm in practice.

The main results of this study are the following:

1. The simplified wind-axes flight dynamics model was very effective at defining minimum waypoint spacing for high-speed cruise conditions. This agrees with (Refs. 17, 23, 24), which demonstrate that at high cruise speeds, a point-mass approximation for helicopter flight dynamics is usually quite accurate.

2. For low-speed flight, the wind-axes flight dynamics model was not applicable for defining minimum waypoint spacing in many important use cases. This was evidenced by the nonlinear simulation results for the low-speed 135° change in ground track angle, where the vehicle can perform "backtracking" maneuvers that do not require minimum waypoint spacing.

3. Anticipating variations in vehicle speed at leg changes is important for capturing change distance variations over complex splines. This was shown by comparing the number and magnitude of violations of flight dynamics-based constraints for constant-speed versus decelerating trajectories that track the exponential Emergency Stop MTE.

Future work involves further exploration of the applicability of our waypoint spacing algorithm to the low-speed maneuverability regime and development of an alternative expression for minimum waypoint spacing that accounts for the ability of the vehicle to backtrack at low speeds and short change distances.

Author contacts:

Bryan Chu: bcchu1@asu.edu
Spring Berman: spring.berman@asu.edu

APPENDIX A

This appendix discusses the change of basis from a cubic polynomial spline in the monomial basis to the B-spline form by way of the Lagrange basis and Bernstein basis. It closely follows the derivation in (Ref. 22) that was performed on quadratic polynomial splines.

A cubic polynomial function of $x \in \mathbb{R}$ can be written in the monomial basis as follows, where $a_i \in \mathbb{R}$:

$$\begin{aligned} f(x) &= a_0 + a_1 x + a_2 x^2 + a_3 x^3 \\ &= \begin{bmatrix} a_0 & a_1 & a_2 & a_3 \end{bmatrix} \cdot \begin{bmatrix} 1 & x & x^2 & x^3 \end{bmatrix}^{\mathrm{T}} \equiv \mathbf{a}^{\mathrm{T}} \cdot \mathbf{m}(x) \end{aligned} \tag{24}$$

The Lagrange basis form of a cubic polynomial can be written as follows, where x_m and b_m, $m = 0,1,2,3$, are the node locations, coefficients of the basis set, and node indices:

$$\begin{aligned} f(x) &= b_0 l_0(x) + b_1 l_1(x) + b_2 l_2(x) + b_3 l_3(x) \\ &= \begin{bmatrix} b_0 & b_1 & b_2 & b_3 \end{bmatrix} \cdot \begin{bmatrix} l_0(x) & l_1(x) & l_2(x) & l_3(x) \end{bmatrix}^{\mathrm{T}} \\ &\equiv \mathbf{b}^{\mathrm{T}} \cdot \mathbf{l}(x), \\ l_j(x) &= \prod_{\substack{0 \le m \le 3 \\ m = j}} \frac{(x - x_m)}{(x_j - x_m)} \end{aligned} \tag{25}$$

The conversion of a cubic polynomial from the monomial basis to the Lagrange basis is readily accomplished by use of the Vandermonde matrix, $\mathbf{V}$. The Lagrange basis functions in $\mathbf{l}(x)$ can be found as follows:

$$\mathbf{l}(x) = (\mathbf{V}^{\mathrm{T}})^{-1} \cdot \mathbf{m}(x), \quad \mathbf{V}^{\mathrm{T}} = \begin{bmatrix} 1 & 1 & 1 & 1 \\ x_0 & x_1 & x_2 & x_3 \\ x_0^2 & x_1^2 & x_2^2 & x_3^2 \\ x_0^3 & x_1^3 & x_2^3 & x_3^3 \end{bmatrix}, \tag{26}$$

and the Lagrange basis coefficients, $\mathbf{b}$, can thus be calculated as follows from the monomial coefficients, $\mathbf{a}$:

$$\begin{aligned} \mathbf{a}^{\mathrm{T}} \cdot \mathbf{m}(x) = \mathbf{b}^{\mathrm{T}} \cdot \mathbf{l}(x) &= \mathbf{b}^{\mathrm{T}} \cdot (\mathbf{V}^{\mathrm{T}})^{-1} \cdot \mathbf{m}(x) \\ \Longrightarrow \mathbf{b}^{\mathrm{T}} &= \mathbf{a}^{\mathrm{T}} \cdot \mathbf{V}^{\mathrm{T}} \end{aligned} \tag{27}$$

The Bernstein basis, which is the basis for Bézier curves, is given by the following, where $\beta_i \in \mathbb{R}$ and $n = 3$:

$$\begin{aligned} f(x) &= \beta_0 B_{0,n}(x) + \beta_1 B_{1,n}(x) + \beta_2 B_{2,n}(x) + \beta_3 B_{3,n}(x) \\ B_{i,n}(x) &= \binom{n}{i} x^i (1-x)^{n-i}, \quad i = 0,1,2,\ldots,n \end{aligned} \tag{28}$$

The relationship between the Bernstein basis coefficients and the monomial basis coefficients can be found by equating the coefficients that multiply the same powers of x, similar to the procedure for determining the relationship between the Lagrange and monomial basis coefficients:

$$\begin{aligned} \beta_0 &= a_0, & \beta_2 &= a_2 + 2a_1 + 3a_0, \\ \beta_1 &= \frac{a_1}{3} + a_0, & \beta_3 &= a_3 + 3a_2 + 5a_1 + 7a_0. \end{aligned} \tag{29}$$

We can similarly obtain the relationship between the B-spline basis coefficients and the monomial basis coefficients. The B-spline representation of a cubic polynomial is as follows, where $C_0, C_1, C_2, C_3 \in \mathbb{R}$ are the spline control points, x_i are the knots of the spline, and $B_{i,3}$ are the B-spline basis functions:

$$\begin{aligned} f(x) &= C_0 B_{-3,3}(x) + C_1 B_{-2,3}(x) + C_2 B_{-1,3}(x) + C_3 B_{0,3}(x), \\ B_{i,d}(x) &= \frac{x - x_i}{x_{i+d} - x_i} B_{i,d-1}(x) + \frac{x_{i+d+1} - x}{x_{i+d+1} - x_{i+1}} B_{i+1,d-1}(x), \\ B_{i,0}(x) &= \begin{cases} 1 & x_i \le x \le x_{i+1}, \\ 0 & \text{otherwise.} \end{cases} \end{aligned} \tag{30}$$

If we assume a uniform knot spacing of $h \equiv x_{i+1} - x_i$ for all $i = 0, \pm 1, \pm 2, \pm 3, \ldots$, then the basis functions for the cubic B-spline form can be simplified to (Ref. 26):

$$B_{i,3}(x) = \frac{1}{6h^3} \begin{cases} (x - x_{i-2})^3, & x_{i-2} \le x \le x_{i-1}, \\ -3(x - x_{i-1})^3 + 3h(x - x_{i-1})^2 + \\ \quad 3h^2(x - x_{i-1}) + h^3, & x_{i-1} \le x \le x_i, \\ -3(x_{i+1} - x)^3 + 3h(x_{i+1} - x)^2 + \\ \quad 3h^2(x_{i+1} - x) + h^3, & x_{i-1} \le x \le x_i, \\ (x_{i+2} - x)^3, & x_{i+1} \le x \le x_{i+2}, \\ 0, & \text{otherwise.} \end{cases} \tag{31}$$

Over a normalized interval spacing such that $h \equiv 1$, the B-spline basis functions can be further simplified to:

$$\begin{aligned} B_{-3,3}(x) &= \frac{1}{6}\left(-x^3 + 3x^2 - 3x + 1\right), \\ B_{-2,3}(x) &= \frac{1}{6}\left(3x^3 - 6x^2 + 4\right), \\ B_{-1,3}(x) &= \frac{1}{6}\left(-3x^3 + 3x^2 + 3x + 1\right), \\ B_{0,3}(x) &= \frac{1}{6}\left(x^3\right), \end{aligned} \tag{32}$$

and thus, Equation (30) can be written as a linear combination of the control points as:

$$f(x) = \frac{1}{6} \begin{bmatrix} x^3 & x^2 & x & 1 \end{bmatrix} \begin{bmatrix} -1 & 3 & -3 & 1 \\ 3 & -6 & 3 & 0 \\ -3 & 0 & 3 & 0 \\ 1 & 4 & 1 & 0 \end{bmatrix} \begin{bmatrix} C_0 \\ C_1 \\ C_2 \\ C_3 \end{bmatrix}. \tag{33}$$

Setting the expressions for $f(x)$ in Equations (24) and (33) equal to each other, we obtain the following relationships between the monomial basis coefficients and the B-spline basis coefficients over the normalized interval $x \in \mathbb{R}$:

$$\begin{aligned}
a_0 &= \frac{1}{6}\left(C_0 + 4C_1 + C_2\right), \\
a_1 &= \frac{1}{6}\left(-3C_0 + 3C_2\right), \\
a_2 &= \frac{1}{6}\left(3C_0 - 6C_1 + 3C_2\right), \\
a_3 &= \frac{1}{6}\left(-C_0 + 3C_1 - 3C_2 + C_3\right).
\end{aligned} \tag{34}$$

This set of linear equations can be easily inverted to find the control point locations in terms of the monomial basis coefficients.

APPENDIX B

This appendix summarizes the derivation of the wind-axes flight model, Equation (2), including the necessary simplifications. The derivation closely follows that in (Ref. 23), with some further developments from (Ref. 17). The following notation is used:

$\mathbf{F}_w$ = Total force on vehicle in wind-axes frame
m = Vehicle mass
$\mathbf{a}_w$ = Vehicle acceleration vector in wind-axes frame
$\mathbf{V}_w$ = Vehicle velocity vector in wind-axes frame
$(\mathbf{i}, \mathbf{j}, \mathbf{k})_w$ = Unit vectors along axes of wind-axes frame
p_w = Vehicle body roll rate in wind-axes frame
q_w = Vehicle body pitch rate in wind-axes frame
r_w = Vehicle body yaw rate in wind-axes frame
$\omega_w = [p_w \; q_w \; r_w]^T$
ϕ_w = Wind-axes bank angle about $\mathbf{V_w}$
γ = Wind-axes velocity flight path angle
χ = Wind-axes velocity ground track angle

The equations of motion for an aircraft relative to a flat, non-rotating Earth in the wind-axes frame are derived from Newton's second law of motion:

$$\mathbf{F}_w = m\mathbf{a}_w = m\frac{d\mathbf{V}_w}{dt} \tag{35}$$

Defining $V = ||\mathbf{V}_w||$, the rate of change of the vehicle velocity vector relative to the wind-axes frame is given by:

$$\begin{aligned} \frac{d\mathbf{V}_w}{dt} &= \frac{d(\mathbf{i}_w V)}{dt} = \mathbf{i}_w\dot{V} + \frac{d\mathbf{i}_w}{dt}V \\ &= \mathbf{i}_w\dot{V} + (\omega_w \times \mathbf{i}_w)V \\ &= \mathbf{i}_w\dot{V} + (\mathbf{j}_w r_w - \mathbf{k}_w q_w)V \end{aligned} \tag{36}$$

The components of the sum of the forces in the wind-axes frame, where the gravitational acceleration vector is expressed in this frame, can then be written as:

$$\begin{aligned} \sum F_{x_w} &= mgn_{xw} - mg\sin\gamma \;=\; m\dot{V} \\ \sum F_{y_w} &= mgn_{yw} = mVr_w \\ \sum F_{z_w} &= mgn_{zw} - mg\cos\gamma \;=\; -mVq_w \end{aligned} \tag{37}$$

The angular rates in the wind-axes frame can be related to the reference rates in the inertial frame as follows:

$$\begin{bmatrix} p_w \\ q_w \\ r_w \end{bmatrix} = \begin{bmatrix} 1 & 0 & -\sin\gamma \\ 0 & \cos\phi_w & \sin\phi_w\cos\phi_w \\ 0 & -\sin\phi_w & \cos\phi_w\cos\gamma \end{bmatrix} \begin{bmatrix} \dot{\phi}_w \\ \dot{\gamma} \\ \dot{\chi} \end{bmatrix} \tag{38}$$

Then, from Equation (37), the components of the vehicle's acceleration in the wind-axes frame are:

$$\begin{aligned} a_{x_w} &= \dot{V} \\ a_{y_w} &= Vr_w \;=\; V(-\dot{\gamma}\sin\phi_w + \dot{\chi}\cos\phi_w\cos\gamma) \\ a_{z_w} &= -Vq_w \;=\; -V(\dot{\gamma}\cos\phi_w + \dot{\chi}\sin\phi_w\cos\phi_w) \end{aligned} \tag{39}$$

Combining Equations (37) and (39), we arrive at:

$$\begin{aligned} \dot{V} + g\sin\gamma &= gn_{xw} \\ V\dot{\gamma} + g\cos\gamma &= gn_{zw}\cos\phi_w - gn_{yw}\sin\phi_w \\ V\dot{\chi}\cos\gamma &= gn_{zw}\sin\phi_w + gn_{yw}\cos\phi_w \end{aligned} \tag{40}$$

Under the assumption that the side-force term n_{yw} is zero, these are the wind-axes equations of motion derived in (Ref. 17) for a single main-rotor tail-rotor helicopter at low speed. The side-force term n_{yw} is zero for tilt-rotors and tandem rotors, and for fixed-wing aircraft that are flying zero sideslip or ball-centered. For single main-rotor tail-rotor helicopters, it is true in general that n_{yw} is zero or negligible when flying ball-centered and performing coordinated turns. Even under steady trimmed conditions, where there is significant sideslip, the wind-axes model in Equation (40) still applies. However, for low-speed dynamic maneuvers where significant changes in sideslip occur, n_{yw} should not be approximated as zero.

Setting $n_{yw} = 0$ and assuming zero flight path angle and zero climb rate ($\gamma = 0, \dot{\gamma} = 0$) in Equation (40), we obtain:

$$\begin{aligned} \dot{V} &= gn_{xw} \\ g &= gn_{zw}\cos\phi_w \\ V\dot{\chi} &= gn_{zw}\sin\phi_w \end{aligned} \tag{41}$$

Finally, the last two equations of Equation (41) can be combined by eliminating the dependence on ϕ_w:

$$\begin{aligned} \cos\phi_w = \frac{1}{n_{zw}} \;&\Rightarrow\; \sin\phi_w = \pm\frac{\sqrt{n_{zw}^2 - 1}}{n_{zw}} \\ \Rightarrow\; \dot{\chi} \;=\; \frac{gn_{zw}}{V}\sin\phi_w \;&=\; \pm\frac{g}{V}\sqrt{n_{zw}^2 - 1} \end{aligned} \tag{42}$$

The expressions for $\dot{V}$ in Equation (41) and $\dot{\chi}$ in Equation (42) comprise the wind-axes model in Equation (2). Note that the sign of $\dot{\chi}$ depends on the value of ϕ_w, which is positive when the aircraft is banking clockwise about $\mathbf{V}_w$ and negative when it is banking counterclockwise.

APPENDIX C

This appendix discusses the computation of the resultant weight-normalized aerodynamic forces along, across, and normal to the flight path, denoted by n_{xw}, n_{yw}, and n_{zw}, respectively, in the nonlinear 6DOF GENHEL rotorcraft simulation. Note that these forces are not to be confused with the body accelerations, as there are many cases in VTOL rotorcraft flight where they are not the same.

The nonlinear 6DOF equations of motion for an aircraft relative to a flat, non-rotating Earth in an inertial reference frame are given by (Ref. 27, Chapter 3.2):

$$\begin{aligned}\dot{u} &= F_x/m - g\sin\theta + rv - qw \\ \dot{v} &= F_y/m + g\sin\phi\cos\theta - ru + pw \\ \dot{w} &= F_z/m + g\cos\phi\cos\theta + qu - pv\end{aligned}$$

$$\begin{aligned}\dot{p} = {} & (I_{zz}M_x + I_{xz}M_z - \{I_{xz}(I_{yy} - I_{xx} - I_{zz})p + \\ & [I_{xz}^2 + I_{zz}(I_{zz} - I_{yy})]r\}q)/(I_{xx}I_{zz} - I_{xz}^2) \\ \dot{q} = {} & [M_y - (I_{xx} - I_{zz})pr - I_{xz}(p^2 - r^2)]/I_{yy} \\ \dot{r} = {} & (I_{xz}M_x + I_{xx}M_z + \{I_{xz}(I_{yy} - I_{xx} - I_{zz})r + \\ & [I_{xz}^2 + I_{xx}(I_{xx} - I_{yy})]p\}q)/(I_{xx}I_{zz} - I_{xz}^2)\end{aligned} \tag{43}$$

The transformation matrix $L_{\phi\theta\psi}$ from the body-fixed frame to the local-level frame is defined as follows, where $c()$ and $s()$ represent the cos() and sin() functions, respectively:

$$L_{\phi\theta\psi} = \begin{bmatrix} c(\psi)c(\theta) & s(\psi)s(\theta) & -s(\theta) \\ c(\psi)s(\theta)s(\phi) - s(\psi)c(\phi) & c(\phi) & s(\phi)c(\phi) \\ 0 & -s(\phi) & c(\phi)c(\theta)c(\phi) \end{bmatrix} \tag{44}$$

Given the translational velocities (u, v, w) from the solution of Equation (43), the local level velocities in the Earth inertial frame can be computed using $L_{\phi\theta\psi}$:

$$\begin{bmatrix} V_{EX} \\ V_{EY} \\ V_{EZ} \end{bmatrix} = L_{\phi\theta\psi} \begin{bmatrix} u \\ v \\ w \end{bmatrix} \tag{45}$$

The ground track angle χ, measured clockwise from true north, can be computed as:

$$\chi = \tan^{-1}\left(\frac{V_{EY}}{V_{EX}}\right) \tag{46}$$

The local level accelerations can be computed by multiplying $L_{\phi\theta\psi}$ by the accelerations in the body-fixed frame:

$$\begin{bmatrix} \dot{V}_{EX} \\ \dot{V}_{EY} \\ \dot{V}_{EZ} \end{bmatrix} = L_{\phi\theta\psi} \begin{bmatrix} F_x/m \\ F_y/m \\ F_z/m \end{bmatrix} \tag{47}$$

The weight-normalized forces along and across the flight path can now be calculated as:

$$\begin{aligned} n_{xw} &= \dot{V}_{EX}\cos\chi - \dot{V}_{EY}\sin\chi \\ n_{yw} &= \dot{V}_{EY}\cos\chi + \dot{V}_{EX}\sin\chi \end{aligned} \tag{48}$$

Assuming that there is no rotor flapping relative to the rotor mast and that the mast is perpendicular to the nose of the aircraft, the weight-normalized force normal to the flight path is given by:

$$n_{zw} = F_z/m \tag{49}$$

In typical dynamic and unsteady maneuvering flight, the forces n_{xw}, n_{yw}, and n_{zw} are not constant. Special cases such as turning at a constant bank angle and slowing to a desired speed at a constant deceleration are exceptions that are used in this study; these maneuvers require tight control of vehicle attitudes in order to maintain constant n_{xw}, n_{yw}, and n_{zw}.

REFERENCES

1. Sujit, P. B., Saripalli, S., and Sousa, J. B., "An Evaluation of UAV Path Following Algorithms," European Control Conference, Zurich, Switzerland, 2013.

2. Rubí, B., Perez, R., and Morcego, B., "A Survey of Path Following Control Strategies for UAVs Focused on Quadrotors," *Journal of Intelligent & Robotic Systems*, Vol. 98, 2020, pp. 241–265.

3. Nelson, D., Barber, B., McLain, T., and Beard, R., "Vector Field Path Following for Small Unmanned Air Vehicles," American Control Conference, Minneapolis, MN, 2006.

4. Park, S., Deyst, J., and How, J., "Performance and Lyapunov Stability of a Nonlinear Path-Following Guidance Method," *Journal of Guidance, Control, and Dynamics*, Vol. 30, (6), 2007, pp. 684–686.

5. Chen, Y., Zeng, W., and Wu, Y., "Vector Field-Based Guidance for Planar Smooth Path-Following of a Small Unmanned Helicopter," *Hindawi: Journal of Robotics*, 2023, pp. 684–686.

6. Curry, R., Lizarraga, M., Mairs, B., and Elkaim, G. H., "L2+, An Improved Line of Sight Guidance Law for UAVs," American Control Conference, Washington, DC, 2013.

7. Cho, N., Kim, Y., and Park, S., "Nonlinear Differential Geometric Path-Following Guidance Law," *Journal of Guidance, Control, and Dynamics*, Vol. 38, (12), 2015.

8. Furieri, L., Stastny, T., Marconi, L., Siegwart, R., and Gilitschensiki, I., "Gone with the Wind: Nonlinear Guidance for Small Fixed-Wing Aircraft in Arbitrarily Strong Windfields," American Control Conference, Seattle, WA, 2017.

9. Chen, Q., and Lu, Y., "An Improved Nonlinear Path-Following Method with On-line Transition Trajectory for Fixed-Wing Unmanned Aerial Vehicles," *International Journal of Advanced Robotic Systems*, Vol. 19, (3), 2022.

10. Chu, B., Keller, J., and Berman, S., "Trajectory Waypoint Spacing for Spline-Based Flight Plans," Vertical Flight Society 6th Decennial Aeromechanics Specialists' Conference, Santa Clara, CA, February 2024.

11. Paduano, J., Wissler, J., Drozeski, G., *et al.*, "TALOS: An Unmanned Cargo Delivery System for Rotorcraft Landing to Unprepared Sites," American Helicopter Society 71st Annual Forum, Virginia Beach, VA, 2015.

12. Petit, P., J., W., B., F., and Greiser, S., "Waypoint Based Online Trajectory Generation and Following Control for the ACT/FHS," AIAA SciTech Forum, 2019.

13. Takahashi, M. D., Fujizawa, B. T., Lusardi, J. A., *et al.*, "Comparison of Autonomous Flight Control Performance Between Partial- and Full-Authority Helicopters," *Journal of Guidance, Control, and Dynamics*, Vol. 45, (5), February 2022, pp. 885–901.

14. deBoor, C., *A Practical Guide to Splines*, Springer-Verlag, New York, NY, 2001.

15. Lane, J. M., and Riesenfeld, R. F., "A Theoretical Development for the Computer Generation and Display of Piecewise Polynomial Surfaces," *IEEE Transactions on Pattern Analysis and Machine Intelligence*, Vol. PAMI-2, (1), January 1980, pp. 35–46.

16. Mork, H., "Spline Approximation Via the Control Polygon," M.S. Thesis, University of Oslo, 2012.

17. Curtiss, H. C., and Price, G., "Studies of Rotorcraft Agility and Maneuverability," Tenth European Rotorcraft Forum, The Hague, The Netherlands, August 1984.

18. Hilbert, K. B., "A Mathematical Model of the UH-60 Helicopter," NASA TM 85890, 1984.

19. Talbot, P. D., Tinling, B. E., Decker, W. A., and Chen, R. T. N., "A Mathematical Model of a Single Main Rotor Helicopter for Piloted Simulation," NASA TM 84281, 1982.

20. Orhan, E. H., "Generic Nonlinear Helicopter Model," https://www.mathworks.com/matlabcentral/fileexchange/72828-generic-nonlinear-helicopter-model, MATLAB Central File Exchange, March 10, 2024.

21. Ivler, C. M., Russell, K., Gong, A., *et al.*, "Toward a UAS Handling Qualities Specification: Development of UAS-Specific MTEs," Vertical Flight Society 78th Annual Forum, Fort Worth, TX, 2022.

22. Ezhov, N., Neitzel, F., and Petrovic, S., "Spline Approximation, Part 2: From Polynomials in the Monomial Basis to B-splines - A Derivation," *Mathematics*, Vol. 9, (18), 2021, pp. 2198.

23. Wood, T. L., *et al.*, "Maneuver Criteria Evaluation Program," USAAAMRDL Technical Report 74-32, 1974.

24. Wells, C. D., and Wood, T. L., "Maneuverability - Theory and Application," *Journal of the American Helicopter Society*, Vol. 18, (1), January 1973.

25. "ADS-33E-PRF, Aeronautical Design Standard Performance Specification Handling Qualities Requirements for Military Rotorcraft," U.S. Army Aviation and Missile Command, Aviation Engineering Directorate, Redstone Arsenal, Alabama, 2000.

26. Munguia, M., "Cubic B-spline: Derivation with two applications," M.S. Thesis, University of Texas-Pan American, 2014.

27. Stengel, R. F., *Flight Dynamics*, Princeton University Press, Princeton, NJ, 2005.

Application of Machine Learning to Maximize Power Density of Propulsion Motors for VTOL Applications

Tadashi Sawata
Senior Technical Fellow
Collins Aerospace
Solihull, UK

Andrei Dinu
Senior Technical Fellow
Collins Aerospace
Solihull, UK

ABSTRACT

Various measures are taken to maximize the power density of electrical machines for aircraft propulsion applications as minimizing the weight penalty is a key to commercializing such technologies. Material selections and setting appropriate electrical and magnetic loading are critical to achieve an optimum balance between power density and other performance indicators. Electromagnetic analysis combined with detailed drive system simulations show that severe nonlinearities in electrical machines can lead to control instabilities. This issue needs to be overcome for further improvements in power density of electric propulsion systems. Simple machine learning solutions can be applied to identify the nonlinear behavior of electrical machines with significant magnetic saturation. The approach involves a combination of off-line learning based on data produced by finite element analysis tools and on-line learning using practical data captured during equipment calibration procedures in the laboratory. The optimal type of learning algorithm can be selected based on the specific characteristics of the motor design.

NOTATION

c	Knot in a hinge function
H	Hinge function
I_d, I_q	d- and q-axis currents
K_t	Torque constant
L_d, L_q	d- and q-axis inductances
L_1, L_2	Number of regression parameters to be optimized
M	Total number of regression parameters to be optimized
T, $\hat{T}$	Torque, approximated torque
t_R	Rotor temperature
Y, $\hat{Y}$	Dependent variable, approximated dependent variable
x	Independent variable
β	Regression parameter
θ	Rotor position
ω	Rotational speed

INTRODUCTION

Maximizing power density, i.e. reducing weight penalty, is one of the most important design targets for electric propulsion drive systems for Vertical Take-Off and Landing (VTOL) and other aircraft. Several examples of high power density machines have been reported in literature (Refs. 1-4). Electrical and magnetic loading of the electric motor must be maximized through materials selection and design optimization while managing thermal performance within manageable limit for a given cooling method. An elevated level of electrical and magnetic loading means that the operation of the motor can be highly nonlinear. Therefore, the conventional assumption of constant parameters such as back-Electromotive Force (EMF) constant, torque constant and inductance would not be adequate, and it would lead to less accurate torque production and increased current and torque ripple that may result in increased vibrations in propellers and airframe and affect its controllability. This paper discusses an application of machine learning algorithm for parametric identification of saturated machines for electric propulsion systems.

DEMONSTRATORS FOR ELECTRIC PROPULSION

Collins Aerospace together with Pratt & Whitney have been developing various electric propulsion drive systems for helicopter and fixed-wing aircraft hybrid-electric propulsion demonstrators including Airbus Helicopters' PioneerLab, Canadian government funded CSAFE, and European Union funded Clean Aviation (Ref. 5) programs. Various electric motors rated from 250 kW to 1 MW have been developed. The motor drive systems have a four-quadrant operation capability meaning that the systems can be used for both motoring and generating modes depending on the requirement for operating conditions of aircraft and/or engine. Figure 1 shows an exterior view of 1 MW electric propulsion

Presented at the Vertical Flight Society's 11th Biennial Autonomous VTOL Technical Meeting, Phoenix, AZ, USA, Feb. 4-6, 2025.

Figure 1. 1 MW electric propulsion demonstrator

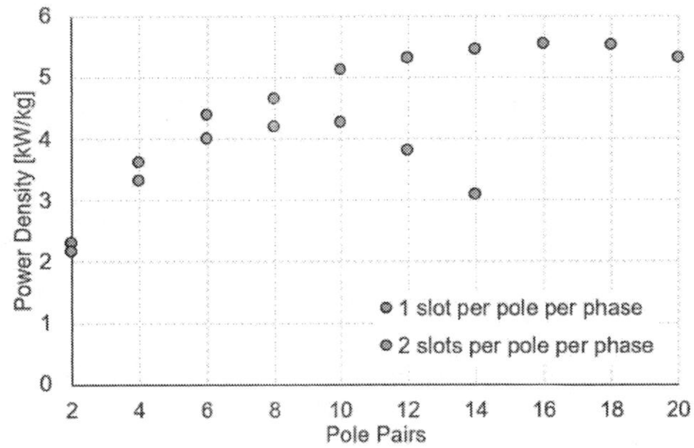

Figure 2. Example of power density vs. pole-pairs.

demonstrator developed and tested at Collins Aerospace, with a rated speed of 12,000 rpm (Ref. 6).

MOTOR DESIGN TO MAXIMIZE POWER DENSITY

Most desirable performance indicators for propulsion electrical machines include high power density, high efficiency and precise and smooth torque output. Various measures to improve the performance are taken but there are also penalties associated with these measures.

Increasing Fundamental Frequency

Increasing the number of poles hence fundamental frequency for a given speed could improve power density by utilizing the electromagnetic gearing effect. Figure 2 shows an example of calculated power density plotted against different numbers of pole-pairs for 500 kW 3,000 rpm direct drive electric propulsion application (Ref. 7). It shows how power density is affected by the number of pole-pairs for two different winding configurations. Although the optimum number differs between the winding configurations, it has a similar trend that power density increases with the increase in the number of pole-pairs until it reaches its peak beyond which power density decreases again. In this example, an optimum number of pole-pairs for the 2 slots/pole/phase design is around 10 corresponding to a fundamental electrical frequency of 500 Hz, and for the 1 slot/pole/phase design is 16 corresponding to 800 Hz. The reason why power density decreases above a certain frequency is increased iron losses (Ref. 8), increased AC losses in copper conductor and magnets, and design practicality in magnetic circuit and winding configuration.

Copper AC losses can be significant in machines with high fundamental frequencies. One of the most effective ways to minimize the AC losses is the use of Litz conductors. Figure 3 shows Finite Element Analysis (FEA) calculated conductor AC loss plots, and Table 1 shows a comparison of calculated AC/DC loss ratios for the 1 MW hybrid-electric demonstration machine (Ref. 6).

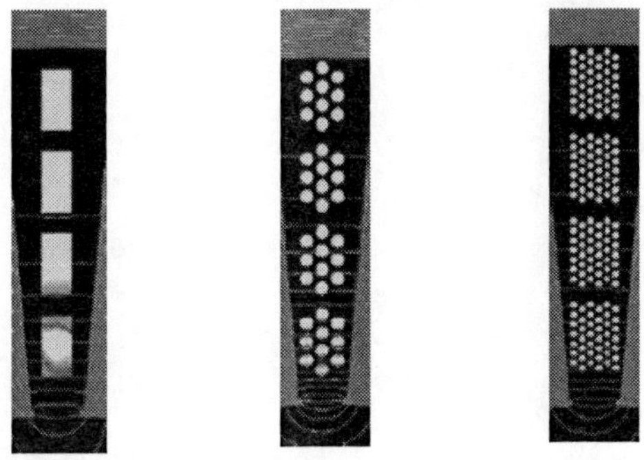

(a) Hairpin (b) 2 mm Litz (c) 1 mm Litz

Figure 3. Conductor AC loss plots.

Table 1. Calculated Conductor AC/DC loss ratios.

Wire	AC/DC loss ratio
Hairpin (bar conductor)	15.2
Parallel conductors 2 mm	14.5
Parallel conductors 1 mm	15.7
Litz 2 mm	2.41
Litz 1 mm	1.02

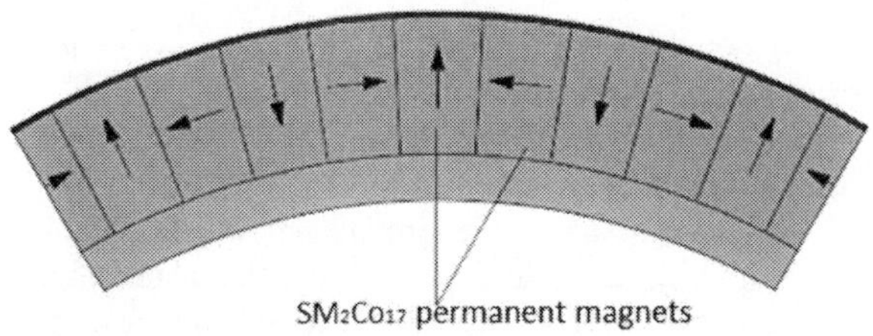

Figure 4. Halbach array PM rotor

Use of Halbach Array Rotor with High Energy Permanent Magnets

A rotor with Halbach array Permanent Magnets (PM) can improve the power density of motors by increasing air gap flux-density as well as providing near sinusoidal EMF

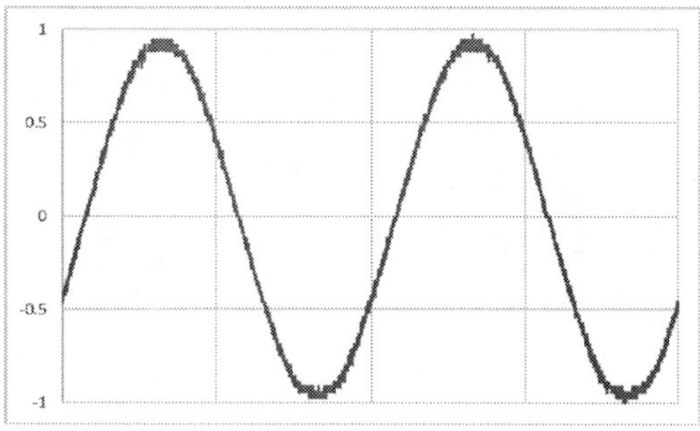

Figure 5. Measured open-circuit EMF of 1 MW machine.

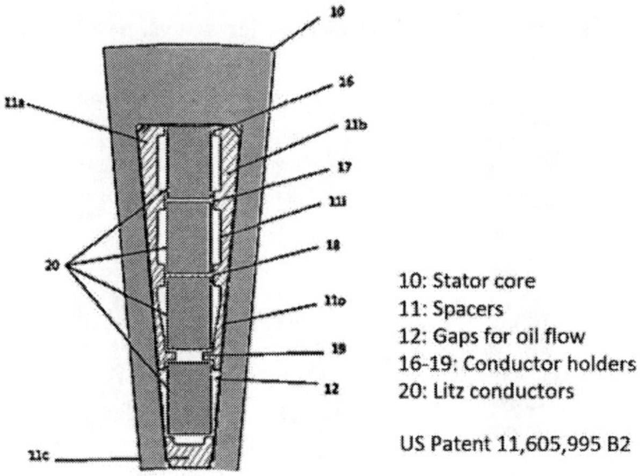

Figure 6. Slot spacers enabling direct slot oil cooling.

waveform minimizing cogging and torque ripples (Ref. 9). Figure 4 shows an example of Halbach array PM rotor with Sm_2Co_{17} PM which consists of pole and transition magnets (Ref. 7). Use of Sm_2Co_{17} over NdFeB material may be beneficial for aircraft propulsion applications due to its superior temperature stability. A measured open-circuit EMF waveform from the 1 MW demonstrator is shown in Figure 5.

Cooling Design

Power densities of electrical machines are determined primarily by their thermal performance (Refs. 10, 11). Various methodologies for thermal designs and analyses for Litz conductors have been published (Refs 12, 13). In the 1 MW machine, conductors are placed in specific locations within slots as modelled in Figure 3 by a patented spacer design shown in Figure 6 (Ref. 14). The design ensures consistency in conductor placement and oil flows in slots thereby optimizing the thermal performance of the machine. It is also expected to improve vibration performance which can be excessive in an engine environment.

Stator Core Material and Operating Point

Cobalt-Iron alloy (48% Co) has superior magnetization properties over widely used silicon steels, with a typical

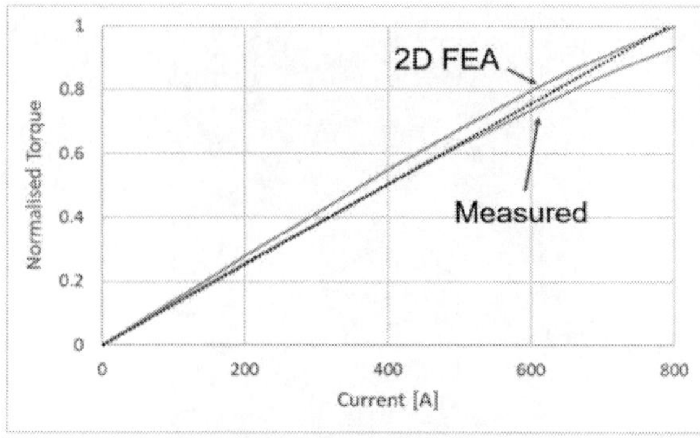

Figure 6. Measured average torque vs. current.

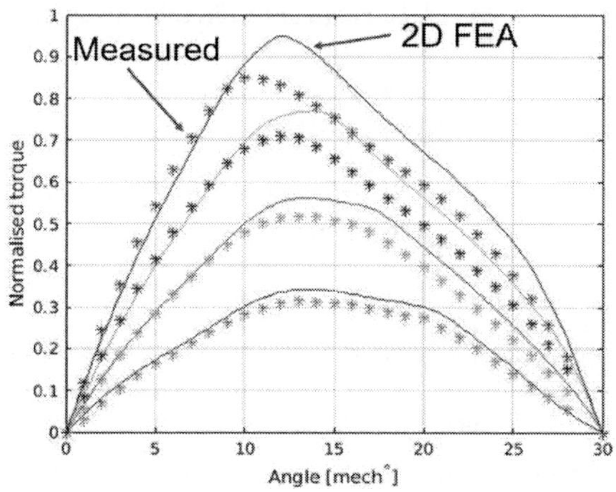

Figure 7. Static torque curves.

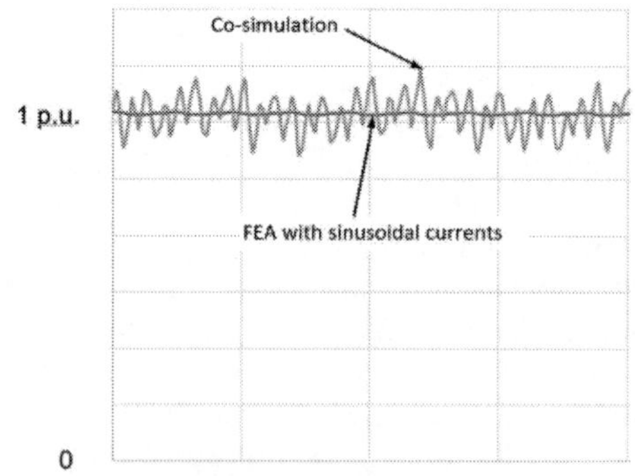

Figure 8. Torque waveforms with sinusoidal and practical currents.

Figure 9. Typical flux-density plot under load.

saturation point of around 2.3 T. Determining operating points of magnetic circuit is a multi-objective optimization problem for power density, efficiency and controllability. Operating at higher flux-densities would result in an improved power density but with a penalty in increased loss hence reduced efficiency, and it may also lead to nonlinearity in torque control and increased torque ripple.

Figure 6 shows measured and FEA calculated average torque vs. phase current curves for the 1 MW machine. The dotted straight line is also plotted so that the saturation effect in the measured data at higher currents can be clearly seen. Figure 7 shows measured and FEA calculated static torque curves when phase windings are excited by DC currents. It can be seen in the static torque curves that saturation causes skew in the shape caused by the armature reaction as well as reducing average torque per ampere. The discrepancies in torque between measured and FEA in Figures 6 and 7 are considered mainly due to the difference in magnetic properties of stator core in the actual machine and the data supplied by manufacturer. It is known that properties of Cobalt-Iron alloy are sensitive to mechanical stress and other manufacturing processes. Investigations are ongoing to identify the sources of discrepancies.

Interaction between Machine Nonlinearity and Controller

Dynamic interactions between saturations in the machine and the controller can affect output torque. Figure 8 shows torque waveform calculated by FEA with ideal sinusoidal phase currents as well the one calculated by a co-simulation of machine FEA and a dynamic model of inverter controller in Matlab/Simulink. Much larger torque ripples are produced in the latter simulation due to Pulse Width Modulation (PWM), dynamics between machine saturation and current regulator etc.

Figure 9 shows a typical flux-density plot under load (rotor rotating in CW direction) in which armature reaction can be seen as saturations in stator teeth in leading edges of magnet poles. Highly saturated regions in teeth and back-iron move as rotor rotates. As phase windings are fixed in stator slots, the effect of saturation on electrical parameters such as L_d and L_q is a function of rotor position as well as current level. Figure 10 shows an example of nonlinearity in torque constant K_t with a function of quadrature and direct axis currents I_q and I_d. It should be noted that other parameters such as the operating temperature would also affect K_t.

Figure 11 shows how control parameters are affected by machine saturations such as K_t, EMFs e_q and e_d, and inductances possibly leading to instabilities in torque T and field wakening controls. Further nonlinearity would be introduced by the variation in winding resistance R with temperature and cross-couplings between d- and q-axis parameters. Potential controller instability was simulated and results are shown in Figure 12, where phase currents (top traces) are gradually increased for higher machine saturations.

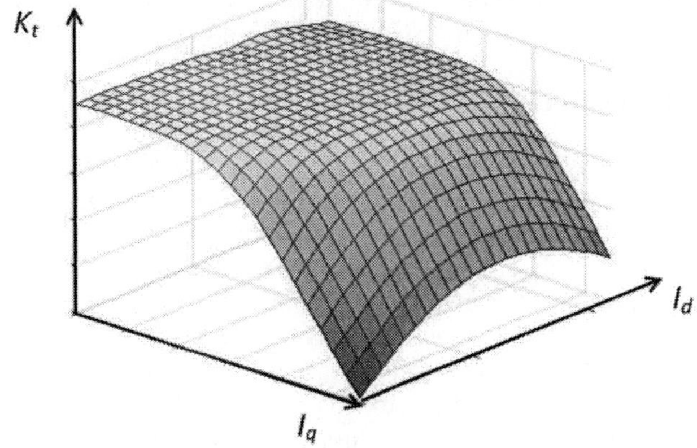

Figure 10. Nonlinearity in K_t.

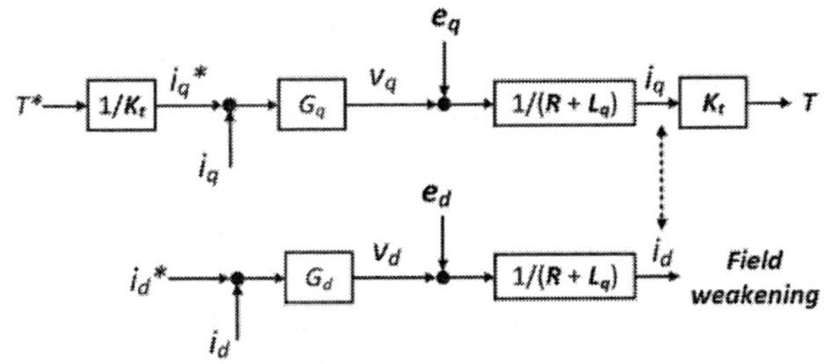

Figure 11. Saturation effect on controller.

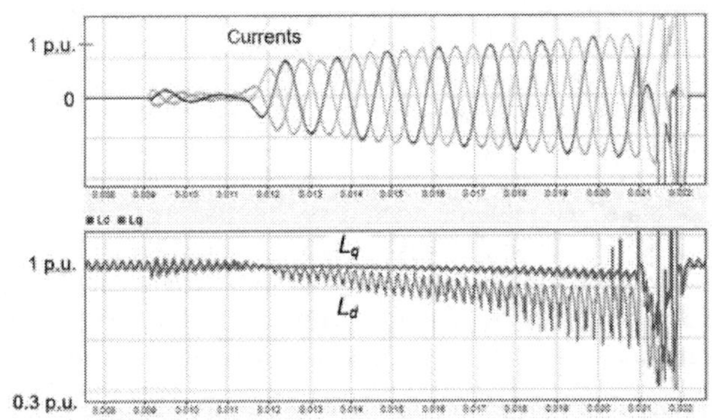

Figure 12. Simulated controller instability.

In the non-salient surface PM machine, L_d and L_q (bottom traces) are comparable values when saturations are low. However, as the armature reaction becomes more pronounced, L_q decreases as well as its value oscillates as rotor rotates. When phase currents exceed significantly their rated values, the controller will no longer be able to regulate currents appropriately.

The above discussion suggests that magnetic saturations in electrical machines need to be limited to a certain level to guarantee the stability of inverter controller. This would limit the use of full potential of materials' capability in the machine and a further improvement in power density. In the following sections, an adaptive controller design to compensate for the machine nonlinearity that could help increase the power density of motor drive systems is discussed.

OPTIMAL CONTROL OF POWER DENSE ELECTRICAL MACHINES

As discussed above, nonlinear behavior is a common consequence of pushing the power density of electrical machines to the limits allowed by the existing materials. While machine torque is linearly proportional to the q-axis current component in an unsaturated surface PM machine, it becomes a non-linear function of several variables in a magnetically saturated machine. These include d-axis and q-axis components of the stator current, rotor temperature, rotor angular position (relative to the stator teeth) and motor speed (which in turn impacts on the eddy currents in the stator laminations). Speed is also very important if the estimated torque is the mechanical torque at the shaft rather than the electromagnetic torque of the electrical machine in which case bearing friction needs to be considered. An additional complication is the variability of magnetic material parameters which causes significant deviation the machine parameters from the theoretical values calculated by FEA.

Supervised machine learning can be applied to capture the nonlinear characteristics of the motor. There are two broad classes of machine learning problems depending on the dimensionality of input space. High-dimensional problems are more suited to artificial neural networks (ANNs) while low-dimensional problems are addressed more effectively by classical regression algorithms. Prognostic and Health Management (PHM) falls in the first category because data points are large vectors containing numerous samples of multiple signals measured over long periods of time. On the other hand, the identification of system parameters (such as an electrical machine) is a low-dimensional regression analysis problem. In this case, data points contain a small number of numerical parameters. Often, input data consists of single samples of phase currents, temperature, speed and rotor angle.

Two sources of data are used to estimate the nonlinear characteristics of the electrical machine. The first source is theoretical data generated by FEA tools. An initial off-line learning algorithm is used to process this data and provide an initial model of the motor nonlinearities. Additional data is collected by subsequent lab measurements. This provides important information about the deviation of the magnetic materials properties from the theoretical values used by FEA tools. The initial motor model is refined by an on-line learning algorithm which processes the live data as it is generated during calibration procedures in the lab.

For reasons of cost optimization, the online learning algorithm is implemented by the same digital controller which governs the normal operation of the motor drive system. Therefore, the computational requirements of the on-line learning task need to be minimized. This means that the mathematical complexity of the algorithm and the number of parameters involved in the learning process need to be restricted. The following sections review a few simple regression solutions and show how the optimal balance between accuracy and computational effort can be identified for practical applications.

Learning Motor Parameters by Multivariate Regression

The general aim of any regression algorithm is to approximate the value $\hat{Y}$ of a dependent variable Y (such as torque, L_d, L_q, etc.) as a function f of the independent variables x_1, x_2, ... x_N. The function also depends on several fixed parameters β_1, β_2, ... β_M as shown in equation (1). Deciding the optimal values of the parameters β is the task of the off-line and the on-line training algorithms.

$$\hat{Y} = f(x_1, x_2, \dots x_N, \beta_1, \beta_2, \dots \beta_M) \tag{1}$$

The regression is considered linear if the function f can be formulated such that parameters β_i are factors in a linear combination of base functions B.

$$\hat{Y} = \sum_{i=1}^{M} \beta_i \cdot B_i(x_1, x_2, \dots x_N) \tag{2}$$

The regression is nonlinear if it contains nonlinear combinations of parameters β_i. For instance, if they are multiplied together or if they are included in the expression of nonlinear base functions such as exponentials, logarithms, etc. A nonlinear regression example is provided below.

$$\hat{Y} = \sum_{i=1}^{M} \beta_i \cdot \prod_{j=1}^{N} e^{\beta_{ij} \cdot x_j} \tag{3}$$

However, the regression is still considered linear if the basis functions are nonlinear in the variable x_1, x_2, ... x_N such as in the example in equation (4).

$$\hat{Y} = \sum_{i=1}^{M} \beta_i \cdot \prod_{j=1}^{N} e^{x_j} \tag{4}$$

The least squares method is the most common type of parameter estimation algorithms applicable to all types of regression analysis (linear and non-linear). The aim is to calculate parameters β such as to minimize an objective function F which is the sum of the squared estimation errors across all the input data points.

$$F(\beta_1, \beta_2, \dots \beta_M) = \sum_j e_j^2 = \sum_j (Y_j - \hat{Y}_j)^2 \qquad (5)$$

The specific algorithm known as the Ordinary Least Squares (OLS) method can be used for linear regression. OLS has a direct non-iterative solution, but this is impractical to implement it in an embedded motor controller because it is mathematically demanding. Furthermore, it requires all input data points to be available from the beginning which makes it unsuitable for on-line learning. Nonlinear regression can only be solved by iterative algorithms. For these reasons, iterative least square algorithms are always used for machine parameters learning. Linear regression is a convex optimization problem therefore, iterative algorithms are guaranteed to converge to the global optimum. However, convergence to a suitable solution is not guaranteed for nonlinear regression because the objective function F will typically have multiple local minima. Nonlinear regression is dependent on the selection of a suitable starting vector of parameters β_i to improve the chance of finding an acceptable local minimum of function F.

Different types of multivariate regression are possible depending on the choice of the basis functions B. Step functions correspond to multivariate partitioning while piecewise linear functions are used in Multivariate Adaptive Regression Splines (MARS). Quadratic and higher order polynomials lead to multivariate spline fitting while Radial Basis Functions (RBF) are used in Support Vector Regression (SVR).

MARS and multivariate splines are most practical for machine learning of machine parameters. For this reason, they presented in more detail here.

Multivariate Splines

A multivariate spline function is a multidimensional extension of a unidimensional spline which in turn is a piecewise polynomial function. The approximate value of an estimated parameter calculated by a one-dimensional spline is:

$$\hat{Y} = \sum_{k=1}^{L} \beta_k \cdot B_k(x)\,. \qquad (6)$$

The basis functions B_k are known as basis splines or B-splines. Each B-spline spans a "support interval" of its variable $\boldsymbol{x}$. The B-spline takes the chosen polynomial expression inside its support interval and is zero outside it. An example of cubic B-spline with support interval $[x_{min}, x_{max}]$ is shown in equation (7).

$$B_k(x) = \begin{cases} Ax^3 + Bx^2 + Cx + D & x \in [x_{min}, x_{max}] \\ 0 & x \notin [x_{min}, x_{max}] \end{cases} \qquad (7)$$

The support intervals of different B-splines can be distinct or can partially overlap. The polynomials are chosen such that the resulting function approximation is continuous and has a certain degree of smoothness. Multivariate B-splines can be constructed as tensor products of one-dimensional B-splines.

$$B_i(x_1, x_2, \dots x_N) = \prod_{j=1}^{N} \sum_{k=1}^{L} B_{ijk}(x_j) \qquad (8)$$

The estimated parameter $\hat{Y}$ is expressed as a linear combination of several multivariate B-splines.

$$\hat{Y} = \sum_{i=1}^{M} \beta_i \cdot \prod_{j=1}^{N} \sum_{k=1}^{L} B_{ijk}(x_j) \qquad (9)$$

It is important to note that the simpler (but less powerful) polynomial regression is achieved if the support interval of each base function is equal to the full range of the independent variable x. In this case, a single polynomial base function is required for each variable x_i and the second summation becomes unnecessary. The regression result becomes a simpler weighted sum of polynomial products as shown in equation (10).

$$\hat{Y} = \sum_{i=1}^{M} \beta_i \cdot \prod_{j=1}^{N} B_{ij}(x_j) \qquad (10)$$

Multivariate Adaptive Regression Splines

MARS is a non-parametric regression technique meaning that it does not make a priori assumptions about the functional form of the relation between the estimated value $\hat{Y}$ and the independent variables $x_1, x_2, \dots x_N$ (see Ref. 15). The general mathematical form of the basis functions is limited to a certain repertoire which is explained below but the number of basis functions and the associated β parameters is decided iteratively while analyzing the available data. MARS is built on the concept of hinge functions. There are two types of hinge functions depending on the sign of their derivative. H_+ functions (Figure 13) have zero or positive derivative. H_- functions have zero or negative derivative.

$$\begin{cases} H_+(x_i, c) = \max(0, x_i - c) \\ H_-(c, x_i) = \max(0, c - x_i) \end{cases} \qquad (11)$$

The base functions involved in the regression can be individual hinge functions, products of hinge functions or

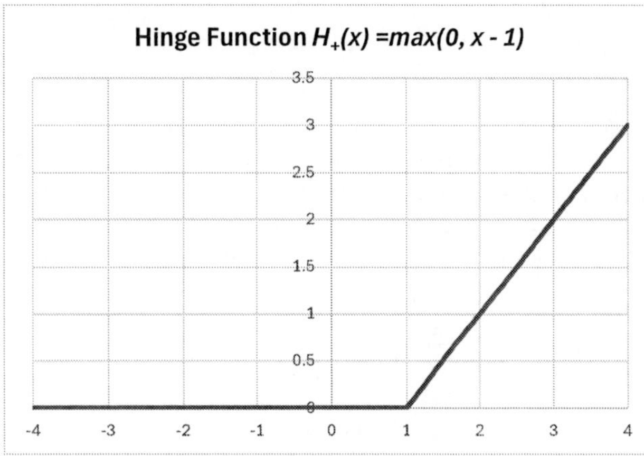

Figure 13. Example of MARS hinge function

simply the constant 1.0 (which is known as the "intercept"). MARS includes two iterative algorithmic stages: the forward stage and the backward stage.

The forward stage starts by calculating the multiplicative parameter β corresponding to the intercept (H_0). This is equal to average of the Y values in all the available data points. Additional pairs of hinge functions (H_+, H_-) are added iteratively with the aim of minimizing the total square-error. Each hinge pair can be included either on its own or by multiplication with other terms already included in the approximation function. Functions H_+ and H_- in a pair share the same constant "c" which is known as the knot and is always selected to be an actual value of the x_i variable from the given data set. The multiplicative constants β are calculated by linear regression after the inclusion of each new term in the expansion. The forward stage is complete when a stopping criterion is reached which can be a maximum number of terms in the expansion or the minimum approximation error. An additional constraint can be placed on the forward stage by specifying a maximum degree of interaction between variables x_i meaning the largest number of hinge functions multiplied in any product term. This reduces the number of possibilities when adding new terms to the expansion which is important because the search of new terms is based on a brute force approach of trial and error.

The backward stage aims to simplify the approximation function generated by the forward stage which is generally too complex and overfits the real model of parameter Y. The backward stage prunes the approximation function in an iterative manner, deleting the least effective term at each step until it finds the optimal solution. Importantly, hinge functions were added in pairs during the forward stage, but only individual hinge functions (or multiplications of individual hinge functions) are removed at this stage. The aim of the backward stage is to achieve the best balance between model complexity and accuracy.

Combining MARS and Multivariate Splines

Motor parameter approximations can benefit from combining concepts from both multivariate spline approximation and from MARS. The result of the MARS algorithm is a formula of the type (12) where H_{ij} are individual hinge functions.

$$\hat{Y} = \sum_{i=1}^{M} \beta_i \cdot \prod_{j=1}^{N} H_{ij}(x_j) \tag{12}$$

This is simpler than the general multivariate spline formula (9), but it is similar in complexity to multivariate polynomial approximation (10). However, MARS is different from polynomial approximation because hinge functions H_{ij} are always piecewise linear (with only two intervals, zero and non-zero). The MARS algorithm allows for hinge functions to be multiplied together and there is no restriction on multiplying hinge functions with different knots c but the same variable x_i. Therefore, some functions may include quadratic terms of the type

$$H_{1+} \cdot H_{2+} = max(0, x_i - c_1) \cdot \max(0, x_i - c_2). \tag{13}$$

However, MARS does not allow knots c_1 and c_2 be equal because this would make the two hinge functions identical. Each possible hinge function is added only once to the function expansion during the forward stage, therefore c_1 and c_2 will be different in MARS. This means that hinge functions cannot be of a quadratic type "$(x-c)^2$" in MARS algorithm. However, the approximation of machine parameters can benefit from using rounded hinge functions of the quadratic type. This is because quadratic hinge functions can be more effective approximators of the smooth variation of electrical machine parameters than the linear hinge functions used by MARS.

Selecting the Optimal Regression Method

A priori knowledge of machine parameters behavior can be used to restrict the configuration space of the regression functions. An iterative search can be executed on a reduced number of most promising configurations which will differ in terms of number and type of base functions (polynomials, linear hinge functions, quadratic hinge functions). This search is similar in nature but more general than the MARS approach which is to identify the optimal solution only among models based on linear hinge functions (see Ref. 16). The parameters β and the knots c of each function can be optimized by versions of the least squares method.

This leads to a regression algorithm which consists of two nested loops. The outer loop iterates over candidate regression functions while the inner loop optimizes the parameters β and c by iterative least-squares method. It needs to be noted that the existence of the knots c causes the regression to become non-linear. This is because they are included in the expressions of both the linear and the quadratic hinge functions which are non-linear. As a result, a simple iterative gradient descent least-squares method will often not converge to an adequate solution. Results are significantly improved by splitting the parameter optimization between an initial stage with fixed c knots and variable β parameters followed by a second stage where all parameters (c and β) are allowed to vary. The initial stage becomes a convex optimization problem once the $\boldsymbol{c}$ knots are fixed. This increases the chances that the second optimization stage will start from a suitable point around the global minimum of the objective function. Therefore, the success of the second optimization is dependent on fixing the c knots to adequate values. In practice, the choice of initial values for the $\boldsymbol{c}$ knots is informed by general technical knowledge of the motor parameters and their dependence on the relevant variables I_d, I_q, θ, ω, etc. For instance, saturation effects are expected at higher currents which means that the $\boldsymbol{c}$ knots for variables I_d and I_q should be distributed closer to the higher end of their respective ranges.

Practical Considerations for Torque Estimation

Restricting the configuration space of the candidate regression functions for motor torque calculation can be achieved based on well-known mathematical properties of non-salient PM machines. The most important property is that torque is zero when current component $I_q = 0$ regardless of the other variables such as I_d, angle θ, speed ω, rotor temperature t_R. Therefore, the torque can be expressed as the product of two regression functions, one depending on I_q and the other depending on all the other variables. The function dependent on I_q will be zero when $I_q = 0$.

A second property is that often there is limited interaction between the variables which influence the torque. This means that the total torque regression function can be approximated as a tensor product of one-dimensional functions of independent variables. The first function (which depends on I_q) generates a rough approximation of the torque while all the other functions provide correction factors. Each correction factor is 1.0 when no correction is needed but takes different values when corrections are required. Therefore, without losing generality, it can be considered that each one-dimensional correction function of variable x_i starts from initial value 1.0 when $x_i = 0$. The zero point of each variable needs to be suitably selected. For instance, the zero point of the temperature parameter may be 20°C. The zero point of the rotor angle may or may not be 0.0 rad.

Conclusions on Motor Parameters Learning

Machine learning algorithms are suitable to identifying the nonlinear behavior of saturated electrical machines. General regression algorithms are available in the literature, but the machine learning performance can be improved by taking into consideration specific physical aspects of electrical machine configuration.

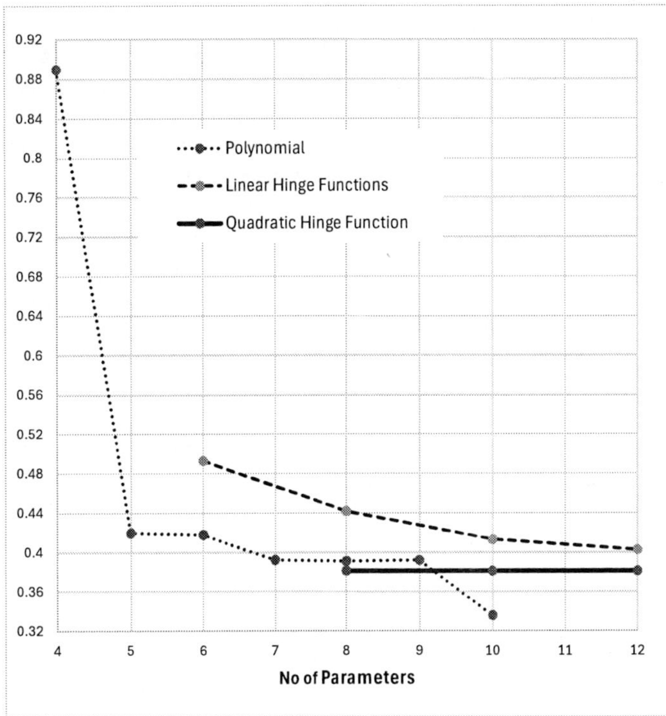

Figure 14. Accuracy of regression solutions

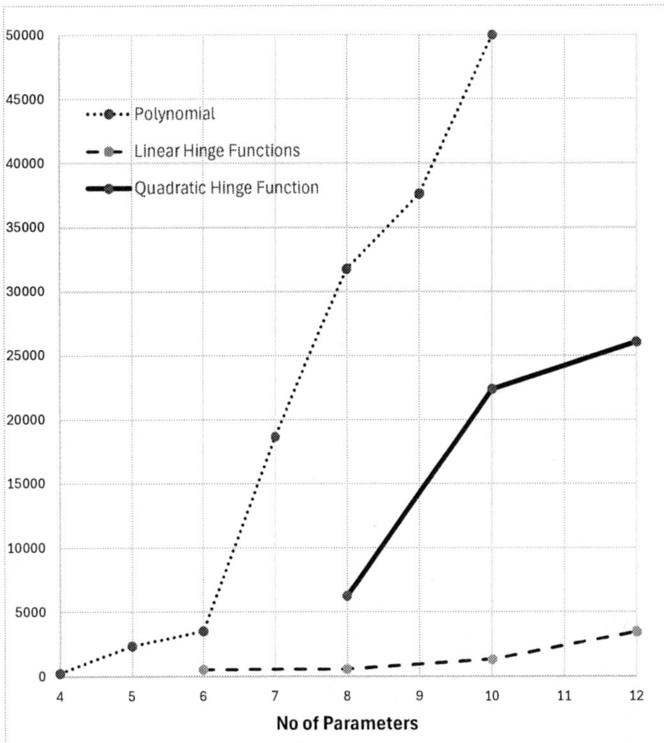

Figure 15. Training effort for each regression solution

Different regression solutions may be selected depending on the machine design and the parameter being estimated (polynomials, linear hinge functions, quadratic hinge functions). The optimal solution is also dependent on the computational capability of the motor drive controller. Figures 14 and 15 indicate results of three types of regression methods for a 1MW electric propulsion method.

The accuracy of the quadratic hinge function approach is generally better than linear hinge functions due to their ability to generate both continuous and derivable approximations which are a better fit to the real behavior of the physical system than piecewise linear solutions generated with linear hinge functions of the MARS algorithm.

At the same time, the quadratic hinge function approach requires less training effort than polynomial approximation, as shown in Figure 15. The training algorithm described here is different from the constructive MARS training algorithm. Unlike MARS, an iterative gradient descent optimization similar to the training of artificial neural networks is used to achieve the optimal accuracy for a fixed regression configuration with a fixed number of training parameters. A specific feature of this training algorithm is the split in two iterative stages with some parameters being iteratively adjusted in both stages while others only in the second stage of the training.

The general framework presented here is very flexible and it is suitable to learning theoretical FEA data as well as practical data collected in the lab. In the first case, off-line learning is implemented on a stand-alone computer and all data is provided to the learning algorithm at once. In the second case, batches of data are used to improve the accuracy of the initial solution generated from the FEA data. The number and the size of the data batches can be adjusted based on the laboratory setup and the capability of the motor drive controller.

GENERAL CONCLUSIONS

Increasing the power density of electrical machines is a key to realizing electric aircraft propulsion feasible and competitive. Various measures from material selection to thermal management were discussed by using an example of prototype 1 MW machine. However, it was shown by detailed simulations that nonlinearities caused by excessive electric and/or magnetic loading can lead to a control instability, which would prevent further improvements in power density. The paper proposed a machine learning algorithm suitable to deal with nonlinear behavior of saturated electrical machines. The proposed method uses an algorithm inspired by MARS where the conventional linear hinge functions are replaced with quadratic hinge functions. The training algorithm similar to the one used for ANN is used providing more flexible and suitable method for the electrical machine application. The method was applied to the 1 MW machine and its validity has been demonstrated, providing a path to further improvements in power density of electric propulsion systems for VTOL and other aircraft applications.

Author contact: Tadashi Sawata tadashi.sawata@collins.com Andrei Dinu andrei.dinu@collins.com.

ACKNOWLEDGMENTS

The authors acknowledge their colleagues at Collins Aerospace for their contributions to the development and characterization of prototype electric propulsion systems.

REFERENCES

1. A. Barzkar and M. Ghassemi, “Components of Electrical Power Systems in More and All-Electric Aircraft: A Review,” IEEE Trans. Transp. Electrif., vol. 8, no. 4, pp. 4037-4053, 2022.

2. E. Sayed et al., “Review of Electric Machines in More-/Hybrid-/Turbo-Electric Aircraft,” IEEE Trans. Transp. Electrif., vol. 7, no. 4, pp. 2976-3005, 2021.

3. P. Alvarez, M. Satrustegui, I. Elosegui, and M. Martinez-Iturralde, “Review of High Power and High Voltage Electric Motors for Single-Aisle Regional Aircraft,” IEEE Access, vol. 10, pp. 112989-113004, Sep. 2022.

4. R.C. Bolam, Y. Vagapov, and A. Anuchin, “A Review of Electrical Motor Topologies for Aircraft Propulsion,” 2020 55th International Universities Power Engineering Conference (UPEC), Sep. 2020.

5. https://www.clean-aviation.eu/programme-overview-and-structure

6. T. Sawata and S. Minshull, "Development of 1 MW Hybrid Electric Propulsion Drive System," IEEE Electrical Systems for Aircraft, Railway, Ship Propulsion and Road Vehicles (ESARS) and International Transportation Electrification Conference, Naples, Italy, 26-29 November 2024.

7. D. Golovanov, D. Gerada, Z. Zu, C. Gerada, A. Page and T. Sawata, "Designating an Advanced Electrical Motor for Propulsion of Electric Aircraft," 2019 AIAA/IEEE Electric Aircraft Technologies Symposium (EATS), Indianapolis, IN, 22-24 August 2019, pp. 1-12.

8. D. Golovanov, D. Gerada, G. Vakil, Z. Xu, M. Degano and C. Gerada, "The Influence of Stator Material on the Power Density and Iron Loss of a High-Performance Starter-Generator for More Electric Aircraft," in 21st International Conference on Electrical Machines and Systems (ICEMS), Jeju, Korea, 2018.

9. Z. Q. Zhu and D. Howe, “Halbach permanent magnet machines and applications—a review,” Proc. Inst. Elect. Eng.—Electric Power Applications, vol. 148, no. 4, pp. 299–308, 2001.

10. M. van der Geest, H. Polinder, J. A. Ferreira, and M. Christmann, “Power Density Limits and Design Trends

of High-Speed Permanent Magnet Synchronous Machines," IEEE Trans. Transp. Electrif., vol. 1, no. 3, pp. 266-276, 2015.

11. C. Dong, Y. Qian, Y. Zhang, and W. Zhuge, "A Review of Thermal Designs for Improving Power Density in Electrical Machines," IEEE Trans. Transp. Electrif., vol. 6, no. 4, pp. 1386-1400, 2020.
12. M. Jaritz, A. Hillers, and J. Biela, "General Analytical Model for the Thermal Resistance of Windings Made of Solid or Litz Wire," IEEE Trans. Power Electron., vol. 34, no. 1, pp. 668-684, Jan. 2019.
13. R. Wrobel, and P. H. Mellor, "Thermal Design of High-Energy-Density Wound Components," IEEE Trans. Ind. Electron., vol. 58, no. 9, pp. 4096-4104, Sept. 2011.
14. P. Sangha, E.G.C. Pocock, and M. Dawson, "US Patent 11,605,995 B2," Mar. 2023.
15. Friedman, J. H. Multivariate Adaptive Regression Splines. The Annals of Statistics, 1991, vol. 19, pp. 1–141.
16. Kartal Koc, E., Bozdogan, H. Model selection in multivariate adaptive regression splines (MARS) using information complexity as the fitness function. *Machine Learning, 2015* vol. 101, pp. 35–58. DOI: 10.1007/s10994-014-5440-5.

Exploring Transition Profiles for the RAVEN Multi-Tiltrotor Aircraft

Tristan Huang
Graduate Research Associate

Brian German
Professor

Daniel Guggenheim School of Aerospace Engineering
Georgia Institute of Technology
Atlanta, GA, U.S.A.

ABSTRACT

The transition between hover and forward flight is a critical consideration for eVTOL aircraft development. This transient and aeromechanically complex maneuver is often analyzed using computationally expensive methods that may not be practical for conceptual design. This paper presents a quasi-steady method capable of optimizing transition paths for multi-tiltrotor aircraft using low- to mid-fidelity models while approximating the transient nature of transition. A 3-degree-of-freedom aircraft model with surrogate models for rotor and airframe aerodynamics was developed along with a novel download model based on momentum theory. The optimization formulation minimizes a thrust-based surrogate for induced power while enforcing trim. This allows estimates for energy-optimal transition paths to be calculated even when accurate rotor power predictions are unavailable during early design phases. Important considerations for operating lift-only rotors through transition are identified and an iteratively refined speed schedule for such rotors is presented. The method presented is applied to analyze the RAVEN multi-tiltrotor aircraft. Transition paths at various fixed angles of attack are first developed to characterize operational stall limits. Optimal paths are then calculated for both accelerating and decelerating transition for RAVEN, with results showing that the aircraft behaves optimally as a multicopter at low airspeed and increasingly like a fixed-wing aircraft at higher airspeeds.

NOTATION

A	Rotor area (ft^2)
A_{proj}	Projected rotor area (ft^2)
a	Acceleration (ft/s^2)
C_L	Lift coefficient
C_D	Drag coefficient
C_m	Pitching moment coefficient
DL	Download (lbf)
D	Drag (lbf)
F	Force (lbf)
M	Moment ($ft-lb$)
m	Aircraft mass ($slugs$)
P	Power
T	Thrust (lbf)
V_∞	Freestream velocity ($KTAS$)
v_{dw}	Downwash velocity (ft/s)
v	Inflow velocity component (ft/s)
α	Angle of attack (°)
β	Rotor blade collective pitch (°)
δ_e	Elevator deflection (°)
δ_f	Flap deflection (°)
θ	Aircraft pitch attitude (°)
ρ	Ambient density ($slug/ft^3$)
τ	Rotor tilt angle (°)
Ω	Rotor angular speed (RPM)

Subscripts

h	Hover
i	Induced
1	Forward tilt rotor
2	Wingtip tilt rotor
3	Rear rotor

Presented at the Vertical Flight Society's 11th Biennial Autonomous VTOL Technical Meeting, Phoenix, AZ, USA, Feb. 4–6, 2025.

INTRODUCTION

Rapid urbanization in the modern world has made the concept of Urban Air Mobility (UAM) increasingly relevant. UAM operations aim to reduce travel times both within and between densely populated urban centers by overflying congested roadways with electric vertical takeoff and landing (eVTOL) aircraft. These aircraft vary in configuration but typically combine multiple tilting rotors with one or more wings to gain both the vertical takeoff and landing capability of rotorcraft and the speed and efficiency of fixed-wing designs. All of these characteristics are crucial in urban environments which have few airports for fixed-wing operations yet demand sufficient range capability and energy efficiency for economic feasibility (Ref. 1). Despite a wealth of understanding about vertical and wing-borne flight existing in aeronautics literature, much less studied is the transition between these flight phases. Most commonly referred to as either "Transition" or "Conversion," this phase of flight can be modeled as a distinct mission segment (Ref. 2). Power requirements and aerodynamic loading are typically highest during this segment, so aircraft sizing during the conceptual and preliminary design phases must satisfy transition performance requirements (Ref. 3). eVTOL aircraft are also highly power-sensitive, which incentivizes minimizing power requirements through the demanding transition regime. Modeling the transition flight profile is therefore crucial to eVTOL aircraft development.

Most recent contributions in this field fall under at least one of three categories: the development of eVTOL aircraft models, the characterization of the transition space, and transition path optimization. A common approach to the first of these is to develop multiple aerodynamics and physics-based sub-models that capture distinct aspects of eVTOL performance. Orndorff et al created five sub-models for flight dynamics, rotor aerodynamics, wing and airframe aerodynamics, electric motor performance, and aeroacoustics (Ref. 4). Dynamics are modeled with a 3-degree-of-freedom (3-DOF) system of equations while each aerodynamic model interpolates rotor performance and aerodynamic coefficients using surrogate models trained on data from first-order aerodynamic approximations. Stoll and Mikić achieved higher fidelity by fitting surrogate aerodynamic models to data from OVERFLOW CFD simulations (Ref. 5). They additionally captured rotor-wing interactions by implementing simple first-order models for download and wing interference.

Once a model is developed, it can be applied to characterize and explore the transition flight regime. A primary objective of such activities is to define operable limits within which eVTOL aircraft can safely fly during the transition segment while maintaining controllability and avoiding structural limits. Kang et al (Ref. 6) define an upper bound as a maximum power constraint and a lower bound as a stall limit. This approach specifies the maximum and minimum rotor tilt angles allowed at each airspeed without exceeding power limits or stalling the main wing, respectively. The resulting transition corridor is shown in Figure 1a. This figure is representative of typical transition corridors that plot tilt rotor shaft angle, or tilt angle, as an airspeed schedule. Stoll and Mikić take a different approach by trimming their Joby S4 model through transition at a range of fixed angles of attack to create a series of trimmed trajectories, or transition paths (Ref. 5). The range chosen of -12° to 12° represents the practical range of allowable angles of attack in transition while maintaining sufficient stall margins. Thus, the transition paths at either extreme of this range of α form the bounds of their transition corridor in Figure 1b.

While transition corridors may provide operational limits, they are relatively large and provide little information on how to best design the transition path itself. More crucially, most eVTOL aircraft are over-actuated such that multiple control solutions exist for any point within a corridor. The task of transition path design is thus inherently an optimization problem.

Modern approaches often employ full-trajectory optimization methods to capture the nonlinear and transient flight dynamics inherent to transition. In these methods, transition trajectories are simulated by modeling the segment as a series of time steps and using an integration scheme coupled with a dynamical aircraft model to solve for aircraft states at each time step (Ref. 4). These trajectories are then optimized by solving a constrained optimization problem, which searches for a control profile minimizing some objective. This is typically the required energy or segment time for transition (Ref. 7). Constraint functions enforce trim, powertrain component limits, and desired initial and end states. Though comprehensive, full-trajectory optimization methods are computationally expensive and can take an hour or more to run (Ref. 4). These methods may be impractical for early-phase aircraft design and transition space exploration, which often require a large number of simulations covering a wide variety of potential aircraft configurations. Instead, a quicker, lower-fidelity method for optimizing transition paths is desired.

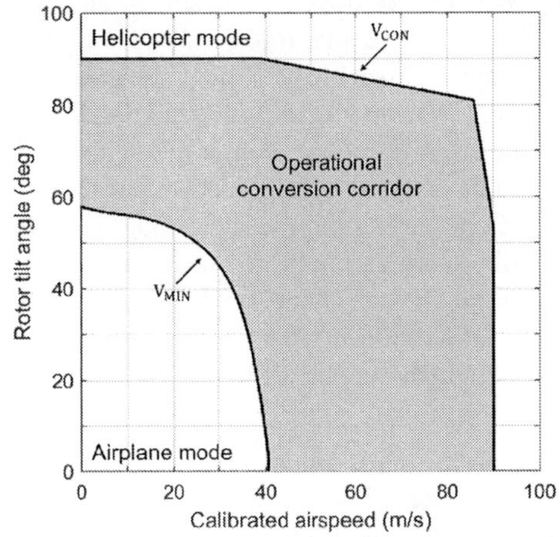

(a) Volante Vision transition corridor (Ref. 6).

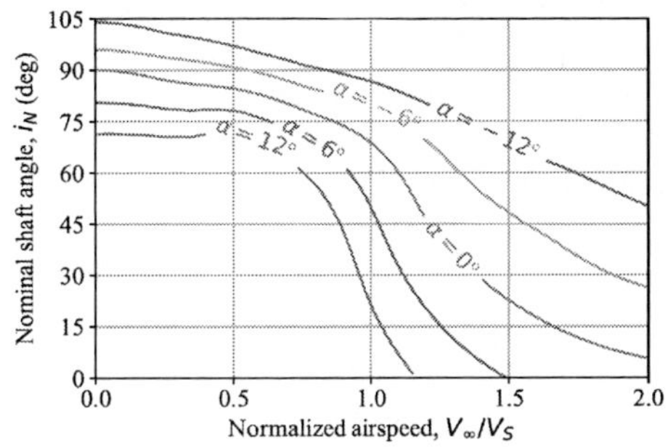

(b) Joby S4 constant-α contours and transition corridor (Ref. 5).

Figure 1: Contributions to transition space characterization.

Quasi-steady methods provide a mid-fidelity capability between simple steady-state analysis and dynamical trajectory optimization formulations. Similarly to dynamical simulations, the transition phase is discretized into a series of shorter sub-segments. Each sub-segment is then evaluated as a steady-state segment, during which the aircraft state is constant. This approach does away with the complex state-space formulations required to model flight dynamics while retaining the fidelity to capture salient aspects of the vehicle state and control profiles. Such a quasi-steady formulation was developed by Sagan et al (Ref. 3). It solves an optimization problem to minimize the total rotor power required to trim at each quasi-steady sub-segment. Flight dynamics are represented using a 3-DOF model and rotor and airframe aerodynamics are modeled by either simple first-order methods or surrogate models fitted to aerodynamic data. Sagan implemented a gradient-based optimizer but noted a prevalence of flight conditions at which multiple or even infinite locally optimal trim solutions exist. Though effective at trajectory optimization, the method necessitates a smooth and accurate rotor power database which may be difficult to develop.

To address the need for versatile, accurate, and computationally inexpensive transition analysis, this paper introduces a quasi-steady methodology capable of both characterizing operational limits for transition flight and designing optimal transition paths. The RAVEN eVTOL aircraft concept (Ref. 8) is modeled as an example application. Models are developed for rotor and airframe aerodynamics, and a download model is developed to capture certain rotor-airframe interactions. A nonlinear constrained optimization problem is then formulated using the Interior Point Optimizer (Ipopt) (Ref. 9) and implemented in JuMP (Ref. 10). The optimizer minimizes a thrust-based approximate surrogate for rotor power. The overall toolchain, consisting of the physics model, optimizer, wrapper, and post-processing scripts, is referred to as the Transition Trim Optimizer (TTO). Optimal transition paths and stall boundaries are developed for the RAVEN aircraft, from which its behavior through transition can be analyzed. An approach to modeling the angular speed profile of lift-only rotors through transition is also presented.

(a) Hover configuration.

(b) Forward flight configuration.

Figure 2: RAVEN eVTOL aircraft configuration.

MODELS

Aircraft Configuration

The development of the Transition Trim Optimizer was motivated by the Research Aircraft for eVTOL Enabling Technologies (RAVEN) activity, a collaboration between Georgia Tech and NASA to develop a large experimental eVTOL aircraft (Ref. 8). The RAVEN aircraft features the 6-rotor distributed electric propulsion (DEP) configuration shown in Figure 2 with four tilt rotors and two fixed lifting rotors. The tilt rotors are mounted in a tractor configuration on two wing sponsons and at the wingtips. The lifting rotors are mounted at the rear of the sponsons aft of the wing. The airframe is a high-wing and conventional tail configuration with a rudder, ailerons, flaps, and an all-moving horizontal tailplane for control.

The RAVEN aircraft is represented as a 3-DOF model assuming symmetry about its X-Z plane, allowing yaw forces and yaw and roll moments to be ignored. This approach requires only three rotors on one side of the aircraft to be modeled, simplifying the trim problem significantly to reduce the computational cost of analysis and thus runtimes. Only the starboard-side rotors are modeled and each port-side rotor is assumed to produce identical thrust to its starboard-side counterpart. The rotors are variable-pitch (collective-controlled) and are presumed to operate at a constant angular speed of 1,850 rotations per minute (RPM) to meet aeroacoustic requirements throughout much of the flight envelope. The only exception occurs during transition, in which the rear rotors spool down to or up from 0 RPM.

The starboard rotors are numbered for identification purposes as follows: rotor 1 refers to the forward sponson-mounted tilt rotor, rotor 2 to the wingtip-mounted tilt rotor, and rotor 3 to the lift-only rear rotor. Modeled control actuators consist of tilt angle, τ, the collective pitch of each rotor, β_1, β_2, and β_3, flap deflection, δ_f, and elevator deflection, δ_e. The all-moving horizontal tailplane serves the role of an elevator and provides pitch control. Aerodynamic actuator control angles, the rotor numbering convention, and primary forces acting on the aircraft are shown in Figure 3.

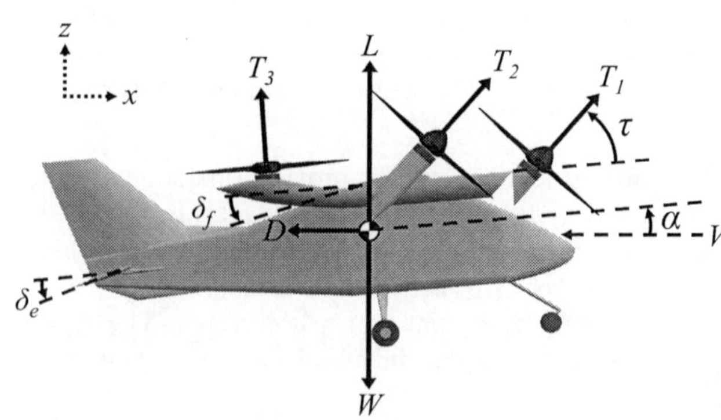

Figure 3: RAVEN aircraft free-body diagram and control actuator deflections.

Aileron and rudder inputs are not modeled to vary during transition since the 3-DOF formulation allows for the roll and yaw moments they control to be ignored.

Rotor Aerodynamics Model

Rotor aerodynamics are captured using surrogate models fitted to aerodynamic force and moment data to reduce computational cost. These models take four lookup parameters as inputs consisting of angular speed, Ω, collective pitch, β, and edgewise and axial inflow velocity components, v_x and v_z.

The surrogate models were trained on an existing aerodynamic database generated using DUST, a mid-fidelity aerodynamic analysis tool for modeling unsteady flows. DUST implements a vortex particle method and is optimized for multirotor configurations relevant to eVTOL research (Ref. 11). Rotor blades were modeled as lifting lines with viscous 2D drag polars implemented to provide estimates for airfoil profile drag. These polars were obtained using XFOIL, a low-fidelity tool for 2D viscous aerodynamics (Ref. 12). Polar extensions for airfoil lift and drag were then applied to cover a wide range of angles of attack. Lastly, the rotor wake was modeled using vortex particles.

Rotor power and torque predictions produced with this approach are considered unreliable. There is low confidence in the viscous corrections due to the limitations of XFOIL and the associated polar extensions. Predictions for viscous forces on the blade and thus profile power are likely compromised. However, the induced power and thrust predictions, being based purely on the rotating lifting line analysis, are considered reliable. An example disk loading plot based on DUST thrust data is shown in Figure 4.

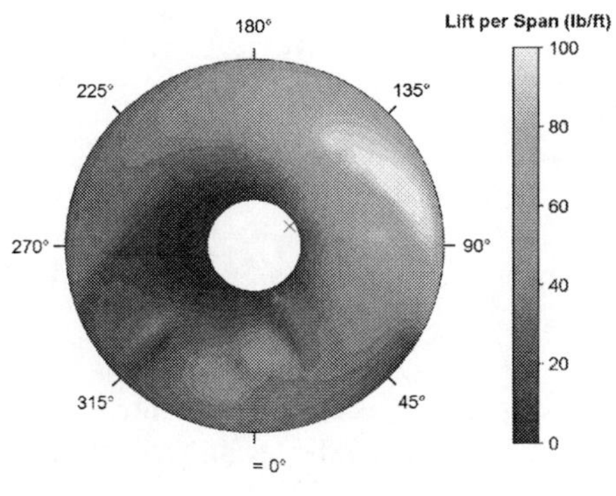

Figure 4: Tilt rotor DUST disk loading plot at $V_\infty = 70$ KTAS.

Blade geometry is based on a 3D model of the rotor selected for RAVEN. DUST was run using this model over rotor conditions corresponding to a set of two 4-dimensional full-factorial grids. One grid covers a range of lookup parameters representing the operating conditions for the tilt rotors and the other covers the operating conditions for the lift-only rear rotor. Using separate grids reduced the number of required DUST runs, as the range of these operating conditions varies greatly between the two types of rotors. Tilt rotors operate across a large variety of axial inflow velocities, requiring a wide range of v_z to be run for their database. The rear rotors, conversely, see less variability in inflow velocity but must operate over a wide range of angular speeds as the rotors spool up or down during transition. This is not a concern for the tilt rotors operating at a constant speed, whose database covers a much more limited set of angular speeds. The resulting tilt rotor aerodynamic database consists of data from 9,984 DUST runs and the rear rotor database includes data from 231,680 DUST runs.

Separate surrogate models were created for tilt and rear rotor aerodynamics. They use 1st-order B-Splines (Ref. 13) to linearly interpolate rotor loads and moments across the entire range of their respective full-factorial grids. Outputs are parameterized by the four lookup parameters. Both surrogate models are further enhanced with the capability for linear extrapolation, expanding the range of rotor operating conditions that can be analyzed. These models produce smooth outputs and gradients, allowing them to be consistently queried by the optimizer. An added benefit of surrogate models is that they can be fitted to databases of any fidelity level, so the rotor aerodynamics model can be improved once higher-fidelity databases become available.

Airframe Aerodynamics Model

The airframe is subdivided into the wing, fuselage, horizontal tail, and vertical tail. Aerodynamic forces and moments on each component are captured using surrogate models fitted to aerodynamic databases specific to that component. Each database consists of a full-factorial grid of aerodynamic force and moment coefficients covering the expected range of operating conditions and control surface deflections for that component. Although this results in a varying number of lookup parameters for each database, each one has at least 5 lookup parameters consisting of angle of attack, sideslip angle, and the three *PQR* aircraft-fixed frame angular rates.

To create the databases, aerodynamic coefficients were first calculated using AVL, a low-fidelity aerodynamic analysis tool that implements an inviscid panel method (Ref. 14). Lifting surfaces were modeled using a vortex lattice method while the fuselage was represented with a slender-body model. The inviscid nature of AVL means it cannot model viscous effects such as parasitic drag. This was instead captured using the OpenVSP parasitic drag tool (Ref. 15), which employs a semi-empirical wetted area method to calculate parasitic drag coefficient, $C_{D,0}$, for each component.

AVL's inability to capture viscous effects also makes it unsuitable for predicting the onset of stall or modeling post-stall behavior. This stall behavior was instead modeled using a method by Viterna and Corrigan (Ref. 16) to calculate post-stall aerodynamic coefficients in conjunction with XFOIL to estimate the angle of attack at which stall occurs. Aerodynamic coefficients were calculated using AVL in the pre-stall regime while Viterna and Corrigan extensions provide coefficients in the post-stall regime. The lift-curve slope for RAVEN's wing in Figure 5 shows both linear pre-stall behavior as well as the onset of stall above $\alpha = 16°$.

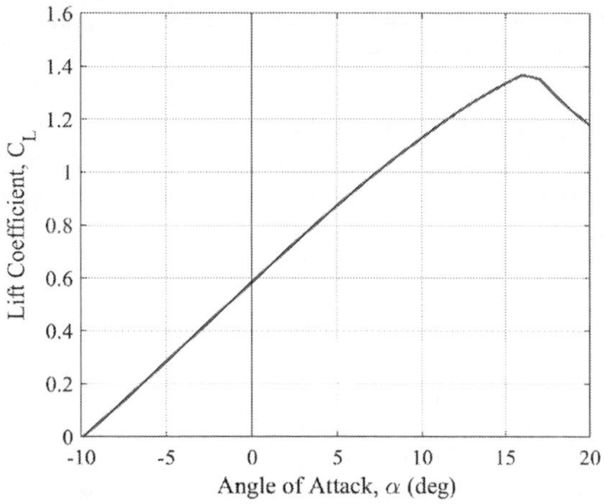

Figure 5: Main wing lift-curve slope for the RAVEN aircraft.

Surrogate models for each component were fitted to their respective databases using 1st-order B-Splines for linear interpolation and modified to support linear extrapolation, as was done for rotor aerodynamics. They output aerodynamic coefficients C_L, C_D, and C_m for lift, drag, and pitching moment, respectively. Each coefficient is parameterized by the component's specific set of lookup parameters. The coefficients are re-dimensionalized using dynamic pressure and the reference wing area. Forces are then superimposed in the aircraft body-fixed reference frame and the net pitching moment due to aerodynamic forces and moments is calculated about the aircraft's center of gravity.

Download Model

A download model based on momentum theory (Ref. 17) was developed to capture the loss in lift due to rotor interference on the airframe and calibrated to align with experimental data. Download is modeled as the drag on the portion of the airframe covered by a rotor's contracted stream tube due to downwash and assumes uniform flow. The velocity of this downwash is considered at the point where the stream tube reaches the airframe. Downwash velocity in hover, $v_{dw,h}$, is defined in Equation 1 as a function of thrust, T, ambient air density, ρ, and rotor area, A, using a download factor of 1.75.

$$v_{dw,h} = 1.75\sqrt{\frac{T}{2\rho A}} \tag{1}$$

Download factor reflects the degree to which the stream tube is contracted at the point it impinges on the airframe. A factor of 2 reflects a fully contracted stream tube and assumes downwash velocity is double the induced velocity, v_i. However, making this assumption was considered to be potentially too conservative and could significantly overpredict download and thus thrust requirements. Instead, download factor was calibrated to achieve agreement between the download model and experimental data for a similar aircraft configuration. This experimental data was obtained from studies characterizing download for the Bell-Boeing V-22 (Refs. 18, 19). The V-22 is configured with wingtip tilt rotors that are a similar distance above the wing when normalized by rotor diameter as RAVEN's wingtip rotors (Ref. 20). As a result, wingtip rotor stream tubes for both aircraft are likely similarly contracted when they impinge on the wing.

Since the V-22 has one wingtip-mounted tilting rotor per wing, only data for RAVEN's wingtip rotor can be compared to calibrate download factor. Assuming a thrust coefficient C_T of 0.012, the ratio of download to rotor thrust DL/T for the V-22 in its hover configuration is about 8% (Refs. 18, 19). Adjusting for RAVEN's larger wing chord relative to its rotor diameter and thus its larger relative projected area results in a target DL/T of 13.6% at this C_T. Download factor was then iteratively adjusted until this target DL/T was achieved, resulting in the selected factor of 1.75. The resulting projected areas for each rotor's stream tube on the airframe are shown in Figure 6.

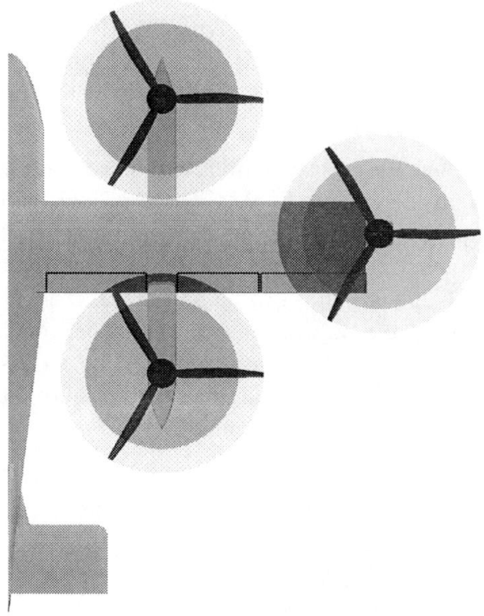

Figure 6: Stream tube projected areas on the RAVEN airframe (shaded in blue) for download factor = 1.75.

Although this approach is less conservative than assuming a fully contracted stream tube, it still constitutes a conservative assumption since both the forward tilt and rear rotors are mounted closer to the wing than the wingtip rotor. Their stream tubes are thus less contracted than that of the tip rotor when they contact the airframe, resulting in lower downwash velocities and thus lower download force than is predicted by the model.

Axial inflow velocity is added to the hover downwash velocity to account for climbing or descending flight. This provides an expression for v_{dw}, which assumes the tilt rotors are in the hover configuration with $\tau = 90°$.

$$v_{dw} = v_{dw,h} + v_z \tag{2}$$

Download is calculated using the projected area in the hover configuration, $A_{proj,h}$, and the drag coefficient for components within this region, $C_{D,proj}$. The former was calculated for each rotor in OpenVSP using a model of the RAVEN aircraft while drag coefficients were obtained from a 1985 report by Jordon et al (Ref. 21) that developed an analytical model of download for the Bell XV-15. Like the V-22 and RAVEN, the XV-15 features an outboard download relief flap. The report found the drag coefficient from above of a wing with a flap length of 0.25c deflected at 60° to be 1.09. Additionally, the drag coefficient of a circular cylinder from above was found to be 1.2. The former coefficient is used to calculate download from the wingtip tilt rotor, which exclusively projects onto the wingtip and outboard flap. The latter is used for the forward tilt rotor and rear rotor as they project over the near-cylindrical sponson mounts.

Reference download DL_{ref} can then be defined as the download force of a tilt rotor in the hover configuration as shown in Equation 3. This expression also defines rear rotor download.

$$DL_{ref} = \frac{1}{2}\rho v_{dw}^2 A_{proj,h} C_{D,proj} \quad (3)$$

As the tilt rotors rotate forward, download decreases as both the projected area and downwash component along the aircraft-fixed z-axis decrease. By calculating projected area for both tilt rotors over a range τ from 0° to 90°, A_{proj} is found to vary proportionally with $sin(\tau)$. Furthermore, geometric relations define the vertical component of downwash velocity as being directly proportional to $sin(\tau)$. Since download DL is directly proportional to A_{proj} and to the square of downwash velocity, it can be defined to vary proportionally with $sin^3(\tau)$. This results in the final tilt rotor download formulation shown in Equation 4 as function of tilt angle and reference download.

$$DL = DL_{ref} sin^3(\tau) \quad (4)$$

It should be noted that the component of download along the aircraft-fixed x-axis, i.e. the drag direction, is not modeled. Unlike the projected area along the z-axis, projected area along the x-axis does not exhibit a variation with tilt angle that can be captured with a simple geometric function. This is acceptable for the mid-fidelity analysis for which the Transition Trim Optimizer is intended.

Furthermore, more complex interactional aerodynamics such as the effects of a rotor's wake on the performance of other rotors and control surface effectiveness are also omitted from the model. Although these interactional effects are important, they are extremely challenging to estimate without a high-fidelity aerodynamics model. Introducing such a model would complicate the approach considerably and require significant additional computational time. Generally, these interactional effects would be expected to deteriorate transition performance. Although the effects are not modeled by the TTO, they should be modeled when a higher-fidelity analysis is desired.

METHODOLOGY

Optimization Problem Formulation

The ideal way to minimize the energy required for transition in a quasi-steady method is to minimize the total shaft power required at every steady-state sub-segment, or operating point, on the transition path. Each of these operating points is associated with a specific airspeed. However, poor confidence in the available power database as discussed in *Rotor Aerodynamics Model* prevents the implementation of a reliable power-objective formulation. Instead, the objective function is defined using rotor thrust terms as an approximate surrogate for shaft power, as there is far greater confidence in the accuracy of the rotor force data. The optimizer minimizes the sum of the square of each rotor's thrust output, known as sum-squared thrust and defined in Equation 5, at each airspeed.

$$\sum_{n=1}^{3} T_n^2 \quad (5)$$

The thrust squared term of T_n^2 captures the nonlinear relationship between induced power, P_i, and thrust. Following momentum theory, induced power is directly proportional to $T^{3/2}$ in hover and to T^2 in forward flight per Glauert's high-speed approximation (Ref. 17), assuming a sufficiently high advance ratio. Thus, minimizing sum-squared thrust also minimizes total induced power. Trends in parasite and climb power requirements in response to varying trim solutions or operating conditions are captured by sum-squared thrust since increased drag and positive climb rate inherently result in increased thrust output in trimmed flight.

Trimmed flight is enforced using equality constraints to ensure equilibrium in each degree of freedom. These constraints require force balances along the aircraft-frame x-axis and z-axis in conjunction with a moment balance about the pitch axis. Constant acceleration terms are included in the force balance constraints to represent user-prescribed nonzero acceleration components a_x and a_z.

The control variables available to the optimizer consist of tilt angle, angle of attack, elevator deflection, and each rotor's collective pitch. Flap deflection is a user input that is held constant through transition and thus not adjusted by the optimizer. Constraint functions place bounds on each control variable to reflect operational or physical deflection limits. Collective pitch is limited to avoid blade stall, and the tailplane's possible range of motion dictates limits on δ_e. Bounds for α and tilt angle can be adjusted based on requirements for the desired analysis, such as to implement a deck angle constraint. An additional set of inequality constraints prevents negative rotor thrust output. The objective and constraint functions collectively comprise the optimization problem formulation in Equation 6.

$$
\begin{aligned}
\min_{\tau,\alpha,\delta_e,\beta_1,\beta_2,\beta_3} \quad & \sum_{n=1}^{3} T_n^2 \\
\text{s.t.} \quad & \sum F_x = ma_x, \\
& \sum F_z = ma_z, \\
& \sum M_y = 0, \\
& 0^\circ \leq \tau \leq 105^\circ, \\
& \alpha_{min} \leq \alpha \leq \alpha_{max}, \\
& -30^\circ \leq \delta_e \leq 30^\circ, \\
& -3^\circ \leq \beta_1 \leq 27^\circ, \\
& -3^\circ \leq \beta_2 \leq 27^\circ, \\
& -3^\circ \leq \beta_3 \leq 20^\circ, \\
& 0 \leq T_1, \\
& 0 \leq T_2, \\
& 0 \leq T_3
\end{aligned} \tag{6}
$$

This formulation is highly nonlinear. To solve such a problem, the Interior Point Optimizer (Ipopt), a gradient-based optimizer tailored for large-scale nonlinear optimization, was selected. Ipopt implements an interior point line search method to find locally optimal solutions (Ref. 9) and is implemented within a broader optimization model that also consists of objective and constraint functions. This broader model is written in JuMP, a mathematical optimization-specific modeling language embedded within the Julia programming language (Ref. 10). Gradient information is provided to the Ipopt optimizer using ForwardDiff, a Julia package that calculates analytical gradients for specified functions using forward mode automatic differentiation (Ref. 22).

A benefit of creating an optimization model in JuMP is the large degree of control it affords over optimizer settings. Convergence criteria are iteratively adjusted to achieve the desired balance between good convergence and low runtime. For example, equilibrium constraint function residuals should be at least 2 to 3 orders of magnitudes smaller than forces on the aircraft at convergence, but targeting even smaller residuals would unnecessarily extend runtime. The impact from a force imbalance on the order of 0.01 lbf is essentially unnoticeable to the 1,000+ lb RAVEN aircraft, so expending significantly more iterations to reduce this imbalance to 1e-6 lbf results in no practical improvement to the trim solution.

The iteration limit, or maximum number of iterations, was also tuned and decreased as the optimization model was refined. A quasi-steady segment at which the aircraft can be successfully trimmed will converge on the order of 10 iterations while segments that do not converge after 200 or so iterations consistently will not converge at all. The maximum iteration count is correspondingly set to 250 to minimize runtime spent on non-converging cases. Tuned optimizer settings are presented in Table 1.

Table 1: Ipopt optimizer settings.

Setting	Value
Maximum iterations	250
Convergence tolerance (relative)	1e-6
Maximum constraint violation (relative)	1e-2

Defining the Transition Segment

Two transition maneuvers are considered. Accelerating transition begins with the aircraft in the hover state and is completed once the aircraft is in the forward flight state. Conversely, decelerating transition begins with the aircraft in forward flight and ends in hover. Both maneuvers are flown purely in level flight with zero climb gradient.

The hover state corresponds to zero airspeed and the aircraft being configured for hover with tilt rotors oriented vertically at $\tau = 90^\circ$. The forward flight state requires that the tilt rotors be fully tilted forward to $\tau = 0^\circ$ and that forward airspeed exceeds the threshold V_{ref} of 85 knots true airspeed (KTAS). V_{ref} is defined as 1.3 times the stall speed V_{so}, which is equal to 65 KTAS with maximum flap deflection, to provide adequate stall margin.

Both maneuvers are modeled with the presumption of a constant $0.1g$ magnitude of acceleration. This acceleration is directed along the direction of flight for accelerating transition and opposite to the direction of flight for decelerating transition. This magnitude was selected to accelerate RAVEN to or from V_{ref} in the target transition time of 45 seconds.

Within the scope of this paper, flap deflection is always set to its maximum value of 30°. This maximizes wing-borne lift and reduces thrust requirements.

Rear Rotor Speed Schedule

The lift-only rear rotor must be spooled down to idle in the forward flight state, as it is no longer needed for lift and would destabilize the aircraft if allowed to continue running (Ref. 23). A speed schedule must be designed to spool the rear rotor down to 0 RPM in accelerating transition and spool it up to the nominal 1,850 RPM in decelerating transition. While the following discussion is focused on the RAVEN aircraft, it is still highly relevant to other eVTOL aircraft designs that utilize lift-only rotors.

Rear rotor angular speed, Ω_3, is scheduled as a function of airspeed, providing commonality with the transition path tilt schedules presented later in this paper. In accelerating transition, the angular speed is held at the nominal 1,850 RPM from hover until the aircraft reaches 50 KTAS. It is then decreased at a constant rate to reach 0 RPM at 85 KTAS, far above stall speed at which point the aircraft is essentially operating as a fixed-wing aircraft.

Spool-down is delayed until 50 KTAS to ensure the rear rotor meets the high thrust requirements at low airspeed without resulting in excessive blade loading or even blade stall. In this low-speed transition regime, low dynamic pressure reduces the wing's effectiveness and forces the rotors to carry a majority of the aircraft's weight. Much of this requirement falls on the rear rotor as the tilt rotors leverage less of their thrust in the lift direction once tilt angle decreases. Additionally, the rear rotor is required to counter the strong nose-up pitching moment produced by the tilt rotors, which both sit forward of the aircraft's center of gravity. The same low dynamic pressure that limits the wing also limits the tailplane's effectiveness. This reduces its ability to assist in trimming aircraft pitch. Trimmed low-speed transition thus requires significant thrust from the rear rotor. If Ω_3 is decreased in this regime, β_3 must increase to maintain the required thrust which increases blade loading, reducing rotor efficiency and increasing the risk of blade stall.

At higher airspeed, rear rotor RPM cannot continue to be held at nominal without destabilizing the aircraft. The rotor blades selected for RAVEN are highly cambered and thus will produce significant thrust if Ω_3 is not decreased during high-speed transition, even when β_3 is lowered. The nose-up pitching moment produced by the tilt rotors becomes far smaller as they enter the forward flight configuration, as very little if any of their thrust is directed in the lift direction at this point. Ω_3 thus must be decreased to reduce rear rotor thrust output and prevent an un-countered nose-down pitching moment.

The finalized rear rotor speed schedule shown in Figure 7 is the product of many iterations and provides the high thrust output required at low airspeed without negatively impacting controllability at high airspeed.

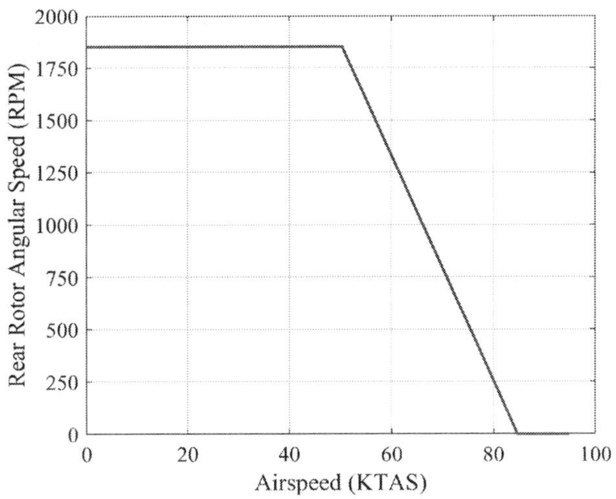

Figure 7: Rear rotor speed schedule for the RAVEN aircraft.

Analysis Methods

The TTO is capable of two main types of analysis. Fixed-α sweeps provide insight into the sensitivity of transition paths to angle of attack and characterize the lower bound of allowable tilt angles. Transition path optimization develops control profiles including tilt angle schedules that minimize sum-squared thrust at quasi-steady points through the transition maneuver.

Fixed-Angle of Attack Sweeps Fixed-α sweeps are produced by calculating the optimal trim solution at a constant angle of attack over the range of airspeeds spanning the transition maneuver. This process is repeated for angles of attack from -20° to 30°, which comfortably encompasses the range of α that the aircraft will operate at. This effectively reduces the number of control variables that can be adjusted by the optimizer by 1, as all control solutions must trim the aircraft at the defined α.

Non-converging operating points indicate conditions for which the aircraft cannot be trimmed and are omitted. A fixed-α path that abruptly ends at $V_\infty = 85$ KTAS, for example, indicates that the aircraft cannot be trimmed above that speed while holding the specified angle of attack.

The lower limit of tilt angles for transition is obtained based on the method by Stoll and Mikić (Ref. 5) by using fixed-α sweeps to derive the stall boundary. A maximum α of 12° is chosen based on the stall onset α for the wing to provide a 25% margin to stall. The fixed-α tilt schedule at $\alpha = 12°$ becomes the lower bound for transition paths. Flying at tilt angles below this region during transition risks stalling the wing and potentially placing the rear rotor in a region of separated flow. Since the transition path flown by RAVEN represents the nominal condition, at no point in transition should the scheduled tilt angle fall below this lower bound.

Transition Path Optimization Optimal transition paths are comprised of the optimal trim solutions at each quasi-steady operating point through the transition segment. Unlike fixed-α sweeps, angle of attack is allowed to vary within its prescribed limits, providing an additional degree of freedom that allows the optimizer to further minimize sum-squared thrust. The paths presented in this paper are specifically calculated over a range of airspeeds from 0 to 95 KTAS in 0.5 KTAS increments. The resulting tilt angle, angle of attack, elevator deflection, and rotor thrusts corresponding to each optimal trim solution are then plotted at these airspeeds to present optimal control schedules.

Each optimal tilt schedule is co-plotted with the stall bound developed from its corresponding fixed-α sweep to confirm it provides ample margin to stall. An upper bound reflecting structural limitations on the airframe will also be shown, though this is not currently used to constrain the optimal path. When the Transition Trim Optimizer was developed, the RAVEN project was still in the conceptual design phase and had not reached a level of maturity at which exact structural constraints could be defined. Instead, placeholder values were selected to align with the expected orders of magnitude for maximum blade bending moment, sponson boom twist, and tilt actuator loading.

Rotor and airframe loads were calculated at each operating point in the transition space over a range of possible aircraft states that could occur in transition. The worst-case loading at each point was then compared against the placeholder constraints to create contours of maximum structural loads on a plot of tilt angle versus airspeed. These constraint contours were combined to create an upper boundary for tilt angle. Once detailed design and structural analysis for RAVEN are completed, the placeholders can be replaced with actual airframe structural limits and the resulting upper bound will become a limit on maximum tilt angle.

During preliminary design, this upper bound can instead be used to inform structural requirements for component selection. eVTOL aircraft such as the RAVEN aircraft are very power-sensitive, so the energy requirement penalty for flying off the optimal path can be very high. It is thus desirable to design an airframe and rotor systems capable of withstanding the loads experienced during optimal transition rather than flying a sub-optimal transition path. The placeholder constraints were iteratively adjusted to produce an upper bound with the desired margin to both the optimal accelerating and decelerating transition paths, resulting in the final values shown in Table 2. Critical components can then be evaluated against these constraints during detailed design.

Table 2: Placeholder structural constraints for the RAVEN aircraft.

Constraint	Limit
Blade root bending moment	500 lb-ft
Sponson boom twist	250 lb-ft
Tilt actuator loading	200 lb-ft

A deck angle constraint is applied to capture passenger comfort requirements, an aspect critical to the success of passenger-carrying UAM operations. Such a constraint has precedent in existing work, such as the deck angle constraint used by Orndorff et al (Ref. 4) for this same reason. The deck angle constraint is implemented in the Transition Trim Optimizer by limiting aircraft pitch attitude, θ, to +/-5°. Since only level flight conditions with no wind are considered for the presented results, θ will be equivalent to α. Thus, α_{min} is set to -5° and α_{max} to 5° in the optimization problem formulation to model the deck angle constraint.

RESULTS AND ANALYSIS

Transition space characterization for the RAVEN aircraft is discussed in Section *Fixed-Angle of Attack Sweeps* and stall boundaries derived from fixed-α sweeps are presented for both accelerating and decelerating transition segments. The two sections following this present optimal accelerating and decelerating transition paths and discuss the factors driving their behavior.

Fixed-Angle of Attack Sweeps

Fixed-α tilt schedules ranging from α = -12° to α = 18° are shown in Figure 8 and illustrate the relationship between angle of attack and optimum tilt angle in trimmed flight. Transition paths fixed to higher angles of attack correspond to lower tilt angles at any given airspeed, as the higher α results in greater wing-borne lift. The tilt rotors in turn are required to produce less lift and can thus tilt further forward to instead provide forward thrust. The inverse is true as well, with lower-α paths requiring higher tilt angles for a fixed airspeed to offset the loss in lift resulting from a less effective wing.

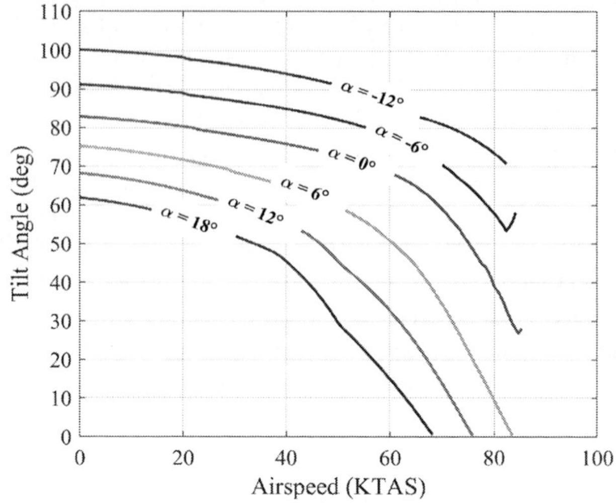

(a) Accelerating transition

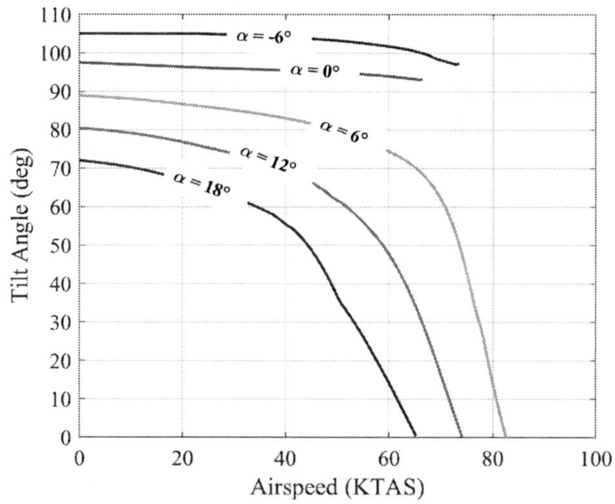

(b) Decelerating transition

Figure 8: Fixed-α sweeps for the RAVEN aircraft.

At a given angle of attack, the RAVEN aircraft favors lower tilt angles in accelerating transition than in decelerating transition. During accelerating transition, this leverages a greater component of the tilt rotors' total thrust toward forward thrust, which is required to both overcome drag and achieve the prescribed $0.1g$ forward acceleration. In decelerating transition, the tilt rotors begin at higher tilt angles as forward thrust requirements are now much lower. Forward thrust is only needed to overcome a portion of drag while the remaining drag is used to decelerate the aircraft. As forward airspeed and drag decrease, the tilt rotors rotate aft past τ = 90° to directly provide the net rearward force required to maintain deceleration.

Transition paths fixed to negative angles of attack consistently terminate prematurely at high speed, indicating the aircraft cannot be trimmed at flight conditions featuring both high V_∞ and low α. High-speed transition is characterized by an increasing reliance on wing-borne lift, something that cannot be met at these negative angles of attack. Lift requirements cannot instead be fulfilled by rotor thrust, as the tilt rotors are needed to overcome drag while the rear rotor's reduced angular speed limits its effectiveness. Thus, the aircraft cannot sustain high-speed trimmed flight without increasing α and transferring lift generation to the wing.

Stall Bound Characterization As demonstrated in Figure 9, limit curves for stall are obtained by isolating the fixed-α tilt schedules for the selected maximum angle of attack of 12°. Hatched lines indicate the aircraft should avoid operating in the lower left-hand corner of the transition space. The low airspeed and tilt angles inherent to this regime reduce both wing effectiveness and the lift-producing component of thrust. Relying on thrust to balance the aircraft's weight would result in excessive power requirements and forward acceleration. Thus, trimmed flight becomes increasingly difficult and eventually impossible at operating points below the stall bound.

Optimal Accelerating Transition

The optimal accelerating transition path for RAVEN is presented in Figure 10. The optimal tilt schedule maintains an ample margin of tilt angle to the stall bound of at least 10°. By considering how this optimal control solution varies with airspeed, it indicates that the aircraft should behave as a multicopter at low speed and increasingly like a fixed-wing aircraft once airspeed exceeds V_{so}.

During low-speed transition and beginning in hover, the tilt rotors are near-vertical and the aircraft favors the most negative allowable angle of attack as shown in Figure 11a. Pitching the aircraft forward in this way allows the fixed rear rotor to provide both lift and forward thrust. Since the wing generates negligible lift due to the low dynamic pressure, rotor thrust is heavily relied upon for lift generation. The tilt rotors therefore maintain a high tilt angle to direct a larger component of their thrust vertically. Low dynamic pressure also results in near-zero elevator effectiveness, so aircraft pitch control is achieved primarily through differential thrust. This is evident as no two rotors provide the same thrust at low forward airspeed. Rear rotor thrust is notably highest, as this is the only rotor positioned aft of the center of gravity and thus must counteract the nose-up pitching moment from both tilt rotors. The elevator does maintain its maximum negative deflection to provide an additional, albeit small, nose-down pitching moment in support of the rear rotor. The limited effectiveness of aerodynamic surfaces in this part of the transition envelope causes the RAVEN aircraft to operate as a multicopter, maintaining pitch trim with differential thrust and pitching forward to accelerate.

As airspeed increases, dynamic pressure rises and wing effectiveness increases. Angle of attack rapidly increases to maximize wing-borne lift as it becomes optimal to transfer

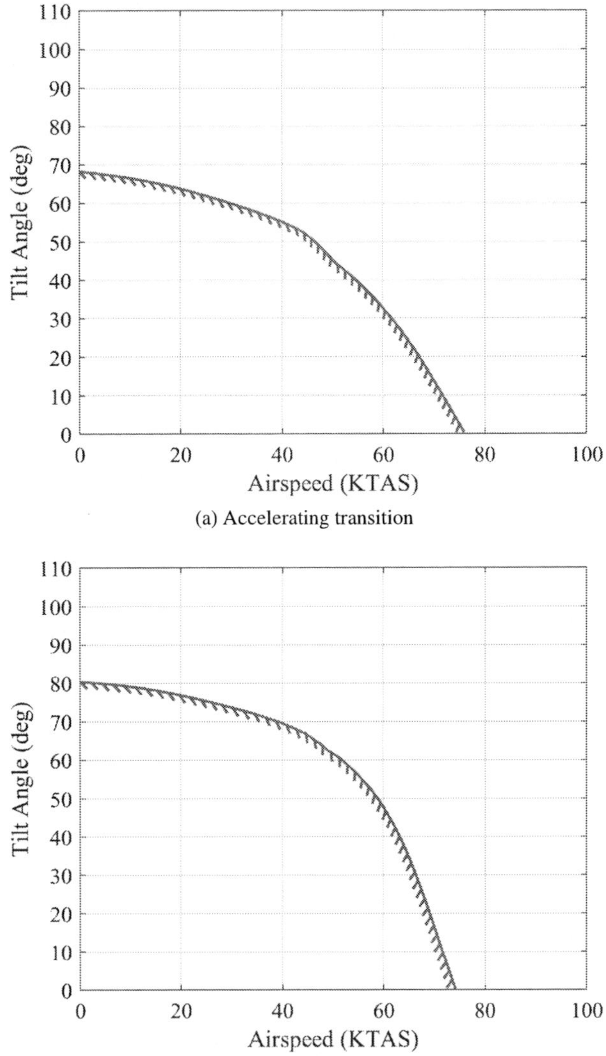

(a) Accelerating transition

(b) Decelerating transition

Figure 9: Transition stall limits for the RAVEN aircraft based on fixed-α tilt schedules at $\alpha = 12°$.

the required lift generation off the rotors and onto the wing. All three rotors reduce their thrust output as the tilt rotors rotate forward as illustrated in Figure 11b. This increases the forward component of their thrust output to counteract rising drag and maintain the required $0.1g$ acceleration. The thrusts from each tilt rotor converge as elevator effectiveness increases and differential thrust is no longer needed for pitch control. Maintaining a balanced thrust distribution is incentivized by the sum-squared thrust objective which has the added effect of minimizing the total induced power from both rotors. At the same time, the rear rotor begins to spool down which reduces its thrust output. Failing to decrease rear rotor RPM would create an aircraft pitch imbalance as the vertically-directed rear rotor thrust would impart a large nose-down pitching moment that can no longer be counteracted by the tilt rotors.

At sufficiently high speed, α becomes a direct function of airspeed. It decreases from its maximum of 5° with increasing V_∞ and thus dynamic pressure to maintain constant lift from

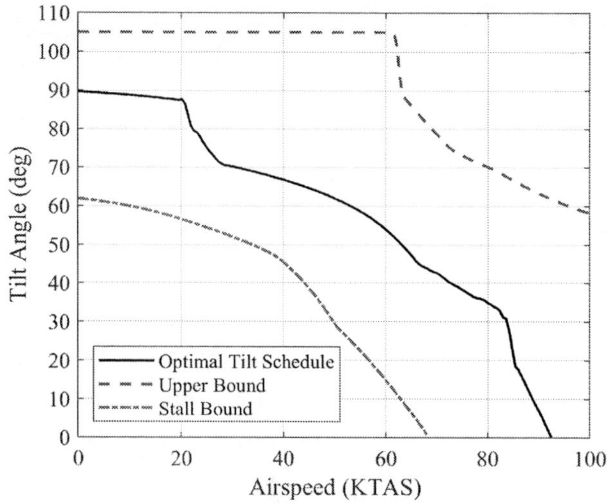

(a) Tilt schedule

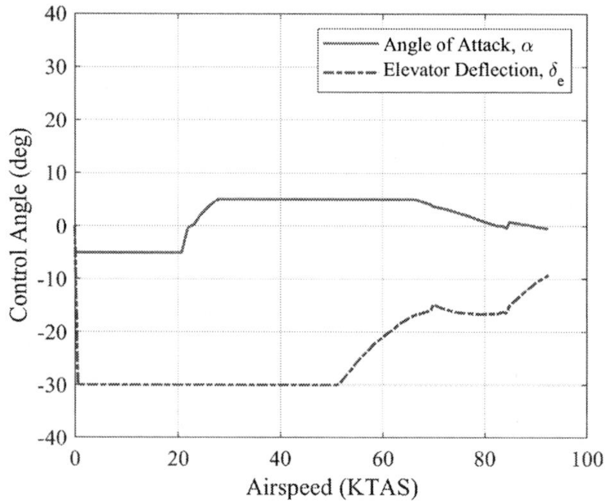

(b) Angle of attack and elevator deflection

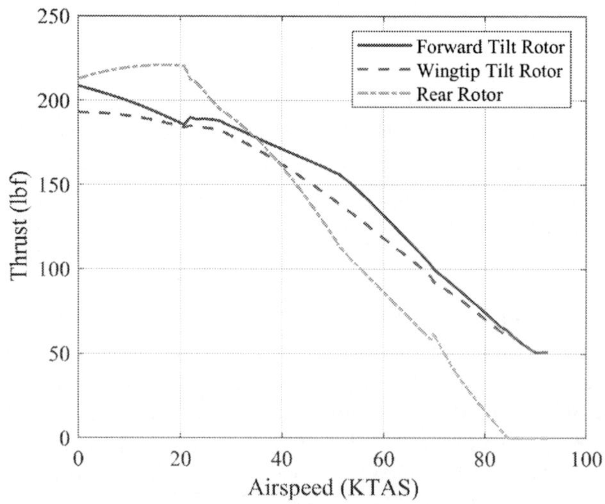

(c) Rotor thrust output

Figure 10: Control solution for the optimal accelerating transition path.

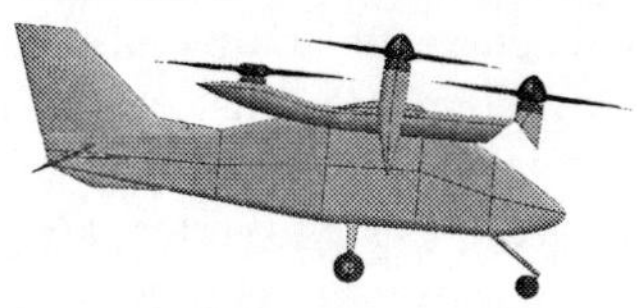

(a) $V_\infty = 20$ KTAS, $\alpha = $-5°, $\tau = 87°$

(b) $V_\infty = 70$ KTAS, $\alpha = 4°$, $\tau = 43°$

Figure 11: Optimal aircraft trim states at specified points in accelerating transition.

the wing, indicating the aircraft has transitioned to wing-borne flight. This behavior continues until the rear rotor fully spools down to 0 RPM and the tilt rotors reach their forward flight configuration at $\tau = 0°$, completing accelerating transition.

Fixed Rear Rotor Collective for High-Speed Transition Earlier during the development of the Transition Trim Optimizer, poor convergence was consistently observed at high speeds above 70 KTAS. This was traced to the rear rotor collective pitch, which would oscillate wildly between successive optimizer iterations. At these airspeeds, Ω_3 is very low since the rear rotor has nearly completed its spool-down. The spool-down limits rear rotor thrust output as intended but also minimizes any effect the control variable β_3 has on the sum-squared thrust objective. Thus, there are likely multiple locally optimal trim solutions at each airspeed in high-speed transition, each with a significantly different β_3 value. These locally optimal solutions prevent the optimizer from converging on a single trim solution.

To resolve this issue, β_3 is fixed to 5° when Ω_3 falls below a threshold speed of 800 RPM. This occurs at 70 KTAS per the speed schedule presented in *Rear Rotor Speed Schedule*. The threshold speed has been iteratively selected and represents the lowest allowable angular speed that consistently prevents poor convergence due to fluctuating β_3. This approach maximizes the time in which all six control variables, including rear rotor collective, can be adjusted by the optimizer without sacrificing convergence in high-speed transition. The 5° fixed collective is selected based on the optimal β_3 immediately before the threshold RPM is crossed. This significantly improves convergence in high-speed transition.

Optimal Decelerating Transition

The optimal decelerating transition path is shown in Figure 12. Like in the accelerating case, the tilt schedule maintains a 10° minimum margin of tilt angle to the stall bound. The aircraft similarly operates as a fixed-wing aircraft at high speed and as a multicopter at low speed. Beginning in high-speed transition above V_{so}, the tilt rotors produce the same thrust output to minimize the objective function because elevator deflection rather than differential thrust provides aircraft pitch control. Compared to high-speed accelerating transition, thrust requirements are lower as less forward thrust is required to achieve the prescribed $-0.1g$ rearward acceleration.

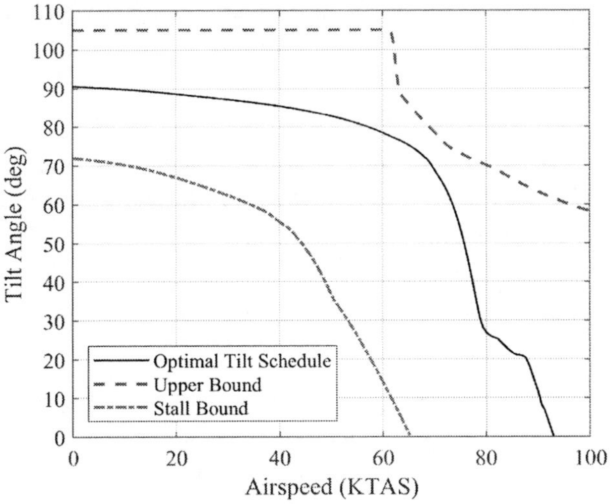

(a) Tilt schedule

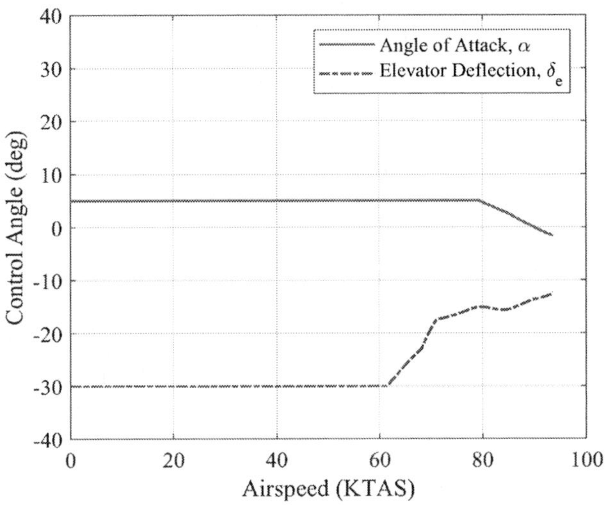

(b) Angle of attack and elevator deflection

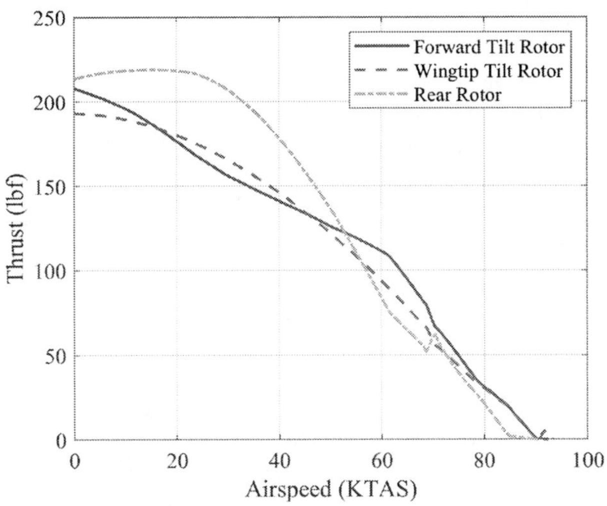

(c) Rotor thrust output

Figure 12: Control solution for the optimal decelerating transition path.

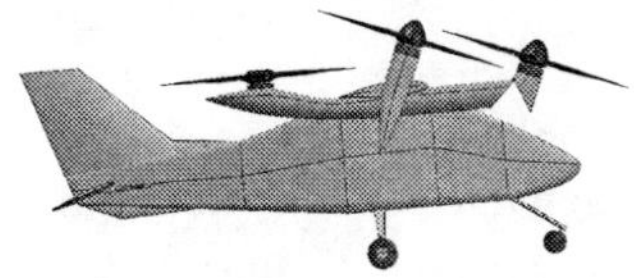

(a) $V_\infty = 70$ KTAS, $\alpha = 5°$, $\tau = 68°$

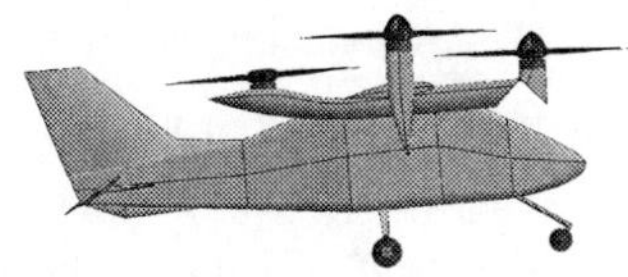

(b) $V_\infty = 20$ KTAS, $\alpha = 5°$, $\tau = 88°$

Figure 13: Optimal aircraft trim states at specified points in decelerating transition.

Angle of attack increases as the aircraft decelerates until it is limited by the deck angle constraint at 5°. This state is shown in Figure 13a. As was done for the accelerating path, β_3 is fixed to 5° for rear rotor speeds below the 800 RPM threshold.

The aircraft pitches backward with decreasing airspeed to maximize angle of attack and thus wing-borne lift while τ sharply increases to raise the tilt rotors near-vertical as shown in Figure 13b. This attitude also leverages the rear rotor for rearward thrust to decelerate. Unlike in accelerating transition, the aircraft continues to maintain this positive α at low speed, following the tendency of multicopters to pitch nose-up when accelerating rearward. Similar to accelerating transition, however, is the use of a differential thrust for trim pitch attitude at these low speeds. This is evident in the varied thrust outputs and higher rear rotor thrust relative to the tilt rotors. These trends continue until RAVEN reaches the hover state with the tilt rotors configured at $\tau = 90°$ with zero airspeed, completing decelerating transition.

CONCLUSIONS

1. A quasi-steady method for characterizing the transition envelope and calculating transition paths for multi-tiltrotor aircraft has been developed. The approach has been implemented for the RAVEN aircraft configuration. Rotor and airframe aerodynamics are modeled using surrogate models, which were fitted to data generated using inviscid methods in conjunction with viscous corrections. Deficiencies in the available rotor power data motivated the development of a thrust-based objective function as an approximate surrogate for power requirements. This approach was implemented within a constrained optimization formulation and solved using a gradient-based optimizer. A download model was created to capture simple aspects of interactional aerodynamics and calibrated to experimental data from similarly configured aircraft.

2. The optimization formulation has three primary limitations. First, it can calculate only local optima and struggles to converge at operating points where multiple similarly optimal trim solutions exist. Second, the reliance on a thrust-based objective prevents optimal trim solutions from being conclusively regarded as power-optimal. The latter can be resolved by obtaining an acceptable rotor power database. Finally, the method as currently implemented does not capture rotor-rotor aerodynamic interactions or rotor-wing interactions except for download. To

model these effects would require a higher-fidelity aerodynamics analysis code.

3. The method implements an approach in which the lift-only rear rotors are spooled down to 0 RPM during the transition to wing-borne flight to prevent them from creating a destabilizing pitching moment with increasing airspeed. However, such a spool-down can only begin at a sufficiently high airspeed, equal to 50 KTAS for RAVEN. During low-speed transition, the lift-only rotors must maintain their nominal RPM to meet lift and pitch attitude control requirements without risking blade stall. These findings are applicable to similarly configured multi-tiltrotor aircraft utilizing lift-only rotors.

4. Fixed-angle of attack sweeps were developed, revealing the aircraft cannot be trimmed in high-speed flight at negative angles of attack. Positive α is required to reach the forward flight configuration and transition to wing-borne flight.

5. During level low-speed transition, the control solution minimizing sum-squared thrust is driven by the lift-only rotors and mimics that of a multicopter. When accelerating, the aircraft pitches forward to leverage its lift-only rotors for forward thrust. Conversely, when decelerating, the aircraft pitches backward to direct the lift-only rotors aft to provide rearward thrust. Aircraft pitch attitude is controlled using differential thrust.

6. For both high-speed accelerating and decelerating transition, the optimal control solution favors high angle of attack to maximize wing-borne lift. Pitch attitude trim is maintained by an increasingly effective elevator rather than differential thrust, allowing tilt rotor thrust outputs to converge for minimal sum-squared thrust.

Author contact:
Tristan Huang tristanhuang2011@gmail.com
Dr. Brian German brian.german@aerospace.gatech.edu

APPENDIX A: AIRCRAFT CONFIGURATION PARAMETERS

Table 3: RAVEN aircraft specifications (Ref. 8).

Characteristic	Value
Gross weight	1140 lb
Wingspan	20 ft
Rotor diameter	5.68 ft
Wing loading	21.5 lb/ft^2
Disk loading	7.5 lb/ft^2

ACKNOWLEDGMENTS

The development of the Transition Trim Optimizer was funded by NASA under the National Institute of Aerospace (NIA) Cooperative Agreement 80LARC17C0004 activity NIA.COOP.05.201201.

At the time of publication of this paper, author Tristan Huang is an employee of The Boeing Company.

REFERENCES

1. Straubinger, A., Rothfeld, R., Shamiyeh, M., Büchter, K. D., Kaiser, J., and Plötner, K. O., "An overview of current research and developments in urban air mobility – Setting the scene for UAM introduction," *Journal of Air Transport Management*, Vol. 87, 2020, pp. 1–12. DOI: 10.1016/j.jairtraman.2020.101852.

2. Silva, C., Johnson, W., Antcliff, K. R., and Patterson, M. D., "VTOL Urban Air Mobility Concept Vehicles for Technology Development," AIAA Aviation 2018 Forum, Atlanta, GA, June 25-29, 2018.

3. Sagan, D. S., Lau, M. T., Krishnamurthy, P. S., Augustine, J. C., and German, B. J., "Minimum-Power Trim of Over-Actuated VTOL Aircraft in Transition," AIAA Aviation 2021 Forum, Virtual, August 2-6, 2021. DOI: 10.2514/6.2021-3216.

4. Orndorff, N. C., Scotzniovksy, L., Sarojini, D., Gill, H., Lee, S., Cheng, Z., Zhao, S., Mi, C., and Hwang, J. T., "Air-taxi transition trajectory optimization with physics-based models," AIAA SciTech 2023 Forum, National Harbor, MD, January 23-27, 2023. DOI: 10.2514/6.2023-0324.

5. Stoll, A., and Mikić, G. V., "Transition Performance Of Tilt Propeller Aircraft," Vertical Flight Society 78th Annual Forum Technology Display, Ft. Worth, Texas, May 10-12, 2022. DOI: 10.4050/F-0078-2022-17443.

6. Kang, N., Whidborne, J. F., Lu, L., and Enconniere, J., "Scheduled Flight Control System of Tilt-Rotor VTOL PAV," AIAA SciTech 2023 Forum, National Harbor, MD, January 23-27, 2023. DOI: 10.2514/6.2023-1530.

7. Chauhan, S. S., and Martins, J. R. R. A., "Tilt-Wing eVTOL Takeoff Trajectory Optimization," *Journal of Aircraft*, Vol. 57, 2020, pp. 93–112. DOI: 10.2514/1.C035476.

8. German, B. J., Jha, A., Whiteside, S. K. S., and Welstead, J. R., "Overview of the Research Aircraft for eVTOL Enabling techNologies (RAVEN) Activity," AIAA Aviation 2023 Forum, San Diego, CA, June 12-16, 2023. DOI: 10.2514/6.2023-3924.

9. Kawajir, Y., Margot, F., Vigerske, S., and Wachter, A., "Ipopt Documentation," `https://coin-or.github.io/Ipopt/index.html`, 2022.

10. Lubin, M., Dowson, O., Garcia, J. D., Huchette, J., Legat, B., and Vielma, J. P., "JuMP 1.0: recent improvements to a modeling language for mathematical optimization," *Mathematical Programming Computation*, Vol. 15, 2023, pp. 581–589. DOI: 10.1007/s12532-023-00239-3.

11. Montagnani, D., Tugnoli, M., Fonte, F., Zanotti, A., Syal, M., and Droandi, G., "Mid-Fidelity Analysis of Unsteady Interactional Aerodynamics of Complex VTOL Configurations," Technical report, Department of Aerospace Science and Technology, Politecnico di Milano, Italy, 2019.

12. Drela, M., "XFOIL: An Analysis and Design System for Low Reynolds Number Airfoils," *Low Reynolds Number Aerodynamics*, Vol. 54, 1989, pp. 1–12. DOI: 10.1007/978-3-642-84010-4.

13. Eilers, P. H. C., and Marx, B. D., "Flexible smoothing with B-splines and Penalties," *Journal of Statistical Science*, Vol. 11, (2), 1996, pp. 89–121. DOI: 10.1214/ss/1038425655.

14. Drela, M., "AVL Overview," `https://web.mit.edu/drela/Public/web/avl/`, 2004.

15. McDonald, R. A., "Advanced Modeling in OpenVSP," 16th AIAA Aviation Technology, Integration, and Operations Conference, Washington, DC, June 13-16, 2016. DOI: 10.2514/6.2016-3282.

16. Viterna, L., and Corrigan, R., "Fixed pitch rotor performance of large horizontal axis wind turbines," *Large Horizontal-Axis Wind Turbines*, Vol. 1, 1982, pp. 69–85.

17. Leishman, J. G., *Principles of Helicopter Aerodynamics*, Cambridge University Press, Cambridge, UK, second edition, 2006, pp. 55–110, 212–272.

18. Young, L. A., Lillie, D., McCluer, M., and Yamauchi, G. K., "Insights into Airframe Aerodynamics and Rotor-on-Wing Interactions from a 0.25-Scale Tiltrotor Wind Tunnel Model," AHS International Aerodynamics, Acoustics, and Test and Evaluation Specialists' Conference, San Francisco, CA, January 23-25, 2002.

19. McVeigh, M. A., "The V-22 Tilt-Rotor Large-Scale Rotor Performance/Wing Download Test and Comparison With Theory," Paper 97, 11th European Rotorcraft Forum, London, UK, September 10-13, 1985.

20. Rosenstein, H., and Clark, R., "Aerodynamic Development of the V-22 Tilt Rotor," 12th European Rotorcraft Forum, September 22-25 1986. DOI: 10.2514/6.1986-2678.

21. Jordon, D. E., Patterson, W., and Sandlin, D. R., "An Experimental and Analytical Method for Approximate Determination of the Tilt Rotor Research Aircraft Rotor/Wing Download," Technical Report NASA-CR-176970, National Aeronautics and Space Administration, San Luis Obispo, CA, 1985.

22. Revels, J., Lubin, M., and Papamarkou, T., "Forward-Mode Automatic Differentiation in Julia," `https://arxiv.org/abs/1607.07892`, 2016.

23. Bridgelall, R., T, A., and Tolliver, D. D., "Introducing an efficiency index to evaluate eVTOL designs," *Technological Forecasting and Social Change*, Vol. 191, 2023, pp. 1–14. DOI: 10.1016/j.techfore.2023.122539.

Aerodynamic Design and Analysis of an Autonomous eVTOL Using Computational Fluid Dynamics

Chunhua Sheng 1
Professor
The University of Toledo
Toledo, Ohio, USA

Qiuying Zhao 2
Associate Professor
The University of Toledo
Toledo, Ohio, USA

ABSTRACT

This study explores the aerodynamic performance of a conceptual multirotor eVTOL aircraft, inspired by the XPENG AeroHT X2 flying car. Computational Fluid Dynamics (CFD) simulations are employed to analyze the aircraft's behavior in both hover and forward flight conditions. A conservative sliding interface technique is employed to capture the relative motion between the rotors and fuselage, while high-order WENO schemes and a hybrid RANS-LES approach ensure accurate modeling of turbulent flows. The results of this research offer valuable insights for the development of future eVTOL designs, providing guidance on achieving efficient, stable, and safe urban air mobility solutions.

NOTATION

C_p	=	power coefficient ($\frac{P}{\rho A V_{tip}^3}$)
C_t	=	thrust coefficient ($\frac{T}{\rho A V_{tip}^2}$)
FM	=	figure of merit ($\sqrt{\frac{T^3}{2\rho A}}/P$)
M_a	=	tip Mach number (V_{tip}/a)
P	=	rotor power
T	=	rotor thrust
V_{tip}	=	rotor tip speed ($R\Omega$)
α	=	rotor/flight tilt angle

INTRODUCTION

As urban areas become increasingly densely populated, cities face growing challenges in transportation, pollution, and the demand for efficient, rapid transit solutions. Electric Vertical Take-Off and Landing (eVTOL) aircraft emerge as a transformative innovation, offering safe, environmentally friendly, and convenient urban air mobility. Unlike conventional aircraft, eVTOLs are specifically designed for vertical takeoff and landing, making them exceptionally well-suited for confined urban environments. This unique capability positions them as ideal solutions for applications such as urban commuting, emergency response, and seamless point-to-point transportation.

There are several prominent eVTOL designs [1], each offering unique advantages and facing distinct engineering challenges. The multi-rotor configuration, the simplest design, utilizes multiple fixed rotors without wings to generate vertical lift and enable maneuverability, with flight control achieved by adjusting rotor RPM. Notable examples include China's Ehang 216 and Xpeng AeroHT X2, Germany's Velocopter Volocity, and France's Airbus CityAirbus. Another design, the lift-cruise type, combines multi-rotors for vertical lift with a dedicated propeller for forward flight, blending VTOL capabilities with conventional aircraft performance. Examples of this type include Beta Technologies' Alia-250C and Wisk's Cora. The most advanced design is the vector thrust or tilt-rotor type, which employs rotatable propellers to transition between lift and cruise modes. Representative models include Joby Aviation's S4, Archer Aviation's Maker, and the Bell Nexus 4EX with ducted fans from the USA. Further details on these eVTOL configurations can be found in Ref. [1].

Achieving optimal aerodynamic performance is critical for all eVTOL designs, as it directly impacts lift generation, drag reduction, stability, and overall operational safety and feasibility in dense urban environments. Aerodynamic optimization presents significant engineering challenges, particularly in managing rotor interactions, ensuring stability during flight mode transitions, and reducing noise levels. Computational Fluid Dynamics (CFD) plays a vital role in addressing these challenges by simulating complex flow fields, analyzing interactions between multiple components, and enabling engineers to refine designs efficiently before physical prototyping and testing.

This study utilizes a high-fidelity 3D unstructured grid computational fluid dynamics (CFD) solver, U2NCLE [2], to examine a conceptual multi-rotor eVTOL operating in hover and forward flight mode. Through computational analysis, this study seeks to identify optimal aerodynamic design

---

Presented at the Vertical Flight Society's 11th Biennial Autonomous VTOL Technical Meeting, Phoenix, AZ, USA, Feb. 4-6, 2025.

parameters that improve the stability, efficiency, and safety of eVTOLs, ultimately supporting their integration into urban air mobility systems.

GEOMETRIC MODEL

Figure 1 illustrates the design of a multirotor eVTOL aircraft featuring four coaxial dual-rotors (eight individual propellers) and a fixed skid landing gear. The airframe measures approximately 17 feet in length, 16.7 feet in span width, and 4.6 feet in height, predominantly constructed from carbon materials to ensure a lightweight yet sturdy structure. The design empty weight of the vehicle is 794 pounds with a payload of 441 pounds. The coaxial rotors have a radius of 3.38 feet, designed to operate at a variable rotational speed from 0 to 2300 RPMs. Table 1 provides the geometric information of the conceptual eVTOL configuration.

The target disk loading for this eVTOL during hover flight is set at 4.3 lbf/ft². To meet the design objectives, a series of CFD simulations is conducted, encompassing both component-level and system-level analyses. These include studies of isolated counter-rotating coaxial rotors in hover, the fully installed eVTOL in hover, and the installed eVTOL in forward flight. These simulations are critical for optimizing aerodynamic performance and ensuring that the design meets the specified targets.

Figure 1. Illustration of a Conceptual eVTOL Configuration.

Table 1. Conceptual eVTOL Geometric Characteristics.

	English	Metric
Body Length	16.96 ft	5.17 m
Body Hight	4.6 ft	1.4 m
Span Width	16.73 ft	5.1 m
No. of Propellers	8	8
Blade Radius	3.38 ft	1.03 m
Solidity	0.03	0.03
Disk Loading	4.30 lbf/ft^2	21 kg/m^2
Empty Weight	794 lbf	360 kg
Payload	441 lbf	200 kg
Max Gross Weight	1,235 lbf	560 kg
Design Cruise Speed	80.78 mph	130 km/h

SIMULATION TOOL

The simulation tool used in this study is U^{2}NCLE [2], a high-fidelity finite-volume CFD solver based on unstructured grid technologies. This solver is designed to simulate three-dimensional unsteady viscous turbulent flows using a hybrid Reynolds-averaged Navier-Stokes (RANS) and Large Eddy Simulation (LES) approach [3]. To effectively capture the turbulent kinetic energy spectrum for hybrid RANS-LES simulations, it is crucial to minimize numerical dissipation errors of the scheme. This is achieved with a high-order Weighted Essentially Non-Oscillatory (WENO) scheme [4], achieving up to sixth-order accuracy.

For time-accurate solutions, the solver employs an implicit Newton's method with second-order temporal accuracy, utilizing symmetric Gauss-Seidel relaxations at each Newton step to solve the linear system of equations. Given the complex geometries and distributed electric propulsion systems typical of eVTOL configurations, advanced computational techniques are employed. These include multi-body relative moving grids and conservative data exchange at sliding interfaces, ensuring high-fidelity simulations of eVTOL systems across various flight scenarios [5].

MODEL SETUP

This study uses advanced computational simulations to evaluate a conceptual eVTOL design, focusing on three key aspects:

1. Hover performance of isolated coaxial dual rotors: This analysis examines thrust, power, and figure of merit (FM) across a range of rotor speeds. The goal is to determine the impact of rotor speed and separation distance on thrust and power consumption, with the most efficient FM point being chosen as the design baseline.

2. Hover performance of an installed eVTOL: This assessment measures the maximum vertical takeoff gross weight (MTGW) achievable at the optimal FM point, providing insight into the eVTOL's best hover capability.

3. Forward flight performance of an installed eVTOL: A trimmed analysis is performed, balancing rotor RPMs, adjusting tilt angles, and assessing performance at a design cruise speed to ensure efficient operation in forward flight.

Computational meshes for both isolated coaxial dual rotors and installed eVTOL were generated, with a mesh size of 7.7 million points for isolated dual rotors and 38.4 million points for a fully installed eVTOL. Table 2 provides the surface and volume mesh statistics for the isolated dual rotors and the installed eVTOL.

Table 2. Conceptual eVTOL Mesh Information.

CFD Mesh Counts	Coaxial Dual-Rotors	Fuselage Body	Installed eVTOL
Surface Mesh	795K	237K	3,283K
Volume Mesh	7,707K	4,490K	38,430K

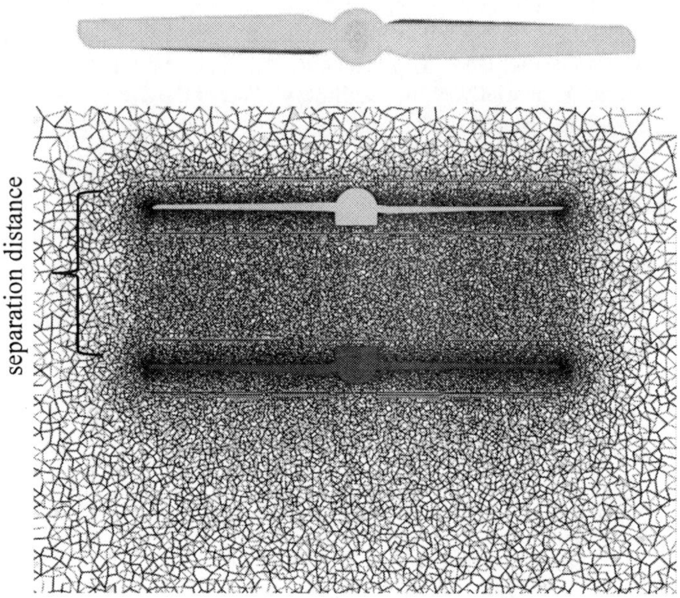

Figure 2. Coaxial Dual Rotors Volume Mesh Presentation.

The mesh for isolated dual rotors comprises two counter-rotating volumes—one for each rotor—and a stationary background volume to simulate the relative motion of the coaxial rotors. The upper rotor rotates counterclockwise, while the lower rotor rotates clockwise. A sliding mesh method is employed to manage the relative motion between the rotors, ensuring smooth flow field transitions across domains through conservative data exchange. Figure 2 shows a volume mesh presentation of the coaxial dual rotors.

The mesh for the installed eVTOL includes eight rotating volumes, one for each rotor, and a stationary volume for the eVTOL body and its components. Figure 3 shows the volume mesh distribution of the installed eVTOL from two cutting planes (top and front views). To accurately capture the complex trailing tip vortices and aerodynamic interactions, a refined mesh resolution is applied around the coaxial dual rotors.

RESULTS AND ANALYSIS

Isolated Coaxial Dual Rotors

The coaxial dual-rotor system is the primary power source under investigation for the current eVTOL design. Initial computational studies focus on isolated coaxial rotors in hover at various rotational speeds (tip Mach numbers). The primary objectives are to evaluate aerodynamic characteristics, including thrust, power, and figure of merit (FM), while assessing the rotor system's overall lifting capability for the vehicle.

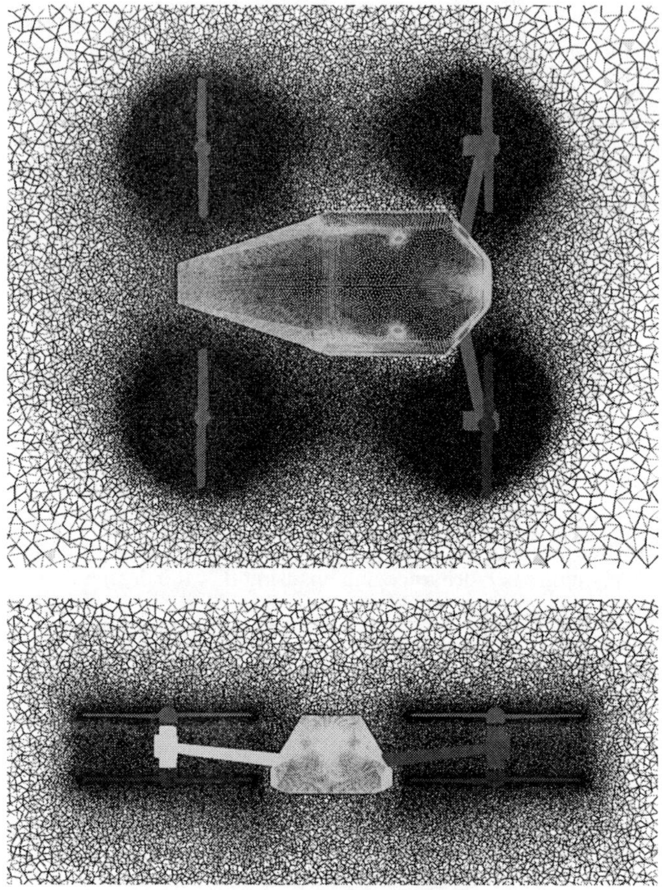

Figure 3. eVTOL Volume Mesh Presentations (top and front views).

The study also explores two rotor separation distances to analyze their impact on aerodynamic performance. As shown in Fig. 2, the dual-rotor system with a shorter separation distance is labeled as Configuration A, while the system with a larger separation distance is labeled as Configuration B. Although rotor separation distances are largely dictated by structural requirements, this analysis is critical for optimizing the vehicle's aerodynamic design and ensuring the rotor system meets performance targets for effective operation.

Figures 4–6 present numerical predictions of the rotor thrust coefficient, power coefficient, and figure of merit across a range of rotor tip Mach numbers for the two configurations. The results reveal notable performance differences in hover, particularly at lower tip Mach numbers ($Ma < 0.35$). For Configuration A (smaller separation distance), the thrust and power coefficients remain relatively stable as the tip Mach number changes. In contrast, Configuration B exhibits a sharp increase in both thrust and power coefficients at tip Mach numbers below 0.35. These variations result in distinct figure-of-merit trends for the upper and lower rotors of the two systems, as illustrated in Figure 6.

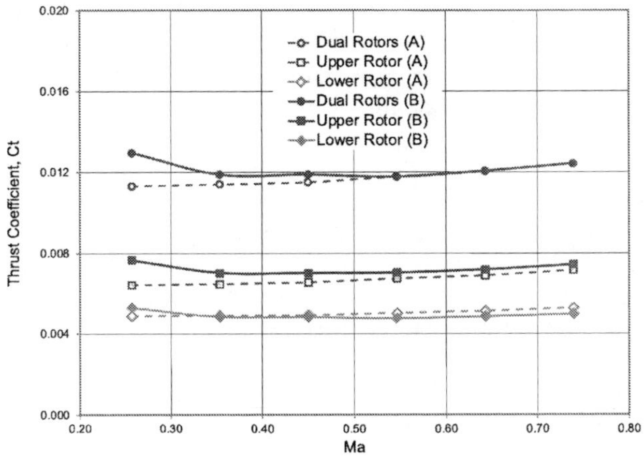

Figure 4. Predicted Thrust Coefficient at Different Tip Mach Numbers for Isolated Dual Rotors.

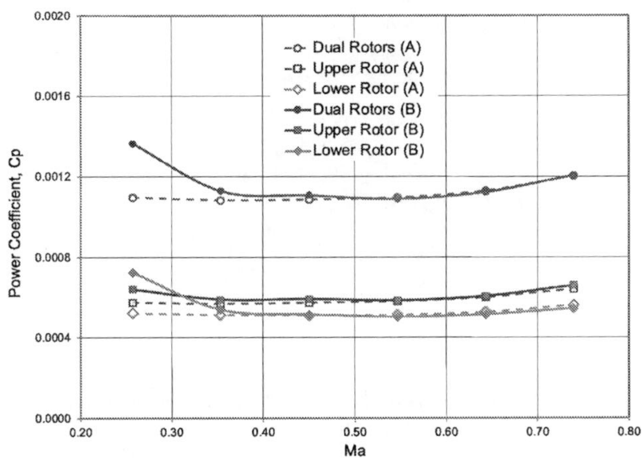

Figure 5. Predicted Power Coefficient at Different Tip Mach Numbers for Isolated Dual Rotors.

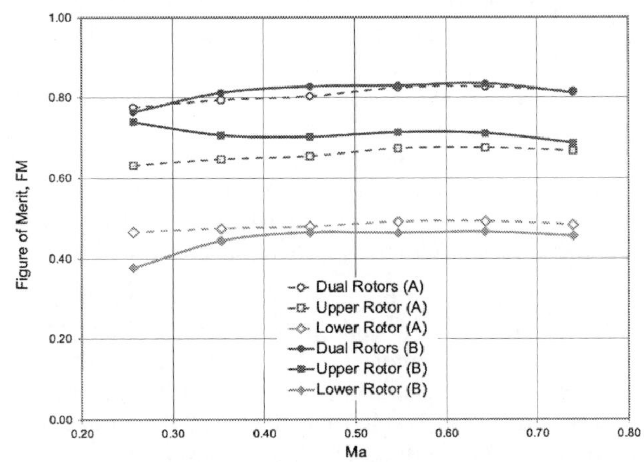

Figure 6. Predicted Figure of Merit at Different Tip Mach Numbers for Isolated Dual Rotors.

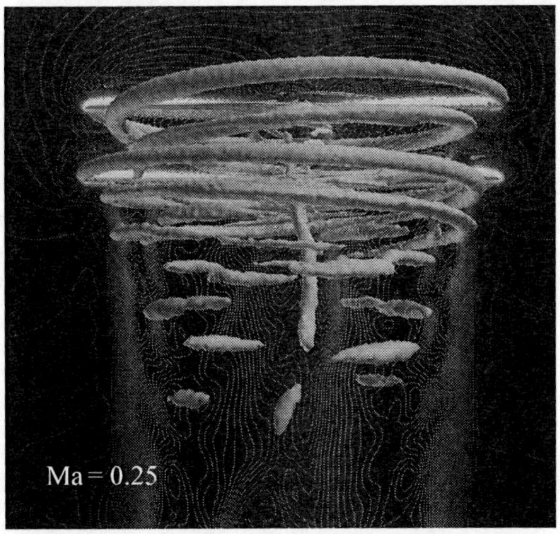

Figure 7. Tip Vortices Generated on Rotor System A with Smaller Separation Distance.

Figure 8. Tip Vortices Generated on Rotor System B with Larger Separation Distance.

These performance differences arise from the aerodynamic interactions between the upper and lower rotors, which are influenced by rotor speed and separation distance. Generally, a shorter rotor separation distance intensifies these interactions. However, a larger separation distance can create a wider range of flow disturbances affecting the lower rotor, leading to greater energy losses for the lower rotor.

Figures 7–8 illustrate the tip vortices generated by the two coaxial rotor systems at a low tip Mach number of 0.25. In rotor system B, which has a larger separation distance, the lower rotor is more significantly affected by the wake of the upper rotor. This is evident from the weaker tip vortices produced by the lower rotor compared to those in rotor system A. This increased sensitivity explains the decline in the lower rotor's figure of merit as rotor speed decreases.

Although the thrust and power coefficients differ notably at low speeds between the two systems, the actual thrust and power generated by the rotors are comparable. Figures 9 and 10 show the dimensional thrust and power predictions for both systems across varying rotor RPMs, indicating that rotor separation distance has minimal impact on lifting capacity and power consumption. Given the geometric constraints of the current eVTOL design, which necessitate a larger rotor separation distance, rotor system B is selected for subsequent installed eVTOL studies. However, operating at low rotor speeds should be avoided, as it can reduce the figure of merit and degrade the performance of the lower rotor.

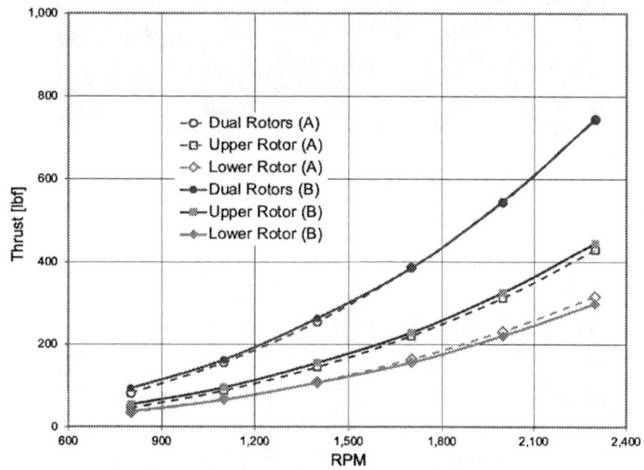

Figure 9. Predicted Thrust at Different Rotor Speeds for Isolated Dual Rotors.

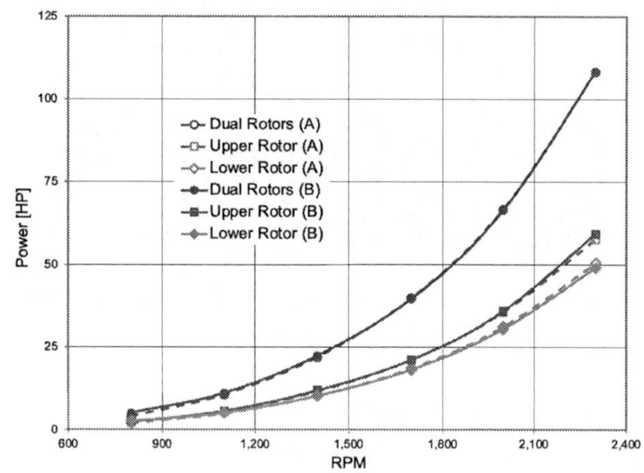

Figure 10. Predicted Power at Different Rotor Speeds for Isolated Dual Rotors.

Installed eVTOL in Hover

The performance analysis of the isolated coaxial rotor system reveals that rotor system B achieves its optimal efficiency (figure of merit) at tip Mach numbers ranging from 0.45 to 0.55, corresponding to rotational speeds of 1,400 to 1,700 RPM. Within this range, a single coaxial rotor system generates lift between 300 and 385 pounds.

The current eVTOL design has an empty weight of 794 pounds and a payload capacity of 441 pounds, resulting in a maximum gross weight requirement of 1,235 pounds. To meet this requirement, a dual-rotor system must produce at least 309 pounds of lift. Based on hover predictions for isolated coaxial rotors, a rotor speed of 1,550 RPM is sufficient to generate the necessary lift. However, due to aerodynamic interference between the coaxial rotors and the eVTOL's body, the actual rotor speed must be slightly increased to compensate for the download effect on the body.

In addition to meeting the maximum gross weight requirement, the multirotor design must minimize the torque generated by all rotors. This is achieved through careful selection of rotor rotation directions, as illustrated in Figure 11, where blue arrows denote clockwise rotation and yellow arrows indicate counterclockwise rotation. The chosen rotor configuration effectively cancels out moments in both the longitudinal and lateral directions of the eVTOL, as demonstrated in the following analysis.

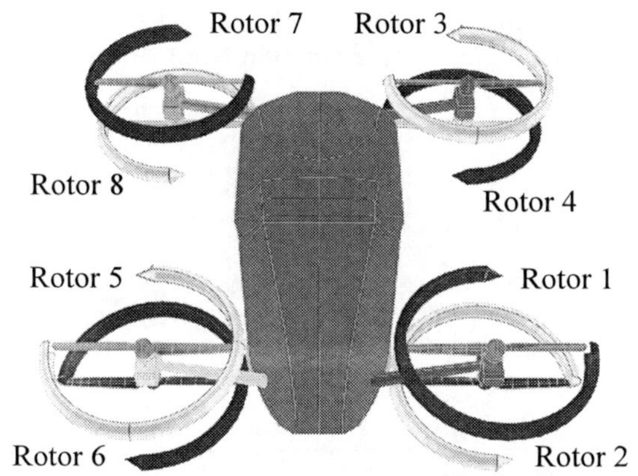

Figure 11. Illustration of Directions of Rotor Rotation in the Installed eVTOL.

To validate these assumptions, computational simulations are conducted on the installed eVTOL, which features four coaxial dual rotors (eight individual rotors). The time-accurate unsteady simulations are performed similarly to those of the isolated coaxial dual rotors, with a time step corresponding to a one-degree change in rotor azimuth angle per time step. Convergence is typically achieved after approximately 12 rotor revolutions when the lift and torque are stabilized into a periodic state. Figure 12 shows the convergence history of the normalized lift and download produced by the eVTOL and its components.

Table 3 summarizes the predicted hover performance of the installed eVTOL, including thrust, torque, and power consumption at three rotor speeds: 1,550, 1,650, and 1,700 RPM. During hover operations, the eVTOL tilt angle is set to zero. The simulation results reveal that at the initial rotor speed of 1,550 RPM, all rotors generate a total lift of 1,123 pounds, which is lower than the lift predicted for the isolated dual rotors at the same rotor speed. Furthermore, a negative download is produced on the eVTOL's body, resulting in a net lift (total gross weight) of 1,090 pounds at 1,550 RPM. This net lift is nearly 12% below the required lift needed to meet the maximum gross weight requirement of 1,235 pounds. However, the torque generated on the eVTOL and its body remains within single-digit values.

Table 3. Predicted eVTOL Hover Performance (α = 0°).

RPM 1,550	Lift [lbf | kg]	Torque [lbf-ft | kg-m]	Power [hp | kw]
eVTOL	1,090 | 495	-1.5 | -20	
Body	-33 | -15	-1.5 | -21	
All Rotors	1,123 | 510		117 | 88
RPM 1,650	**Lift [lbf | kg]**	**Torque [lbf-ft | kg-m]**	**Power [hp | kw]**
eVTOL	1,167 | 530	-0.5 | -7.1	
Body	-29 | -13	-0.5 | -6.6	
All Rotors	1,195 | 542		125 | 93
1,700 RPM	**Lift [lbf | kg]**	**Torque [lbf-ft | kg-m]**	**Power [hp | kw]**
eVTOL	1,261 | 572	-1.0 | -14	
Body	-17 | -7.5	-0.5 | -6.7	
All Rotors	1,278 | 580		131 | 98

The shortfall in net lift generated by the installed eVTOL is primarily due to aerodynamic interference caused by the eVTOL's structures on the rotor systems. Figure 13 illustrates the rotor tip vortices produced by the coaxial rotor systems, showing that most of the vortex structures are concentrated around the dual rotors. These vortices exhibit a complex pattern but break down and dissipate after a few rotor revolutions due to interactions with the surrounding structures. This interference is the primary factor leading to the degraded hover performance of the installed eVTOL compared to the isolated rotor case.

To meet the lift requirement of the eVTOL in hover, increasing the rotor speed is necessary. At 1,650 RPM, the total net lift improves to 1,167 pounds, representing a 6.6% increase over the previous rotor speed. However, this still falls 5.5% short of the target gross weight. Only at 1,700 RPM does the eVTOL achieve its design target for total gross weight, generating a total net lift of 1,261 pounds. At this speed, the upper rotors contribute approximately 60% of the total lift, while the lower rotors contribute 40%. The required rotor power in hover at this speed is 125 horsepower (93 kW).

While the eVTOL meets its maximum gross weight requirement at 1,700 RPM (corresponding to a rotor tip Mach number of 0.55), this operating condition may result in elevated noise levels. For urban air mobility applications, it is critical to minimize noise, which typically requires maintaining a rotor tip Mach number below 0.45. This discrepancy underscores a key design challenge: balancing lift requirements with noise constraints. To address this, future efforts should focus on advancing rotor system design to enhance lifting capability at lower rotor speeds. Additionally, optimizing the fuselage design to reduce rotor-vehicle interactions could improve aerodynamic efficiency and further lower noise levels, ensuring the eVTOL is both effective and environmentally suitable for urban operations.

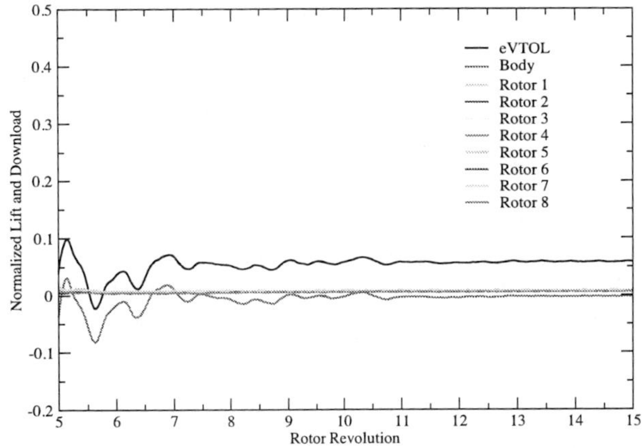

Figure 12. Lift and Download Convergence Histories of Installed eVTOL in Hover.

Figure 13. Simulation of Installed eVTOL in Hover.

Installed eVTOL in Forward Flight

The final phase of the eVTOL simulations focuses on evaluating its aerodynamic characteristics during level forward flight. In this phase, the eVTOL must generate sufficient lift to support its weight and payload while overcoming drag from the fuselage and its components, all while maintaining the designated cruise speed of 80.78 miles per hour (130 km/h). Achieving this balance requires precise control of the rotor systems to ensure stable and efficient flight.

Since the current eVTOL aircraft features the rotor systems with a fixed blade pitch angle, rotor speed is the primary control variable for adjusting flight status. By varying the speeds of the individual rotors, the aircraft can effectively maneuver, enabling forward movement, turns, and other flight adjustments. This capability is critical for following desired flight paths and maintaining stability under varying flight conditions.

To achieve this, the front and rear coaxial rotor systems can operate at different RPMs. This differential rotor speed configuration generates the necessary nose-down pitch moment and propulsive force, allowing the eVTOL to transition from hover to forward flight and maintain its cruise speed. This approach highlights the importance of rotor speed control in enabling the eVTOL's versatility and adaptability across different flight regimes.

Simulations for the installed eVTOL in level forward flight are conducted at three rotor speeds (1,350, 1,450, and 1,550 RPM) and two flight tilt angles (5° and 10°). In all scenarios, the rotors operate at the same speed, following the rotational pattern illustrated in Fig. 11. This investigation aims to evaluate the influence of rotor speed and flight tilt angle on the eVTOL's aerodynamic performance under cruise-level flight conditions.

By systematically varying these parameters, the study seeks to identify optimal operating conditions that balance lift, drag, and propulsion efficiency. The rotor speed directly affects the thrust generated by the rotors, while the tilt angle determines the distribution of forces between lift and forward propulsion. Understanding the interplay between these variables is critical for optimizing the eVTOL's performance, ensuring stable and efficient flight at the designated cruise speed of 80.78 miles per hour (130 km/h).

Figure 14 illustrates the convergence history of the normalized lift and download produced by the eVTOL and its components during forward flight. Unlike the relatively smooth lift profile observed during hover, the lift generated by the rotors in forward flight exhibits noticeable fluctuations. These fluctuations are influenced by the combined effects of rotor speed and flight tilt angle, which alter the aerodynamic forces acting on the rotors and fuselage.

Figure 15 shows the rotor tip vortices generated by the coaxial rotor system in forward flight. In this phase, the interaction between the vortices from the front and rear rotors becomes more pronounced, leading to complex flow patterns and increased aerodynamic interference. This interaction significantly amplifies the lift fluctuations observed in Figure 14, highlighting the challenges of maintaining stable and efficient performance during forward flight.

Tables 4 and 5 summarize the predicted lift, propulsive force (or drag), and power consumption of the eVTOL at flight tilt angles of 5° and 10°, respectively. The simulation results reveal that both rotor speed and flight tilt angle significantly influence the aircraft's lift and propulsive force during forward flight. Notably, the lift generated in forward flight is substantially greater than that in hover, even at the same rotor speed. This increase is attributed to the combined effects of wind speed and flight tilt angle, which alter the rotor's effective pitch angle and, consequently, its aerodynamic characteristics.

Additionally, during forward flight, a portion of the rotor thrust is redirected to generate propulsive force, which is essential for overcoming the drag produced by the vehicle's body. This redirection highlights the critical role of the flight tilt angle in balancing lift and propulsion. By adjusting the tilt angle, the eVTOL can optimize the distribution of thrust between lift and propulsive force, ensuring sufficient lift to support the vehicle's weight and payload while effectively countering drag.

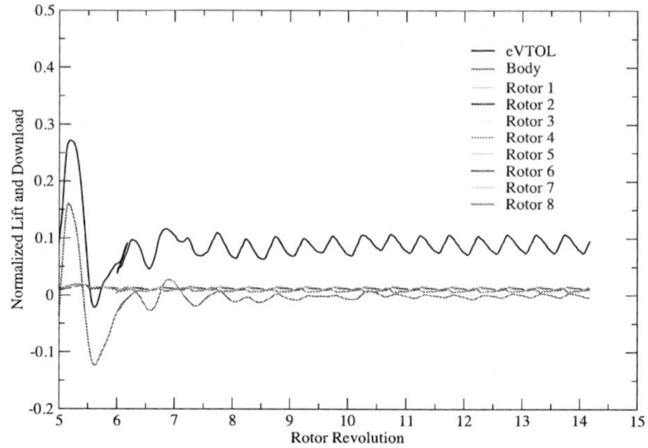

Figure 14. Lift and Download Convergence Histories of Installed eVTOL in Level Forward Flight.

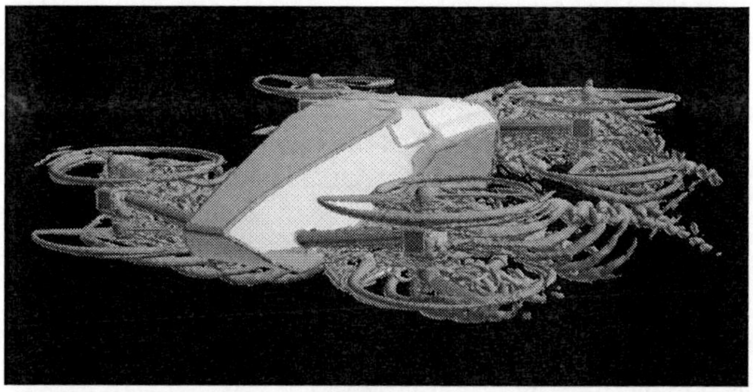

Figure 15. Simulation of Installed eVTOL in Forward Flight.

At a 5° tilt angle, both lift and propulsive force generated by the rotors increase with rotor speed. For instance, at 1,350 RPM, the eVTOL generates 1,385 pounds of lift, which increases to 1,579 pounds at 1,550 RPM. Similarly, the propulsive force produced by the rotors rises from 4.6 pounds at 1,350 RPM to 29 pounds at 1,550 RPM (with the negative drag value indicating propulsive force). However, even at the higher rotor speed of 1,550 RPM, the propulsive force remains insufficient to fully overcome the drag generated by the eVTOL during cruise forward flight.

The simulations also reveal that the interaction between the front rotor's wake and the rear rotors creates unfavorable aerodynamic conditions at 1,450 RPM. At this specific rotor speed and tilt angle combination, the rotors generate drag instead of propulsive force, highlighting the sensitivity of the system to rotor speed and flight tilt angle.

When the flight tilt angle increases to 10°, the overall lift and drag generated by the eVTOL decrease, leading to improved propulsive force and a more balanced distribution of lift compared to the 5° tilt angle. For instance, the propulsive force generated by the rotors increases significantly to 80 pounds at 1,350 RPM and 129 pounds at 1,550 RPM. Despite this improvement in propulsive force, the net drag on the eVTOL remains substantial, exceeding 137 pounds under these flight conditions.

Table 4. Predicted eVTOL Forward Flight Performance ($\alpha = 5°$).

RPM 1,350	Lift [lbf | kg]	Drag [lbf | kg]	Power [hp | kw]
eVTOL	1,385 | 628	242 | 110	
Body	-7.4 | -3.3	245 | 111	
All Rotors	1,386 | 629	-4.6 | -2.1	102 | 76
RPM 1,450	**Lift [lbf | kg]**	**Thrust/Drag [lbf | kg]**	**Power [hp | kw]**
eVTOL	1,537 | 698	258 | 117	
Body	-2.6 | -1.2	260 | 118	
All Rotors	1,425 | 647	8.6 | 3.9	118 | 88
RPM 1,550	**Lift [lbf | kg]**	**Drag [lbf | kg]**	**Power [hp | kw]**
eVTOL	1,579 | 717	202 | 92	
Body	6.5 | 2.9	230 | 105	
All Rotors	1,573 | 714	-29 | -14	111 | 83

Table 5. Predicted eVTOL Forward Flight Performance ($\alpha = 10°$).

RPM 1,350	Lift [lbf | kg]	Drag [lbf | kg]	Power [hp | kw]
eVTOL	1,021 | 463	202 | 92	
Body	-106 | -48	280 | 127	
All Rotors	1,121 | 509	-80 | -37	101 | 75
RPM 1,450	**Lift [lbf | kg]**	**Drag [lbf | kg]**	**Power [hp | kw]**
eVTOL	1,227 | 575	156 | 71	
Body	-73 | -33	272 | 124	
All Rotors	1,342 | 609	-116 | -53	118 | 88
RPM 1,550	**Lift [lbf | kg]**	**Drag [lbf | kg]**	**Power [hp | kw]**
eVTOL	1,470 | 667	137 | 62	
Body	-62 | -28	276 | 125	
All Rotors	1,540 | 699	-139 | -63	135 | 101

The most favorable flight condition occurs at a rotor speed of 1,450 RPM, where the eVTOL generates 1,227 pounds of lift and experiences a drag of 156 pounds. In this configuration, the rotor power consumption is 118 horsepower (88 kW). While this condition represents an improvement in propulsive efficiency, the persistent drag highlights the ongoing challenge of optimizing aerodynamic performance during forward flight.

These findings underscore the importance of identifying the optimal rotor speed and flight tilt angle to ensure stable operation. By carefully balancing these parameters, the eVTOL can achieve efficient performance during level forward flight, meeting the demands of urban air mobility applications. This analysis provides insights for refining the eVTOL's design and control strategies, ultimately enhancing its aerodynamic stability and operational efficiency.

CONCLUSIONS

This study investigated the aerodynamic design of a conceptual multirotor eVTOL aircraft, with a focus on both hover and cruise-level forward flight conditions. Computational analyses were performed to evaluate the influence of several factors, including rotor speeds (RPMs), rotor separation distances, and flight tilt angles, on the thrust and power output of the conceptual vehicle. The simulation results provided key insights for further design optimization:

1. Hover performance and noise considerations: For the current eVTOL configuration, a rotor speed of 1,700 RPM is required to meet the gross weight target during hover. However, this operating condition results in slightly elevated noise levels. Optimization of the coaxial rotor profiles is necessary to improve hover performance while balancing noise reduction, ensuring compliance with urban air mobility requirements.

2. Drag reduction in forward flight: Forward flight analysis reveals that the majority of drag is generated by the eVTOL's body, significantly impacting overall performance. Substantial improvements in the aerodynamic design of the fuselage and other structural components are essential to reduce profile drag and enhance forward flight efficiency.

3. Rotor speed and tilt angle optimization: Both rotor speed and tilt angle critically influence the eVTOL's lifting capability and propulsive force during cruise-level forward flight. Selecting an optimal combination of these parameters is crucial to ensure efficient operation, minimize adverse aerodynamic effects, and achieve stable forward flight performance.

4. CFD tool for design and optimization: The current CFD tool serves as a valuable resource for preliminary design assessments and further optimizations. By enabling high-fidelity simulations of complex eVTOL's aerodynamic interactions, the tool can significantly mitigate the risks and costs associated with eVTOL prototyping, accelerating the development process.

In conclusion, the performance of the eVTOL can be enhanced by optimizing rotor speed, tilt angle, and rotor system design to achieve efficient, low-noise, and high-lift flight capabilities, making it suitable for urban air mobility.

Author Contact: Chunhua Sheng 1 chunhua.sheng@utoledo.edu, Qiuying Zhao 2 qiuying.zhao@utoledo.edu

REFERENCES

1 eVTOL Aircraft Directory, https://evtol.news/aircraft.
2 Sheng C., A preconditioned method for rotating flows at arbitrary Mach number, *Modeling and Simulation in Engineering,* ISSN: 1687-5591, Vol. 2011, January 2011, Article ID 537464, doi:10.1155/2011/ 53746.
3 Sheng C., Zhao Q., and Khan R., Assessment of hybrid RANS-LES methods for simulating viscous flows, 2023 AIAA SciTech Forum, 23-27 January 2023, National Harbor, MD.
4 Zhong D., Sheng C., A new method towards high-order WENO schemes on structured and unstructured grids, Computers and Fluids 200 (2020) 104453.
5 Sheng C., Zhao Q., Hybrid RANS-LES simulations for HVAB rotor in hover and descent flight, 2025 AIAA SciTech Forum, 6-10 January 2025, Orlando, FL.

Airfoil Table Generation Comparison Utilizing XFOIL and UNS2D Flow Solvers

Gianmarco Sahragard-Monfared
Aerospace Engineer
NASA Ames Research Center
Aeromechanics Office
Moffett Field, CA, USA

Isaac Cavazos
Student
California State University, Sacramento
Dept. of Mechanical Engineering
Sacramento, CA, USA

Joseph Gabriel Estrada
Student
California State University, Sacramento
Dept. of Mechanical Engineering
Sacramento, CA, USA

Dalton Hall
Student
California State University, Sacramento
Dept. of Mechanical Engineering
Sacramento, CA, USA

Sarah Maadarani
Student
California State University, Sacramento
Dept. of Mechanical Engineering
Sacramento, CA, USA

Min Kaung Myat
Student
California State University, Sacramento
Dept. of Mechanical Engineering
Sacramento, CA, USA

ABSTRACT

The flow solvers XFOIL and UNS2D were utilized within the airfoil table generator AFTGen to generate airfoil performance tables, and these results were compared to determine accuracy at low Mach numbers that may be relevant to Urban Air Mobiltiy (UAM) applications. Airfoil tables were generated for the NACA 0012 airfoil as well as a set of three asymmetrical airfoils developed in this study from which the rotor blade CSUS 001 was comprised. To determine which solver more accurately represented real-world aerodynamics, the airfoils were experimentally tested in the CSU 2-foot by 2-foot wind tunnel. The results of XFOIL and UNS2D were compared to the experimental results. It is important to note that due to differences in the Reynolds numbers between simulations and experimental testing, the magnitude of the section lift and drag coefficient varies between the two methods. However, observations on the accuracy of XFOIL and UNS2D could still be made based on the trends of the data. For the symmetrical NACA 0012 airfoil, it was found that XFOIL and UNS2D accurately predicted the trends of the section lift and drag coefficients. However, XFOIL predicted slightly lower section drag coefficients than UNS2D for the tested angles of attack. For the asymmetrical CSUS 001 airfoils, XFOIL could not predict a steady to decreasing section lift coefficient phenomenon at negative angles of attack observed in both the UNS2D and experimental results. Similarly, XFOIL was unable to predict an increased section drag coefficient at negative angles of attack for the CSUS 001 airfoils as observed in the UNS2D and experimental results. Additionally, 3D computational fluid dynamics was utilized through RotCFD to compare the performance of a rotor blade using the CSUS 001 airfoils versus a rotor blade using the NACA 0012 airfoil.

NOTATION

A	surface area of rotor
α	angle of attack, degrees
C	airfoil chord length
c_d	section drag coefficient, c_d=$(2F_D)/(\rho v^2 A)$
c_l	section lift coefficient, c_l=$(2F_L)/(\rho v^2 A)$
c_m	section moment coefficient, c_m=$(2F_M)/(\rho v^2 Al)$
C_P	power coefficient
C_T	thrust coefficient
C_T/σ_s	blade loading, $C_T/\sigma_s = C_T/(N_B*c/(2\pi R))$
CFD	computational fluid dynamics
$Deg.$	collective angle
F_D	drag force, Newtons, $F_D = c_d * \rho(v^2/2)A$
F_L	lift force, Newtons, $F_L = c_l * \rho(v^2/2)A$
F_M	moment force, Newtons, $F_M = c_m * \rho(v^2/2)Al$
FM	figure of merit, $FM = C_T^{(3/2)}\sqrt{2})/C_P$
H	shape parameter at the end of wake
l	reference length, meters
M	Mach number
m	maximum camber
M_{Tip}	tip Mach number
N_B	number of blades
P	pressure, pascals
p	location of maximum camber
q	total pressure (Bernoulli)
r	radial station of specified airfoil
R	overall radius of entire blade
Re	Reynolds number, $Re = \rho vl/\mu$
ρ	fluid density around airfoil, kilograms per cubic meter
σ_s	solidity of the blade, dimensionless, $\sigma_s=(N_B*c)/(\pi R)$
T	temperature, Kelvin

Presented at the Vertical Flight Society's 11th Biennial Autonomous VTOL Technical Meeting, Phoenix, AZ, USA, Feb. 4-6, 2025.

t	maximum thickness, percent of chord
θ	momentum thickness at end of wake
θ_i	momentum thickness downstream
u	edge velocity at end of wake, meters per second
v	fluid velocity around airfoil, meters per second
V	freestream velocity, meters per second

INTRODUCTION

Urban Air Mobility (UAM) efforts are currently driving research initiatives aimed at enhancing the accessibility of current air vehicles, such as helicopters, for public use and emergency response systems. Rotor blade design is a critical aspect of rotorcraft performance, with the geometric aspects influencing flight characteristics. The lift-to-drag ratio is the balance of two forces correlated to the blade's total lift force and drag force. By maximizing the lift-to-drag ratio, it is possible to increase the efficiency of rotorcraft and make them more practical for urban air mobility. This can be achieved by utilizing software that mimics real-life conditions under which the rotorcraft will perform.

A similar study to the one presented in this work was performed by Kallstrom (Ref. 1) utilizing AFTGen (Ref. 2) to compare the flow solvers XFOIL (Ref. 3) and OVERFLOW (Ref. 4). Kallstrom observed that XFOIL is limited in angle of attack range and thus cannot accurately model stall and post-stall conditions. Kallstrom also stated that OVERFLOW does not perform well at low Mach numbers, with the best performance occurring at Mach numbers of 0.4 and above. However, it is important to acknowledge that AFTGen limits the capability of higher fidelity flow solvers such as OVERFLOW. Additionally, Kallstrom noted that the solving time of XFOIL is much faster than OVERFLOW. Therefore, this study investigates if UNS2D (Ref. 5 and 6), a flow solver made specifically for use with AFTGen, performs better than XFOIL at low Mach numbers. This study utilizes an NACA 0012 airfoil as well as custom asymmetric airfoils (used to generate the CSUS 001 rotor blade) to compare XFOIL and UNS2D. The success of each airfoil generation tool was determined by comparing its predictions with experimental data acquired through wind tunnel testing. Additionally, the aerodynamic performance of CSUS 001 was compared to a rotor blade generated from the NACA 0012 airfoil utilizing the computational fluid dynamics (CFD) tool RotCFD.

In the design process, CFD simulations are utilized to anticipate and reduce uncertain factors such as fluid behavior around the blade, rotor performance, and boundary condition influences. This enables the most cost-effective evaluation of design iterations as needed and helps shape future changes to achieve the desired blade performance characteristics.

For this study, a list of functions and specifications were created in tandem with the goal of creating an optimized blade design: a novel blade geometry is developed; the blade radius must fit within a 2-foot by 2-foot wind tunnel; the blade design must deliver a 30:1 lift-to-drag ratio; and the rotors should achieve higher figure of merit, in the tested collective range, than NACA 0012 blades.

One of the primary ways to manipulate an airfoil profile for an effective design that produces a favorable lift-to-drag ratio is to align the airfoil profile to that of a symmetrical or asymmetrical airfoil shape. The symmetrical airfoil profile produces no lift at an angle of attack of zero, is heavy and stable, and has no camber due to the identical lower and upper surfaces. The asymmetrical airfoil creates a good lift-to-drag ratio even at low angles of attack and produces a lift force when oriented at an angle of attack of zero due to its camber.

A study by Thakur and Choudhary (Ref. 7) compares the characteristics of an asymmetrical airfoil profile with the symmetrical airfoil profile of the NACA 0012. At an angle of attack of zero, it was determined that the symmetrical airfoil had a lower section lift coefficient compared to the asymmetrical airfoil due to its camber (Ref. 7).

APPROACH

An asymmetrical blade was utilized due to the increased lift-to-drag ratio at low angles of attack compared to a symmetrical airfoil. Then a viable airfoil shape with that characteristic was created and validated with the 2D flow solvers XFOIL and UNS2D within AFTGen. After verifying that the blade geometry and performance had successfully converged at various angles of attack and Mach number, the next step was the utilization of the 3D solver to test the blades in an isolated rotor hover configuration within RotCFD. This allowed for efficiency assessment by looking at factors such as power coefficient, C_P, and thrust coefficient, C_T. Once computational analyses and simulations were complete, physical testing in a wind tunnel was performed to validate the aerodynamic coefficients and rotor performance data obtained from the 2D and 3D simulations.

Furthermore, to ensure the consistency of this process, testing of the NACA 4-series airfoil and the custom designed CSUS 001 airfoil were done concurrently. The NACA 0012 airfoil was chosen due to the large amount of public domain experimental data available for comparison.

The wind tunnel wind speed and size limitations restricted the tested Reynolds number to 1.39×10^5 at a Mach number of 0.1. Due to a current inability of the flow solvers to converge at low Reynolds numbers, an Re/M of 3.24×10^6 was used for UNS2D and XFOIL analysis. Despite the two order of magnitude difference in Reynolds number between experimental testing and flow solver analysis, observations could still be made based on data trends.

DESIGN

Airfoil shape

Developing an airfoil design involves using simulations to understand the physical principles and aerodynamic characteristics of an asymmetrical airfoil profile and consideration of the aspects of basic airfoil behaviors such as generating aerodynamic lift, drag, and pressure variance at different angles of attack. As part of the design modification process, exploring airfoil databases such as the NACA 4-digit series asymmetrical airfoil shape is essential in obtaining the desired airfoil applications. In the final CSUS 001 blade design, three airfoils with various thicknesses such as 11%, 13%, and 15% thickness-to-chord ratio with a camber of 9% were placed at three radial stations along the span of the blade, as detailed in Table 1.

Table 1: CSUS 001 Airfoil Specifications at Various Radial Stations

Airfoil	Thickness	Camber	Chord Length (inches)	Radial Station (r/R)
Airfoil 1	15%	9%	2.5	0.445
Airfoil 2	13%	9%	2.5	0.771
Airfoil 3	11%	9%	2.5	1.0

MATLAB code

MATLAB was utilized to generate the geometry of the custom airfoil. The NACA series four airfoil equations were modified and used as the basis for the airfoil. The MATLAB script allowed the user to specify the maximum thickness of the airfoil, the location of maximum thickness, the camber of the airfoil, the location of maximum camber, as well as the number of points to be generated. From these inputs MATLAB plotted the airfoil and wrote the X and Y coordinates into a text file that could then be used as input for airfoil table generators within AFTGen. The equation used to generate the uncambered upper surface of the airfoils is shown in Equation 1.

$$y = 5t\,(.51x^{1/2} - .3291x - .1562x^2 + .0789x^3 - 0.1015x^4) \quad (1)$$

Where t is the maximum thickness as a percentage of the chord. The uncambered bottom surface of the airfoil was created using Equation 2.

$$y = 5t\,(.61x^{1/2} - .6291x - .1562x^2 + .3789x^3 - 0.2020x^4) \quad (2)$$

Where t is the same value as in the top surface equation. The value x is then calculated between zero and one on an equally spaced vector with as many points as specified by the user. After the uncambered airfoil shape has been generated and plotted, the code then applies the cambering equation, Equations 3 and 4, to both the top and bottom surfaces. This equation is identical to that used in the NACA series 4 airfoils.

$$y = \frac{m}{p}(2px - x^2)\; 0 \le x \le p \quad (3)$$

$$y = \frac{m}{(1-p^2)}\left((1-2p) + 2px - x^2\right) p \le x \le 1 \quad (4)$$

Where m is the maximum camber and p is the location of maximum camber.

Changes

The earliest concept of the blade design originally consisted of five distinct airfoils starting at 13% thickness-to-chord ratio near the hub and decreasing until it reached 5% thickness-to-chord ratio at the end of the blade. This design had the same camber of 9% throughout all design with airfoils placed at five radial stations along the blade. However, as previously stated, the number of airfoils decreased to three. This was a result of the airfoil generation within the CFD solvers provided and the realization that an airfoil with a thickness of 5% would be difficult to print for physical testing. Figure 1 illustrates the final design of the CSUS 001 rotor blade.

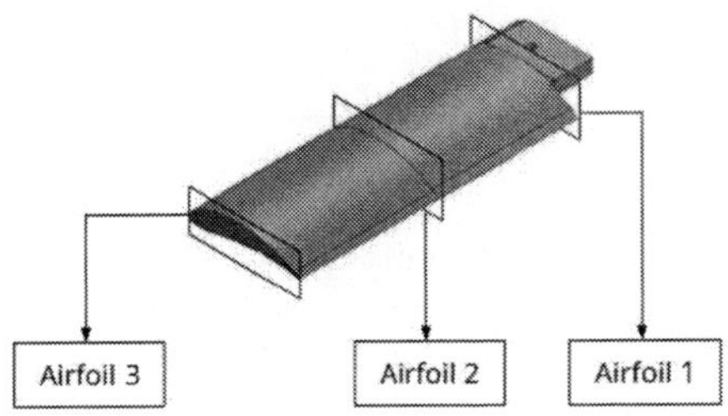

Figure 1: CSUS 001 Rotor Blade

COMPUTATIONAL METHODS

As stated previously in the approach section, 2D and 3D solvers were utilized to determine the aerodynamic properties of the airfoils being tested and generated using AFTGen between the angles of attack of -10 and 10 degrees and Mach numbers ranging from 0.1 to 0.3. Developed by Sukra-Helitek, AFTGen provides a GUI for various flow solvers, including XFOIL, ARC2D, OVERFLOW, UNS2D, MSES, and UNS2D (Ref. 2).

The 3D software RotCFD was used to "test" the three airfoils used in the CSUS 001 rotor blade, and the NACA 0012 airfoil to determine their efficiency in a rotorcraft configuration. The simulation setup was appropriately adjusted to obtain converged results for the coefficient of thrust (CT) and total coefficient of power (CP). Detailed information on the input settings for RotCFD and rotor specifications can be found in the RotCFD Testing section.

XFOIL

XFOIL is an airfoil analysis program developed by Prof. Mark Drela at MIT and Harold Youngren at Aerocraft, and it is able to quickly analyze inviscid and viscous flows around a 2D airfoil for subsonic flight conditions. The XFOIL solver utilizes an integral-based approach to determine the section lift coefficient (c_l), section drag coefficient (c_d), and section moment coefficient (c_m) of the blade. This integral considers the values from both ends of the panel being assessed and iteratively calculates the aerodynamic coefficients until convergence (Ref. 3).

To obtain converged values for the airfoil aerodynamic coefficients, the solver first finds the pressure distribution over the panels or surface of the airfoil, which utilizes the user-inputs for the flow conditions provided. Subsequently, this allows the solver to solve for the velocity and allows for the integral approach of the pressure along the airfoil's surface to calculate the airfoil's aerodynamic properties. The integral approximation approach can be described by utilizing Equations 5, 6, 7, and 8 (Ref. 3).

$$c_l = \frac{L}{q} = \int c_p d\underline{x} \tag{5}$$

$$c_m = \frac{M}{q} = \int -c_p[(x - xref)dx + (y - yref)dy] \tag{6}$$

$$c_d = \frac{D}{q} = \int 2\theta_i = 2\theta\left(\frac{u}{V}\right)^{\frac{H+5}{2}} \tag{7}$$

$$q = 0.5 * r * V^2 \tag{8}$$

Where q is the total pressure (assuming that there is no static pressure), θ is the momentum thickness at the end of the wake, u is the edge velocity at the end of the wake, H is the shape parameter at the end of the wake, V is the freestream velocity, θ_i is the momentum thickness at "downstream infinity," and $\underline{x}$ and $\underline{y}$ are a function of angle of attack (Equations 9 and 10) (Ref. 3):

$$\underline{x} = x *cos\ \cos(\alpha) + y * si\,n(\alpha) \tag{9}$$

$$\underline{y} = y *cos\ \cos(\alpha) + x * si\,n(\alpha) \tag{10}$$

The number of panels and panel density ratio utilized in XFOIL for this study were 425 and 0.18, respectively.

UNS2D

UNS2D is a mid-fidelity flow solver developed by Sukra-Helitek for generation of 2D aerodynamic coefficients of an airfoil in AFTGen. Rather than utilizing a panel around the surface of the airfoil like XFOIL, UNS2D uses a hybrid of an O-grid near the surface of the airfoil and an unstructured triangular grid farther from the airfoil surface (Ref. 5 and 6). The mathematical model of UNS2D results in a longer run time for the aerodynamic properties within c81 tables to reach convergence at various Mach numbers and angles of attack, elaborated as follows.

UNS2D's approach to obtaining values used for c81 tables utilizes the unsteady-state Navier-Stokes equations in 2D space (Ref. 5 and 6). These equations govern the motion of fluid flow and include the partial derivative terms in the continuity equation and momentum equations. As implied, rather than looking at it in a 3D space, UNS2D utilizes the partial derivatives of pressure, time, and the two directions in the cartesian coordinates generally noted as the x-direction and y-direction.

The grid generation in UNS2D discretizes the domain around the airfoil into a mesh of elements. Serving as a framework to apply the Navier-Stokes equations, the equations are solved at the nodes to calculate the vector field variables such as velocity, pressure, and density (Ref. 5 and 6).

ROTCFD

The Rotorcraft Computational Fluid Dynamics (RotCFD) software, developed by Dr. R Ganesh Rajagopalan, Sukra Helitek, Inc., is a reliable GUI environment to analyze the behavior of the flow around a blade in three-dimensions. Based on the solver interface, the governing equations of the flow are unsteady, incompressible Navier Stokes equations. Regarding the simulation, the first-time step is run with incompressible Navier Stokes equations; the flow can be considered incompressible as the local Mach numbers for the cells fall below 0.25 (Ref. 8). Mach numbers greater than 0.25 are influenced by the velocity of the object, speed of sound in the medium, density of the medium, temperature, and atmospheric conditions (Ref. 8). In these cells, the future time step is calculated using the compressible Navier Stokes equations. These equations incorporate the fundamental principles of mass conservation, momentum conservation, and energy conservation to analyze the fluid motion around the blade. Under the assumption of incompressible flow, several simplifications can be considered regarding the fluid properties. These simplifications involve isothermal temperature, constant volume regardless of the pressure applied, a continuous medium, and Newtonian behavior: constant viscosity across a range of shear rates and maintaining a constant temperature. In this study, the solver considered both compressible and incompressible flow conditions by utilizing the Navier Stokes equation at the local Mach Number for a particular cell. The grid generation panel in RotCFD uses a Cartesian Octree grid to surface geometry and creates the planes along X, Y, and Z.

NACA 0012 and CSUS 001: XFOIL vs UNS2D COMPARISONS

The following section presents a comparison of the results obtained from the flow solvers XFOIL and UNS2D of the NACA 0012 airfoil and the airfoils used in the CSUS 001 rotor blade. Moreover, a common approach was taken before running these solvers, such as setting appropriate inputs to obtain reasonable and precise results with low non-convergence rates. Table 2 conveys the input settings for the flow solvers. The Re/M of 3.24 x 10^6 was chosen because it was the lowest value that could be used while avoiding significant non-convergence. A low Re/M was preferrable to match the wind tunnel Re as closely as possible.

Table 2: Flow Conditions and Primary Inputs for XFOIL and UNS2D

α Range	[-10, -8, -6, -4, -2, 0, 2, 4, 6, 8, 10]
M_{Tip} Range	[0.1, 0.2, 0.3]
Re/M	3.24 x 10^6

CSUS 001 consists of three airfoils with thicknesses of 11%, 13%, and 15% (Table 1). The solvers provided the aerodynamic performance c81 tables for each CSUS 001 airfoil and the NACA 0012 airfoil, which consist of section lift coefficients (c_l), section drag coefficients (c_d), and section moment coefficients (c_m) at the corresponding angle of attack (α) and M_{Tip} number, which refers to the speed of the tip of the blade relative to the speed of sound in the ambient air.

NACA 0012

As observed in Figure 2, for the NACA 0012 UNS2D result, the section lift coefficient at both extremes of the angle of attacks tested is slightly below ± 1 while the XFOIL section lift coefficient at ± 10 angle of attack is slightly above ± 1. There is little discrepancy between the UNS2D and XFOIL NACA 0012 section lift coefficient results. Figure 3 illustrates the section drag coefficient of NACA 0012 with the UNS2D result showing a maximum section drag coefficient of 0.0375. For XFOIL the section drag coefficient shows a maximum value of 0.014, less than half of the UNS2D result. The NACA 0012 section moment coefficient observed in Figure 4 shows similar results between UNS2D and XFOIL between an angle of attack of ± 5. Outside of this angle of attack range the UNS2D and XFOIL results diverge.

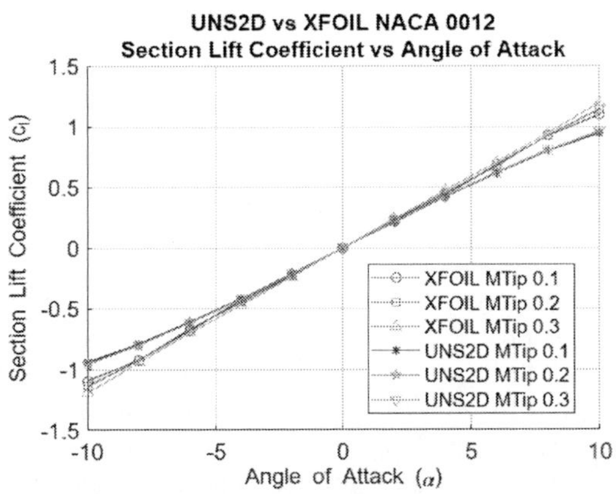

Figure 2: Section Lift Coefficient vs Angle of Attack (NACA 0012 UNS2D and XFOIL)

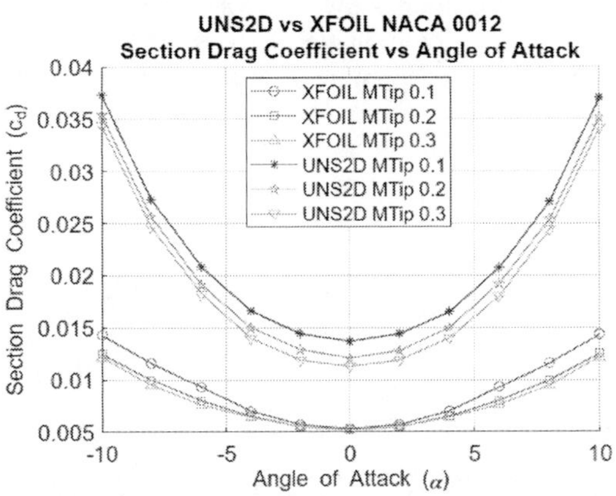

Figure 3: Section Drag Coefficient vs Angle of Attack (NACA 0012 UNS2D and XFOIL)

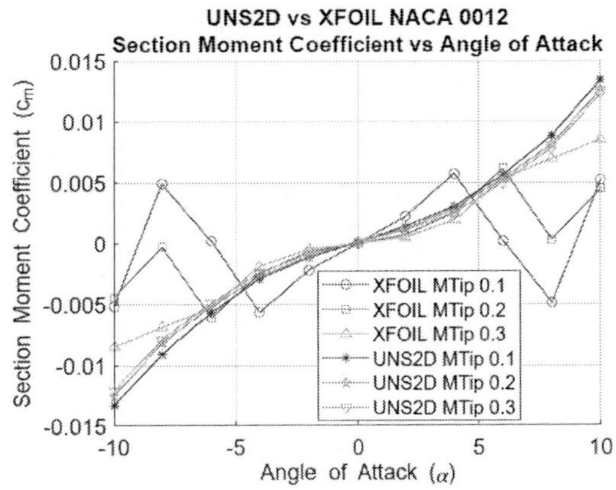

Figure 4: Section Moment Coefficient vs Angle of Attack (NACA 0012 UNS2D and XFOIL)

CSUS 001

For the CSUS 001 rotor, each airfoil profile will be assessed to determine the aerodynamic characteristics as the CSUS 001 rotor is composed of all three of these airfoils. The section lift coefficients for the CSUS 001 airfoil produced by UNS2D, Figures 5, 6, and 7, show a positive linear relationship with the angle of attack, except between the angle of attack range of -10 to -6 in which a steady to decreasing section lift coefficient is observed. On the other hand, the XFOIL results do not show the same negative angle of attack behavior observed in the UNS2D results. Instead, the section lift coefficient versus angle of attack result for XFOIL is a quadratic relationship, almost resembling a linear line. However, there is little discrepancy between UNS2D and XFOIL section lift coefficient values, especially for Airfoil 3 for which the results are almost identical.

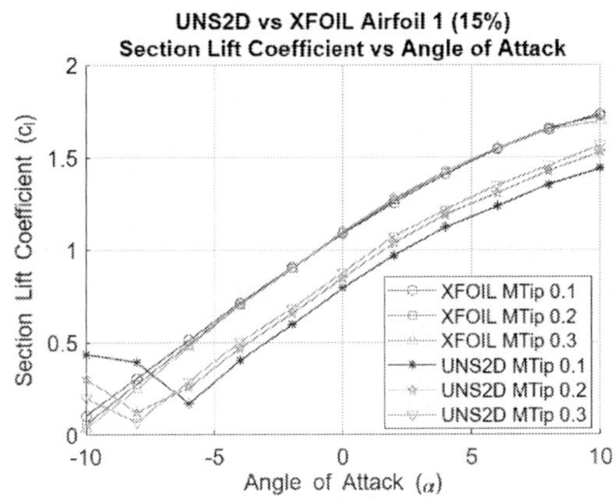

Figure 5: Section Lift Coefficient vs angle of Attack (CSUS 001 Airfoil 1 UNS2D and XFOIL)

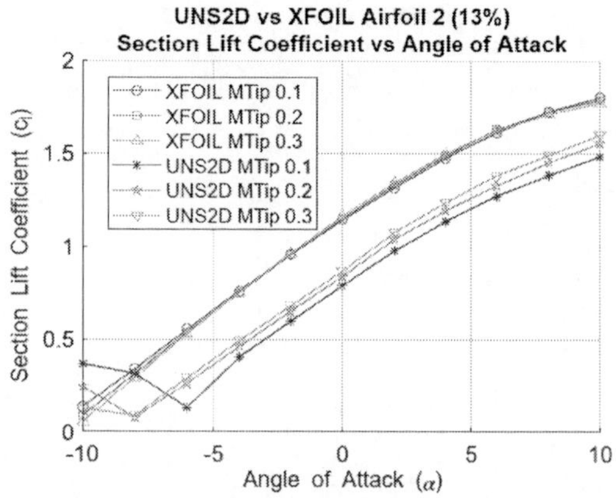

Figure 6: Section Lift Coefficient vs Angle of Attack (CSUS 001 Airfoil 2 UNS2D and XFOIL)

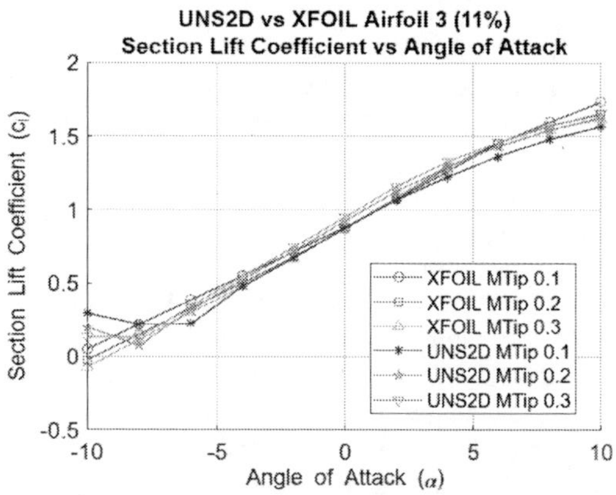

Figure 7: Section Lift Coefficient vs Angle of Attack (CSUS 001 Airfoil 3 UNS2D and XFOIL)

As seen in Figures 8, 9, and 10, the CSUS 001 UNS2D dataset consists of a pattern where the section drag coefficient is similar to NACA 0012 in the sense that there is a higher section drag coefficient between the angles of attack of -10 to -4 for all three tip Mach numbers. The XFOIL section drag coefficient shows a steady increase from an angle of attack of -10 to 10. Throughout Airfoils 1-3 and the three tip Mach numbers, the XFOIL section drag coefficient is at least half of the UNS2D section drag coefficient result.

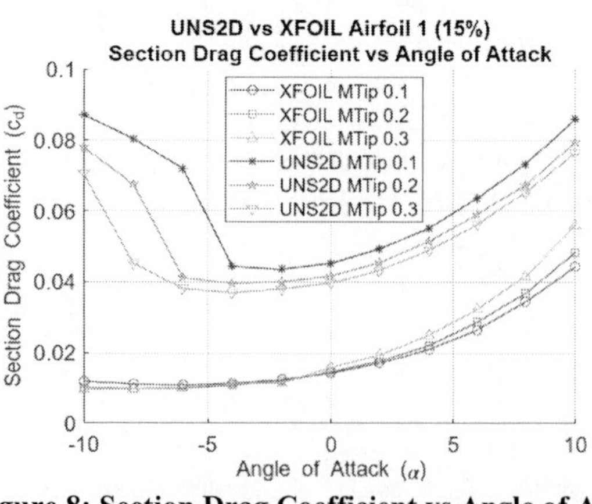

Figure 8: Section Drag Coefficient vs Angle of Attack (CSUS 001 Airfoil 1 UNS2D and XFOIL)

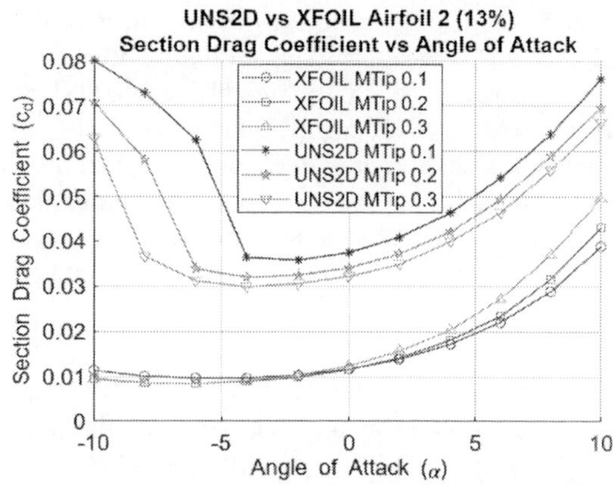

Figure 9: Section Drag Coefficient vs Angle of Attack (CSUS 001 Airfoil 2 UNS2D and XFOIL)

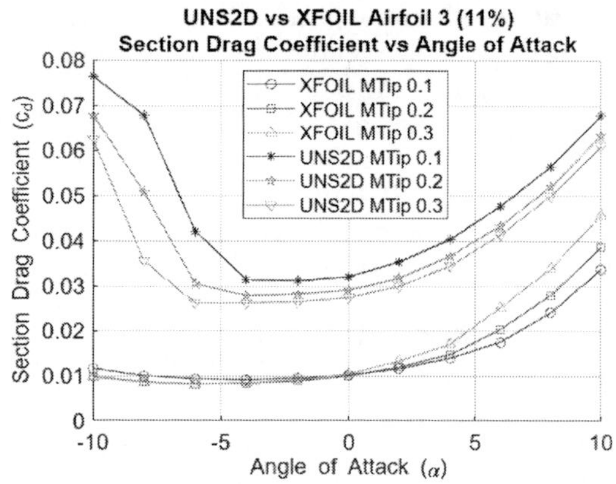

Figure 10: Section Drag Coefficient vs Angle of Attack (CSUS 001 Airfoil 3 UNS2D and XFOIL)

Figures 11, 12, and 13 illustrate the relationship between section moment coefficient and angle of attack for the CSUS 001 airfoils resulting from UNS2D and XFOIL. For Airfoils 1 and 2, the results of both solvers agree at an angle of attack of ten. However, UNS2D predicts higher section moment coefficients at lower angles of attack. All three airfoils show opposite solver trends between -10 and 0 angle of attack with UNS2D predicting a generally negative slope and XFOIL predicting a positive slope. However, for Airfoil 3, the XFOIL section moment coefficient values surpass those predicted by UNS2D at around -4 to -3 angle of attack while the Airfoil 1 and 2 UNS2D section moment coefficients remain higher than the XFOIL predictions at all except the highest angles of attack investigated.

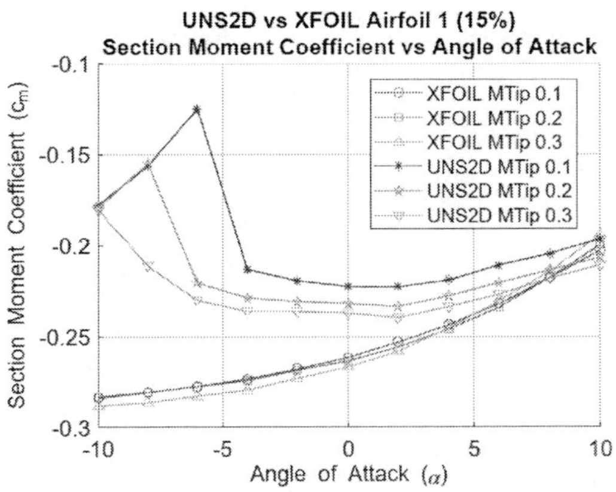

Figure 11: Section Moment Coefficient vs Angle of Attack (CSUS 001 Airfoil 1 UNS2D and XFOIL)

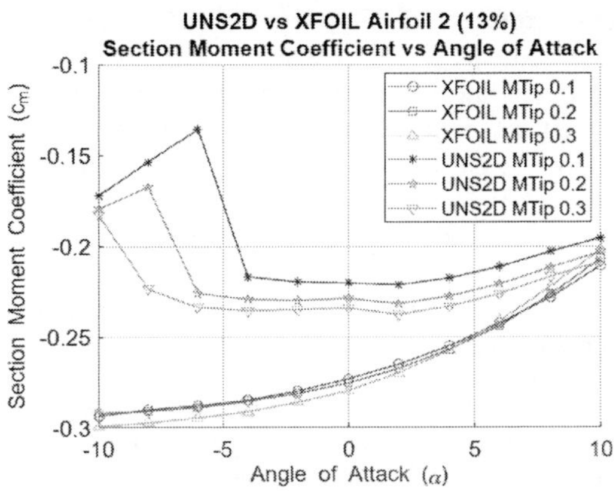

Figure 12: Section Moment Coefficient vs Angle of Attack (CSUS 001 Airfoil 2 UNS2D and XFOIL)

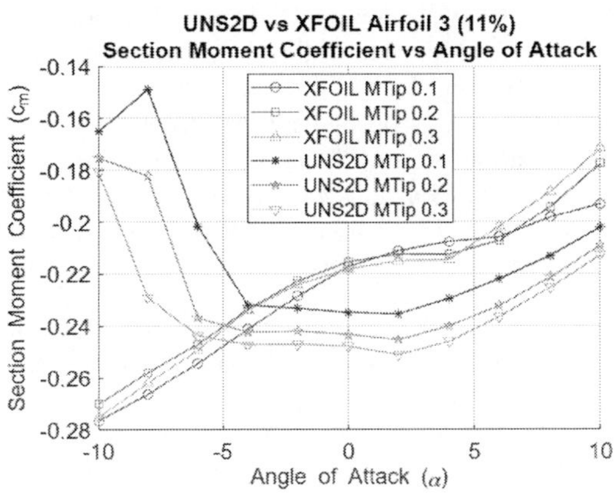

Figure 13: Section Moment Coefficient vs Angle of Attack (CSUS 001 Airfoil 3 UNS2D and XFOIL)

ROTCFD TESTING

Approach

RotCFD software was employed for the analysis of rotor performance through computational fluid dynamic simulations. RotCFD provides visual representations of rotor performance under various conditions (Ref. 9). Rotor performance data, such as the coefficient of thrust and total coefficient of power, can be examined at each collective pitch angle. This visual feedback plays a crucial role in observing the convergence of the solver to a solution. To achieve accurate simulation results, appropriate grid refinement based on the blade geometry, rotor refinement, and setting appropriate time step size is important. These steps involve calculations such as determining the cutout radius of the blade, measuring the distance from the center of rotation to the unnormalized blade tip, and measuring the distance from the center of rotation to each radial station along the blade. The RotCFD inputs utilized are displayed in Table 3.

Table 3: RotCFD Input Settings and Rotor Specifications

Rotor Refinement	11
Time Steps	3000
Tip Mach Number	0.1, 0.2, 0.3
Cutout Radius (r/R)	0.389
Radius	0.26543 m
Number of Blades	3
Radial Station Input (r/R) of Airfoil 1 – Thickness 15%	0.445
Radial Station Input (r/R) of Airfoil 3 – Thickness 11%	1.0
Gen. Rotor Refinement	9
Boundary X Cells	5
Boundary Y Cells	5
Boundary Z Cells	5

A common approach in CFD for analyzing rotorcraft flows is to model a rotor as a distribution of momentum sources in the flow. The rotor geometry and the flow behaviors relate to the momentum provided by the rotor. The rotor model calculates rotor performance such as the coefficient of thrust (C_T), total coefficient of power (C_P), and figure of merit (FM). Airfoil tables generated by XFOIL and UNS2D solvers in AFTGen were utilized as the input to determine the rotor performance. The simulations were performed based on NACA 0012 and the CSUS 001 asymmetrical airfoils with various thicknesses: 11%, 13% and 15%, respectively. To comprehensively evaluate the rotor performance across the different hover conditions, a parametric analysis was performed. The simulations involved a range of angles of attack from –10 degrees to +10 degrees. Additionally, the simulations examined a tip Mach number range of 0.1 to 0.3 in 0.1 intervals.

For the Mach 0.1 case, increasing the time step led to better convergence results due to the low speed of flow and incompressibility condition. In some cases, low angles of attack led to convergence issues, and this occurred due to several conditions, such as insignificant pressure differences across the blade as the angle between the rotor blade and incoming flow is small, characteristics of weak flow, and boundary layers.

A smaller time step was required for the Mach 0.2 and Mach 0.3 cases to obtain sufficient results and improve simulation speed. Based on the RotCFD simulation theory, a Mach number higher than 0.25 is considered an incompressible flow. In essence, the incompressible flow has higher velocity due to escalating external forces, the influence of airfoil geometry, and the behavior of the flow. The decision to set a reasonable time step to obtain convergence results depends on the flow speed, and in other scenarios, the complexity of fluid can affect the convergence results.

Results and Discussion

The following section represents the results generated by using RotCFD. The relationship between the proportion of ideal power to the real power used in hover, known as the figure of merit (Equation 11), and blade loading (C_T/σ_S), which is the coefficient of thrust of the rotor divided by the relative blade area, known as solidity (Equation 12), are discussed in the following sections. Utilizing the coefficient of thrust and total power coefficient results of NACA 0012 and the asymmetrical airfoil with various thicknesses, the blade's figure of merit can be determined. The calculated solidity (σ_S) value based on the blade area and disk area is 0.228 for both blades.

$$\mathrm{FM} = \frac{\frac{C_T{}^{3/2}}{\sqrt{2}}}{C_P} \tag{11}$$

$$\sigma_s = \frac{(N_B)(c)}{(\pi R)} \tag{12}$$

NACA 0012

In this section, the data from RotCFD, which incorporates both XFOIL and UNS2D c81 tables, was analyzed to evaluate the coefficient of power (C_P) and coefficient of thrust (C_T). By analyzing the coefficient of thrust produced with its associated coefficient of power, quantitative assessment of the rotor efficiency for the blade geometry can be performed. Ideally, the blade geometry or solver with a lower coefficient of power associated with a larger coefficient of thrust would be the most efficient. As seen with the RotCFD results from XFOIL and UNS2D in Figure 14, the UNS2D result conveys the highest performance and efficiency at a tip Mach number of 0.3 for producing a coefficient of thrust of 5.7 x 10^{-3} and corresponding coefficient of power of 4.5 x 10^{-4}. Based on the XFOIL result, the highest performance and efficiency is produced at a tip Mach number of 0.3, but the highest coefficient of thrust of 5.7 x 10^{-3} was achieved at a coefficient of power of 3.7 x 10^{-4}. This indicates that XFOIL results demonstrate a slightly better performance and efficiency for the given design space with a C_T over C_P value of 12.7. For transparency, outliers in Figure 14 have not been removed.

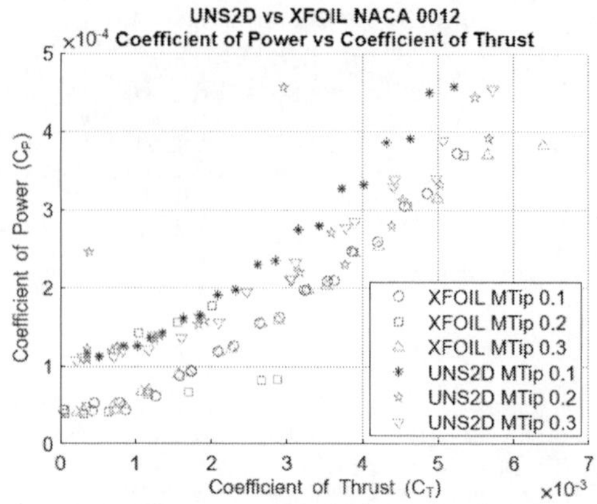

Figure 14: Coefficient of Power vs Coefficient of Thrust (NACA 0012 UNS2D and XFOIL)

In this section, figure of merit, which describes the efficiency of a rotor, is compared against blade loading, which defines how the aerodynamic forces are distributed along the blade, by examining the figure of merit versus (C_T/σ_S) plots based on the results obtained from XFOIL and UNS2D. This analysis indicates that the rotor blade achieves an optimal balance between the generation of lift and efficiency at different tip Mach numbers. Furthermore, it can ensure that the rotor is efficiently generating thrust and functioning optimally by avoiding factors limiting the optimum blade loading.

At $M_{Tip} = 0.1$, Figure 15 shows that the maximum hover figure of merit is about 0.75, which occurs at a blade loading of 0.021 based on XFOIL. On the other hand, the maximum figure of merit is 0.58 at a blade loading of 0.023 for $M_{Tip} = 0.1$ based on UNS2D.

Results from both XFOIL and UNS2D for M_{Tip} values of 0.2 and 0.3 followed a similar pattern as stated above. The optimal blade loading for XFOIL was found to be more efficient compared to UNS2D. Specifically, for $M_{Tip} = 0.2$, the optimal blade loading occurred at a figure of merit of 0.75 for XFOIL, whereas for UNS2D, the highest figure of merit is 0.69. For $M_{Tip} = 0.3$, the optimal blade loading occurred at a figure of merit of 0.81 for XFOIL and 0.73 for UNS2D.

These XFOIL and UNS2D results indicate that the figure of merit is slightly higher at $M_{Tip} = 0.3$ than the other tested tip Mach numbers. This illustrates that the rotor is more efficient at higher speeds. While XFOIL results show more efficiency than UNS2D results, UNS2D data meets the fundamental principle of a rotor's figure of merit not exceeding 0.8.

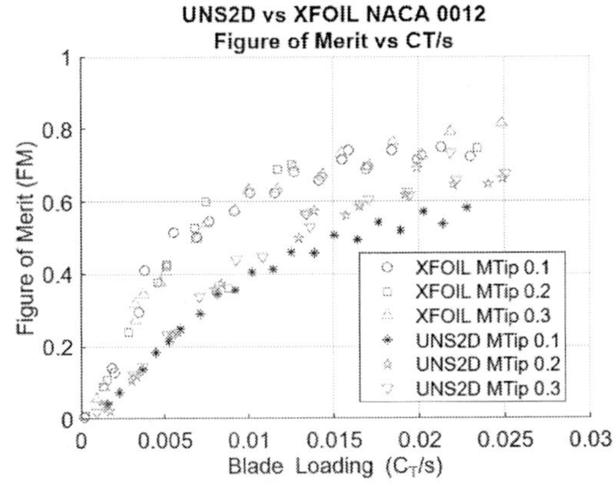

Figure 15: Figure of Merit vs Blade Loading (NACA 0012 UNS2D and XFOIL)

Another measurement of performance of the blade geometry on the rotorcraft hub configuration is the measurement of C_T/σ_s versus the collective (deg.). By looking at the datasets generated by RotCFD from the XFOIL and UNS2D c81 tables, the peak performance across different collective angles can be assessed. As seen from the datasets in Figure 16, the deliverable thrust based on the blade geometry increases as the collective angle increases between the ranges of -10 to 10. Figure 16 shows that XFOIL and UNS2D have a minimum blade loading value of -0.021 at a collective angle of -10 and a maximum blade loading value of 0.024 at a collective angle of 10. Figure 16 also illustrates that, regardless of which solver is used, the lift produced by the blade geometry increases with angle of attack. Since the collective angle range tested was between -10 and 10 for this study, the maximum performance in this design space is at a collective angle of 10.

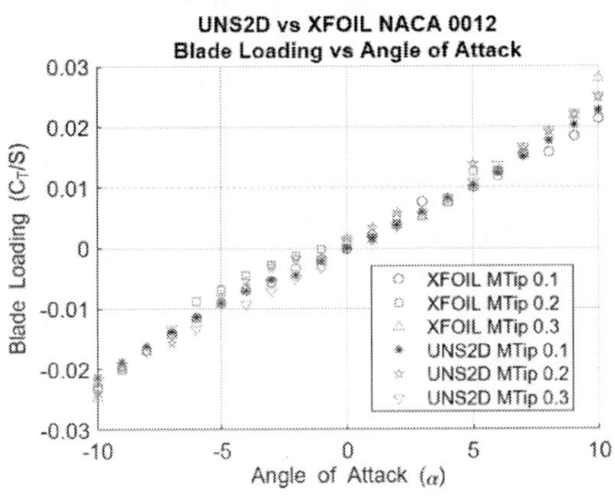

Figure 16: Blade Loading vs Collective (NACA 0012 UNS2D and XFOIL)

CSUS 001

When looking at the coefficient of power versus coefficient of thrust for the custom-designed blade in Figure 17, it is apparent that it differs from the NACA 0012 blade. Furthermore, the values within UNS2D have some datasets between the ranges where the coefficient of thrust is around zero to 4×10^{-3} showing a reverse effect in terms of coefficient of power. This can be attributed to UNS2D's interpretation of the values of coefficient of power, which shows that at some point in the negative angle of attack range, the coefficient of thrust attributed to those values requires a larger coefficient of power. While the initial data points show an increase in efficiency, in the sense that the power needed to deliver a certain thrust force is lower, the efficiency then decreases again where the coefficient of power needed for a larger coefficient of thrust increases more than that of the actual coefficient of thrust value. Both sets of data show that the efficiency and performance of the custom-designed blade are outmatched by the performance of the NACA 0012 blade data. The maximum C_T over C_P value of the CSUS 001 blade was 11.4. Additionally, both data sets and graphs show that the efficiency decreases with a higher angle of attack.

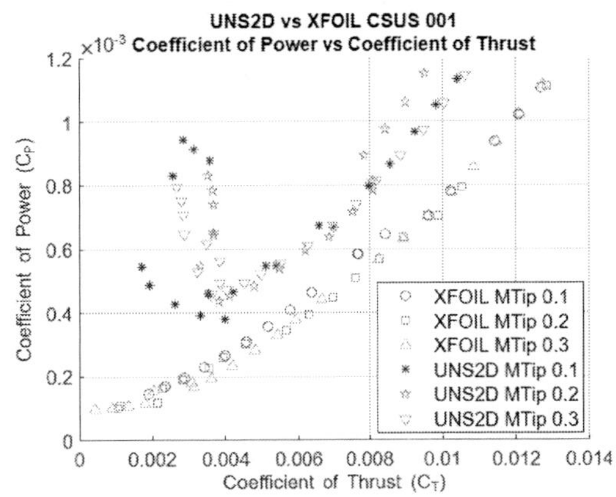

Figure 17: Coefficient of Power vs Coefficient of Thrust (CSUS 001 UNS2D and XFOIL)

The analysis of the figure of merit versus blade loading of CSUS 001 aligns with the optimal blade loading interpretation for NACA 0012. When comparing the data at $M_{Tip} = 0.1$ for XFOIL and UNS2D results, as observed in Figure 18, XFOIL shows a higher hover figure of merit of about 0.95, with a corresponding blade loading of 0.042. On the other hand, UNS2D has a figure of merit of around 0.66 at a blade loading of 0.046.

According to the XFOIL result, the optimal blade loading for $M_{Tip} = 0.2$ occurs at a figure of merit of around 0.984, with a corresponding blade loading of 0.044. Meanwhile, the UNS2D result shows that the optimal blade loading occurs at a figure of merit of 0.66 and a blade loading of 0.036.

At $M_{Tip} = 0.3$, the XFOIL result indicates the decreasing figure of merit pattern after the optimal blade loading point due to stall and profile drag. However, the UNS2D result shows that the data ended at the optimal blade loading point. The optimal blade loading point for the XFOIL can be examined at the figure of merit of 0.943, and the corresponding blade loading point is 0.045. On the other hand, the UNS2D result shows that the optimal blade loading point corresponds to a figure of merit of 0.675 and blade loading of 0.046.

These analyses based on the optimal blade loading for both XFOIL and UNS2D results indicate that the rotor is more efficient at $M_{Tip} = 0.2$ regarding the XFOIL result, while the UNS2D result shows that the rotor is more efficient at $M_{Tip} = 0.3$. However, based on the fundamental principle of a rotor's figure of merit, the UNS2D result would be more reliable for examining the efficiency of rotor operation. The unrealistically high figure of merit values produced by XFOIL for the CSUS 001 blade indicates that the UNS2D result for both NACA 0012 and CSUS 001 is more accurate to real-world expectations.

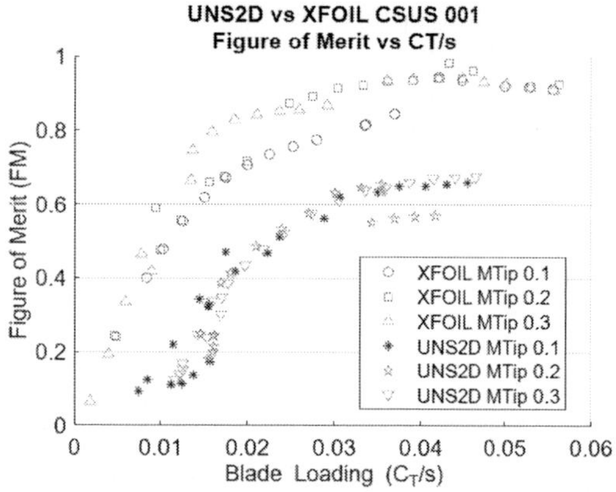

Figure 18: Figure of Merit vs Blade Loading (CSUS 001 UNS2D and XFOIL)

Performance of the CSUS 001 blade is examined in Figure 19 by examining blade loading versus collective. According to the XFOIL and UNS2D generated datasets for CSUS 001, the lift produced by the blade geometry increases as the collective angle increases between the range of –10 to 10.

XFOIL data indicates a minimum blade loading value of 0.0019 at a collective angle of –10 and a maximum blade loading value of 0.056 at a collective angle of 10. On the other hand, UNS2D data shows that a minimum blade loading occurs at 0.0075 and corresponds to a collective angle of –6 and the maximum blade loading value is 0.046 at a collective angle of 10.

The test data generated by both solvers within the examined collective angle range of –10 to 10 degrees indicates that the blade design achieves a balance between thrust generation and efficiency. The optimal point for the CSUS 001 blade was achieved at a collective angle of 10. The XFOIL solver suggests $M_{Tip} = 0.2$ for peak efficiency, while the UNS2D solver points towards $M_{Tip} = 0.3$.

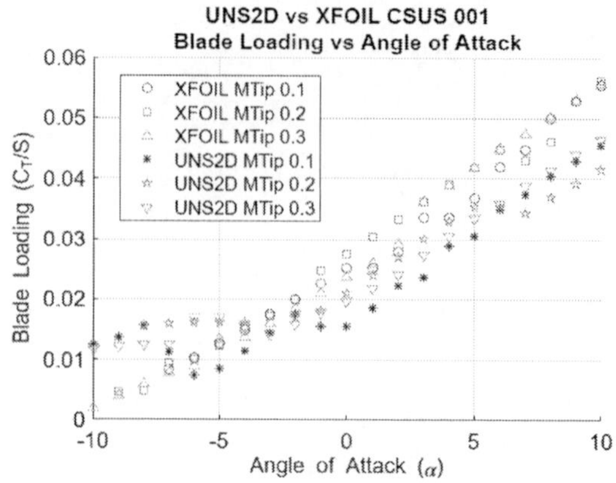

Figure 19: Blade Loading vs Collective (CSUS 001 UNS2D and XFOIL)

EXPERIMENTAL TESTING

To confirm the section lift and drag coefficient values generated in AFTGen by XFOIL and UNS2D were accurate, they were compared to experimental data for the respective airfoils. Using the three airfoils developed in this study and NACA 0012, four test sections were 3D printed with Polylactic Acid (PLA). The testing was done using the California State University (CSU), Sacramento, 2-foot by 2-foot wind tunnel. Figure 20 depicts the wind tunnel experimental setup.

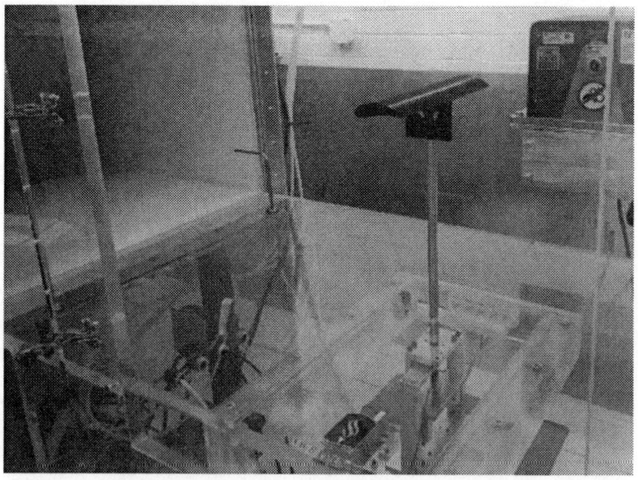

Figure 20: Experimental Wind Tunnel Testing Setup

The experimental data obtained from the wind tunnel was collected with a load cell that effectively measures the lift and drag of an attachment inside the wind tunnel. To ensure that the data gathered was precise, the base values for the lift force and drag force were observed without the 3D printed airfoil attachments, making sure to do this each time the attachment was changed to ensure consistency in subsequent tests. For the experimental tests with the airfoil attachments inside of the wind tunnel testing area, the airfoils printed had an extrusion at the bottom of the airfoil to allow for a pivot screw to be attached and a secondary screw to allow for the angle of attack to be changed. A digital angle finder was utilized to determine the angle of attack of each airfoil within the wind tunnel, where the data was observed with the angle of attack varying incrementally by 1 degree from -10 to 10 degrees. The final output load values were subtracted from the static loads and converted to section lift and drag coefficients utilizing Equations 13 and 14. These were then compared to the c_l and c_d values from the c81 tables that were created in AFTGen. Section moment coefficient was not experimentally measured.

$$c_l = \frac{(2 * F_l)}{(\rho * v^2 * A)} \tag{13}$$

$$c_d = \frac{(2 * F_D)}{(\rho * v^2 * A)} \tag{14}$$

Additionally, the values for the wind tunnel such as pressure, temperature, humidity, density, dynamic viscosity, and wind speed were measured to allow for the determination of the Reynolds number within the wind tunnel (Table 4 and Equation 15).

Table 4: Wind Tunnel Airflow Data

Pressure	$101000\ Pa$
Temperature	$296.56\ Kelvin$
Humidity	54.2 %
Air Density	$1.18\ kg/m^3$
Dynamic Viscosity	$1.83 * 10^5\ Pa * s$
Freestream Velocity	$34\ m/s$
Reynolds Number	$1.39 * 10^5$

$$Re = \left(\frac{\rho * V * c}{\mu}\right) \tag{15}$$

A freestream velocity of 34 m/s was the highest achievable speed of the wind tunnel. The highest freestream velocity was used to maximize the tested Re. Additionally, utilizing this freestream velocity meant the airfoils were being tested at a Mach number of 0.1 which was ideal for comparison to flow solver results. Given the dimensions of the wind tunnel testing chamber, the maximum achievable Re using a chord of 2ft was 1.34×10^6. However, this Re is still an order of magnitude below the flow solver Re at a Mach number of 0.1. Additionally, a 2ft chord rotor blade was unfeasible to fabricate due to 3D printer build plate dimension limitations.

3D Printing

Manufacturing of the extruded airfoil profiles for wind tunnel testing was done using a Sovol SV06 3D printer and Ultimaker Cura slicer. For these prints, universal settings were used and are as follows: the infill was set to 100%, with a print quality of 0.2 mm, print speed was left at the default of 50 mm/s, tree supports were used with raft adhesion method to increase plate contact and prevent the prints from toppling over, z-hop was also activated to prevent the nozzle from colliding with the print during traveling. A finer quality was considered but decided against it because the surface finish seemed to diminish, as well as the increased time demand for finer quality prints. All other unmentioned parameters were left at the slicer default. All specimens produced generated a smooth finish.

COMPUTATIONAL VS EXPERIMENTAL COMPARISONS

In the interest of comparing computational and experimental data, the airfoil table data generated in AFTGen was compared to the experimental data.

Regarding the section lift coefficient versus the angle of attack for the NACA 0012 airfoil, as seen in Figure 21, the experimental results yield a larger section lift coefficient at angles of attack below 2 degrees and smaller section lift coefficient at angles of attack above 2 degrees compared to the UNS2D and XFOIL computational results. Additionally, as seen in Figure 22, the measured section drag coefficient versus angle of attack in the experimental data is larger at ± 10 degrees compared to the UNS2D and XFOIL computational results.

For Airfoils 1-3 in the CSUS 001 blade design, comparing the section lift coefficient versus angle of attack from UNS2D with the experimental data, as seen in Figures 5, 6, 7, and 21, the experimental results show the same pattern where the section lift coefficient has a stall behavior at around -6 to -10 degrees and increases between the angles of -6 to 10 degrees. It is important to note that this section lift coefficient phenomenon at a low angle of attack was not captured by XFOIL. However, the results of the experimental section lift coefficient are just short of the expected section lift coefficient from the computational data. Similar to the section lift coefficient results, XFOIL was not able to predict the increased section drag coefficient at negative angles of attack, as observed in the UNS2D and experimental results.

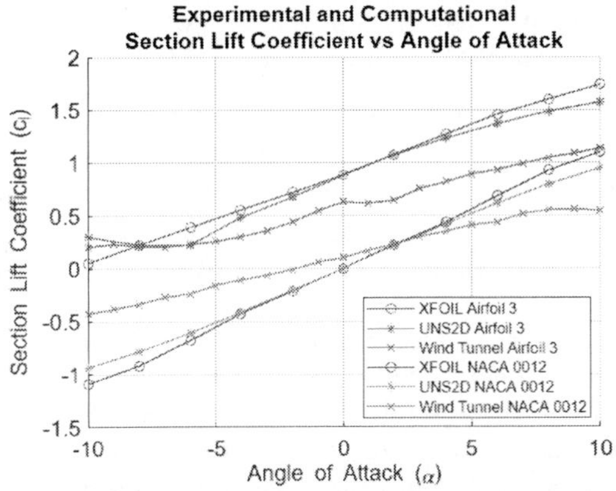

Figure 21: Section Lift Coefficient vs Angle of Attack (UNS2D, XFOIL, and Experimental)

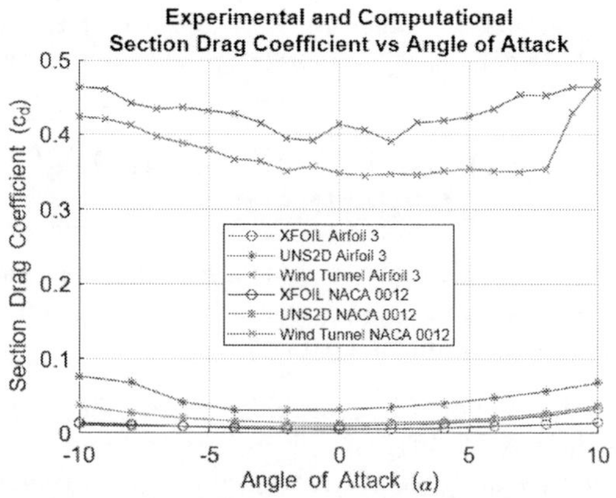

Figure 22: Section Drag Coefficient vs Angle of Attack (UNS2D, XFOIL, and Experimental)

Looking at both the XFOIL and UNS2D interpretation of section drag coefficient versus angle of attack in Figure 22, neither of the NACA 0012 or CSUS 001 datasets have a magnitude similar to the experimental data. This is likely the result of different Reynolds number conditions in the computational and experimental portions of this study. It is important to note that XFOIL and UNS2D were run with a Reynolds number of 3.24 x 10^7 at a Mach number of 0.1, whereas the experimental runs in the 2ft x 2ft wind tunnel were run with a Reynolds number of 1.39 x 10^5 at a Mach number of 0.1. As demonstrated experimentally by Dafnis, et al. (Ref. 9), Reynolds number and section drag coefficient are inversely related while Reynolds number and section lift coefficient are directly related. The magnitude of these relationships is also relevant as an increase in Reynolds number significantly decreases section drag coefficient while section lift coefficient only slightly increases. As illustrated in Figures 21 and 22, the same Reynolds number relationships were observed in this study. Therefore, observations on the accuracy of XFOIL and UNS2D could still be made based on the trends of the data despite the significant different in Reynolds number.

CONCLUSION

This study compared the accuracy of the 2D flow solvers XFOIL and UNS2D. It was observed that although XFOIL is more time efficient than UNS2D, XFOIL does not provide accurate results for asymmetrical airfoils at negative angles of attack. Furthermore, the inaccuracy of the XFOIL airfoil tables led to unrealistically high figures of merit in the RotCFD simulations. While XFOIL is an excellent tool for quick airfoil table generation for symmetrical airfoils at any angle of attack and asymmetrical airfoils at positive angles of attack, it cannot confidently be used for asymmetrical airfoils at negative angles of attack.

Author contact:
Gianmarco Sahragard-Monfared
G.Monfared@nasa.gov

ACKNOWLEDGEMENTS

Special thanks are given to the late Professor Ganesh Rajagopalan and his wife, Professor R. S. Nappinnai, for granting the use of the RotCFD software. Gratitude and appreciation are also expressed to Kristen Kallstrom and Lauren Wagner from the NASA Ames Research Center for their support throughout this study.

REFERENCES

1. Kallstrom, K., "Exploring Airfoil Table Generation using XFOIL and OVERFLOW," Aeromechanics for Advanced Vertical Flight Technical Meeting, Transformative Vertical Flight 2022, San Jose, CA, 2022.
2. "AFTGen Final Report," Sukra Helitek, Inc., 2020.
3. Drela, M., "XFOIL: An Analysis and Design System for Low Reynolds Number Airfoils," Low Reynolds Number Aerodynamics, edited by T. J. Mueller, 1989. DOI: https://doi.org/10.1007/978-3-642-84010-4 1.
4. "OVERFLOW Module for AFTGen Application User Manual," Sukra Helitek, Inc., 2020.
5. "UNStructured 2D Solver Application RotCFD 0.9.15 (Beta) Build 402 K04 Theory," Sukra Helitek, Inc., 2018.
6. "UNS2D Module for AFTGen Application User Manual," Sukra Helitek, Inc., 2021.
7. Thakur, G., and Choudhary, N., "Comparative study of symmetrical Vs asymmetrical Vs semi-symmetrical airfoils," WEENTECH Proceedings in Energy, 09 2020, pp. 193–205. DOI: 10.32438/WPE.0602164.
8. Rajagopalan, R., Baskaran, V., Hollingsworth, A., Lestari, A., Garrick, D. P., Solis, E., and Hagerty, B. P., "RotCFD - A Tool for Aerodynamic Interference of Rotors : Validation and Capabilities," American Helicopter Society Future Vertical Lift Aircraft Design Conference, San Francisco, CA, 2012
9. Ballmann, J., Dafnis, A., Korsch, H. J., Buxel, C., Olivier, H., Braun, C., Baars, A., and Boucke, A., "Experimental Analysis of high Reynolds Number structural Dynamics in ETW," 46th AIAA Aerospace Sciences Meeting and Exhibit, 2008. DOI: 10.2514/6.2008-841.

Sensitivity Analysis and Aerodynamic Modeling of eVTOL Low-speed Predicted Handling Qualities

Agata Rylko
PhD Candidate
Politecnico di Milano
Milan, Italy

Lorenzo Favaro
MSc Candidate
Politecnico di Milano
Milan, Italy

Giuseppe Quaranta
Full Professor
Politecnico di Milano
Milan, Italy

ABSTRACT

This study explores aerodynamic modelling and the application of Sensitivity Analysis (SA) to the low-speed predicted Handling Qualities (pHQs) of a small-scale eVTOL, to identify the impact of uncertainty in modeling. The methodology integrates the statistical simulation tool Dakota with a Flight Simulation Model (FSM) to assess the propagation of input parameter uncertainties on pHQs: Attitude Quickness, Dynamic Stability, and Bandwidth. The aerodynamic analysis, conducted using the DUST aerodynamic model, determines the correct approach for computing dynamic stability derivatives. The results highlight the dependency of these derivatives on reduced frequency, emphasizing the need to consider this in FSM development and uncertainty analysis. A hybrid SA framework that combines the Morris One-at-a-Time (MOAT) method for preliminary analysis with Variance-Based Decomposition (VBD) using a Kriging meta-model for efficiency is implemented. Key parameters whose uncertainty affects pHQs, such as mass, system time delay and aerodynamic parameters, are identified using the SA. This work demonstrates how to implement a robust SA process for uncertainty analysis, to support informed decision-making in assessing the credibility of flight simulation models to be used for certification by simulation.

NOTATION

Symbols

C_{j_0}	Zero attitude coefficient
C_{j_α}	Derivative due to angle of attack
C_{j_q}	Derivative due to pitch rate
$\mathbf{c}$	Wing chord
I_y	Moment of inertia with respect to y-axis
k	Reduced frequency
K_T	Thrust coefficient
M	Mass
q	Pitch rate
S_i	First-order sensitivity index
S_{Ti}	Total effect index
T_u	Thrust derivative due to velocity
V_{lon}	Longitudinal velocity input
μ	Mean of the Elementary Effects
μ^*	Modified Mean of the Elementary Effects
ω_n	Natural frequency
ω_{BW}	Bandwidth
σ	Standard Deviation of the Elementary Effects
τ_d	Rotor's delay
τ_p	Phase delay
θ	Pitch angle
ζ	Damping

Presented at the Transformative Vertical Flight 11th Biennal Autonomous VTOL Technical Meeting, Phoenix, AZ, USA, Feb 4–6, 2025.

Acronyms

ACAH	Attitude Command/Attitude Hold
EASA	European Union Aviation Safety Agency
EE	Elementary Effect
eVTOL	electric Vertical Take-off and Landing
FFT	Fast Fourier Transform
FSM	Flight Simulation Model
FTM	Flight Test Maneuver
GP	Gaussian Process
HQ	Handling Qualities
LUT	Lookup Table
MC	Monte Carlo
MHQRM	Modified Handling Qualities Rating Method
MOAT	Morris One-at-a-time
MTE	Mission Task Element
pHQ	Predicted Handling Qualities
qLPV	quasi-Linear Parameter Varying
RoCS	Rotorcraft Certification by Simulation
SA	Sensitivity Analysis
TAS	True Air Speed
TPR	Transient Peak Ratio
VBD	Variance-based Decomposition
VPM	Vortex Particle Method

INTRODUCTION

eVTOL manufacturers are addressing new technical challenges alongside regulators and standards bodies to achieve certification. Their close collaboration is essential to ensure that new technologies, such as distributed propulsion and advanced flight control systems, meet stringent safety and performance requirements within ambitious time frames. In particular, to support compliance demonstration with EASA SC VTOL.2135 and VTOL.2145 (Ref. 1) using the EASA Modified Handling Qualities Rating Method (MHQRM) MOC.VTOL.2135 (Ref. 1) as Means of Compliance, EUROCAE released a new guideline, ED-295 (Ref. 2). The approach to collect evidence presented therein is based on newly defined Flight Test Manuevers (FTM), similar to the Mission Task Elements (MTEs) that were introduced in (Ref. 3). The MHQRM requires to investigate numerous flight conditions; some may present challenges to flight safety or be difficult to carry out these tests, or more simply the process may be too expensive to be done entirely through flight testing. Therefore, the ED-295 explicitly allows the use of simulation to complement flight tests.

To achieve the goal of applying Simulation by Certification at a desired level of decisiveness and impact, the approach outlined in the Rotorcraft Certification by Simulation (RoCS) (Ref. 4), as well as in previous publications (Ref. 5) and (Ref. 6), is followed. This approach requires first building the Flight Simulation Model (FSM) within the Domain of Validation, enabling simulation results to be compared against flight test data. During the validation process, it must be ensured that errors remain within the required tolerances, and uncertainties are assessed to evaluate the credibility of the simulation.

A similar methodology is adopted in the current research, consisting in the following steps:

1. For each FTM select a set of Predicted HQs to validate the model.
2. Identify the uncertainty associated with each relevant parameter, such as model inputs or numeric parameters.
3. Evaluate the error and uncertainty of each selected Predicted HQ metric to ensure that the error is constrained and the uncertainty is quantified.
4. Analyze the impact of simulation uncertainty on the evaluation of HQ ratings.
5. Conduct the FTM in the simulator to assess the HQ rating, taking into account the influence of uncertainty in this evaluation.

It is important to underline that the term "Predicted Handling Qualities" (pHQ) equivalent to "Flying Qualities", is intended as a set of objective characteristics of the aircraft that influence significantly the HQs, as defined in Ref. (Ref. 3). These characteristics, that can be quantified in numerical terms through a specific test, are different from the "Assigned HQ", which must instead be determined by the test pilots using the Cooper-Harper HQ rating Scale by performing specific MTEs.

In the current research, the focus is on FTM VTOL Deceleration to Hover. The chosen Predicted HQs, that is bandwidth, dynamic stability and attitude quickness in pitch, are used as fundamental indicators of the model's sufficiency for intended use, that is for flight simulation trials based on a designated FTM.

Building on the aerodynamic model developed in the previously published work (Ref. 6), which investigated the effect of propellers wake interference on the vehicle static derivatives and propeller-propeller interaction on the generated thrust, the current research addresses steps (2)-(4) of the outlined strategy. The paper begins with a sensitivity analysis aimed at quantifying the input parameters uncertainty propagation within the FSM. Based on engineering judgment, uncertainty ranges are assigned to key FSM input parameters identified as critical to the fidelity of the model. These ranges are systematically propagated through the FSM to assess their impact on the pHQ metrics. After the sensitivity analysis, the focus shifts to addressing an unresolved issue from the previous study: the dependency of dynamic stability derivatives on the reduced frequency. The derivatives are computed taking into account the variation in reduced frequency, and their resulting range of uncertainty is identified. Finally, the ranges of the identified aerodynamic parameters—dynamic derivatives from the current study and static derivatives from the previous one—are incorporated into the FSM which is utilized to recalculate the variations in pHQ metrics. This recalculation serves to verify the outcomes of the sensitivity analysis by focusing exclusively on aerodynamic parameters, providing a deeper understanding of their influence on the pHQs.

VTOL DESCRIPTION

The small unmanned eVTOL under investigation was designed and manufactured by the Aerospace Systems and Control Laboratory (ASCL) of the Department of Aerospace Science and Technology at the Politecnico di Milano (DAER) (Refs. 7–10), and is depicted in Figure 1. It utilizes 8 three-bladed propellers for vertical lift, each operating with variable RPM and fixed pitch. In helicopter mode, the aircraft is controlled by the differential thrust of all vertical propellers. For horizontal thrust, the aircraft is equipped with 2 two-blade propellers. The wings are rectangular without twist, while the empennage is H-shaped. The main geometry and mass properties of the eVTOL are detailed in Table 1. The eVTOL center of gravity is located in its vertical plane of symmetry and at 25% of the wing chord. This system has the scale of a drone and not the scale of an eVTOL capable of carrying passengers. However, it has been considered a valid benchmark for this work, since it resembles a typical lift + cruise eVTOL configuration. For the development of the FSM, it was chosen not to scale up the vehicle, to allow a direct comparison of the simulation results with future flight tests.

Table 1. eVTOL Characteristics.

Parameter	Value	Unit
Max Gross Weight	7	kg
Wing and fuselage span	3	m
Wing chord	0.26	m
Wing incidence	4.5	deg
Horizontal stabilizer span	0.721	m
Horizontal stabilizer chord	0.13	m
Horizontal stabilizer arm	1.08	m
Horizontal stabilizer incidence	0	deg
Vertical propeller diameter	0.178	m

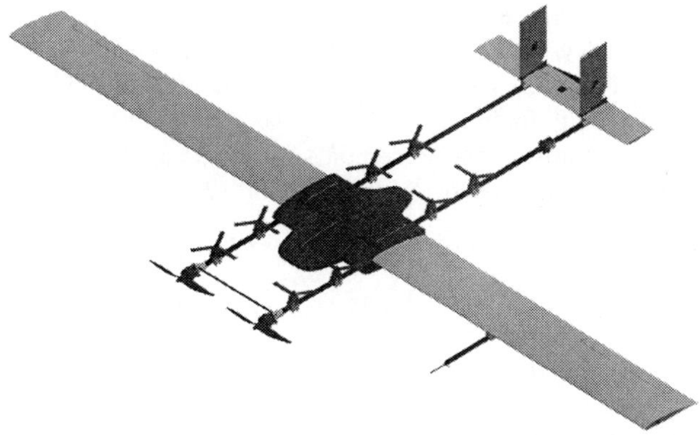

Figure 1. eVTOL overview.

METHODOLOGY AND MODELS DESCRIPTION

Sensitivity and uncertainty analysis techniques

The main features of the SA techniques employed in this work are outlined in this section. The first selected technique is Morris One-at-a-time (MOAT) which is able to distinguish between factors that have Negligible, Linear and Additive or Nonlinear/Interaction effects on the model output. The effect of each parameter, which can assume only a discrete number of values called "levels", is investigated by varying the parameter individually, along a set of "trajectories". A trajectory consists of $k+1$ points, where k is the number of specified model inputs. At each simulation point of the trajectory, only one input factor is varied with respect to the previous one (exactly "one-at-a-time"), in a derivative-like approach. In this way, for each subsequent couple of simulations, an Elementary Effect (EE) is defined as the incremental ratio produced by the changed variable. In the end, the EE distribution is averaged and two main sensitivity measures are obtained: the Mean of EE, μ, and their standard deviation, σ. A large μ is assigned to a factor whose overall effect on the output is high. In contrast, a variable with a high σ-level either undergoes nonlinear effects, increasing the variance in the results or significantly interacts with other parameters.

This work uses the Modified Mean μ^*, proposed in (Ref. 11), as an improved mean metric. μ^* averages the absolute values of the EE distribution, ranking parameters by importance and avoiding Type II errors, such as missing key parameters due to cancellation effects between EEs of opposite signs.

MOAT is primarily a method for preliminary analysis and it is not able to isolate possible interaction effects that may arise between FSM inputs from non-linearities. A more quantitative operating framework is offered by Variance-based Decomposition (VBD) approaches which, as (Ref. 12) states, help in ranking "the factors according to the amount of output variance that is removed when we learn the true value of a given input factor". To do so, a fundamental distinction shall be made between "first-order" and "total-effect" contributions. A first-order index acts as an "importance measure" of the single parameter taken alone and, for a purely additive model, represents exactly the percentage of the total output variance caused by each factor. Its quantitative representation is given by Sobol's main-effect index S_i, associated with the ith input factor. When the "overall" influence on the output cannot be solely attributed to the sum of all first-order effects, an interaction between two or more variables exists. Taking this into account, for a k-dimensional model, the total variance associated with a generic output Y could then be defined as

$$V(Y) = \sum_i V_i + \sum_i \sum_{j>1} V_{ij} + \ldots + V_{12\ldots k} \quad (1)$$

as described in (Refs. 12, 13), which accounts for all the possible forms of interactive effects. However the so-called "curse of dimensionality" makes "high-order" effects computation unpractical for the SA purpose. A well-established criterion is to compute the Total Effect index S_{T_i} instead, which groups the first-order effects of a factor with all of its interactive contributions. S_{T_i} not only is a standard SA index in the literature but has also shown good agreement with MOAT's μ^* and therefore allows for a comparative analysis between methods.

Since the amount of MC simulations needed to perform a VBD on the full FSM would make the process prohibitively expensive, a surrogate model is built for each of the pHQ response functions of interest, running a Monte Carlo (MC) analysis to generate a samples set of input/output relations and used to build a factor-response mapping that "emulates" the full behaviour of the eVTOL. The approach followed in this work employs Kriging interpolation (Ref. 14): building on a limited dataset of simulations performed on the true Simulink FSM, it uses the spatial correlation in the sample points to predict the value assumed by the response function at a new point, building an estimation of the unknown true response surface. This offers a fast computer emulator to be run in place of the true model to perform the MC simulations required to decompose the variance and compute Sobol Indices.

In this work, the approaches described in (Refs. 15–17) are taken as guideline. To reduce the number of FSM executions, the problem is first tackled qualitatively, performing a screening exercise through MOAT method. A VBD quantitative analysis follows, starting with a Monte Carlo simulation set. It employs Latin Hypercube sampling strategy, which ensures a better coverage of the input space for classic random sampling strategies. Next, a Kriging metamodel is created to perform the VBD at low computational cost. Finally, MOAT results and VBD-based Sobol Indices are compared.

VTOL Flight Simulation Model description

The FSM, developed in Matlab/Simulink environment at Politecnico di Milano (Refs. 7–10), is built upon a "stitched quasi-Linear Parameter Varying" (qLPV) model. Whereas a simple LPV model is a linear system whose state-space equations show a continuous dependence on time-varying scheduling parameters, a quasi-LPV system occurs when those same scheduling parameters are also part of the states of the system, eventually resulting in non-linear feedback. When a qLPV system is coupled with non-linear equations of motion, the full model is said to be a "stitched qLPV" (Refs. 18, 19). The simulator structure is divided into two fundamental sections: an Airframe block containing VTOL's dynamics description and its integration within the external environment, and the Control Unit, further divided into fixed-wing and multicopter flight control systems.

The function of the Airframe Block is to simulate VTOL's 6DOF rigid-body dynamics by integrating its non-linear equations of motion, taking into account aerodynamic, propulsive and gravitational contributions. Regarding aerodynamics, the used model is linear, accounting for contributions from all aerodynamic coefficients and stability derivatives. Neither stall nor propeller-airframe interaction effects are accounted for by the FSM. As a consequence of the qLPV modelling approach, the aircraft's trim states are computed through interpolation of a lookup table (LUT), pre-compiled along VTOL's airspeed envelope. Similarly, aerodynamic coefficients and stability derivatives are scheduled in an airspeed-dependent LUT, . These values have been initially estimated using OpenVSP software for aerodynamic modelling. Propulsive terms computation follows an analogous path. Employing TAS information and the throttle percentage command coming from the Control Unit, vertical rotors' angular speed, thrust and torque are obtained using three-dimensional LUT, containing data obtained during experimental wind tunnel tests. After applying mixer matrix conversion, multicopter thrust and moments become available, already set in the body frame.

The Control Unit for the VTOL mode is set in a cascade control arrangement, made of four loops: body angular rates, attitude angles, inertial velocity and inertial position. While velocities and rates use full PID controllers, position and attitude rely only on proportional ones. The cascade structure proves to be particularly useful for piloting mode definition, as managing the loops, provides different levels of assistance to the pilot. To investigate pHQs , the eVTOL is always operated in Position Mode, using longitudinal and lateral inputs to control the ground speeds, V_{lon} and V_{lat} respectively, vertical inputs to control the vertical velocity V_{vert}, and a separate command for the yaw rate, expressed as yaw setpoint. Consequently, position controllers remain always disconnected throughout the pHQ simulations, except for the altitude one.

Dakota and FSM integration

A statistical simulation tool was required to adapt the existing flight simulation model for SA purposes. Thanks to its flexibility and ease of use the choice naturally fell on Dakota, a state-of-the-art software for uncertainty quantification, model calibration and optimization. It offers various variance-based/sensitivity analysis methods to be used in Design of Computer Experiments practice, parameters studies, together with meta-modeling capabilities for the construction of powerful surrogate models. Moreover, it can be easily coupled with external computational codes such as the eVTOL FSM, by means of text files, treating the physics-based simulation models as simple black-boxes. More about Dakota, sensitivity and uncertainty analysis capabilities can be found in the related manual and software training documents (Refs. 14, 20–22).

DUST Aerodynamic model description

As a more accurate aerodynamic modelling approach, the computational tool called DUST developed by Politecnico di Milano, was used. DUST is open-source software that allows for the modelling of solid bodies such as thick surface panels, thin vortex lattice elements, and lifting line elements. In this study DUST version with the reformulated Vortex Particle Method (VPM) was used, which provides a more realistic wake modeling, through the reformulated vortex stretching estimation, described in (Ref. 23). The model developed in (Ref. 6) was used in configuration without the propellers, composed of wing and horizontal stabilizer. For the fixed lifting surfaces, vortex lattice elements were used, without aerodynamic tables correction (inviscid flow).

Dynamic stability derivatives due to pitch rate, C_{L_q} and C_{Mm_q} were calculated based on the single point method outlined in (Refs. 24, 25), by imposing a sinusoidal oscillatory pitching motion to the eVTOL. According to this method, the dynamic stability derivative, expressed as a coefficient dependent on the angle of attack, can be expressed as follows:

$$c_{jq} = \frac{\Delta c_{jq}(t)|_{t=t_{\alpha_0}}}{\alpha_A k} \qquad (2)$$

where α_A is the amplitude of the angle of attack for pitching oscillation, k is the reduced frequency, and the term $\Delta c_{jq}(t)|_{t=t_{\alpha_0}}$ corresponds to the operation of measurement of the thickness of the aerodynamic force or moment hysteresis ellipse around α_0 (see Figure 2).

SENSITIVITY ANALYSIS RESULTS AND DISCUSSION

To approach the pHQ definition problem systematically, the previously described SA techniques are merged together into

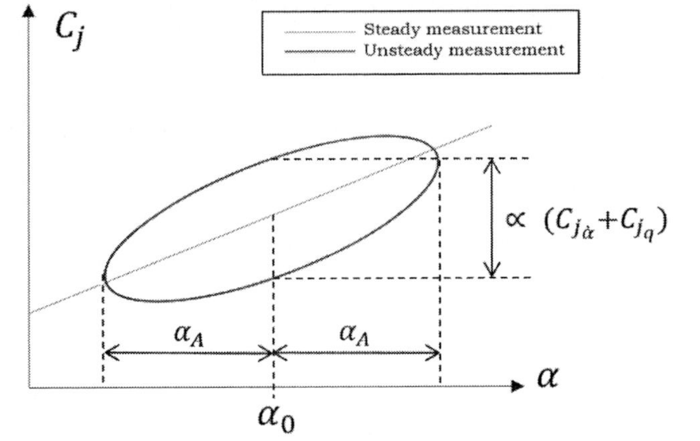

Figure 2. Graphical representation of the extraction of the dynamic derivative, from (Ref. 24).

Table 2. SA uncertain input variables.

Parameter	Value	Range	Unit
M	7	$\pm$ 0.5	kg
I_y	0.7237	$\pm$ 10%	$kg \cdot m^2$
K_T	$6.96 \cdot 10^{-8}$	$\pm$ 15%	-
T_u	1	$\pm$ 5%	-
τ_d	0	+ 0.03	s
C_{L_0}	0.7745	$\pm$ 20%	-
C_{D_0}	0.0183	$\pm$ 20%	-
$C_{M_{m_0}}$	-0.0240	$\pm$ 20%	-
C_{L_α}	5.1270	$\pm$ 20%	-
C_{D_α}	0.2558	$\pm$ 20%	-
$C_{M_{m\alpha}}$	-0.5402	$\pm$ 20%	-
C_{L_q}	7.6136	$\pm$ 20%	-
C_{D_q}	0.7179	$\pm$ 20%	-
$C_{M_{mq}}$	-16.169	$\pm$ 20%	-

a unique, comprehensive simulation framework. This allows for the repeatability of the tasks and makes the results on different pHQs comparable.

Table 2 shows the main FSM factors to whom an uncertainty range has been assigned, which are composed of three main subgroups: eVTOL's mass and inertia parameters (namely M and the pitch inertia moment I_y), aerodynamic coefficients and stability derivatives related to the longitudinal dynamics, and vertical rotors' characteristics. The latter include rotors' thrust coefficient K_T, thrust's dependence upon the longitudinal speed (modeled as a gain on the airspeed information fed by the FSM Sensor Unit to the thrust-related LUT, as not to influence thrust dependence upon throttle command) and an artificially added delay in rotors' response, formulated as a simple time-delay. The reason behind its addition comes from a series of experimental tests (Ref. 9) that indeed highlighted a significant delay A small amount of delay (0.02 s) was manually inserted in the system, allowing for the characterization of its effect on eVTOL's dynamic behaviour. It should be noted that the parameters list was originally longer, including parameters possibly related to the coupling between the pitch and the roll axes . However, a SA factor screening based on ADS-33 3.3.9.2 requirement ("Roll-due-to-pitch Coupling for Aggressive Agility") showed that there was no significant inter-axes coupling and all roll-related factors were consequently discarded.

Due to the differing nature of the factors, uncertain ranges are assigned based on different logic: aerodynamic and stability terms are derived from their nominal values, with a 20% increase/decrease applied. Similarly, I_y, K_T and T_u are defined in percentage terms, limiting their range of variability to avoid excessive interference with pHQ-related manoeuvres. The rotor delay, τ_d, as previously stated, follows the same logic. Since "there is insufficient information concerning the quantity of interest to specify either a fixed value or a precisely known probability distribution" (Ref. 26), all of these factors are treated as epistemic uncertain parameters, with their distributions generically assumed to be uniform. The mass has been varied in both directions by a maximum of 0.5 kg, accounting for the possibility of missions with different payloads; again, a uniform distribution has been used.

In the following paragraphs, a brief description of the required test manoeuvres and of the main adopted metrics precedes the SA results.

Moderate-amplitude pitch Attitude Changes (Attitude Quickness)

The first pHQ to be analyzed refers to ADS-33 Requirement 3.3.3, Pitch Attitude Quickness: the idea behind this criterion is "to characterize the helicopter's ability to achieve rapid, precise attitude changes when performing sharp, moderate amplitude manoeuvres" (Ref. 27).

However, given the dynamics and control system architecture of the eVTOL under study, a re-interpretation of the requirement was needed, for two main reasons. The first one regards the input specification: ADS-33 suggests the application of a step in the longitudinal cyclic for ACAH Response Types or a pulse for Rate Response Types. In the case of the eVTOL, when operated in Position Piloting Mode, this translates into a ΔV_{lon} input, thus making the identification of the Response Type challenging, not being directly linked to either of the previous categories. However, given that when a positive control input in the longitudinal speed is applied to the FSM, the drone pitches down to achieve the α_{trim} required by the new condition and when the command is released, the eVTOL returns to its previous state, in a fashion similar to what an ACAH-flown helicopter would do, a step was chosen as the best input option to perform the test.

The second challenge is associated with the correct definition of the output metrics: ADS-33 asks for the attitude change to "be made as rapidly as possible from one steady attitude to another" (Ref. 3) and requires the extraction of two quantities from the outputs' time histories: the ratio of the peak angular rate reached after application of the step to the peak θ change and the minimum attitude excursion experienced by the vehicle before reaching the new steady attitude. In the case

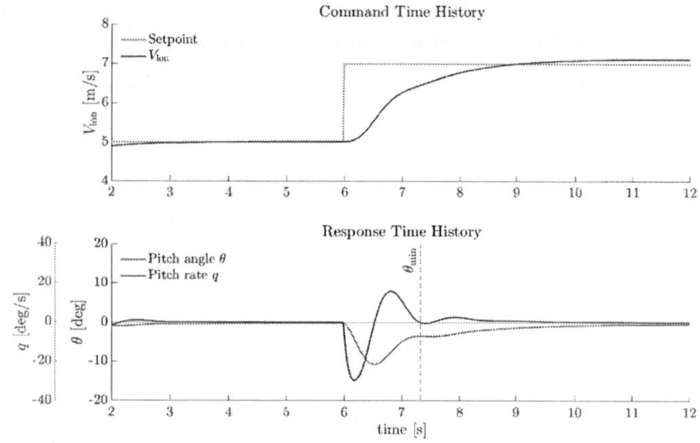

Figure 3. Attitude Quickness, Test Manoeuvre.

of this study the latter measure, mainly due to the small trim pitch angle the eVTOL flies at, does not directly fit the definition provided by ADS-33: in fact, when a step is applied to the FSM after its maximum attitude change the vehicle finds a new trim state without any undershoot, not allowing for a clear $\Delta\theta_{\text{min}}$ to be measured. The problem is overcome by looking at the recorded pitch-rate time history: the minimum attitude change is then taken as the angle reached by the vehicle when the on-axis angular rate approaches a stationary point for the second time after the application of the controller input.

Figure 3 shows the input/output time histories: ΔV_{lon} has been chosen as not to undergo stall effects (over all of the SA simulations $\Delta\theta_{\text{pk}}$ does not exceed 13 deg); this brings with it a limitation on the $\Delta\theta_{\text{min}}$, which indeed always remains below required minimum attitude change of 5 deg . Nevertheless, this does not affect the SA, which is still able to highlight the main trends.

Figures 4 and 5 display the Modified Mean μ^* against the standard deviation σ of the computed Elementary Effects, as in (Ref. 16). The parameters located in the lower right-hand area behave mostly monotonically, if not almost linearly, and appear to be highly significant in driving model's behaviour, but without interactions with the other uncertain variables. On the other hand, the variables placed in the upper left-hand region show a high non-linear/interactions effect. The x-axis representation quantitatively captures the outputs' variability. . Since μ^* is derived from the absolute value of the mean of the Elementary Effect (EE), it also indicates how much the output varies, on average, when varying the corresponding input of one Δ level. To compare factors and assess the relative importance, the criterion based on μ^* is selected. Alternatively, the Euclidean distance from the origin in the $\mu^* - \sigma$ plane could be employed (Ref. 15).

Figure 4 shows the results of MOAT application to $\Delta\theta_{\text{min}}$ index: at first glance, it is immediately clear that almost all factors behave monotonically, with little or no non-linearities/interactions. Secondly, a large number of them seem to affect the output measure, from mass to aerodynamic parameters. Aerodynamics plays indeed a role of primary importance: all lift-related coefficients are present, with a predominance of C_{L_0} and C_{L_α}. Pitching-moment stability derivatives appear too, together with drag-related ones (this constitutes a unicum among all the performed analyses, as pHQ rarely show dependence on drag, as emerges from the following paragraphs). Among the others, I_y and K_T prevail.

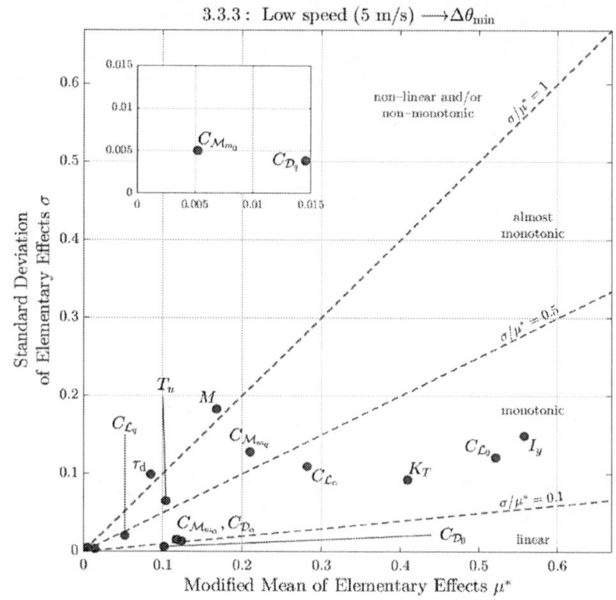

Figure 4. Attitude Quickness, $\Delta\theta_{\text{min}}$ Results.

Figure 5 illustrates how the behaviour changes drastically when focusing on the quickness metrics, making the distinction between parameters clearer. The eVTOL is mainly guided by mass and rotor-related factors, with response delay τ_d assuming a bigger role. As for aerodynamics, only C_{L_0} and C_{L_α} are visible. The other parameters are relegated to smaller orders of magnitude.

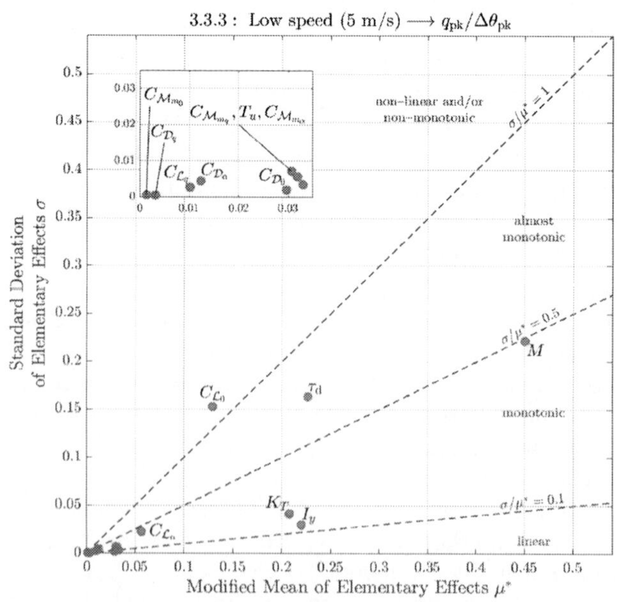

Figure 5. Attitude Quickness, $q_{\text{pk}}/\Delta\theta_{\text{pk}}$ Results.

MOAT results are now complemented with Kriging ones. Firstly a 500-sample MC simulation is performed, from which Dakota builds two different GP surrogates, one for each pHQ sensitivity measure. The metamodel is hence available to apply a VBD technique, from which Sobol indices are computed. The results are divided into two subsets: the S_i related to the first-order effects and the overall S_{T_i} related to the Total Effects, thus dividing each factor's own effect from its interaction with the others.

In Figure 6 Sobol Indices for both pHQ metrics are shown. Focusing on the minimum attitude change, good agreement with MOAT results is visible: both methods identify I_y, K_T and aerodynamic terms as the most significant ones, confirming the utmost importance of C_{L_0}, $C_{L\alpha}$ and $C_{M_{mq}}$. The small cluster of C_{D_0}, $C_{D\alpha}$ and $C_{M_{m\alpha}}$ is also correctly identified by both methods, with some small differences in the $\mu^* - S_{T_i}$ ranking as the variable importance decreases. Even though most of the variance is explained by the Main Effects, M and τ_d mainly contribute to the output through higher-order terms. This can be ascribed to their interactive effect: this can be ascribed to the lower end of the mass uncertainty range that, when combined with a sufficiently high level of rotors' delay, eventually drives the system into a sustained oscillation, hence causing a high level of variance in the results. The same combined effect reflects also into the pHQ specification of Figure 7: some outliers are in fact produced in the upper part of the MC output distribution, in both $\Delta\theta_{\text{min}}$ and $q_{\text{pk}}/\Delta\theta_{\text{pk}}$, which arise when both the mass and the delay simultaneously reach the extremes of their prescribed ranges.

Regarding the quickness index, VBD confirms MOAT results for the first 5 input factors (mass parameters, rotors' features and C_{L_0}). Although both methods correctly identify $C_{L\alpha}$, T_u and $C_{M_{m\alpha}}$, some discrepancies affect the lower-end part of the ranking, as $C_{M_{m0}}$ and C_{D_q} gain some positions with respect to the previous MOAT results, especially thanks to higher order effects (as shown by the white-coloured part of the histogram): even if these parameters are of much smaller of importance for the others (both MOAT and VBD locate them at a lower order of magnitude) and do not influence the factor prioritization process significantly, some caution should be placed in considering the effects of the aerodynamic parameters.

Figure 7 projects the MC results on the ADS-33 requirement plane: most of the variability is visibly displaced along the y-axis with the quickness values ranging from 2.5 to 3.25 ($\pm 13\%$). Furthermore, they exceed the typical limits the document prescribes for helicopters, being way above the Level 1 threshold which is due to the small dimensions of the vehicle.

Mid-term Response to Pitch Control Inputs (Dynamic Stability)

In this section, Requirement 3.3.2.3 is analyzed: the pHQ refers to the vehicle's dynamic stability and investigates eVTOL's poles, as extracted from the pitch-rate free response to a pulse in the longitudinal controller. In Figure 8 the main time

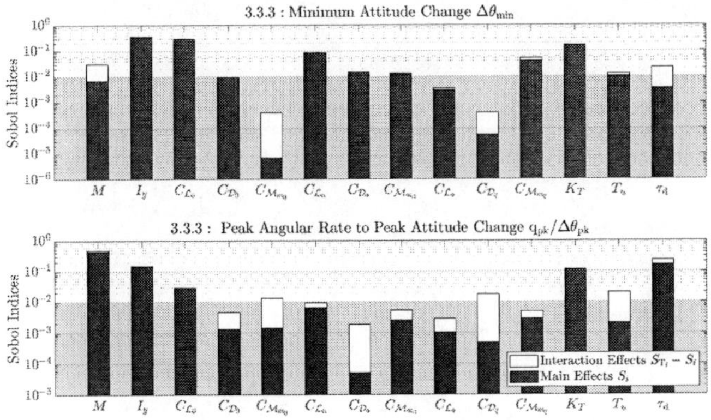

Figure 6. Attitude Quickness, VBD Results.

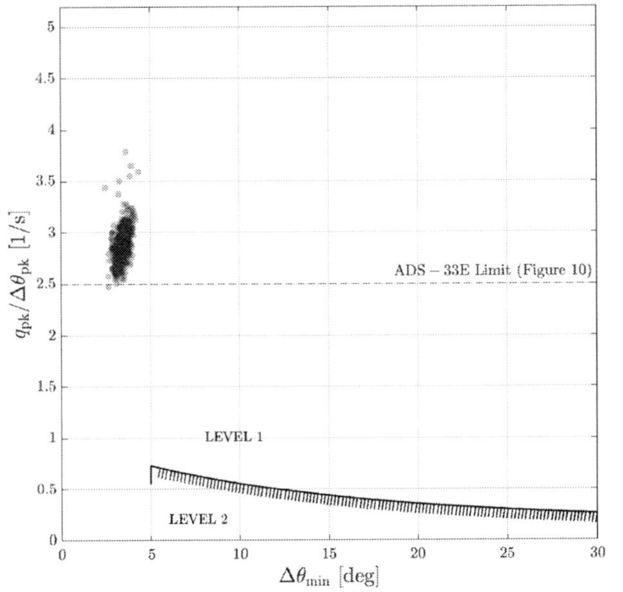

[1] The applied levels are derived from helicopter standards and may require revision to align with eVTOL-specific characteristics.

Figure 7. Attitude Quickness, pHQ ADS-33 Results.[1]

histories requirements are shown. Since ADS-33 states that "the portion of the time history used to calculate the damping ratio should occur after the input is completed", the first oscillation - driven by the pulse in the longitudinal speed - is neglected. Furthermore, it is important to measure the damping ratio from the free response only, not the forced response: this raised some concerns about the effect of the onboard attitude and rate controllers, which rapidly take the aircraft back to the previous trim state. To limit the effect of the control system on the response, the damping ratio ζ has been measured - using the Transient Peak Ratio (TPR) criterion, as suggested by ADS-33 itself (Ref. 3), employing the first peak and the first valley only. The second positive peak has been used only as a mean to compute the characteristic period of the oscillation and its natural frequency ω_{n}.

The MOAT results exhibit two main trends: first of all, aerodynamics assumes greater importance than in the Quickness framework, as C_{L_0} and $C_{M_{mq}}$ figure among the factors with the highest μ^* for the damping ratio ζ, to which C_{L_q}, $C_{D\alpha}$ and

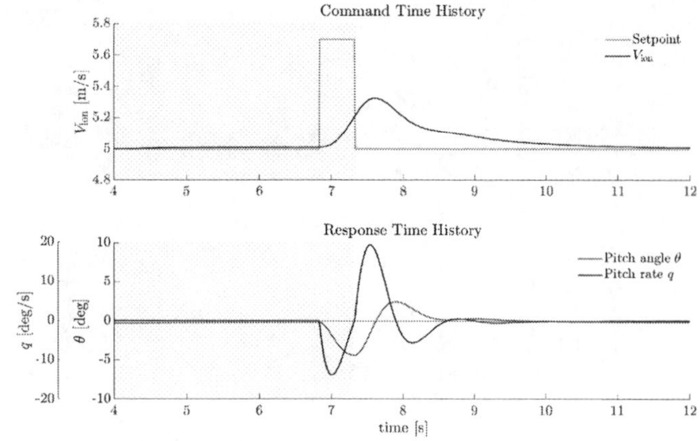

Figure 8. Dynamic Stability, Test Manoeuvre.

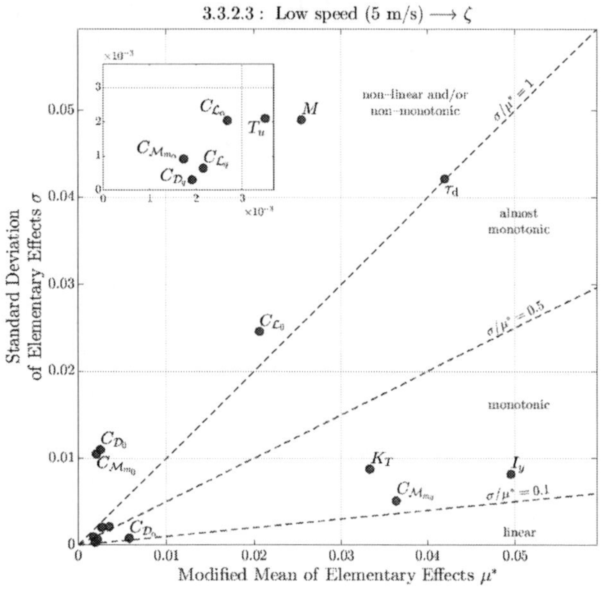

Figure 9. Dynamic Stability, ζ Results.

$C_{M_{m\alpha}}$ are added for the natural frequency. The second one is the high level of standard deviation associated with M and τ_d in the ζ results: this may suggest the existence of non-linearities or, possibly, an interactive effect between the variables (Figures 9 and 10).

Sobol Indices in Figure 11 confirms the MOAT results and, as the MOAT itself already predicted, indicate a strong Total order effect coming from the mass factor. Analyzing the MC training set, a behaviour analogous to the one described for the minimum attitude change in the previous section was recognized: when combined, mass and rotors' delay eventually drive the system into a Level 2 ζ specification.

Figure 12 shows how the aircraft flies over the Level 1/2 margin, with a predominance of Level 1 simulations. As MOAT highlighted in Figures 9 and 10, while the damping oscillates around the $\zeta = 0.35$ line of about ± 0.5, the frequency has a much wider variability and furthermore is much higher than the typical helicopter values expected by ADS-33 (far from the axes origin).

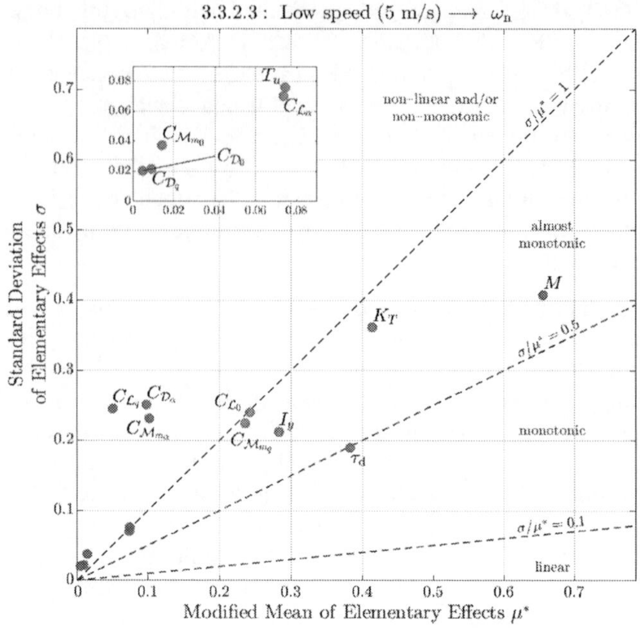

Figure 10. Dynamic Stability, ω_n Results.

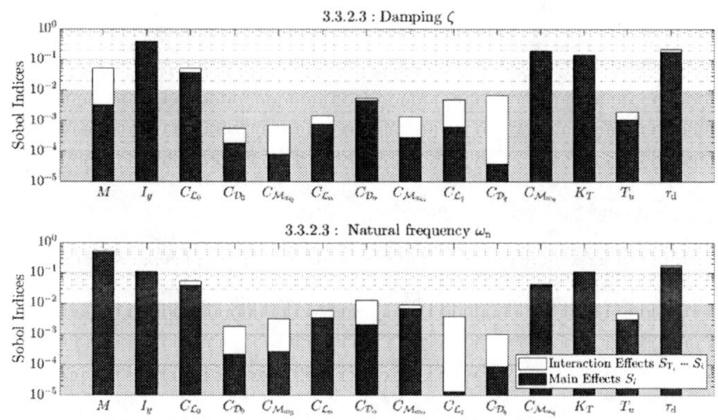

Figure 11. Dynamic Stability, VBD Results.

It is important to note that the DS values for this vehicle shown in Figure 12 lie well above the typical scale of the ADS-33 graph, with $\omega_n = 5.61$ rad/s corresponding to a natural oscillation frequency of 0.9 Hz. This represents a very high frequency, attributed to the small size of the vehicle.

Short-term Response to Pitch Control Inputs (Bandwidth)

The last pHQ subjected to analysis is 3.3.2.1, referred to as Bandwidth, which investigates aircraft's behaviour in the frequency domain, over a frequency range of interest. The reasoning behind the application of this requirement is "to determine if there is sufficient phase margin to allow the pilot to close the attitude loop [...] without threatening stability" (Ref. 28). The test course requires the pilot to apply a frequency sweep input in the longitudinal controller and to register the pitch angular rate response, which are processed by means of a Fast Fourier Transform (FFT) to obtain Bode plots. The two quantities of interest are ω_{BW} (bandwidth) - taken as

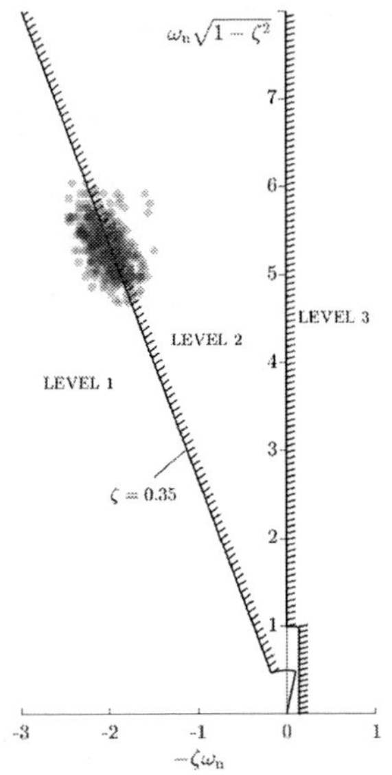

[2] The applied levels are derived from helicopter standards and may require revision to align with eVTOL-specific characteristics.

Figure 12. Dynamic Stability, pHQ ADS-33 Results.[2]

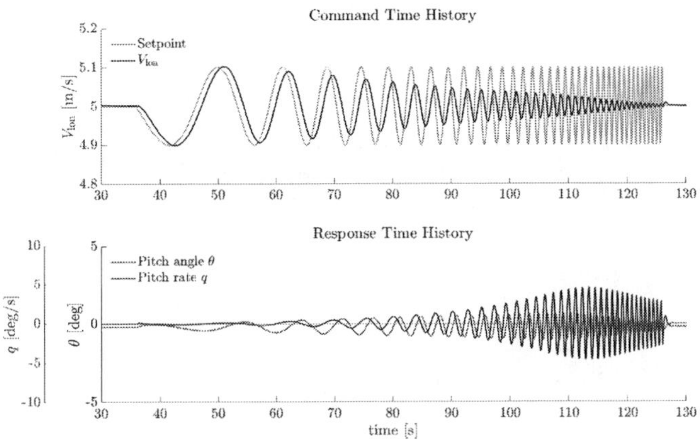

Figure 13. Bandwidth, Test Manoeuvre.

$\omega_{BW_{phase}}$ because of the ACAH Response Type (Ref. 3) - and τ_p (phase delay).

Following the instructions of (Ref. 28), a 90 s long automated sweep is generated , with a frequency content linearly increasing from 0.3 rad/s to 12 rad/s (approximately from 0.05 Hz to 2 Hz) that is the pilot frequency of interest. Figure 13 illustrates the reference sweep in the longitudinal speed (green curve) and the related q-rate response.

However, setting a 90 s input naturally leads to inherently time-demanding simulations which, in the context of a MOAT or MC analysis, could easily become prohibitively expensive. To overcome this problem, a different approach is explored for the purposes of SA application. Instead of analyzing the full non-linear system response, the criterion is instead applied to a linearized FSM . The main idea is to linearize the longitudinal dynamics near a trim condition point and to generate the related Bode plot employing Matlab®/Simulink® *Model Linearizer* routine. By doing so, the time cost is significantly brought down and the main trends in eVTOL's bandwidth be-

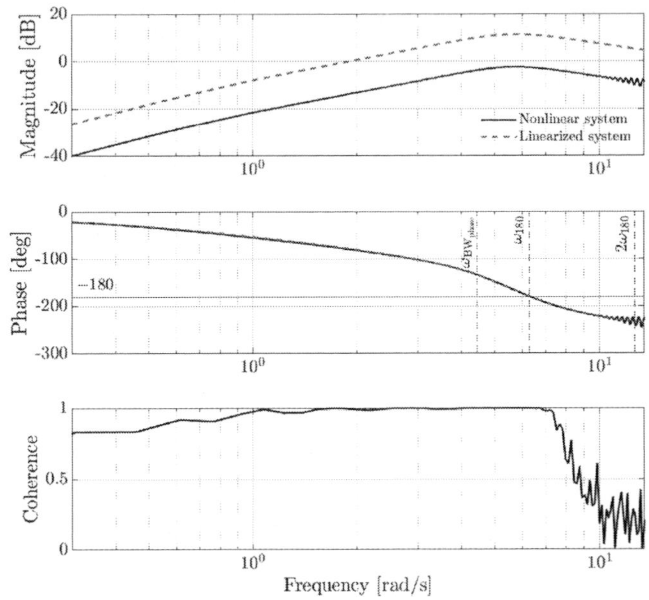

Figure 14. Closed-loop Bode plot for $q/\Delta V_{lon}$.

haviour are still captured. To check for the applicability of the method, FFT-based response of the full non-linear system is compared to the linearized one (around a $V_{lon} = 5$ m/s trim point), as shown in Figure 14 .

Even though some discrepancies are present between the magnitude Bode plots, the phase behavior shows to be valid inside of the specified frequency range. It should be made clear that the agreement between models holds as long as the sweep applied to the non-linear model stays within reasonably small amplitudes. As the applied sweep's amplitude increases, differences may arise. A coherence plot is also shown for the FFT-based results, as ADS-33 indicates a minimum value of 0.6 to check for the goodness of the non-linear system sweep results.

The discussion of Bandwidth SA results is straightforward, following the main considerations already made for the previous pHQ. In both SA indices (from Figures 15 and 16) - aside from the typical prevalence of the mass, inertia and thrust coefficient - aerodynamics exhibits a noticeable effect, mainly due to C_{L_0} and $C_{M_{m_q}}$. At a lower level, at least for the bandwidth metrics, α-related derivatives must be also accounted for. Sobol Indices confirm the MOAT results , as shown in Figure 17.

At last, the MC results are compared with the expected limits: Figure 18 shows a cluster of points well above the threshold between the levels, placing the eVTOL in Level 1. $\omega_{BW_{phase}}$ shows a wider range of variability for τ_p: when parameters are varied in the MOAT and MC simulations the Bode phase curve undergoes greater variation in the range between the bandwidth and the crossover frequency (ω_{180}), shifting bandwidth values all over the 4-5 rad/s interval. Since the phase curve is less sensitive to variations for $\omega > \omega_{180}$, τ_p instead remains almost unchanged after ω_{180}, resulting in a much lower variance in phase delay results.

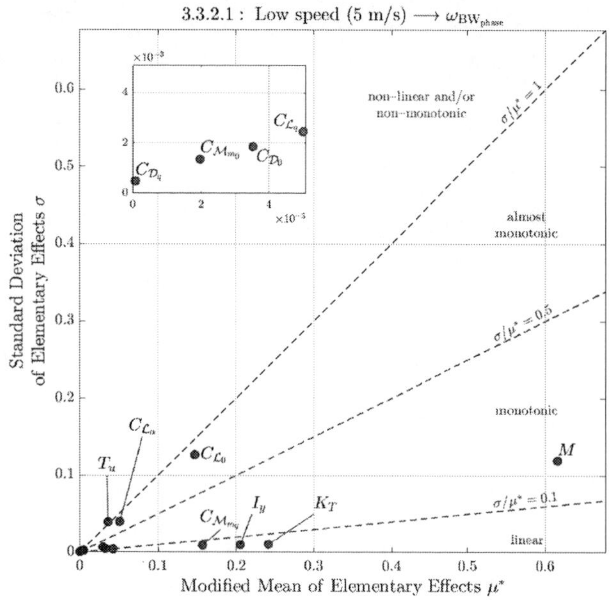

Figure 15. Bandwidth, $\omega_{BW_{\text{phase}}}$ Results.

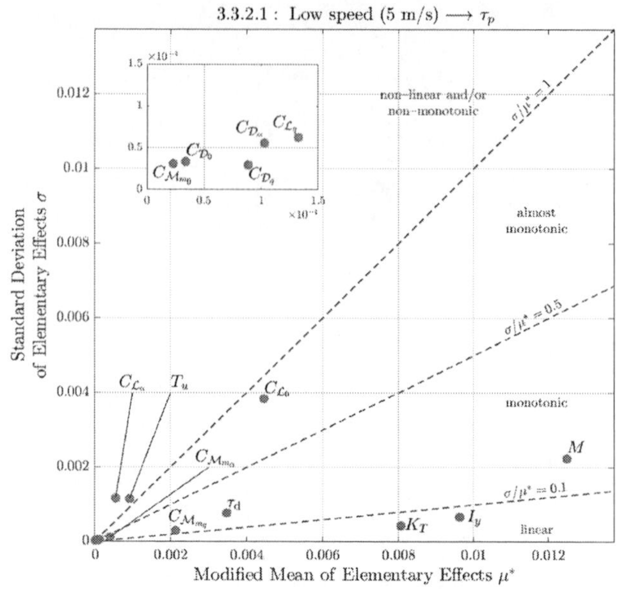

Figure 16. Bandwidth, τ_p Results.

AERODYNAMIC ANALYSIS RESULTS AND DISCUSSION

Dynamic stability derivatives for an isolated wing based on the Theodorsen model

In the previous work (Ref. 6), the derivatives due to pitch rate C_{L_q} and C_{Mm_q}, computed according to the methodology reported in (Refs. 24, 25) were found to be dependent on the oscillation frequency. To understand this unexpected dependency, a comparative study with a simple isolated rectangular wing, using Theodorsen theory was conducted. The eVTOL wing has a relatively high aspect ratio $b/c = 11.5$, so the 3D flow effects have less influence on the unsteady aerodynamics and Theodorsen's theory can provide a useful approximation of the aerodynamic response. A detailed description of the model and the derivation of the equations can be found in (Ref. 29). The methodology allows to compute the static and dynamic derivatives for pitching and plunging sinusoidal oscillations, to evaluate their dependency on frequency. In this section only the dynamic derivatives are reported, computed as follows, for the pitching oscillation $\alpha = \alpha_0 sin(\omega t)$:

$$C_{L_q} + C_{L_{\dot{\alpha}}} \quad = \quad \frac{1}{k}\frac{\Im(\mathsf{C_L})}{\alpha_0} \tag{3}$$

And for the heave oscillation $h = h_0 sin(\omega t)$:

$$C_{L_{\dot{\alpha}}} \quad = \quad -\frac{1}{k^2}\Re(\mathsf{C_L})\frac{c}{2h_0} \tag{4}$$

Where k is the reduced frequency $k = \frac{c\omega}{2U_\infty}$.

Figure 19 presents the results obtained for different pitching and heave oscillation frequencies, in function of reduced frequency, which range corresponds to frequencies between 0.1 and 3 Hz. The calculations were performed for a wing with

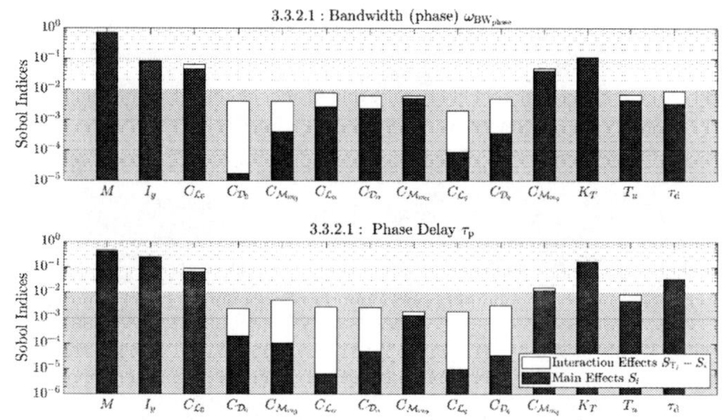

Figure 17. Bandwidth, VBD Results.

a non-symmetric profile Selig 2046, for incompressible and inviscid flow, in the range of $\alpha = \pm 5$ deg. The amplitude of sinusoidal pitching motion was set to 2 deg for all simulations. The heave amplitude was imposed so that the resulting amplitude of the angle of attack oscillation was equal to 2 deg.

The lift dynamic derivative, $C_{L_q} + C_{L_{\dot{\alpha}}}$, exhibits a strong dependency on the reduced frequency, which becomes less pronounced at higher reduced frequencies. The lift dynamic derivative due to heave motion, $C_{L_{\dot{\alpha}}}$, shows a similar dependency on the reduced frequency, is always negative, and cannot be neglected. The lift derivative due to pure pitch rate, C_{L_q}, was determined as the difference between the aforementioned quantities and corresponds to a value that could be obtained by simulating a phugoid motion where $\alpha =$ const. It is positive, indicating that an increase in pitch rate (nose-up motion) results in an increase in overall lift and it decreases at higher reduced frequencies, due to the growing influence of unsteady flow effects.

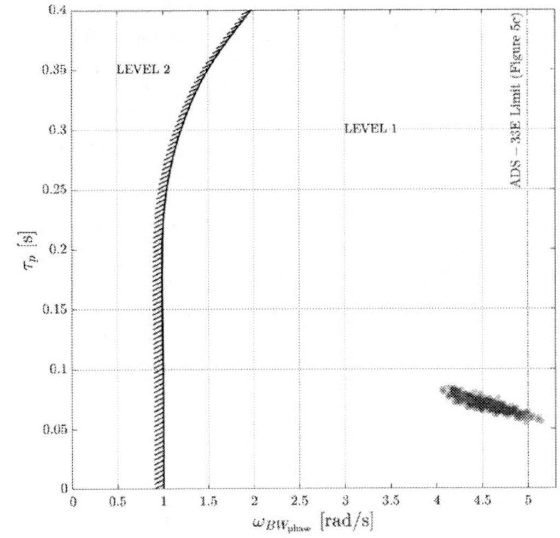

[3] The applied levels are derived from helicopter standards and may require revision to align with eVTOL-specific characteristics.

Figure 18. Bandwidth, pHQ ADS-33 Results.[3]

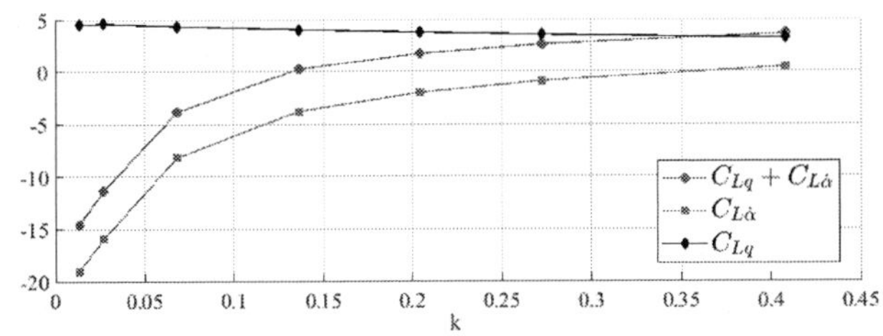

Figure 19. Lift derivative due to pitch rate vs reduced frequency, for pitch oscillation and pure heave case, obtained based on Theodorsen theory.

Stability derivatives results using DUST model

In this section, the results of the calculation of lift and pitching moment derivatives in DUST are reported. Figure 20 shows the lift and pitching moment dynamic derivatives, $C_{L_q} + C_{L_{\dot{\alpha}}}$ and $C_{Mm_q} + C_{Mm_{\dot{\alpha}}}$ respectively, for an isolated wing, obtained from pitching sinusoidal oscillations conducted using DUST. It can be seen that the lift dynamic derivative has the same dependency on the frequency as in Figure 19 and is positive. The pitching moment dynamic derivative is negative, meaning that the wing is self-stabilizing.

Both methods exhibit similar trends in the computation of the dynamic stability derivatives, but there still are quantitative differences in the magnitude of derivatives. Figure 21 represents the lift and pitching moment dynamic derivatives, for a wing and horizontal tail. It was obtained by oscillation of 2 deg about a 0 mean angle of attack. It can be seen that the horizontal tail provides a negative contribution to the pitching moment derivative due to pitch rate. A negative C_{Mm_q} is essential for longitudinal dynamic stability, as it ensures the damping of oscillatory modes like the short-period mode or phugoid mode. As expected, for different velocities, but maintaining the same k, the computed derivatives have the same values.

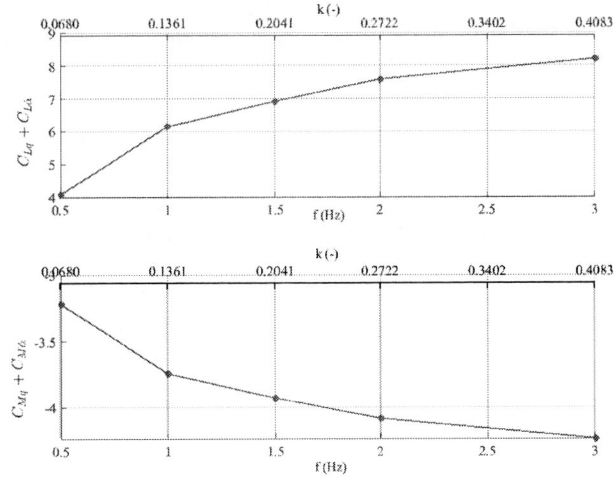

Figure 20. Lift and pitching dynamic derivatives for an isolated wing, for 6 m/s.

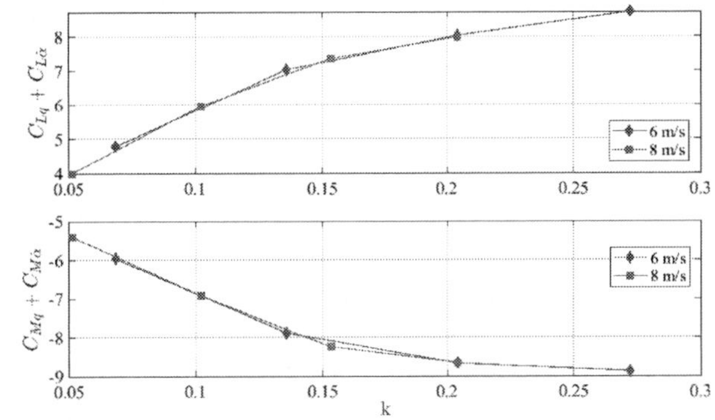

Figure 21. Lift and pitching dynamic derivatives for a wing and horizontal stabilizer, for 6 and 8 m/s.

Moderate-amplitude pitch Attitude Changes (Attitude Quickness)

Based on the MOAT and VBD results presented in the previous section, significant aerodynamic parameters were selected to perform comparative Attitude Quickness simulations (see Table 3). The initial ranges of aerodynamic parameters, as listed in Table 2, were refined by studying the influence of propeller wake on the static stability derivatives and the effect of viscosity on the computation of C_{L_α} (Ref. 6), as well as the variation in dynamic derivatives due to frequency, as discussed in the previous subsection.

The maximum value of C_{L_0} represents the value without propeller wake interference, while the minimum value accounts for lift loss on the wing and negative lift on the horizontal stabilizer. Although C_{Mm_0} would also change due to propeller wake interference, it is kept constant in this study, as it was found to be negligible in the previous sensitivity analysis. The minimum value of C_{L_α} corresponds to the lift coefficient of the wing and horizontal tail for inviscid flow, while the maximum value was computed in (Ref. 6), considering low Reynolds numbers for the vehicle scale and flight velocity. The maximum and minimum values of C_{L_q} and C_{Mm_q} were computed

Table 3. Aerodynamic parameters values for Attitude Quickness.

Case:	1	2	3	4	5
C_{L0}	0.4524	**0.6268**	**0.6268**	**0.6268**	**0.6268**
$C_{M_{m0}}$	-0.0240	-0.0240	-0.0240	-0.0240	-0.0240
$C_{L\alpha}$	5.115	5.115	**7.48**	**7.48**	**7.48**
$C_{M_{m\alpha}}$	-0.5402	-0.5402	-0.5402	-0.5402	-0.5402
C_{L_q}	5.9649	5.9649	5.9649	**7.9898**	**7.9898**
$C_{M_{mq}}$	-6.9263	-6.9263	-6.9263	-6.9263	**-8.6516**

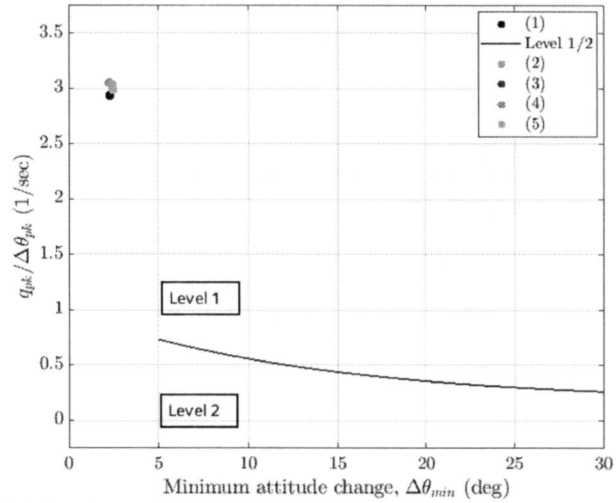

[1] The applied levels are derived from helicopter standards and may require revision to align with eVTOL-specific characteristics.

Figure 22. Attitude Quickness results variability due to uncertainty of aerodynamic parameters.[1]

for $k = 0.1$ and $k = 0.2$, respectively, corresponding to oscillation frequencies of $0.7 - 1.5$ Hz.

The manoeuvres were conducted in the same manner, with an initial trim condition in forward horizontal flight at 5 m/s, giving a step input into the horizontal velocity V_{lon}. The results of Attitude Quickness are presented in Figure 22. No delay was introduced ($\tau_d = 0$). The changes to the aerodynamic parameters were implemented gradually, one parameter at a time, in each simulation.

As anticipated in SA, this analysis confirms that the attitude quickness parameter exhibits low sensitivity to variability in aerodynamic parameters. Even when these parameters are modified, the quickness value at the uppermost position on the y-axis is only 4% higher than the baseline value, indicating a limited influence. The variability in the minimum attitude change further reinforces this observation. Despite simultaneous changes to multiple aerodynamic parameters, the variability is constrained within a narrow range, spanning from -4% to $+7\%$.

This consistency implies that the aerodynamic parameters exert only a marginal effect on attitude quickness, at the investigated low speed and small amplitudes, and are predominantly governed by other factors, such as mass or delay, or are inherently robust to aerodynamic variability. This behaviour might

Table 4. Aerodynamic parameters values for Dynamic Stability.

Case:	1	2	3	4
C_{L0}	0.4524	**0.6268**	**0.6268**	**0.6268**
$C_{M_{m0}}$	-0.0240	-0.0240	-0.0240	-0.0240
$C_{L\alpha}$	5.115	5.115	5.115	5.115
$C_{M_{m\alpha}}$	-0.5402	-0.5402	-0.5402	-0.5402
C_{L_q}	5.9649	5.9649	5.9649	5.9649
$C_{M_{mq}}$	-6.9263	-6.9263	**-8.6516**	**-16.169**

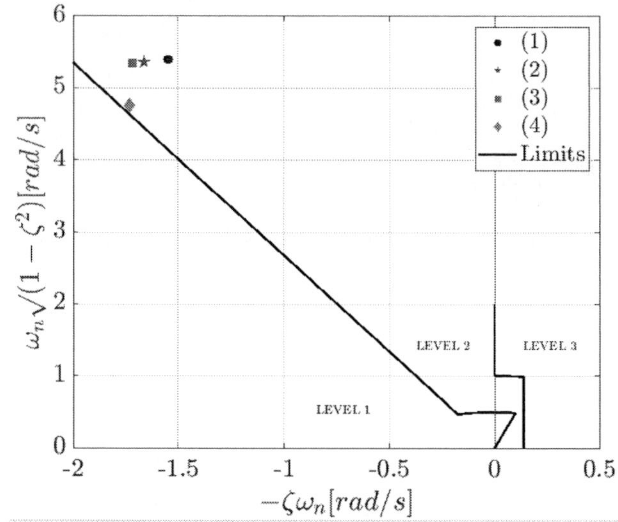

[1] The applied levels are derived from helicopter standards and may require revision to align with eVTOL-specific characteristics.

Figure 23. Dynamic Stability variability due to uncertainty of aerodynamic parameters.[1]

change at higher speeds and higher pitch angles, where dynamic stall effects should be considered.

Mid-term Response to Pitch Control Inputs (Dynamic Stability)

Following the strategy outlined in the previous subsection, the most significant aerodynamic parameters for Dynamic Stability, identified from the SA analysis—namely C_{Mmq} and C_{L0}—were selected to perform a quantitative comparison of Dynamic Stability results (see Table 4). Again, the simulations were performed in the same way, introducing a pulse of $V_N = 0.7$ m/s of duration of 0.5 s during a trimmed horizontal flight at 5 m/s. The damping and natural frequency were computed, as described in the previous section dedicated to Dynamic Stability, considering the first peak and the first valley of pitch rate for ζ, and including the second positive oscillation peak for ω_n.

In contrast to Attitude Quickness, Dynamic Stability demonstrates greater sensitivity to aerodynamic parameters (see Figure 23), such as C_{L0} and $C_{M_{mq}}$. The lift coefficient C_{L0} visibly influences damping, particularly shifting the pHQ closer to Level 1 under the studied conditions. This indicates that ignoring the effects of propeller wake interference, which decreases the zero aerodynamic lift coefficients, can lead to overestimated stability predictions. Moreover, as previously stated, the small vehicle's dimensions result in higher and more spread ω_n values, which exceed typical helicopter scales.

After computing the dynamic derivatives in this subsection, it was observed that the obtained C_{Mmq} is significantly higher than the values reported in recent studies (Refs. 7–10). This value was also used in the sensitivity analysis conducted in the previous section (see Table 2). Due to the high sensitivity of the DS to C_{Mmq}, a comparative simulation, referred to as (4) in Table 4, was performed, which demonstrates further increment in ζ and a decrease in ω_n. The discrepancies between the results could not be explained, either by tracing back the design changes of the aircraft or by analyzing the employed reference frames.

Short-term Response to Pitch Control Inputs (Bandwidth)

The bandwidth and phase delay trials described in this section were conducted using the non-linear FSM, and with a greater sweep amplitude compared to the simulations presented in the sensitivity analysis section. The sweep was performed with an amplitude of $V_{lon} = 0.5$ m/s, resulting in a maximum pitch angle of approximately 5°. It was verified that this amplitude does not affect the computation of bandwidth and phase delay.

Table 5. Aerodynamic parameters values for Bandwidth and phase delay.

Case:	1	2	3	4	5
C_{L_0}	0.4524	**0.6268**	**0.6268**	**0.6268**	0.4524
$C_{M_{m_0}}$	-0.0240	-0.0240	-0.0240	-0.0240	-0.0240
$C_{L\alpha}$	5.115	5.115	**7.48**	**7.48**	5.115
$C_{M_{m\alpha}}$	-0.5402	-0.5402	-0.5402	-0.5402	-0.5402
C_{L_q}	5.9649	5.9649	5.9649	5.9649	5.9649
$C_{M_{mq}}$	-6.9263	-6.9263	-6.9263	**-8.6516**	**-16.169**

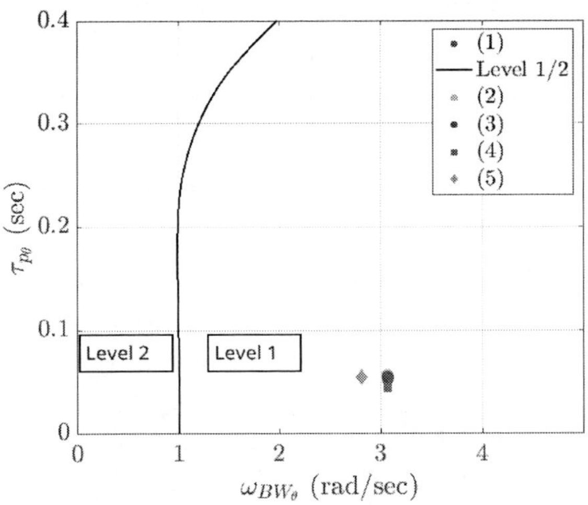

[1] The applied levels are derived from helicopter standards and may require revision to align with eVTOL-specific characteristics.

Figure 24. Bandwidth and phase delay variability due to the uncertainty of aerodynamic parameters.[1]

The parameters selected from the previous MOAT and VBD analyses are C_{L_0}, $C_{L\alpha}$, and C_{Mm_q} (see Table 5). Figure 24 shows the bandwidth and phase delay results for different aerodynamic data sets. Similarly to Attitude Quickness, Bandwidth and Phase Delay are not significantly sensitive to aerodynamic parameters. Changes in C_{L_0} and $C_{L\alpha}$ (Cases 2 and 3) result in a minimal increase in τ_d, while C_{Mm_q} causes a decrease in τ_d of approximately 17%. However, it is worth noting that the phase delay is generally very low because the model does not include a delay in propeller response, considering only the intrinsic delay of the control system.

After significantly modifying the derivative C_{Mm_q} to cross-check with previous computations, a dependency of ω_{BW} also becomes evident.

CONCLUSIONS

The applied SA analysis demonstrated that a valid framework can be built coupling the FSM with an external statistical tool, such as Dakota. The chosen methods , MOAT and emulator-based VDB, show satisfactory agreement in identifying the most influential factors, with a relatively small number of simulations.

Concerning Attitude Quickness, the SA showed that most of pHQ variance is to be ascribed to changes in the quickness parameter $q_{\text{pk}}/\Delta\theta_{\text{pk}}$: mass, inertia, and rotor-related factors play a primary role, while aerodynamics contributes to the problem only secondarily through C_{L_0} and C_{L_α}. Minimum attitude change $\Delta\theta_{min}$ depends on a much wider set of factors, among which aerodynamics coefficients and stability derivatives prevail, but shows little to no variability. A possible cause for this behavior can be sought in the limited pitch angles , constrained by the applied linear aerodynamic model, or by the intrinsic robustness of this pHQ on FSM aerodynamic parameters. The aerodynamic analysis that follows the SA, confirms that the Attitude Quickness is relatively insensitive to aerodynamic parameter variations, under low-speed and small-amplitude conditions.

Dynamic stability is situated at the boundary between Levels 1 and 2, with Level 2 values mainly caused by the coupled $M - \tau_d$ effect on ζ. Dynamic Stability exhibits higher sensitivity to aerodynamic parameters, particularly C_{L_0} and $C_{M_{mq}}$, in comparison to Attitude Quickness. Variations in the lift coefficient C_{L_0} have a pronounced impact on damping, bringing the pHQ closer to Level 1 under the analyzed conditions. Failing to account for the reduction in C_{L_0} caused by propeller wake interference may lead to an overestimation of stability.

Bandwidth analysis revealed that for the SA, the linearization approach can satisfactorily substitute the application of an automated sweep to the non-linear FSM, to significantly reduce the computational effort. However, high-amplitude commands should be applied too, to explore possible discrepancies arising further from the linearization point. The MC analysis indicated that most of the variability is to be attributed to $\omega_{BW_{\text{phase}}}$, which is mostly sensitive to M, K_T, I_y and C_{Mmq}. In the aerodynamic analysis, bandwidth and phase delay were found to be rather non-sensible to aerodynamic parameters alone, with C_{Mmq} having the highest impact on bandwidth, which confirms the SA results.

Overall, among aerodynamic parameters, C_{L_0} and $C_{M_{mq}}$ have the most significant impact on the analyzed pHQs. In order to better characterize their effect, an aerodynamics-only SA could be conducted, using the coupled Dakota/FSM approach.

Having identified the predominant uncertain factors and building on the preliminary SA results, a deeper characterization of inputs' ranges and probability distributions may be considered, to resize the effect of mass, inertia, and rotor-related parameters (which in some cases may absorb the majority of output variance, thus overshadowing aerodynamics) and to account for possible range-dependent behaviours or non-linearities which may occur beyond the intervals' limits employed in this study. Similarly, pHQ manoeuvres could be executed at different V_{lon}, as to investigate the dependency of aerodynamics on the test velocity.

Given the width of Dakota capabilities, different MOAT settings could be studied, as well as different meta-modelling techniques may be introduced in the SA framework to compute, for example, second-order interactions between parameters.

The aerodynamic analysis using DUST simulations demonstrated that the trend of the lift dynamic derivative due to pitching, $C_{Lq} + C_{L\dot{\alpha}}$, corresponds to the trend predicted by the Theodorsen model. Therefore, the methodology for obtaining stability derivatives due to pitch rate is proven to be correct, but the dependency on the reduced frequency should be accounted for when building the FSM, or at least in the later stage of RCbS, during uncertainty analysis. To obtain pure C_{Lq} and C_{Mmq} derivatives, a phugoid maneuver should be simulated in DUST.

The research presented in this paper demonstrates how to perform a robust and automated sensitivity analysis, as an essential component of uncertainty analysis, and how to propagate model input uncertainties onto the final pHQ metrics. This activity supports the model user in gaining a deeper understanding of the uncertainty in FSM predictions, thereby enabling correct decisions within the certification process and building model credibility with the authority.

Author contact:
Agata Rylko - agata.rylko@polimi.it
Lorenzo Favaro - lorenzo.favaro@mail.polimi.it
Giuseppe Quaranta - giuseppe.quaranta@polimi.it

ACKNOWLEDGMENTS

The work in this study done by Agata Rylko was carried out within the 38th cycle of doctoral study in aerospace engineering, in the research field "Policymaking for Urban Air Mobility acceptance" and received funding from PIANO NAZIONALE DI RIPRESA E RESILIENZA (PNRR). The work in this study done by Giuseppe Quaranta was carried out within MOST -- Sustainable Mobility National Research Center and received funding from the European Union Next-Generation EU (PIANO NAZIONALE DI RIPRESA E RESILIENZA (PNRR) – MISSIONE 4 COMPONENTE 2, INVESTIMENTO 1.4 – D.D. 1033 17/06/2022, CN00000023). This manuscript reflects only the authors' views and opinions, neither the European Union nor the European Commission can be considered responsible for them.

REFERENCES

1. Anon., *Third Publication of Means of Compliance with Special Condition VTOL, MOC-3 SC-VTOL Issue 2*, Cologne, Germany, EASA, 2023.

2. Anon., *ED-295, Guidance on VTOL Flight Control Handling Qualities Verification*, EUROCAE, 2024.

3. Anon., *Proposed Revisions to Aeronautical Design Standard – 33E (ADS-33E-PRF) Toward ADS-33F-PRF*, U.S. Army Combat Capabilities Development Command, Redstone Arsenal, AL, September 2019.

4. Lu, L., van't Hoff, S., Padfield, G., Podzus, P., Quaranta, G., and White, M., *Rotorcraft Certification by Simulation and Analysis; an Introduction to the Principles and Practices*, Springer, to be published, 2023.

5. Rylko, A., van't Hoff, S., Lu, L., Padfield, G., Podzus, P., White, M., and Quaranta, G., "Rotorcraft Flight Simulation to Support Aircraft Certification: A Review of the State of the Art with an Eye to Future Applications," 49th European Rotorcraft Forum, Bückeberg, Germany, September 2023.

6. Rylko, A., Quaranta, G., and M., L., "Aerodynamic Model Development and Fidelity Assessment for eVTOL Predicted Handling Qualities," 50th European Rotorcraft Forum, Marseille, France, September 2024.

7. Battaini, N., *Design and dynamic modeling of a VTOL UAV*, Master's thesis, Politecnico di Milano, Italy, 2019.

8. Martinelli, E., *Modelling, control, integration and testing of an eVTOL drone*, Master's thesis, Politecnico di Milano, Italy, 2022.

9. Martello, N. C., *Modelling and integration of an eVTOL UAV*, Master's thesis, Master Thesis, Politecnico di Milano, Italy, 2021.

10. Marchesi, M., *Structural redesign, simulation, control, and state estimation of an eVTOL UAV*, Master's thesis, Politecnico di Milano, Italy, 2024.

11. Campolongo, F., Cariboni, J., and Saltelli, A., "An effective screening design for sensitivity analysis of large models," *Environmental Modelling & Software*, Vol. 22, (10), 2007, pp. 1509–1518.

12. Saltelli, A., Tarantola, S., Campolongo, F., and Ratto, M., *Sensitivity Analysis in Practice. A Guide to Assessing Scientific Models*, Wiley, 2004.

13. Saltelli, A., Ratto, M., Andres, T., Campolongo, F., Cariboni, J., Gatelli, D., Saisana, M., and Tarantola, S., *Global Sensitivity Analysis*, Wiley, 2008.

14. Dalbey, K., Eldred, M., Geraci, G., Jakeman, J., Maupin, K., Monschke, J., Seidl, D., Swiler, L., Tran, A., Menhorn, F., and Zeng, X., *Dakota, A Multilevel Parallel Object-Oriented Framework for Design Optimization, Parameter Estimation, Uncertainty Quantification, and Sensitivity Analysis: Version 6.14 Theory Manual*, Sandia National Laboratories, Albuquerque, NM, 2021.

15. Campolongo, F., and Saltelli, A., "Sensitivity analysis of an environmental model: an application of different analysis methods," *Reliability Engineering & System Safety*, Vol. 57, (1), 1997, pp. 49–69.

16. Menberg, K., Heo, Y., and Choudhary, R., "Sensitivity analysis methods for building energy models: Comparing computational costs and extractable information," *Energy and Buildings*, Vol. 133, 2016, pp. 433–445. DOI: 10.1016/j.enbuild.2016.10.005.

17. Gan, Y., Duan, Q., Gong, W., Tong, C., Sun, Y., Chu, W., Ye, A., Miao, C., and Di, Z., "A comprehensive evaluation of various sensitivity analysis methods: A case study with a hydrological model," *Environmental Modelling & Software*, Vol. 51, 2013, pp. 269–285. DOI: 10.1016/j.envsoft.2013.09.031.

18. Nabi, H., de Visser, C., Pavel, M., and Quaranta, G., "A quasi-Linear Parameter Varying (qLPV) Approach for Tiltrotor Conversion Modeling and Control Synthesis," Vertical Flight Society 75th Annual Forum & Technology Display Proceedings, Philadelphia, PA, May 13–16, 2019.

19. Marcos, A., and Balas, G., "Development of LinearParameter-Varying Models for Aircraft," *Journal of Guidance, Control and Dynamics*, Vol. 27, 2004, pp. 218–228. DOI: 10.2514/1.9165.

20. Adams, B., Bohnhoff, W., Dalbey, K., Ebeida, M., Eddy, J., Eldred, M., Hooper, R., Hough, P., Hu, K., Jakeman, J., Khalil, M., Maupin, K., Monschke, J., Ridgway, E., Rushdi, A., Seidl, D., Stephens, J., Swiler, L., Tran, A., and Winokur, J., *Dakota, A Multilevel Parallel Object-Oriented Framework for Design Optimization, Parameter Estimation, Uncertainty Quantification, and Sensitivity Analysis: Version 6.16 User's Manual*, Sandia National Laboratories, Albuquerque, NM, 2022.

21. Anon., *Dakota Sensitivity Analysis and Uncertainty Quantification, with Examples*, Sandia National Laboratories, 2014.

22. Anon., *Dakota Software Training, Surrogate Models*, Sandia National Laboratories, 2016.

23. Cocco, A., Saetti, U., and Savino, A., "Aeroelastic Load Evaluation During Tiltrotor Transition Using a Comprehensive Mid-Fidelity Approach," 50th European Rotorcraft Forum, Marseille, France, September 2024.

24. Granata, D., Savino, A., and Zanotti, A., "Numerical Evaluation of Aircraft Aerodynamic Static and Dynamic Stability Derivatives by a Mid-Fidelity Approach," *Aerospace*, Vol. 11, (3), 2024. DOI: 10.3390/aerospace11030213.

25. Granata, D., Savino, A., and Zanotti, A., "Numerical Calculation of a Tiltrotor Aircraft Aerodynamic Stability Derivatives," 50th European Rotorcraft Forum, Marseille, France, September 2024.

26. Mamino, M., *Comparison of Uncertainty Quantification and Validation Methodologies on a Civil Tiltrotor Flight Simulation Model*, Master's thesis, Politecnico di Milano, Italy, 2023.

27. Pavel, M., and Padfielf, G., "Progress in the development of complementary handling and loading metrics for ADS-33 manoeuvres," American Helicopter Society 59th Annual Forum Proceedings, Phoenix, AZ, May 6–8, 2003.

28. Anon., *Test Guide for ADS-33E-PRF*, U.S. Army Research, Development, and Engineering Command, Moffet Field, CA, July 2008.

29. Rodden, W. P., and Giesing, J. P., "Application of oscillatory aerodynamic theory to estimation of dynamic stability derivatives," *Journal of Aircraft*, Vol. 7, (3), 1970, pp. 272–275.

A Future for Rotorcraft that Engages with New Application Domains

Larry A. Young[1]
NASA Ames Research Center
Moffett Field, CA 94035

ABSTRACT

The objective of this paper is to present ideas related to possible new application domains to consider for future rotorcraft research and development efforts. Some of these application domains have been points of discussion in an earlier body of work at NASA Ames Research Center, but some are anticipated to be wholly new. The goal is to aid in securing a robust and expansive rotorcraft research community for the next four decades at least. To achieve that goal and to pursue the identified new application domains, it will be essential to engage in collaborative networking with other research communities such as intelligent systems developers and roboticists.

NOMENCLATURE

A	Rotor disk area, ft^2; $A=\pi R^2$
CATS	Climate-adjusted air transportation systems
C_P	Rotor, or proprotor, or fan power coefficient, nondim.; $C_P = P/\rho A V_{tip}^3$
C_T	Rotor, or proprotor, or fan thrust coefficient, nondim.; $C_T = T/\rho A V_{tip}^2$
DRER	Disaster relief and emergency response
DOF	Degree-of-freedom, nondim.
L/D	Lift to drag ratio, nondim.
L/D_e	Effective lift to drag ratio for rotorcraft, $L/D_e = WV/P$
P	Vehicle total power,
PAV	Planetary aerial vehicles
PAX	Number of passengers, nondim.
R	Rotor radius, ft
RAR	Rotorcraft as robots
RIA	Runway independent aircraft
SAS	System analysis as applied to autonomy
SMR	Single main rotor
S	(Fixed-) wing planform area, ft^2
UFM	Universal flying machines
V	Cruise velocity, ft/s (m/s
W	Takeoff gross weight of vehicle, lb_f
α	Angle of attack, Deg.
β	Sideslip angle, Deg.
Γ	Wing dihedral/anhedral angle, Deg.
$\Delta D_e/L$	Delta effective drag to lift ratio, nondim.

INTRODUCTION

This paper describes a possible future for rotorcraft and the rotorcraft research community that would emphasize new mission application domains that respond to emerging, or largely unaddressed by aviation, societal problems. This paper emphasizes

---

[1] Presented at the 11th Biennial Autonomous VTOL Technical Meeting, Phoenix, AZ, Feb 4-6, 2025. Work of the US Government and not subject to Copyright protection.

that vehicle autonomy is an integral aviation capability necessary for the future.

The overarching objectives of the paper are to:

(1) present a perspective of engineering innovation for early career engineers,

(2) plant the seeds for possible research and innovation within the rotorcraft research community for the next two or three decades, and

(3) inspire cross-disciplinary and community collaboration.

Societal problems should drive new fields of study and new application domains. Not all societal problems have technological solutions. Not all technology solutions have an aerospace/rotorcraft component to them. However, there are far more societal problems that could be at least partially addressed by rotorcraft and rotary-wing technologies than have been imagined to date, and this paper addresses this gap.

BACKGROUND

Examples of some societal problems that could be partially addressed by rotorcraft/aviation are: disaster relief and emergency response (i.e., responding within the 'golden hour' no matter where, or how far, or under what circumstances); terrestrial and planetary science (i.e., science everywhere, all the time, for everyone); wildlife conservation/preservation (i.e., to minimize/prevent the sixth mass extinction, aka 'great die-off,' of life on our planet); improving the livability of cities and strengthening rural communities.

The following areas of research will be discussed: (1) Planetary rotorcraft (vertical lift planetary aerial vehicles, Refs. 1-3); (2) 'Rotorcraft as robots,' (aerobots as a ubiquitous part of society, Refs. 4-5); (3) Disaster relief and emergency response using autonomous vertical lift aerial vehicles, Refs. 6-8; (4) 'Resilience in aviation systems,' where this includes urban and regional rotorcraft networks, novel modular and scalable aerial vehicle designs, and accessibility and standard of living considerations in emerging aerospace applications, Refs. 9-10; (5) 'Universal flying machines,' i.e. static and reconfigurable structures composed of flying elements; machines with rotors/thrusters as actuators versus conventional servo-motors and linear-actuators; (6) High-speed rotorcraft to close the productivity gap between rotorcraft and commercial aircraft, Refs. 11-13; (7) Transforming or morphing hybrid multi-modal aerial vehicles including variable geometry vehicles and modular and/or distributed, homogeneous or heterogeneous, vehicles, Refs. 13-14; (8) Smart, multifunctional, variable-geometry (rotating and fixed-frame) structures and/or subsystems, Refs. 15-16; (9) Opportunities for space and aeronautics synergy (in addition to planetary rotorcraft); (10) Advances in emerging areas of system analysis, including 'system analysis as applied to autonomy', sizing analysis of hybrid multi-modal systems, and new societal metrics to factor into design (e.g. livability indices); and (11) Asymmetry in aircraft/rotorcraft design for enhanced functionality and capability.

The common set of themes for these eleven research areas are listed as follows:

- Autonomous system technology
- Robotics and GNC (guidance, navigation, and communication) advances

- All-electric and hybrid-electric propulsion, battery and/or fuel-cell, and power electronics advances.
- Variable geometry and/or modular rotorcraft subsystems or vehicles
- Hybrid air/ground or air/sea mobility

THE NEED FOR NEW APPLICATION DOMAINS

For the rotorcraft research community to grow in the future, it is necessary to think beyond the well-established aviation markets for commercial and military helicopters and tiltrotor aircraft. Even with the recent emergence of the eVTOL urban air mobility (UAM) application, there is still considerably more work needed within the rotorcraft community to consider other application domains. But, even more important than that, there are clearly unaddressed or under-addressed societal problems that could be addressed, in part, by rotary-wing technologies and/or rotorcraft vehicles.

The bulk of the early work cited in this paper is that of the Aeromechanics Office at NASA Ames Research Center. This does not mean, though, that other researchers in the past have not touched upon or worked in similar nascent application domains. What is perhaps most pressing now versus back then, at the time of the earlier work, is that many of the societal problems noted back then have only unfortunately worsened with time.

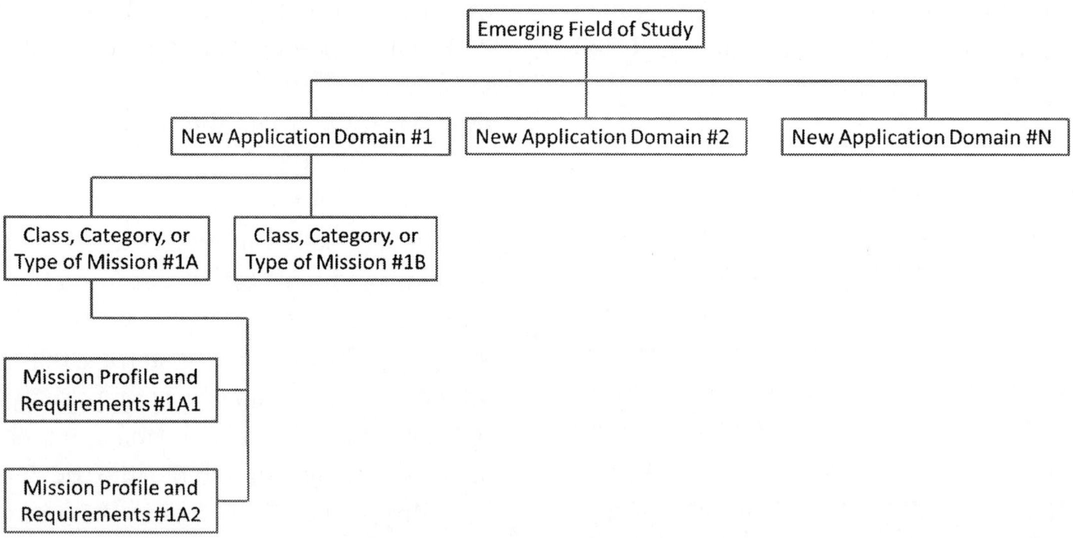

Figure 1. Generic Flow Chart of Emerging Fields of Study vs. New Application Domains vs. New Types of Missions

Note that there is some inevitable confusion between an 'emerging field of study' versus a 'new application domain.' In a few cases in this paper, a 'new application domain' is listed/summarized in the text but, admittedly, what is being summarized is more one of an 'emerging field of study' (to which, as shown in Fig. 1, there can be many application domains that can fall under a given field of study). The distinction between the two will hopefully be made clearer as one proceeds through the paper.

As a body of work builds up with time, there begins to be developed a certain level of formalism in the treatment of that area of research. With increasing formalism, it consequently becomes more efficient to generate new ideas and technical approaches in the emerging field of study and its associated application domains.

The fields of study – and associated new application domains suggested – are not intended to be comprehensive. Instead, this paper should be a considered a call to action to consider the future of the rotorcraft research and development community.

RECOMMENDED NEW APPLICATION DOMAINS

The following discussion summarizes the recommended new fields of study and new rotorcraft application domains.

Vertical lift planetary aerial vehicles

NASA Ames has held for twenty-five-plus years a leadership role in planetary rotorcraft, e.g., Refs. 1-3. Ingenuity and Titan Dragonfly are just the beginning though. New types of planetary aerial vehicles should be developed. For example, the planet Venus might be the next planetary rotorcraft opportunity. Or, alternatively, planetary aerial vehicles for the outer giant planets might be worth exploring (for likely non-rotorcraft aerial vehicles). And consider this, a propeller is effectively a rotor: Mars/Venus airplanes could be investigated by the rotorcraft community (and not just by the fixed wing airplane community). Synergy with, and expansion of, terrestrial field science campaigns could be an outgrowth of planetary rotorcraft work.

Figure 2 presents an illustrative example of a flow chart for the planetary aerial vehicle field of study (with emphasis, because of the necessity for brevity, on Mars).

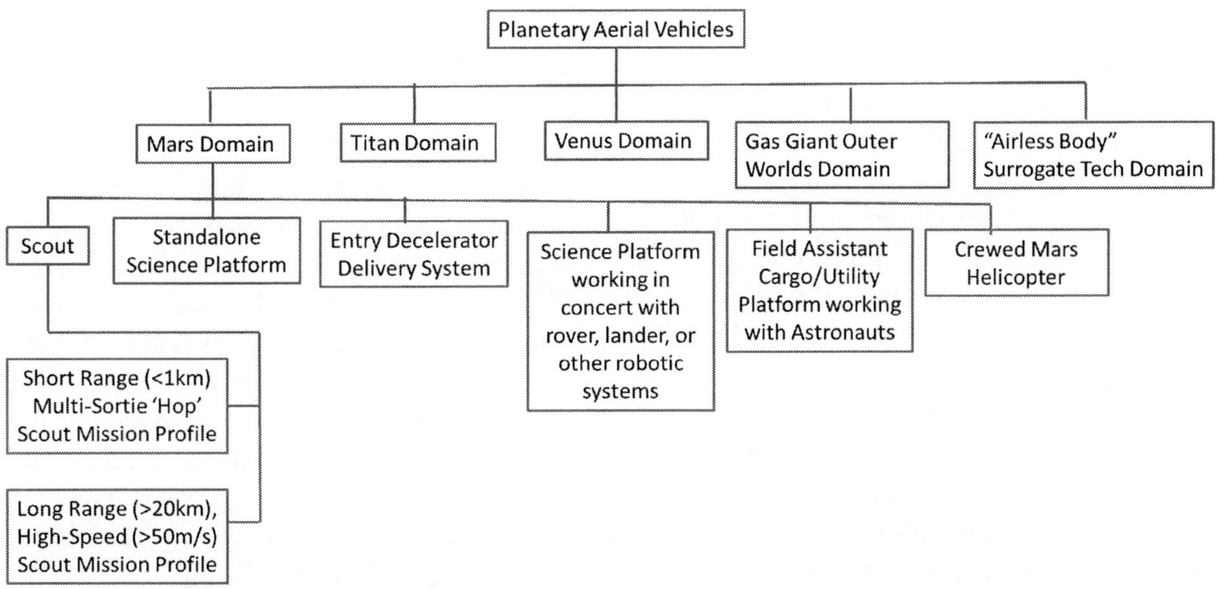

Figure 2. Representative Example of a Flow Chart for Planetary Aerial Vehicles Field of Study, Application Domains, and Types of Missions

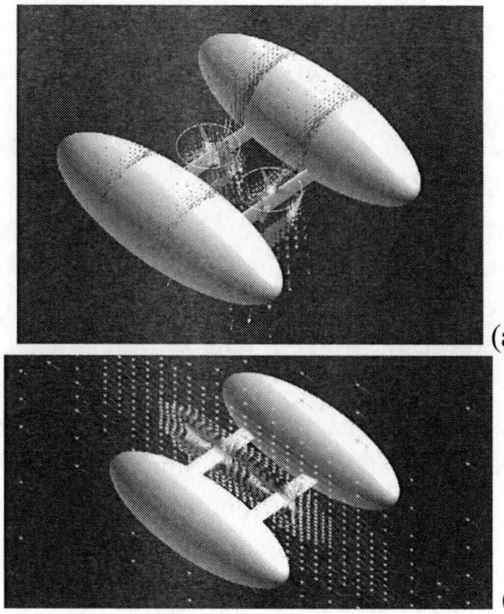
(a)
(b)

Figure 3. Venus flyers (hybrid rotorcraft for near-surface flight, Ref. 17), (a) hover and (b) forward flight

Flight on Venus, near its surface will be very difficult because of the tremendous pressures and high temperatures near the surface. But the problem of flight becomes more tractable the higher the altitudes to be flown at during a planetary mission. Despite the major environmental challenges it poses, rotary-wing flight near the surface should not be abandoned though. Figure 3 is one notional vehicle configuration (twin hulls internally pressurized with tandem tilting wings and propellers in this example) for such near-surface flight, see Ref. 17. The high temperatures near the surface might limit mission duration near the surface to a few hours, because of the onboard electronics heat death, but undertaking and delivering such a Venus flyer mission would be a major planetary science accomplishment.

An alternate type of Venus mission is the use of rotary-wing decelerators for descent into the atmosphere of Venus (Fig. 4). This is perhaps less challenging than the VTOL flyer because most of the descent will be at altitudes where the density, pressure, and temperatures are less severe than at the surface and 'heat death' of the onboard electronics can be delayed.

Figure 4. Notional Venus Descent Rotary-Wing Decelerator (Ref. 18)

Rotorcraft as robots

The emerging field of 'rotorcraft as robots' (RAR) provides the aspirational goal to think beyond just autonomous helicopters (i.e., doing more than merely replacing the pilot). Rotorcraft that are surface- and object-interactive and multi-modal; and rotorcraft that have effector/actuator capabilities, see Fig. 5. Additionally, this includes robotic rotorcraft that interact with other robotic systems for complete missions and even complete vehicle life cycles. 'Rotorcraft as robots' could have a profound influence on how future terrestrial field science, environmental, and biodiversity monitoring will be conducted in the future.

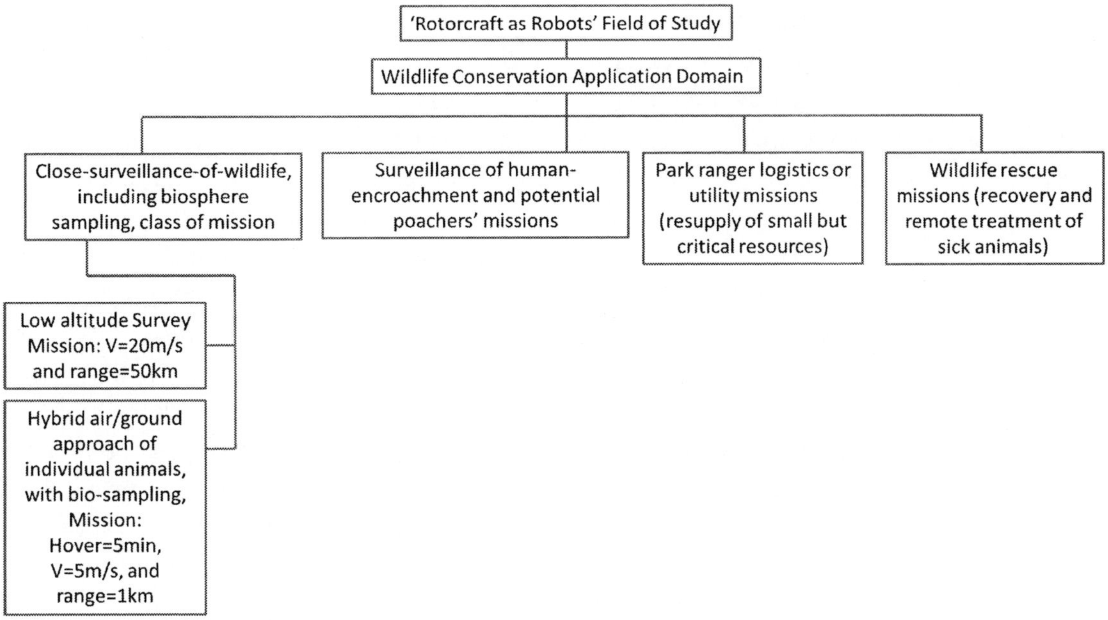

Figure 5. 'Rotorcraft as robots' supporting wildlife conservation (one example of a multitude of RAR applications)

Figure 6 is a notional artwork of RAR assets used for wildlife conservation.

Figure 6. RAR potential serving wildlife conservation missions; AI-generated artwork from Microsoft Copilot software

Figure 7 is one potential RAR hybrid air/ground mobility robotic system. In this case, the ground mobility is provided by passive legged locomotion (motive force provided by propellers) with propellers that are tiltable to switch between one mode (propeller rotational axes are horizontal) where the propellers aid ground mobility and switch to another mode where the propellers are tilted vertically and support air mobility.

Figure 7. CAD/CFD model of robot with tiltable side-by-side rotors integrated with Jansen walking machine

'Rotorcraft as Robots,' though perhaps not generally well known by that title, is perhaps one of the most widely developed of the new fields of study noted in this paper. Electrical engineers, computer scientists and technologists, and roboticists have significantly embraced the use of readily available commercial off the shelf (COTS) microelectronics, micromechanical control actuators, and multirotor drones or UAVs to investigate novel aerial robotic systems, e.g., Refs. 4 and 19. Early work at NASA Ames Research Center helped pioneer some of these investigations but this type of research saw most of its subsequent advances from nontraditional (with respect to rotorcraft) research communities. Despite some noteworthy work by academic researchers, e.g., Refs. 32-33, this application domain is largely unexplored by the rotorcraft community. The 'rotorcraft as robots' field could benefit from rotorcraft expertise in defining practical mission CONOPS and fieldable systems.

Disaster relief and emergency response using autonomous vertical lift aerial vehicles

Ever since the first practical helicopter, helicopter use for search and rescue has been the rotorcraft community's first and most important application domain.

Disaster relief and emergency response (DRER) is the first and best use of helicopters, but the same will be true for autonomous vertical lift aerial vehicles. Such vehicles can potentially span two or more orders of magnitude of gross weight, payload, range/endurance, and speed. Such DRER aerial vehicles will be used to monitor, warn, aid, rescue, and rebuild (MWARR). Cross-mission and application synergies (between defense, commercial delivery, and DRER) are possible, and likely necessary, to arrive at large, cost-effective, and highly responsive networks.

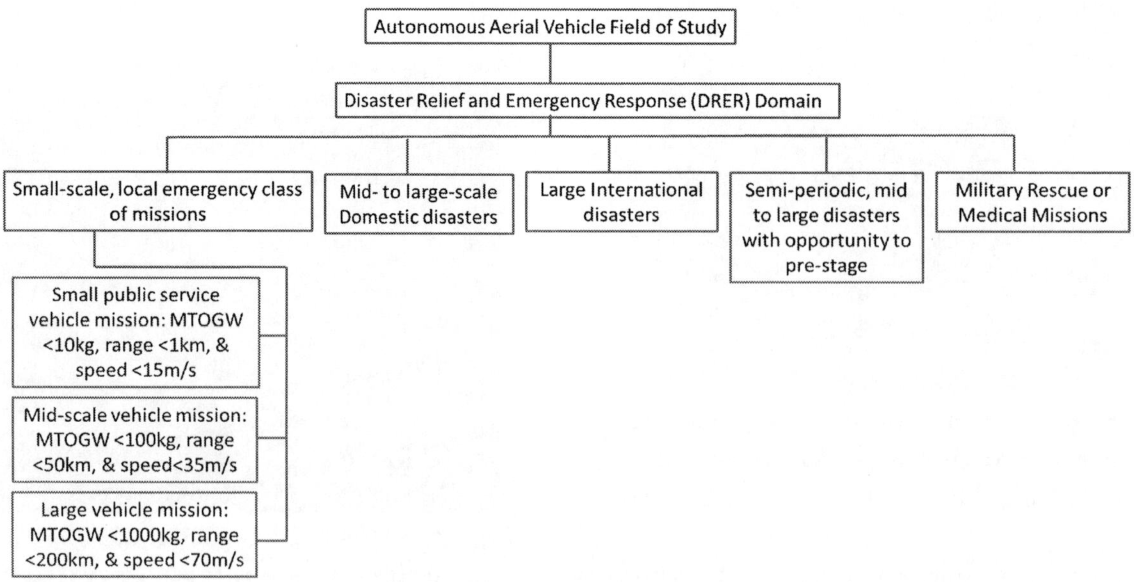

Figure 8. Autonomous aerial vehicles and disaster relief and emergency response applications and missions

There has been research at NASA Ames Research Center into autonomous aerial vehicles – and how they could support DRER missions – since the 2000 timeframe, e.g., Ref. 20. In recent years, there has been increased interest in DRER missions for rotary-wing UAVs, e.g., Ref. 21.

Figure 8 summarizes some potential DRER application domains and sample classes of missions. A large spectrum of vehicle sizes, mission capabilities, and new technologies are possible for DRER. Currently, the NASA technology portfolio for DRER is focused on wildfire response and more on UAV traffic management (UTM) considerations and not so much on vehicle-centric research and development activity, e.g., Ref. 29. Hopefully, this will change in the future; there are substantial opportunities to consider novel vehicle designs for DRER missions, see, for example, Fig 9a-b.

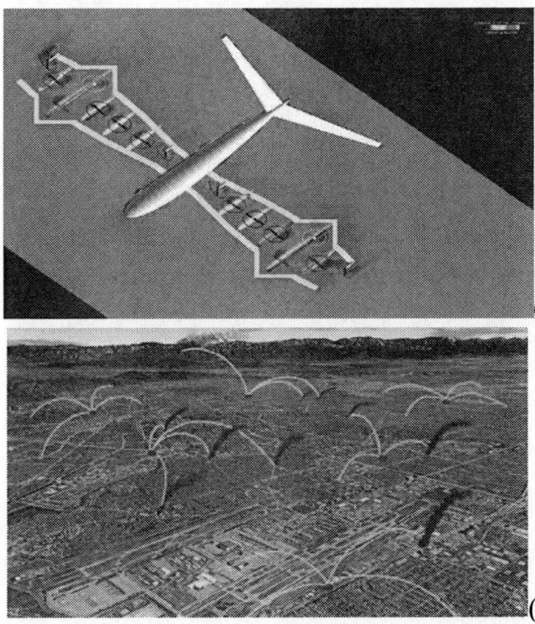

(a)

(b)

Figure 9. DRER: (a) damage surveys and (b) urban emergency aid networks

The introduction of rotary-wing UAVs, or drones, to support DRER and other public service missions will likely be enhanced by using these aerial assets in a dual-purpose (commercial and public service vehicles), coordinated manner.

Resilience in aviation systems

The aviation sector is a very complex but generally successful enterprise within the United States. It spans a spectrum of stakeholders that ranges from: the traveling public, to aircraft researchers and developers, to airport operators, to airlines and other aircraft operators, to the public/government regulators and officials. Overall, the aviation sector is critical to the smooth operation of society. After a series of demonstrable major disruptions to aviation over multiple decades, it is appropriate to consider how to enhance the overall resilience of the aviation sector. This will entail developing research programs that promote system-of-systems engineering solutions that focus on vehicle design advances and novel airspace system architectures.

To accomplish this goal, it is necessary to consider engineering solutions to make the aviation sector more resilient to anticipated and potential future crises.

1. How can the aviation fleet be quickly 'right sized' and/or re-configured to meet the challenges of rapid changes in the economic operational environment?
2. How can aviation be refined to improve economic opportunities for urban and rural/regional communities in the US?
3. How can aviation respond to changes stemming from environmental challenges, including climate change?

4. How can aviation evolve to respond to the ever-increasing challenges of disaster relief and emergency response?

The following are some foundational principles that can aid in addressing the above questions:

I. Vertical lift and short takeoff and landing (STOL) aerial vehicles, particularly those having hybrid-electric propulsion, have the potential to expand into and fill several subsectors of the aviation sector.
II. Autonomous systems and robotics technology also potentially have a profound influence on aviation.
III. Aviation can and must expand to new markets and to new applications to meet critical societal needs. This, in turn, will economically sustain aviation during difficult times.
IV. Modular and distributed vehicles and/or subsystems could radically change aircraft design and operations.
V. Wholly new types of amphibious and multimodal vehicles and networks also have an opportunity to transform urban and regional aviation markets.
VI. A broad spectrum of vehicle sizes and capabilities will be required for building aviation resilience (e.g., disaster relief and emergency response, Refs. 5-20.

Figures 10-11 are some examples of how runway independent aircraft (VTOL, VSTOL, and STOL aircraft) will be a vital component of enhanced aviation resilience.

Figure 10. Runway independent aircraft (RIA) for aviation resiliency (Ref. 11-13 and 22-23)

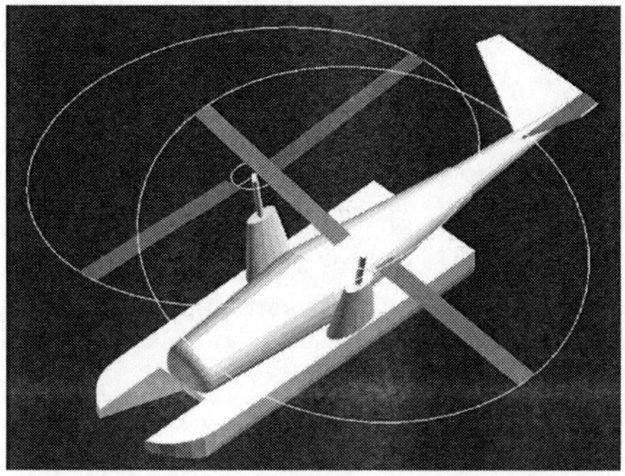

Figure 11. An emerging possible role for amphibious rotorcraft/aircraft (Ref. 24)

Figure 12 illustrates a possible new generation of STOL or V/STOL aircraft to maximize mission flexibility by being able to operate from vertiports, short or long runway airports, and even operate amphibiously. It emphasizes the use of distributed and heterogenous rotors and, further, tilts the wings and rotor nacelles independently to enable this V/STOL flexibility.

Figure 12. Hybrid tiltwing/tiltrotor aircraft with heterogeneous rotors and propellers (Ref. 23)

Universal flying machines

Universal flying machines (UFM) are conceived of as modular rotorcraft 'elements' physically joined to form temporary reconfigurable structures and/or on-demand 'constructs' (i.e., modular machines or vehicles). This work lies at the intersection between 'swarms' and modular systems. In that regard, UFM can be considered large-scale flying Lego™ bricks or Tinkertoy sets. Equally important, though, UFM could also be used to build simple or modest machines that make use of rotors/thrusters as motive-force 'actuators' in 'constructs' versus motors and linear actuators.

A considerable body of work exists with regards to swarms of robots, aerial or otherwise. However, less work has been focused on the notion of multiple robots structurally intertwined/semi-integrated (if only for a short time with or without limited aggregate, or ensemble, versus individual robotic subsystem functionality/actuation). The term "construct" is introduced in this paper as a general term to be applied to all these collective intertwined/semi-integrated multi-robot assemblies. Also, for the purposes of this paper the individual robotic systems will be referred to as "elements." When aerial mobility is an important attribute of either the elements or the construct, or both, then also for the purposes of the paper, the elements and constructs are interchangeably referred to as universal flying machines (UFM), see Fig. 13. Nothing in this terminology precludes swarms of robots forming and decomposing from swarms to constructs back and forth upon need. Nothing in this notion of construct prohibits the intertwined/semi-integrated multiple robots be passive or stationary during integration, nor does it exclude the possibility that the integrated construct has unique aggregate/ensemble mobility, enhanced, or transformed, actuation or manipulation capabilities, or other types of collective versus individual element robotic functionality. Ideally this formation and/or decomposition of constructs from elements should happen without external assistance either manually from people or from independent robotic assembly systems. This ideal, though, is not an absolute requirement and there are many applications/missions where this requirement is unnecessary. Additionally, it is also acceptable to consider constructs whose formation or decomposition is a one-way or nonreversible process. An inherent, intrinsic requirement of the construct, and consequently the universal flying machine, concept is the need for transitory formation and decomposition of constructs from elements.

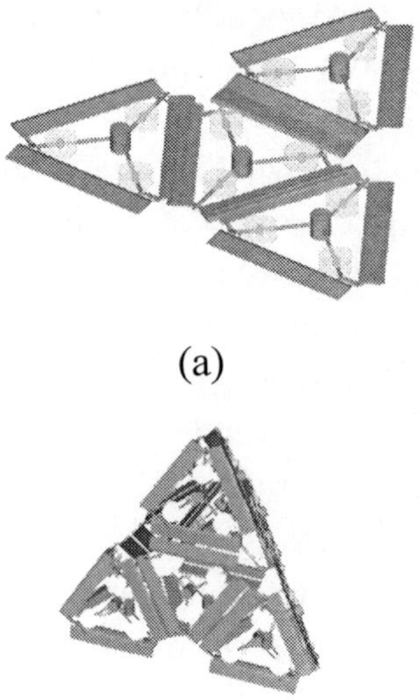

(a)

(b)

Figure 13. UFM elements that could reconfigure to form structures, aggregate assemblies or constructs, and even simple machines

High-speed (and high-productivity) rotorcraft

The NASA Aeronautics Research Mission Directorate's RVLT project has substantially invested over the past five plus years on research into the development of urban air mobility (UAM), aka eVTOL, vehicles. This parallels a very large development effort by an emerging eVTOL industry, see Ref. 25. Such eVTOL vehicles are on track to be certified by the FAA in the next couple of years and then become commercially available. This then begs the question of 'what next?' for rotorcraft research, see Fig. 14. The one possible answer is to revisit past interest into rotorcraft that can support regional commercial markets, and, further, (re)emphasize higher speed flight, while at the same time emphasizing the need to incorporate into these regional high-speed rotorcraft sustainable aviation attributes. High-speed (>275 knots cruise), regional (range>500 nautical miles (nm)) rotorcraft will build in needed reliance in the aviation system of the future; providing sustainable aviation capability, in parallel, is also a worthwhile cause.

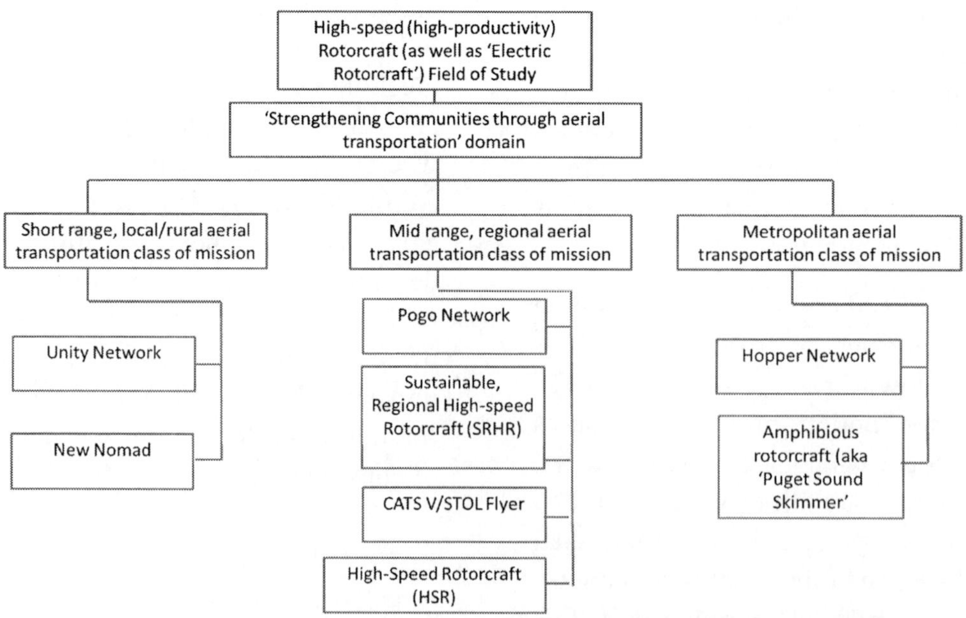

Figure 14. High-speed (and/or high productivity) rotorcraft

One possibility to investigate is the sustainable, regional, high-speed rotorcraft (SRHR) concept that seeks to address the following problem: how to best arrive at a class of rotorcraft that has higher speed, has regional-scale range capability, retains VTOL capabilities, but also embodies technologies and vehicle/system designs that emphasize sustainable aviation goals. The overall problem can be broken down into two parts: why do regional high-speed rotorcraft need to be developed and, secondly, if needed, why, and how much, should they embody sustainable aviation capabilities? The NASA rotorcraft community in the past has devoted a considerable amount of research into the potential benefits of regional rotorcraft supporting the commercial passenger transport role; this includes the Short Haul Civil Tiltrotor (SHCT) and the Runway Independent Aircraft efforts of the early 2000's. NASA also in the past has a large body of work considering the technical feasibility of high-speed rotorcraft, with the primary focus of that effort on tiltrotor aircraft. This past work on civil tiltrotor, high-speed rotorcraft, and 'runway independent aircraft' would suggest focusing on high passenger counts and speed to make a regional rotorcraft more economically competitive against regional jets and turboprop aircraft. After the definition of the overall problem, the next step is to define the overall proposed effort objective to identify those technologies and conceptual designs that make a modest but significant impact on reducing the carbon footprint of this proposed new class of SRHR; more specifically, the objective is to have an SRHR hybrid-electric vehicle provide 10-20% of the total mission energy expended to come from electrical power. Preliminary work, Ref. 11, has studied novel SRHR vehicle concepts that targeted twenty-five percent of VTOL and hover power to be derived by electric propulsion running in parallel with turboshaft engine mechanical power supplied to the rotors; cruise power was provided solely by the turboshaft engine in this early study; the net effect of this design trade study sizing analysis was energy provided by electric propulsion to be about 5-10% of the total mission energy. (Only limited effort is currently being expended within NASA as to applying sustainable aviation goals to regional, high-speed rotorcraft.)

Figure 15. Tilting ducted-fan sustainable regional high-speed rotorcraft concept (Refs. 11-12); low-speed helicopter-mode

Three different hybrid-electric propulsion concepts should be studied in the future: (1) battery (and/or hydrogen-based fuel-cells) and electric motors partially offsetting turboshaft mechanical power to the rotors during vertical takeoff and landing and hover (with turboshaft power only during cruise); (2) turboshaft-driven dual-use electric motor and generator for discharging/charging battery storage inflight for on-demand rotor power augmentation; (3) turboshaft-driven generators providing full flight power to electric motors, which are the sole means of delivering rotor power. New or refined analysis tools as well as new technologies and propulsion architectures will need to be explored. Some preliminary work by industry is looking to extend their UAM/eVTOL vehicles to regional markets but these vehicles are comparatively small (4-6PAX) and do not represent the size of vehicle necessary for regional commercial passenger transport (>50PAX).

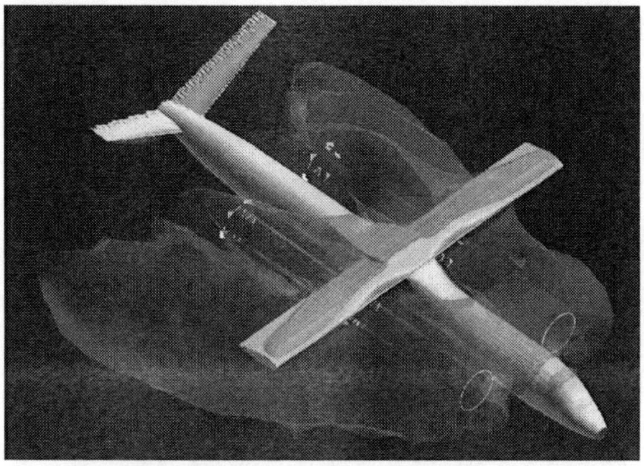

Figure 16. One notional high-speed rotorcraft: stopped cycloidal rotor aerial vehicle (Ref. 11)

Transforming/morphing hybrid multi-modal aerial vehicles

Another field of study for future rotorcraft are 'transformers' (extreme variable geometry, aka 'morphing,' or hybrid multi-modal) rotorcraft; for example, refer to Fig. 17, a notional variable geometry tiltrotor aircraft with variable sweep and longitudinal positioning.

Though there are nontrivial downsides in terms of vehicle weight, there are also significant potential mission capability advantages of variable-geometry morphing rotorcraft as well as hybrid multi-modal vehicles.

Though these types of vehicles have primarily been examined for fixed-wing aircraft, this design space is largely unexplored for rotorcraft. See Ref. 13 for some more advanced tiltrotor aircraft concepts.

(a)

(b)

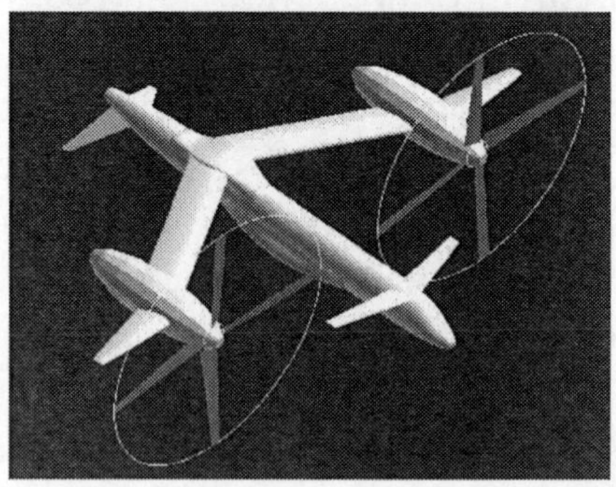
(c)

Figure 17. Notional variable-geometry tiltrotor aircraft (Ref. 13): (a) hover configuration, (b), transition or conversion and (c) cruise configuration

Smart, multifunctional, variable-geometry structures and/or subsystems

The field of study of smart multifunctional variable-geometry structures (especially variable-geometry rotors) and/or subsystems is at least a forty-year effort within the rotorcraft community. It is suggested that this whole field of study might be rebranded as the 'Babbage[2] program.' Why this field of study has renewed relevance in recent years is twofold. First, continued advances in low-cost COTS microelectronics and micromechanical systems. Second, the potential promise of 'electric rotorcraft' whereby providing for electric-propulsion for aerial vehicles also enables the potential for electrically powering active flow control devices on the airframes and rotors of such aerial vehicles as well as powering variable-geometry devices. Previously, enabling such devices would result in too much delta increases in power, weight, and complexity (and cost) to be implemented onto a production aircraft.

This field of study continues to have a lot of potential and could greatly benefit from having a unifying foundation to build upon. Use of the 'Babbage program' label is one small part of building this unifying foundation (from an advocacy perspective at least). 'Babbage' would include embedding smart, multi-functional, multi-actuated, micro-mechanical systems in aircraft and rotorcraft structures. It should be noted that work on embedded systems for rotorcraft has been studied for nearly forty years. Such past work includes: free-pitching or indexed rotor blade tips; rotor blade-embedded servo-flaps; on-demand deployment of blade airfoil slats; variable/actuated blade airfoil camber; on-demand trim tabs for blade airfoils; and last but not the least, active and passive blade twist. Despite these earlier studies the full potential of 'smart,' multifunctional, with embedded-actuator, rotor and vehicle structures has still not been adequately explored. This is why a new field of study centered around the 'Babbage' concept will, hopefully, expedite synthesis on old and new concepts and technologies into a unified technical framework. Further, beyond smart, embedded rotor-blade actuators, new rotor design and manufacturing approaches should continue being developed to go beyond the SOA of composite rotor blade construction to include a combination of additive manufacturing, advance composite layup, and multi-part sensors, processors, and micro-mechanical systems. These might even include the speculative addition of 'clockwork'-like Legos™ and other multifunctional structural building-blocks and mechanical systems.

Figure 18. Indexed-Tip Proprotor (Ref. 26)

[2] In recognition of Charles Babbage, the inventor of the first general purpose mechanical computer, the 'Difference Engine.'

Space and aeronautics synergy (in addition to planetary rotorcraft)

Rotorcraft expertise could be applied to surrogate lander development as well as Vertical Motion Simulator (VMS) simulations of lunar and planetary landers (some of which has already occurred in past work). Further, rotorcraft expertise could also be applied to rotary-wing decelerators (for Earth reentry or planetary entry). Rotorcraft expertise could also be applied for flight dynamics and simulation work for development of cold-gas thruster 'drones' for airless planetary bodies. Rotorcraft expertise could be applied to the design of aeroassist/aerobrake 'coaxial rotor' orbiters (rarified gas analysis), e.g., Ref. 27. Finally, rotorcraft expertise could also be brought to bear on the development of propeller-driven planetary fixed-wing airplanes, e.g., Ref. 28).

Figure 19 further summarizes some of these opportunities for aeronautics and space community synergy.

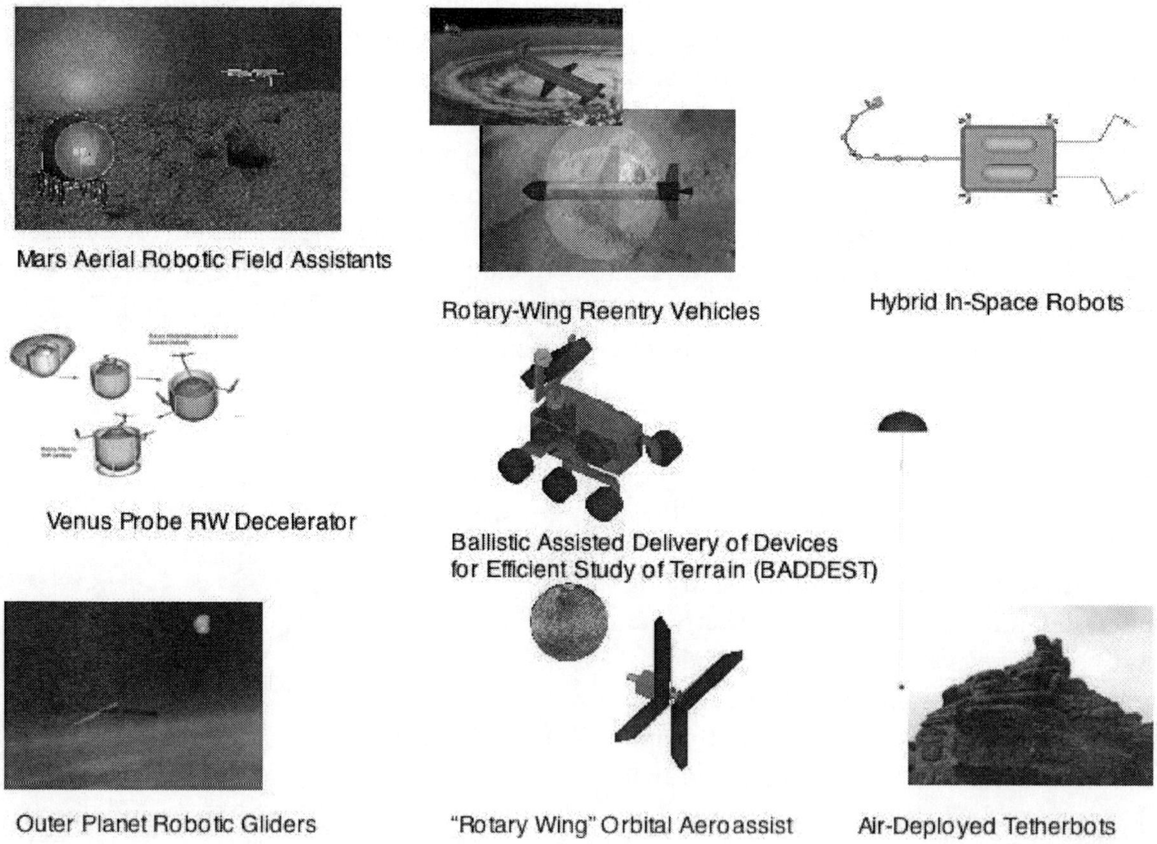

Figure 19. Possible space and aeronautics synergy

Emerging areas of system analysis

Future rotorcraft system analysis efforts will need to begin to try addressing the following analysis problems: autonomy, intelligent systems, and robotics design considerations; livability metrics included in conceptual design refinement/optimization (how are these aircraft going to make our cities and/or rural communities more livable); accessibility considerations in vehicle design; complete life-cycle and system-of-systems considerations; sustain aviation design requirements; factoring in wholly new application domains and technical disciplines and disparate/hybrid technologies.

Figure 20 reflects one possible approach to defining and implementing a 'system analysis as applied to autonomy' methodology.

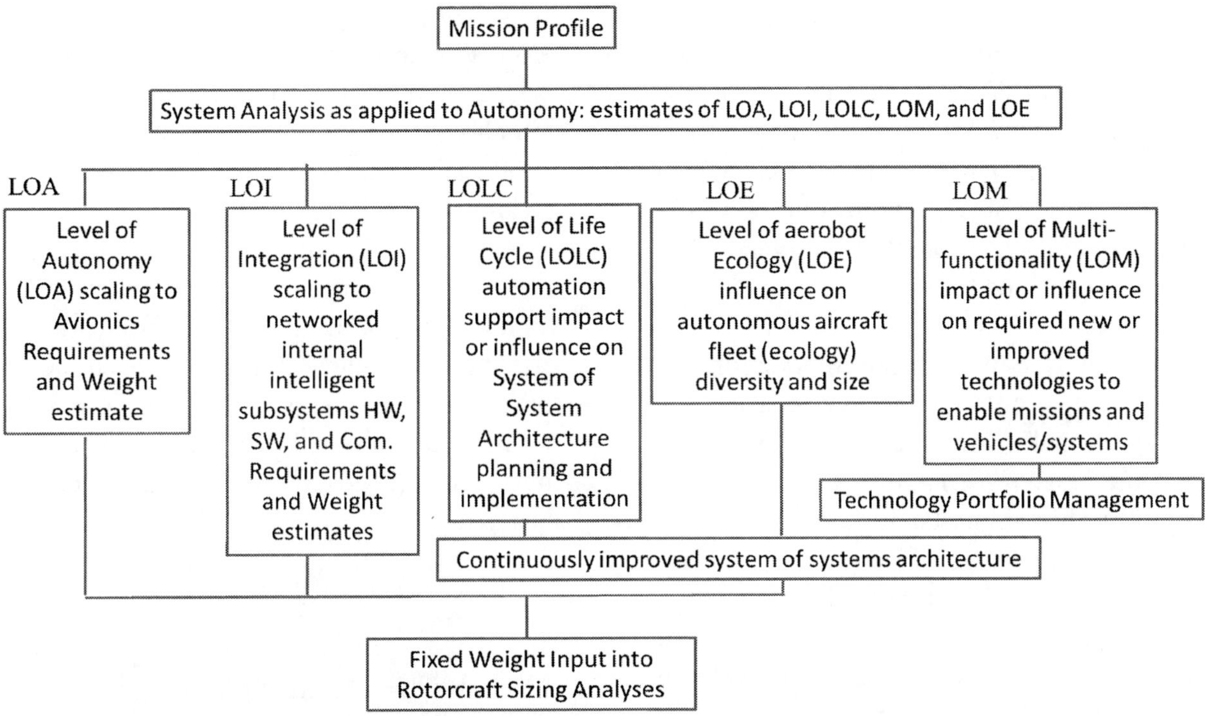

Figure 20. New system analysis approaches required: System Analysis as applied to Autonomy (SAS)

The aerospace community needs to actively fight against the perception that aircraft design – and systems analysis and engineering – are only evolutionary in nature from this point forward. To combat this perception, it is necessary to not only explore new aerial vehicle concepts and new technologies, but it is also necessary to develop (as emphasized in the theme of this paper) new fields of study and application domains that respond to largely unaddressed societal problems. But this, in turn, dictates the development of new system analysis and system engineering tools that encompass new performance requirements, constraints, and success criterion metrics that are novel in nature as compared to the tools and metrics typically used in aerospace analysis.

Asymmetry in aircraft/rotorcraft design

After several decades of CTOL and VTOL aircraft development, most (both commercial/civilian and military) aircraft have begun to look all the same. For rotorcraft, most vehicles are single main rotor and tail rotor helicopters. For commercial aircraft, the conventional tube and wing configuration with four or fewer turbofan engines dominates. Periodically, new types of commercial aircraft designs have been proposed and studied – the recent blended-wing-body (BWB), Ref. 30, and the truss-braced-wing (TBW), Ref. 31, concepts come to mind – and, of course, there are tandem and coaxial helicopters (and tiltrotor aircraft) flying in addition to SMR helicopters. But there is a continued need for aircraft design innovation.

This study considers the implications of intentionally building in geometric (or configurational), structural, dynamic, and mechanical asymmetry into aerial vehicle designs. This suggests, for example, early asymmetric aircraft such as the Blohm and Voss BV141 and Rutan/Scaled Composite's Boomerang aircraft. This is not just about making aesthetic choices in aerial vehicle configurations though. The technical focus of this field of study should be on assessing whether intentional asymmetry in vehicle designs can enable unique flight performance benefits that otherwise symmetrical designs cannot readily achieve. The goal of this field of study should be to promote the development of a unified aerial vehicle design philosophy that embraces asymmetry. The initial steps toward developing such a design philosophy lies in summarizing a few specific vehicle conceptual designs (some previously reported in the literature and some for the first time) that embrace asymmetry. Accordingly, this paper summarizes some of the asymmetric vertical lift aerial vehicle concepts that could be explored. Most of these aerial vehicle concepts exhibit configurational asymmetry and not the other three types possible. (The four types of asymmetries are: configurational, structural, dynamic, and mechanical.)

A few key questions need to be considered:

- What is asymmetry?
- Why and when should asymmetry be adopted in aerial vehicle configurations (with special emphasis on vertical lift aerial vehicles)?
- How can be asymmetry be embodied as a rotorcraft design philosophy?
- What is the potential power of asymmetry as an enabler of wholly new vehicle designs, missions, and applications?

At this introductory point of discussion of 'asymmetry in design,' it is more crucial to define an expansive design space than to deep dive into analysis of individual conceptual designs.

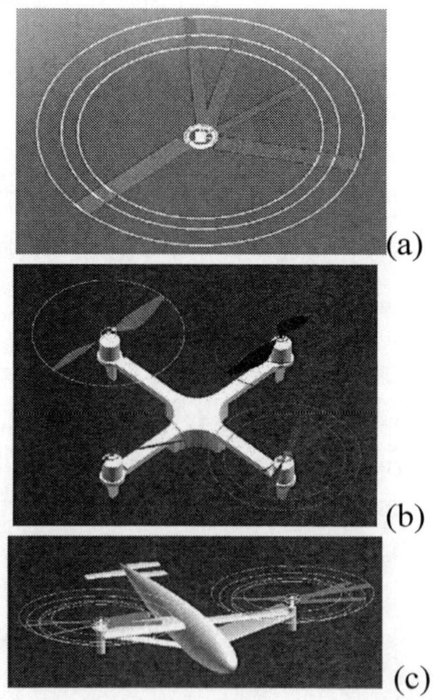

Figure 21. Asymmetry: (a) 'Thistle' rotor, (b) multirotor with 'Thistle' rotors, and (c) 'Nettles' aerial vehicles

UNIFYING THEMES INHERENT IN THESE NEW APPLICATION DOMAINS

As noted earlier, the unifying themes for these new application domains are twofold. First, expanding research communities' conception of what rotorcraft and aerospace, in general, might aspire towards as to addressing societal problems beyond that of the traditional/well-known transport of people, goods, and services. And second, advances in key technologies are also paving the way for innovation: i.e., autonomous and/or intelligent system technologies; robotics, and GNC (guidance, navigation, and communication); all-electric and hybrid-electric propulsion, battery and/or fuel-cell, and power electronics; variable geometry and/or modular rotorcraft subsystems or vehicles.

Figure 22 outlines one possible approach to methodically proceed from identified societal problem to vehicle concepts and technologies addressable by the rotorcraft community.

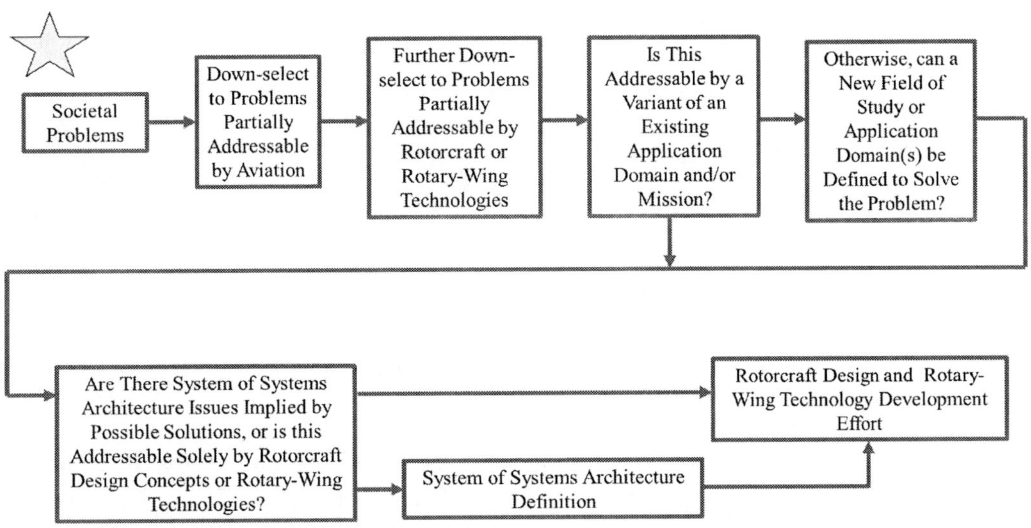

Figure 22. One notional approach to go from societal problems to realizable systems architecture, or vehicle design, or technology investigations

In many of the presented suggestions for new fields of study and/or novel application domains for rotorcraft, there is inevitably going to be some overlap between the presented 'fields' and 'application domains.' Figure 23 illustrates some of that inevitable overlap.

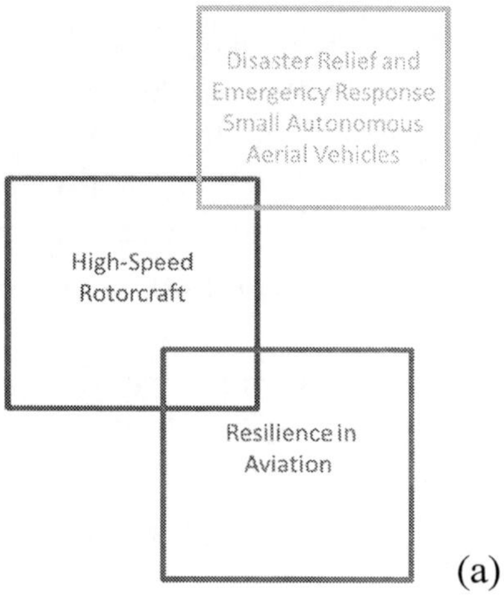

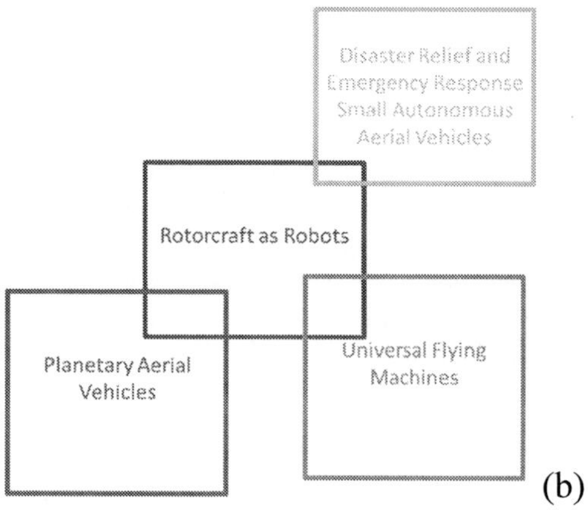

Figure 23. Certain inevitable overlapping between fields of study

MOVING FORWARD TO ENABLE NEW APPLICATION DOMAINS

New mission and vehicle concepts (not just the ones summarized in this paper) are required to identify new application domains. Novel technologies will need to be developed as well to help realize the new mission and vehicle concepts necessary for future application domains.

As much as this is a paper summarizing a perspective of what the future of rotorcraft research and development might entail, this paper is a call to the rotorcraft community to think expansively and innovatively beyond the traditional rotorcraft missions and markets and to consider additional future opportunities.

There are at least two methods of defining new missions for future rotorcraft: a classic mission-centric approach and, alternatively, a vehicle- or technology-centric approach. Figure 24 is a simple illustration of those two approaches and how, in engineering practice, there can be a switching back and forth between those approaches.

Figure 24. Mission- versus vehicle-centric approaches to defining new missions for future rotorcraft.

REACHING OUT AND BUILDING NEW RESEARCH COMMUNITIES

As noted early, much of the work identified in this paper has stemmed from over two decades worth of research at NASA Ames Research Center. In addition to that body of work, in parallel, other researchers both within the rotorcraft community and outside of it (e.g., robotics researchers), have been also exploring similar alternate application domain research themes.

For the rotorcraft community to successfully engage with some of the emerging fields of study and application domains summarized in this paper, it will be essential to network and collaborate with nontraditional aerospace research communities and other potential future stakeholders. Besides the obvious rotary-wing expertise that the rotorcraft research community could bring to these new applications, the unique mission-focus of aerospace technologists and developers may very well be the critical missing element necessary to address key societal problems that might be more amenable to aviation solutions than currently recognized.

CONCLUDING REMARKS

Applying rotorcraft technologies into new application domains will be an essential element of future growth in the rotorcraft community (industry, academia, and government). Several application domains are suggested in this paper, along with representative sample mission and vehicle concepts.

This paper seeks to inspire a new generation of rotorcraft researchers and developers to consider mission application domains outside the well-explored conventional uses of rotary-wing technologies. The result could potentially see a significant expansion in the overall size of the rotorcraft aviation sector. Exploring these novel mission application domains will likely require collaboration between the rotorcraft community and other research communities, such as roboticists and information technologists, and new/emerging end-user communities.

Author contact: Larry Young,
larry.a.young@nasa.gov

ACKNOWLEDGMENTS

This work is for the aerospace engineering innovator. This work is also for the NASA and U.S. Army rotorcraft research community at NASA Ames Research Center. In particular, the inspiration of Mr. Robert Stroub (Ret. NASA), Mr. Edwin Aiken (Ret. NASA), Dr. William Warmbrodt (Ret. NASA), Dr. Chee Tung (Ret. U.S. Army), and Dr. Robert Ormiston (Ret. U.S. Army) is gratefully acknowledged. The development of the mid-fidelity CFD software tool, RotCFD, by the late Prof. Ganesh Rajagopalan of Iowa State University is gratefully acknowledged. Finally, in memoriam, Anne McCracken.

REFERENCES

[1] Cornelius, J., Schmitz, S., Young, L.A., et al, "Dragonfly – Aerodynamics during Transition to Powered Flight," VFS Annual Forum, May 2021.

[2] Grip, H., Scharf, D., Malpica, C., Johnson, W., Mandić, M., Singh, G., and Young, L.A., "Guidance and Control for a Mars Helicopter," AIAA Science and Technology Forum and Exposition (SciTech 2018), Kissimmee, Florida, January 8-12, 2018.

[3] Young, L.A., Aiken, E.W., Gulick, V., Mancinelli, R., and Briggs, G.A., "Rotorcraft as Mars Scouts," 2002 IEEE Aerospace Conference, Big Sky, MT, March 9-16, 2002.

[4] Young, L.A., "Rotorcraft as Robots," Vertical Flight Society webinar, https://www.youtube.com/watch?v=A843t70amxA, presented August 21, 2021; last accessed 7-27-2024.

[5] Young, L.A., "Aerobots as a Ubiquitous Part of Society," AHS Vertical Lift Aircraft Design Conference, San Francisco, CA, January 18-20, 2006.

[6] Young, L.A., "Smart Precise Rotorcraft InTerconnected Emergency Services (SPRITES)," AIAA Science and Technology Forum and Exposition (SciTech 2018), Kissimmee, Florida, January 8-12, 2018.

[7] Young, L.A., "Rotorcraft and Enabling Robotic Rescue," Heli-Japan 2010: AHS International Meeting on Advanced Rotorcraft Technology and Safety Operations, Ohmiya, Japan, Nov. 1-3, 2010.

[8] Young, L.A., "Future Roles for Autonomous Vertical Lift in Disaster Relief and Emergency Response," Heli-Japan 2006: AHS International Meeting on Advanced Rotorcraft Technology and Life Saving Activities, Nagoya, Japan, November 15-17, 2006.

[9] Young, L.A., "Urban Aerial Mobility Networks using Amphibious Vertical Takeoff and Landing Vehicles," 9th Biennial Autonomous VTOL Technical Meeting & 8th Annual Electric VTOL Symposium, Jan. 26–28, 2021.

[10] Alonso, J.J., Arneson, H.M., Melton, J.E., Vegh, J.M., Walker, C., and Young, L.A., "System-of-Systems Considerations in the Notional Development of a Metropolitan Aerial Transportation System: Implications as to the Identification of Enabling Technologies and Reference Designs for Extreme Short Haul VTOL Vehicles with Electric Propulsion," NASA TM 2017-218356, September 2017.

[11] Young, L.A., "Novel Conceptual Designs for Stopped-Rotor Aerial Vehicles and Other High-Speed Rotorcraft," 6th Decennial VFS Aeromechanics Specialists' Conference, Santa Clara, CA, February 6-8, 2024.

[12] Young. L.A., "Novel Tilting Ducted-Fan Aerial Vehicle Configurations," 10th Autonomous VTOL Technical Meeting and 10th Annual Electric VTOL Symposium, Mesa, AZ, January 24-26, 2023.

[13] Young, L.A., "What is a Tiltrotor? A Fundamental Reexamination of the Tiltrotor Aircraft Design Space," AHS International Technical Meeting on Aeromechanics Design for Transformative Vertical Flight, San Francisco, CA, January 16-18, 2018.

[14] Young, L.A., "Conceptual Design Aspects of Three General Sub-Classes of Multi-Rotor Configurations: Distributed, Modular, and Heterogeneous," Sixth AHS International Specialists Meeting on Unmanned Rotorcraft Systems, Scottsdale, AZ, January 20-22, 2015.

[15] Young, L.A., "Performance, Inflow, and Tip Loss Characteristics of Rotors with Discontinuous Steps in Twist and Lift at their Blade-Tips," NASA/TM-2022-0017246, February 2023.

[16] Stroub, R.H., Young, L.A., Keys, C.N., and Cawthorne, M.H., "Free-Tip Rotor Wind Tunnel Test Results," *Journal of the American Helicopter Society*, Vol. 31, No. 3, July 1986.

[17] Young, L.A. and Aiken, E.W., "Vertical Lift Planetary Aerial Vehicles: Three Planetary Bodies and Four Conceptual Design Cases," 27th European Rotorcraft Forum, Moscow, Russia, September 11-14, 2001.

[18] Young, L.A., Briggs, G., Aiken, E., and Pisanich, G., "Rotary-Wing Decelerators for Probe Descent through the Atmosphere of Venus," 2nd International Planetary Probe Workshop, NASA Ames Research Center, Moffett Field, CA, August 2004 (NASA CP-2004-213456, April 2005).

[19] Young, L.A., "Rotorcraft as Robots," Soon to be Published NASA TM.

[20] Aiken, E. W., Young, L. A., Ormiston, R. A., "Future Directions in Rotorcraft Technology at Ames Research Center," 56th Annual Forum of the American Helicopter Society, Virginia Beach, VA, May 2000.

[21] Conley, S., Peters, N., Aires, J., Wagner, L., Kallstrom, K., Withrow-Maser, S., Wright, S., Radotich, M., Chianello, C., Ortiz, G., Ewing, G., Anderson, A., Klokkevold, K., Kluch, E., "VTOL Analysis for Emergency Response Applications (VAERA) – Identifying Technology Gaps for Wildfire Relief Rotorcraft Missions," Vertical Flight Society Sixth Decennial Aeromechanics Specialists' Conference, Santa Clara, CA, February 2024.

[22] Radotich, M., "Conceptual Design of Tiltrotor Aircraft for Urban Air Mobility," Presented at the VFS Aeromechanics for Advanced Vertical Flight Technical Meeting, San Jose, CA, Jan 25-27, 2022.

[23] Young, L.A., "Building Enhanced Resilience into Aviation," Transformative Vertical Flight 2025, 11th Biennial Autonomous VTOL Technical Meeting, Phoenix, AZ, February 4-6, 2025.

[24] Young, L.A., "Urban Aerial Mobility Networks using Amphibious Vertical Takeoff and Landing Vehicles," 9th Biennial Autonomous VTOL Technical Meeting & 8th Annual Electric VTOL Symposium, Jan. 26–28, 2021.

[25] Vertical Flight Society eVTOL newsletter; https://evtol.news/; last accessed December 15, 2024.

[26] Young, L.A., "Performance, Inflow, and Tip Loss Characteristics of Rotors with Discontinuous Steps in Twist and Lift at their Blade-Tips," NASA/TM-2022-0017246, February 2023.

[27] Young, L.A., Aiken, E.W., "Exploration: Past and Future Contributions of the Vertical Lift Community and the Flight Vehicle Research and Technology Division," AIAA 1st Space Exploration Conference, Orlando, FL, January 30-February 1, 2005.

[28] Young, L.A., Pisanich, G., and Ippolito, C., "Aerial Explorers," 43rd AIAA Aerospace Sciences Meeting, Reno, NV, January 10-13, 2005.

[29] Xue, M., et al, "Advanced Capabilities for Emergency Response Operations Project;" https://www.nasa.gov/directorates/armd/aosp/acero-wildfire/; last accessed January 16, 2025.

[30] Gelzer, C., "X-48B Blended Wing Body;" https://www.nasa.gov/aeronautics/x-48b/; last accessed January 16, 2025.

[31] O'Shea, C.A., "Next Generation Experimental Aircraft Becomes NASA's Newest X-Plane;" https://www.nasa.gov/news-release/next-generation-experimental-aircraft-becomes-nasas-newest-x-plane/; last accessed January 16, 2025.

[32] Ackerman, E., "Aerial-Biped Is a Quadrotor With Legs That Can Fly-Walk," IEEE Spectrum, August 13, 2018; https://spectrum.ieee.org/aerial-biped-quadrotor-with-legs; last accessed January 16, 2025.

[33] Ackerman, E., "Pogo Stick Microcopter Bounces off Floors and Walls," IEEE Spectrum, April 12, 2024; https://spectrum.ieee.org/jumping-robot-quadrotor; last accessed January 16, 2025.

Novel Control of a Propeller Driven Rotor as a Swashplateless System

Robert Brown
Aerospace Engineer,
Naval Surface Warfare Center
Carderock Division,
Bethesda, MD, USA

Inderjit Chopra
Director, Alfred Gessow
Rotorcraft Center,
University of Maryland,
College Park, MD, USA

ABSTRACT

A novel method of swashplateless rotor control is presented for a propeller driven rotor. A propeller driven rotor consists of small electric motors and propellers attached to the rotor blade to spin the main rotor. The control methodology involves varying the propeller RPM around one rotor revolution to generate rolling and pitching moments for the main rotor. A small test stand was fabricated using 2.7 ft diameter 3D printed rotor blades to investigate this concept with various tip propeller geometries. The test stand was utilized to characterize changes in motor commands to rotor rolling and pitching moments. Next, in an effort to reduce the required power for the propeller driven rotor, the propeller placement along the rotor blade span was varied. Placing the propeller at the mid span location on the main rotor blade resulted in a significant reduction of 50% in the electrical power loading coefficient C_{Pe}/σ compared to a tip propeller placement. Finally, the propeller driven rotor control strategy was experimentally compared with a 6 ft diameter, shaft driven rotor that used a conventional swashplate system. A comparison of the control authority derivatives for the different configurations showed that the propeller driven rotor control strategy resulted in a similar amount of control authority as a conventional swashplate system.

I. INTRODUCTION

A propeller driven rotor, shown in Figure 1, uses compact electric motors and propellers, which are attached to the rotor blade, to spin the main rotor. Electric power is transfered from the fixed frame through a high-powered slip ring to the electric motor on the rotor blade tip using wires embedded inside the rotor blades.

A propeller driven rotor utilizes recent advancements in electric aviation to create an alternative method of spinning the main rotor compared to the conventional, shaft driven method, and offers several potential advantages. First, large mechanical transmissions and gears associated with a conventional shaft driven single main rotor design are replaced with efficient direct-driven electric motors. Second, the propeller driven rotor requires no anti-yaw device - eliminating the need for a tail rotor and as a result further reducing vehicle mechanical complexity. Third, this concept appears attractive for distributed electric propulsion. Finally, this design places high RPM actuators capable of generating propulsive forces in the rotating frame, and could offer potential stability and control advantages.

Presented at the Vertical Flight Society's 11th Biennial Autonomous VTOL Technical Meeting, Phoenix, AZ, USA, Feb. 4-6, 2025.

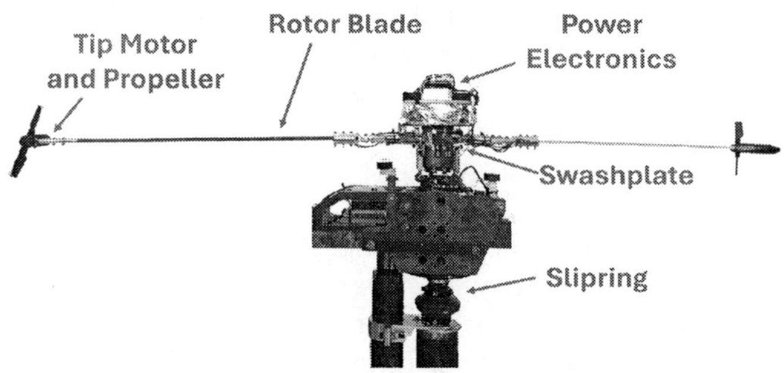

Figure 1: Propeller driven rotor system.

Tip Driven Rotorcraft: State-of-the-Art

The idea of a tip-driven rotor system dates back to one of the earliest attempts at vertical flight, by the Brennan Helicopter, shown in Figure 2a. Tested in 1916, the Brennan Helicopter utilized propellers located at the tip of the rotor blade. Since then, there have been several other attempts that have utilized tip-driven rotors. In the 1950s, the tip-driven design largely focused around hot or cold gas jets, such as the Doblhoff WNF 342 (Ref. 1). Arguably the most famous of these tip-driven rotorcraft was the Fairey Rotordyne, shown in Figure 2b, which used tip-pulsejet engines to rotate the rotor in hover and flew as an autogiro with wings for high speed forward flight.

(a) Brennan Helicopter (1916)

(b) Fairey Rotordyne (1962)

Figure 2: Early examples of tip driven rotorcraft.

The present authors have investigated the feasibility of the propeller driven rotor and have compared the performance of a propeller driven rotor to a conventional, shaft driven rotor (Refs. 2, 3). During these studies, it was observed that increasing the rotor collective angle also increased the vertical component of the propeller thrust, and thus the total thrust of the propeller driven rotor system increased. The tip propeller had a rotational speed 25 times faster than the main rotor, indicating it could potentially be used to generate rotor rolling and pitching moments. This concept is investigated in this paper.

Additionally, for the design to be an attractive alternative for rotorcraft design, the propeller driven rotor should require roughly the same amount of power as a conventional rotor. Some key findings from (Refs. 2, 3) are that adding the motor nacelle to the blade tip resulted in a doubling of the rotor power coefficient. As such, aerodynamically efficient propellers, streamlined nacelle designs, and alternative propeller locations along the main rotor blade span need to be investigated in an effort to reduce required power for the propeller driven rotor.

Literature Review of Swashplateless Micro Air Vehicles

A literature review of similar swashplateless rotorcraft found two notable examples. First, the 38 gram *Samara II* (Refs. 4, 5), shown in Figure 3, flight vehicle utilized a single rotor blade with a brushless motor and propeller attached at the blade tip, and a trailing edge flap. The brushless motor and propeller generated the torque to spin the rotor blade, and a cyclic variation was applied to the trailing edge flap for rotor control.

Figure 3: *Samara II* micro air vehicle from the University of Maryland. Image taken from (Ref. 5).

The second notable concept was the Singapore University's *Monocopter* (Ref. 6), shown in Figure 4. The 29 in wingspan, *Monocopter* design was similar to the *Samara II* aircraft, with the addition that the motor command also varied with rotor blade rotation. In (Ref. 7), the *Monocopter* completely eliminated the trailing edge flap and controls the aircraft solely through propeller RPM variations.

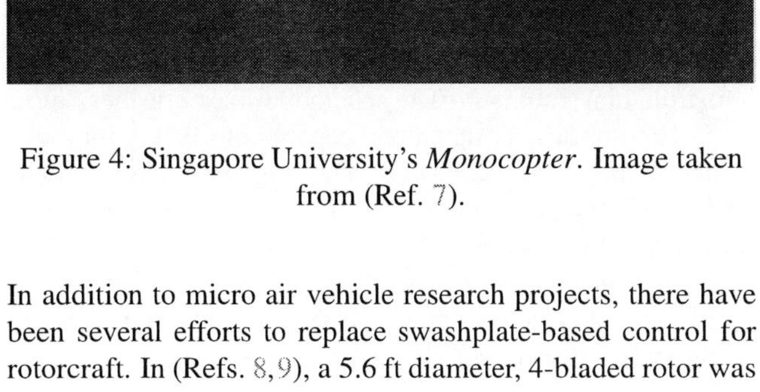

Figure 4: Singapore University's *Monocopter*. Image taken from (Ref. 7).

In addition to micro air vehicle research projects, there have been several efforts to replace swashplate-based control for rotorcraft. In (Refs. 8, 9), a 5.6 ft diameter, 4-bladed rotor was modified with a trailing edge flap, located at 75%-89% of the rotor radius. A brushless DC motor, embedded in the rotor blade, actuated the trailing edge flap for roll and pitch control. Experiments demonstrated the feasibility of the trailing edge flap based control in both hover testing and wind tunnel testing up to a rotor advance ratio of 0.278. This work demonstrated the potential for high frequency brushless motors to augment rotorcraft control systems.

While these works documented the concept of swashplateless rotorcraft, there are several key areas missing from these studies. First, there is a shortcoming on the amount of control moments that these strategies can generate. Secondly, there is a lack of a comparison with conventional swashplate control. Next, the scalability of the proposed design is not discussed. Finally, these studies did not investigate placing the actuators at alternative locations on the rotor span.

Proposed Propeller Driven Rotor Control Strategy

A conventional rotorcraft uses a swashplate mechanism to adjust the rotor blade angle around the rotor azimuth. The propeller driven rotor control strategy takes advantage of the large difference between the main rotor rotational speed (250 RPM) and the tip propeller rotational speed (5000 RPM) to generate rolling and pitching moments as needed for primary control of the aircraft. The tip motor and propeller are rigidly attached to the tip of the main rotor blade with a slight upward tilt. As such, a component of the tip propeller thrust acts in the vertical direction of the rotor disk, shown in Figure 5.

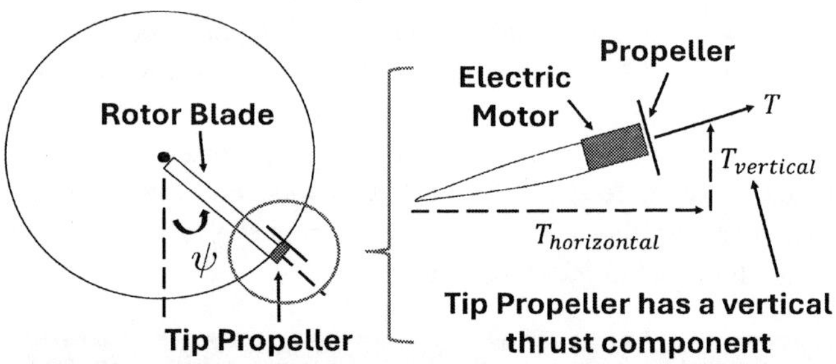

Figure 5: Top down view of rotor plane. The tip propeller has a component of vertical thrust that was used to generate control moments for the rotor.

The propeller driven rotor combines the 1/rev excitation of a swashplate from a conventional rotor with the fast electronics of a quadrotor. For stable flight, a quadrotor adjusts motor commands to augment propeller RPM to create a rolling/pitching moment. The scheme utilizes brushless motors as high frequency actuators, capable of quickly increasing/decreasing rotational speed. The propeller driven rotor's control moments are generated by increasing the propeller rotational speed on one side of the rotor disk while reducing the rotational speed on the opposite side of the rotor disk. Figures 6 and 7 shows how rotor rolling and pitching moments are generated, respectively.

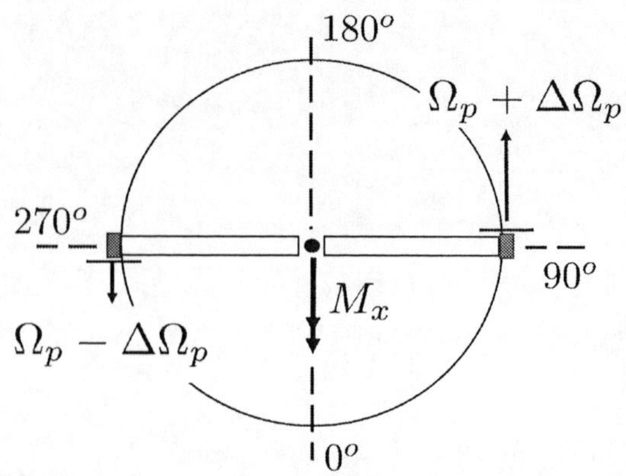

Figure 6: Rolling moment strategy.

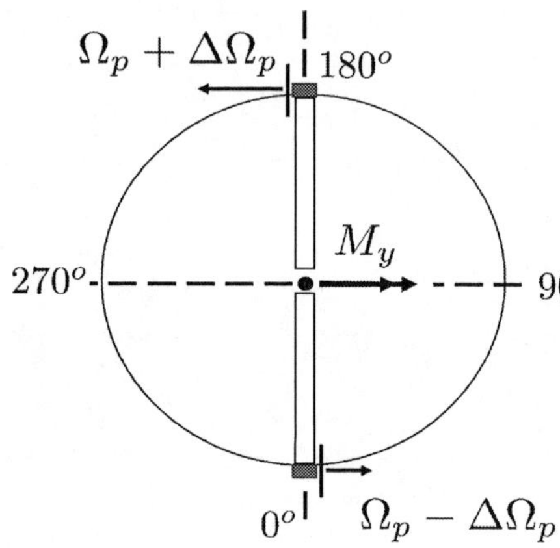

Figure 7: Pitching moment strategy.

The strategy works by adjusting the propeller speed, and thus the vertical portion of propeller thrust, at specific locations around the rotor azimuth. Consider the propeller is spinning at some nominal rotational speed $\Omega_{p,0}$. A positive rolling moment was generated by increasing the propeller speed at $\psi = 90$ and decreasing the propeller speed at $\psi = 270$. This corresponds to adding a $\Omega_{p,1s} \sin \psi$ portion to the propeller speed, thus:

$$\Omega_p(\psi) = \Omega_{p,0} + \Omega_{p,1s} \sin \psi$$

A similar strategy can be considered for the pitching moment. For a positive pitching moment, the propeller speed must decrease at $\psi = 0$ and increase at $\psi = 180$. This results in a $-\Omega_{p,1c} \cos \psi$ portion being added to the nominal propeller speed $\Omega_{p,0}$. Dropping the minus sign for convince, the final strategy results is the propeller speed variation with rotor azimuth ψ as:

$$\Omega_p(\psi) = \Omega_{p,0} + \Omega_{p,1s} \sin \psi + \Omega_{p,1c} \cos \psi \quad (1)$$

where $\Omega_{p,1s}$ and $\Omega_{p,1c}$ represent the longitudinal and lateral propeller RPM variations, respectively.

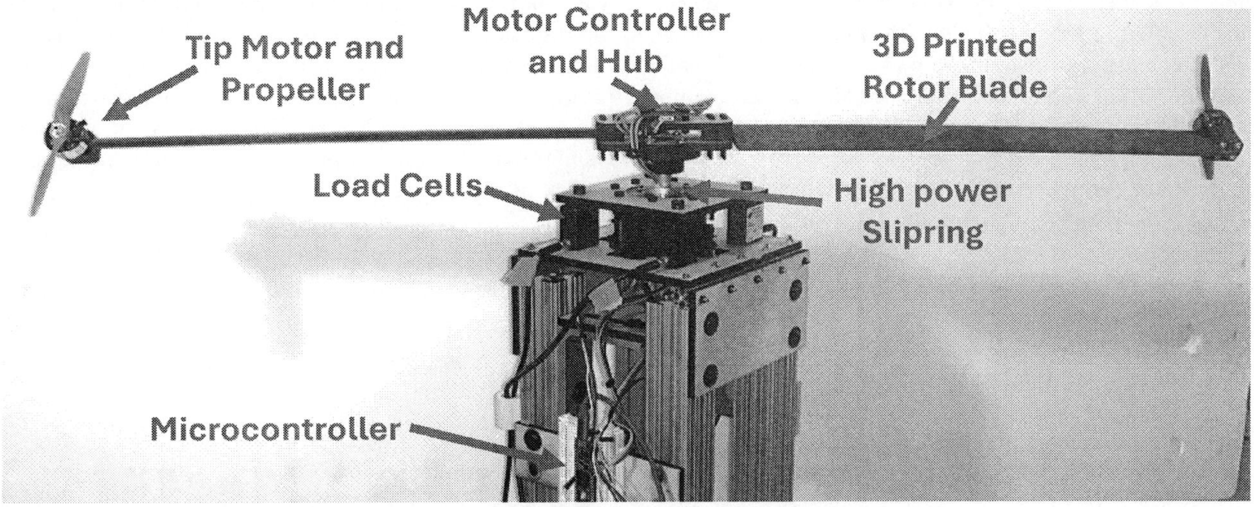

Figure 8: Propeller driven rotor test stand.

Technical Approach

The study had three research objectives. First, to implement the swashplateless propeller driven rotor control scheme and measure the resulting control authority. Second, to investigate impact of propeller placement along the main rotor blade on propeller driven rotor performance. Finally, compare the propeller driven rotor control methodology to a conventional swashplate system. Two test stands were utilized to complete these research objectives:

- **Propeller Driven Rotor Test Stand** - A 2.7 ft diameter propeller driven rotor test stand was fabricated to examine the propeller driven rotor control concept. The propeller driven rotor test stand was used for propeller driven rotor control testing and for the propeller placement study. The propeller driven rotor test stand does not have a swashplate.
- **Conventional Rotor Test Stand** - A 6 ft diameter shaft driven rotor test stand that features composite carbon fiber blades and a swashplate. The rotor swashplate enables cyclic pitch variation and allows the study to compare the propeller driven rotor's control methodology to a conventional swashplate control methodology.

The remainder of the paper is as follows. Section II describes the propeller driven rotor test stand and details on the propeller driven rotor control performance. Section III presents the results of placing the propeller at alternative locations along the main rotor span, and their impact on main rotor thrust, power, and propeller speed. Section IV details the conventional rotor test stand and the swashplate control results, and compares the swashplate and the propeller driven rotor control performance. Lastly, Section V presents the summary, conclusions, and future work.

II. PROPELLER DRIVEN ROTOR CONTROL EXPERIMENTS

A propeller driven rotor test stand, shown in Figure 8, was fabricated to investigate the propeller driven rotor control strategy. This propeller driven rotor test stand featured a high powered 12 channel slipring, four force loads cells, DC power measurements, a rotor azimuth magnetic encoder, and a microcontroller. A commercial-off-the-shelf 5x5 propeller, shown in Figure 9, was used for all propeller driven rotor experiments. A microcontroller reads the rotor azimuth location and adjusts the motor commands accordingly. Additionally, the stand can be modified to enable shaft driven rotor testing.

There were three test objectives with the propeller driven rotor test stand. First, conduct an initial experiment for the propeller driven rotor control scheme outlined in Section I. Secondly, investigate reducing propeller driven rotor performance with the propeller mounted at various locations along the main rotor blade span. Finally, compare the propeller driven rotor to the shaft driven rotor performance for the various rotor configurations. Table 1 summarizes characteristics of the propeller driven rotor test stand.

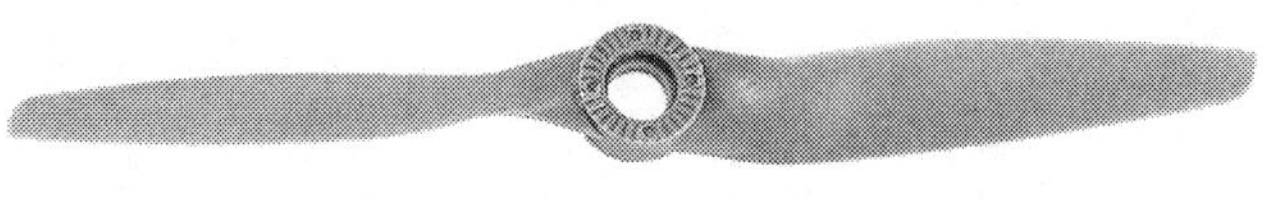

Figure 9: Five in (5x5) diameter propeller that was used for all propeller driven rotor experiments.

Figure 10 shows the a schematic of the propeller driven rotor test stand setup. A microcontroller reads the rotor azimuth angle and adjust the motor throttle settings at a rate of over 1000 times per second. All sensors and the microcontroller were connected to a LabView DAQ and computer for real time data viewing and saving. The motor controller was placed on the hub in the rotating frame to reduce the number of channels that were required to transmit high powered current. The microcontroller connected to the motor controller in the rotating frame via signal wires that went through the slipring.

Table 1: Rotor, propeller, and powertrain characteristics for the propeller driven rotor test stand.

Rotor Radius R	16.2 in
Rotor Chord c	2.5 in
Rotor Aspect Ratio R/c	6.4
Number of Rotor Blades	2
Rotor Airfoil	NACA0012
Rotor Solidity σ	10%
Rotor Collective θ_0	8 deg
Rotor Twist	None
Propeller	5x5, shown in Figure 9
Propeller Radius	2.5 in
Propeller Chord	0.4 in
Number of Propeller Blades	2
Propeller Solidity	10%
Motor	EMAX Eco II 1700 Kv
Motor Controller	T-Motor Velox 50A

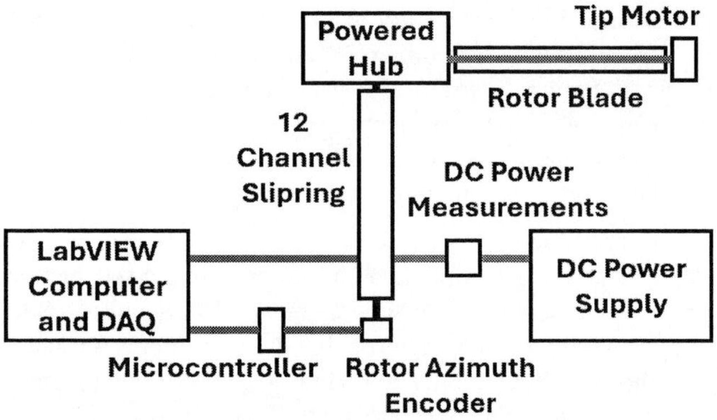

Figure 10: Wiring diagram for the propeller driven rotor test stand.

3D Printed Main Rotor Blades

The next phase included the fabrication of the main rotor blade using 3D printing. Due to the large number of designs, 3D printing the blades was preferable to using composite blade designs (making CNC molds for each design, etc.) and resulted in significant time savings. Using 3D printers allowed the blade design to be systematically varied in a quick and efficient manner. The blade tip nacelle configuration, as well as the motor/propeller placement along the blade span were altered. Figure 11 shows the blade, tip motor, and tip propeller.

Three printers were investigated for 3D printing rotor blades: the Bambu Labs X1 Carbon, the Prusa MK4, and the Flashforge Creator 3 Pro. A small blade section was printed on each printer, and after comparing the surface quality, print times, and print options of all printers it was decided to utilize the Bambu Labs X1 Carbon printer as it resulted in the highest quality blade section.

The 3D printed rotor blades were printed in two sections per blade and reinforced with a 4 mm diameter solid carbon fiber beam as a spar. The blades were made of PLA and had a 1 mm thickness. Spin tests revealed that the 3D printed blades were able to spin at over 1500 RPM, with a tip speed of 65 m/s or 212 ft/s.

Figure 11: The 14 inch span, 3D printed propeller driven rotor blade.

Propeller Tip Angle δ

For propeller driven rotor control strategy development, the vertical portion of propeller thrust is directly related to the control moment magnitude. As such, the angle δ between the rotor blade chord and the tip propeller thrust vector was adjusted. This propeller tip angle δ is shown in Figures 12. A larger propeller tip angle δ tilts the propeller thrust in the vertical direction, and thus causes a larger moment with a slight reduction of propulsive thrust.

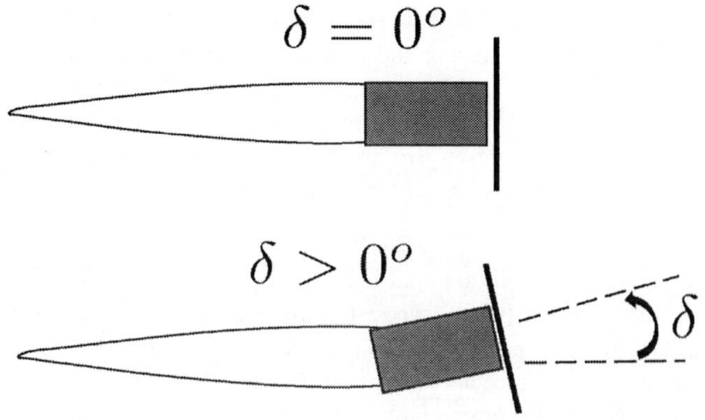

Figure 12: Propeller tip angle δ increases the vertical component of propeller thrust and thus creates a larger moment.

Open Loop Motor Control

Due to the limited number of slipring channels on the propeller driven rotor test stand it was not possible to implement closed-loop propeller RPM control. Instead, the open loop motor throttle commands u were adjusted. The motor throttle u is the Pulse Width Modulated value to the motor controller, and has been normalized between 0-100%. A microcontroller was added to the propeller driven rotor test stand to control the input throttle setting to the motor. The motor throttle consists of a nominal portion u_0 and components for the longitudinal u_{1s} and lateral u_{1c} variations. Additional details on the motor controller and brushless motor performance can be found in (Ref. 10).

Instead of directivity controlling propeller speed, such as in equation 1, the microcontroller adjusted the motor throttle based on the rotor azimuth ψ as:

$$u(\psi) = u_0 + u_{1c}\cos\psi + u_{1s}\sin\psi \qquad (2)$$

The values of u, u_0, u_{1c}, and u_{1s} are normalized between 0-100% of the motor's throttle range. For example, Figure 13 shows the motor throttle u over one rotor azimuth with a u_0 value of 30% and a u_{1s} variation of 5%. Additionally, there is some time delay between the input signal and propeller speed, which can result in some cross-axis coupling with the moments. The time delay is comprised of the motor response, the propeller dynamics, and the rotor response.

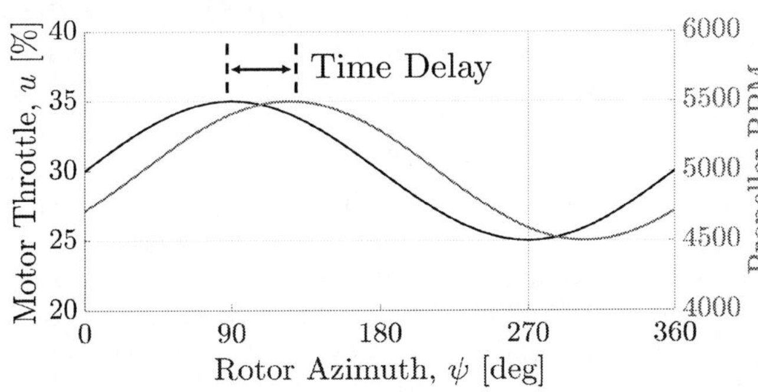

Figure 13: Motor throttle variation over one rotor azimuth with $u_0 = 30\%$ and $u_{1s} = 5\%$. Some time delay is expected between the motor throttle and the propeller speed.

Propeller Driven Rotor Experiments

The propeller driven rotor with various propeller tip angles δ of 0°, 10°, and 20° was characterized in two experiments. In the first experiment, an RPM sweep of the main rotor was performed and the main rotor's thrust, electrical power, and propeller RPM were recorded. This sweep was accomplished by gradually increasing the nominal motor throttle u_0 while the u_{1s} and u_{1c} values were set to 0%. In the second experiment, the rotor's rolling and pitching moments were characterized by changing either u_{1s} or u_{1c} independently. The main rotor was held at a constant rotational speed of 250 RPM while either u_{1s} or u_{1c} were adjusted. The resulting change in rotor moments, rotor thrust, and DC electrical power were recorded.

Exp. 1 - Propeller Driven Rotor RPM Sweep

The 2.7 ft diameter propeller driven main rotor blades with propeller tip angles δ of 0°, 10°, and 20^owere tested up to a main rotor rotational speed up to 700 RPM. The motor input throttle u_0 was increased and the resulting rotor thrust, rotor electrical power, and propeller speed were recorded. All tests were conducted at a main rotor collective of 8 deg. These results are shown in Figure 14.

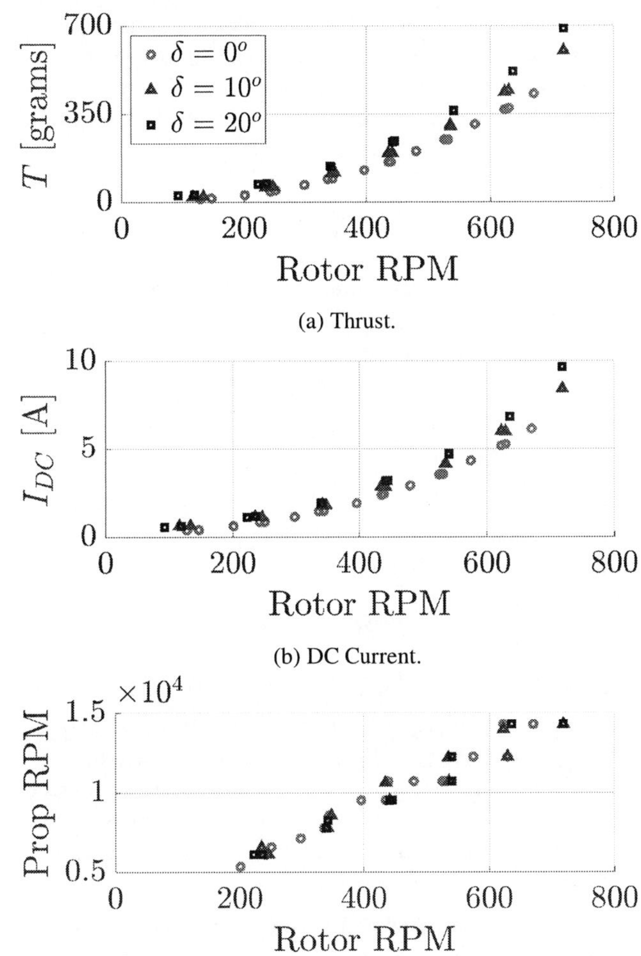

(a) Thrust.

(b) DC Current.

(c) Propeller RPM.

Figure 14: Propeller driven rotor tip results with different δ angles at a main rotor collective of 8 deg.

The primary comparison metrics for the 2.7 ft diameter propeller driven rotor tests are the thrust loading coefficient C_T/σ, electrical power loading coefficient C_{Pe}/σ, and the propeller/rotor RPM slope m.

The thrust loading coefficient C_T/σ is defined as:

$$\frac{C_T}{\sigma} = \frac{1}{\sigma}\frac{T}{\rho A(\Omega R)^2}$$

The electrical power P_{DC} is defined as:

$$P_{DC} = V_{DC} I_{DC}$$

Due to the limited number of slipring channels, propeller mechanical power could not be measured. Instead, the DC electrical power to the slipring was measured in the fixed frame and will be used as a comparison metric, denoted as C_{Pe}.

$$P_{DC} = C_{Pe}\rho A(\Omega R)^3 + P_{DC,o} \tag{3}$$

where C_{Pe} is the electrical power coefficient that contributed to spinning the rotor whereas $P_{DC,o}$ is the power component to overcome the static drag of the rotor. The static drag comes from frictional forces in the bearings, slipring, and viscous rotor drag. From this relation, the static DC current I_o required to overcome this drag is defined as:

$$I_o = \frac{P_{DC,o}}{V_{DC}}$$

The propeller mechanical power P_p is related to the electrical power P_e through the powertrain efficiency η_{PT}, which includes the electric motor, motor controller, and wiring losses:

$$P_p = \frac{P_{DC}}{\eta_{PT}}$$

From the experimental test data, it can be seen that the main rotor RPM Ω and the propeller RPM Ω_p are a linear relationship, which can be written as:

$$\Omega_p = m\Omega + \Omega_{p,offset}$$

where m is the slope between rotor RPM and propeller RPM, called the propeller/rotor RPM slope. The propeller RPM offset $\Omega_{p,offset}$ is the required propeller RPM to overcome the static friction of the propeller driven rotor test stand. Figure 15 shows a sketch of how $\Omega_{p,offset}$ and the propeller/rotor RPM slope m are defined. The terms $P_{DC,o}$, I_o, and $\Omega_{p,offset}$ were directly related to one another.

Table 2 presents the measured performance metrics for this test. The rotor thrust loading coefficient C_T/σ increased by 78% between the $\delta = 0^o$ to 20° for a 24% increase in an electrical power loading C_{Pe}/σ. This increased thrust was expected as a portion of the propeller thrust was directed in the vertical direction. This vertical propeller thrust was used to generate control moments in the following experiment. Additionally, Table 2 shows an increase in I_o and the propeller RPM offset $\Omega_{p,offset}$ with an increase in the propeller tip angle δ. This was expected as the larger tip angle δ results in more propeller thrust directed upwards at a cost of less propulsive thrust being used to rotate the main rotor.

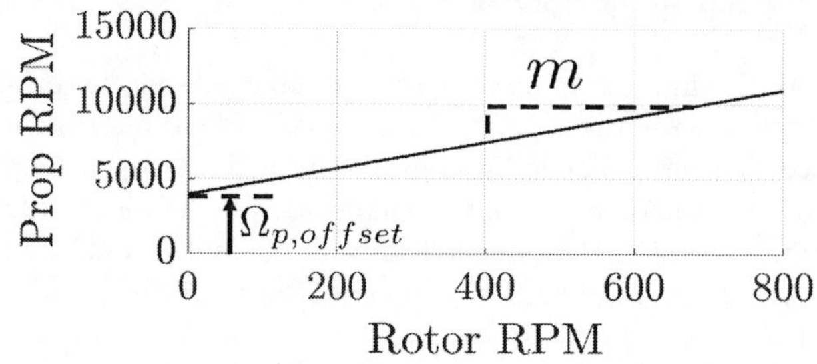

Figure 15: Definition of propeller offset $\Omega_{p,offset}$ and the slope m between rotor RPM and propeller RPM.

Table 2: Propeller driven rotor performance with various propeller tip angles δ.

δ	0°	10°	20°
C_T/σ	0.071	0.094	0.127
C_{Pe}/σ	0.0455	0.0495	0.0566
$I_o[A]$	0.63	0.89	0.88
m [Propeller RPM/Rotor RPM]	17.7	17.2	17.0
$\Omega_{p,offset}$ [Propeller RPM]	2075	2250	2320
T/P_{DC} at 500 Rotor RPM [N/W]	0.0569	0.0599	0.0623

Exp. 2 - Control Characterization

After the propeller driven rotor RPM sweep performance sweep was conducted, the control experiment characterized the variation in motor commands to rotor moments. The tip motor's input throttle u_o was set such that the main rotor spun at 250 RPM, and the control inputs u_{1s} and u_{1c} were varied from 0 to 10% of the motor's throttle range.

Figure 16 shows the change in electrical power while using the propeller driven rotor control scheme. Electrical power was found to increase by 24.5%, 21.7%, and 22.9% for tip propeller angles δ of 0°, 10°, and 20°, respectively. Figures 17 and 18 on the next page show the rolling and pitching moments with change in u_{1s} and u_{1c}, respectively. This control scheme resulted in satisfactory on-axis control, with little coupling between the axes. Additionally, increasing the tip angle linearly increased the control moments.

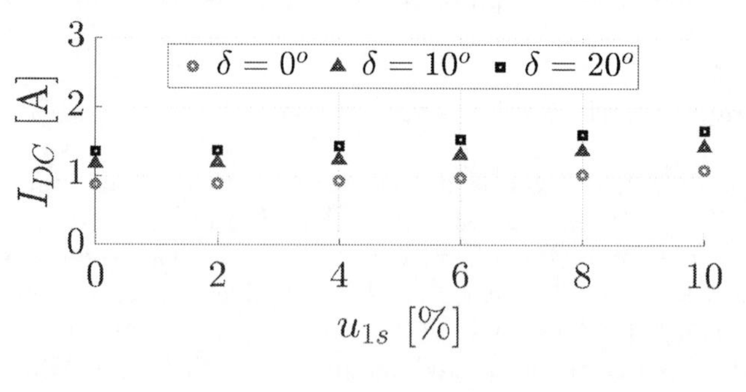

Figure 16: Change in DC current for a u_{1s} variation at 250 rotor RPM. A similar trend was observed for u_{1c}.

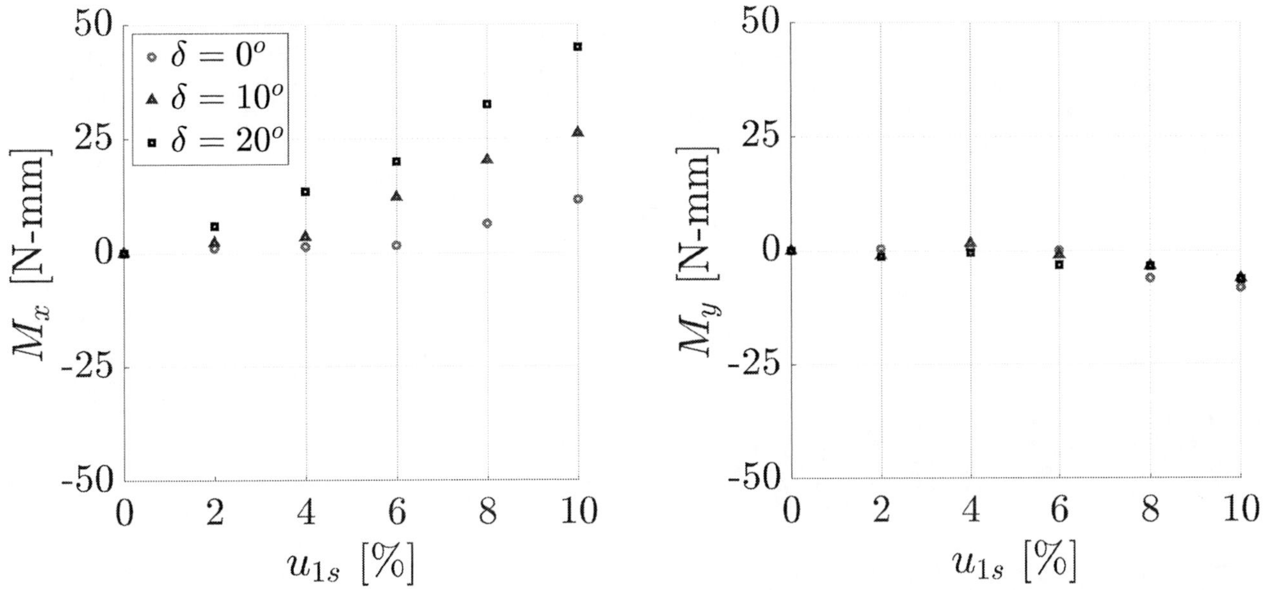

(a) On-axis rolling moment response to u_{1s} variation.

(b) Off-axis pitching moment response to u_{1s} variation.

Figure 17: Rolling moment M_x and pitching moment M_y variation with u_{1s} at 250 main rotor RPM with a fixed main rotor collective of 8 deg.

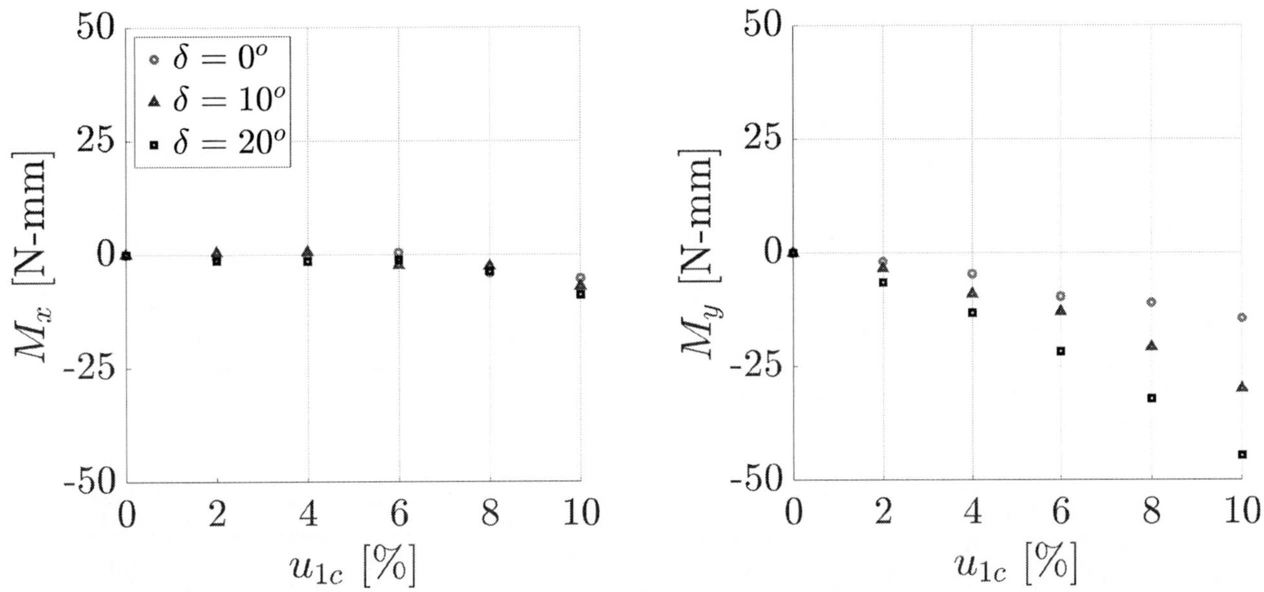

(a) Off-axis rolling moment response to u_{1c} variation.

(b) On-axis pitching moment response to u_{1c} variation.

Figure 18: Rolling moment M_x and pitching moment M_y variation with u_{1c} at 250 main rotor RPM with a fixed main rotor collective of 8 deg.

III. PROPELLER PLACEMENT ALONG THE BLADE SPAN

A second objective was to evaluate the propeller driven rotor performance with the propeller placed at various locations on the main rotor span. The design of the 3D printed rotor blade was modified to have the propeller placed at the 1/2 span, 3/4 span, and at the tip of the rotor blade. Figure 19 shows the 3D printed rotor blades. All blades utilized the same propeller and had the rotor collective θ_0 fixed at 8 deg, similar to the experiments in Section II.

A rotor blade in hover has a change in velocity proportional to the distance from the hub, shown in Figure 20. A propeller mounted on the main rotor blade will experience this incoming velocity as a propeller climb velocity v_c.

Figure 19: 2.7 ft diameter main rotor blades showing the propeller at the 1/2 span, 3/4 span, and tip locations on the rotor blade.

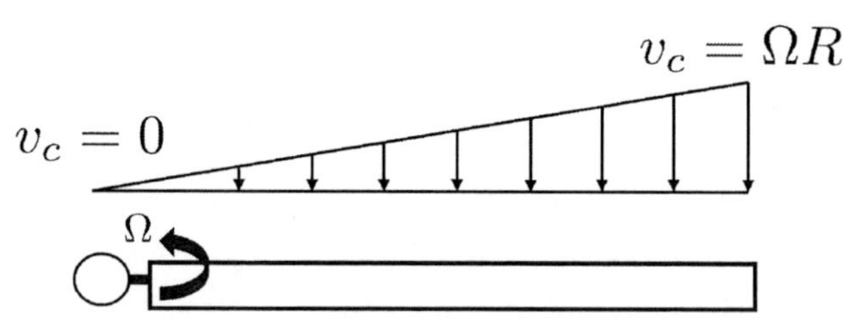

Figure 20: A main rotor blade in hover has a tangential velocity proportional to its distance from the hub.

The term x_p, shown in Figure 21, represents the non-dimensional location along the rotor span, relative to the hub, that the propeller was placed. An $x_p = 1/2$ represents placing the propeller at the half span location. Note that the propeller climb velocity v_c also increased with an increased x_p position, as $v_c = x_p(\Omega R)$. It was assumed that the propeller radius was small relative to the main rotor radius such that the climb velocity v_c was uniform across the propeller span.

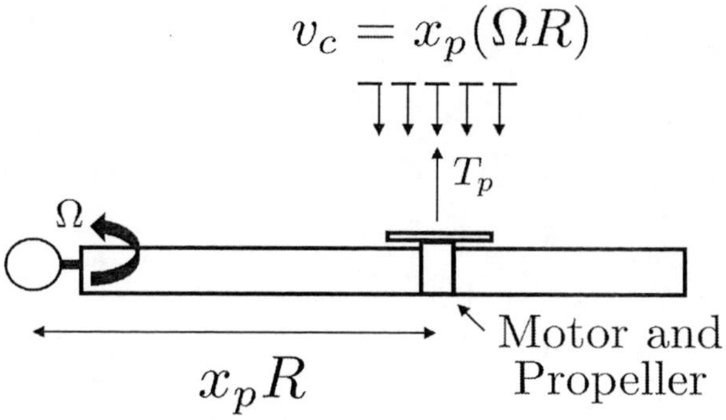

Figure 21: Definition of propeller location x_p.

Figure 22 shows the change in propeller climb velocity for the different propeller placements. The rational for placing the propeller at certain locations on the rotor span is as follows:

- **Rotor Blade Tip** - The propeller is furthest from the rotor shaft and thus would produce the largest moment arm. However, placing the motor and propeller nacelle at the tip results in a high drag penalty. Previous studies found that this drag penalty can double the profile drag of the rotor.

- **Rotor 1/2 Span** - The incoming climb velocity v_c of the propeller is reduced, which could have impacts on the propeller performance. However, the moment arm is also reduced so propeller thrust and power could increase.

- **Rotor 3/4 Span** - This acts as an in-between case for the propeller placed at either the blade tip and half span locations.

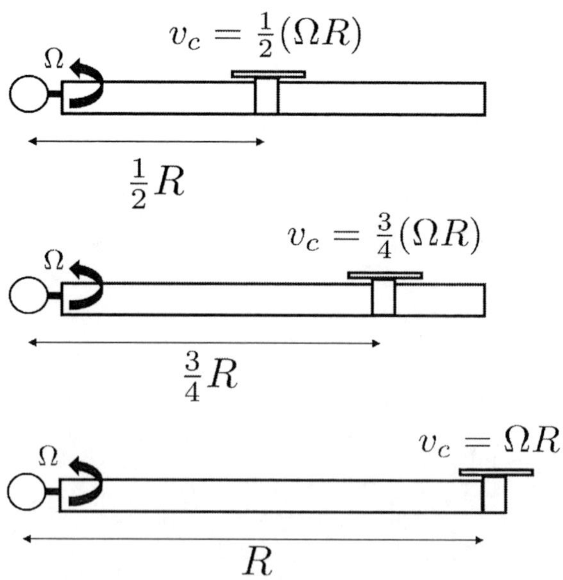

Figure 22: Climb velocities for the 1/2 span, 3/4 span, and tip mounted propellers.

Propeller Placement Experimental Results

Figure 23 shows the propeller driven rotor's thrust, electrical power, and propeller RPM under x_p values of 1/2, 3/4, and 1. Table 3 shows the measured performance metrics for the test.

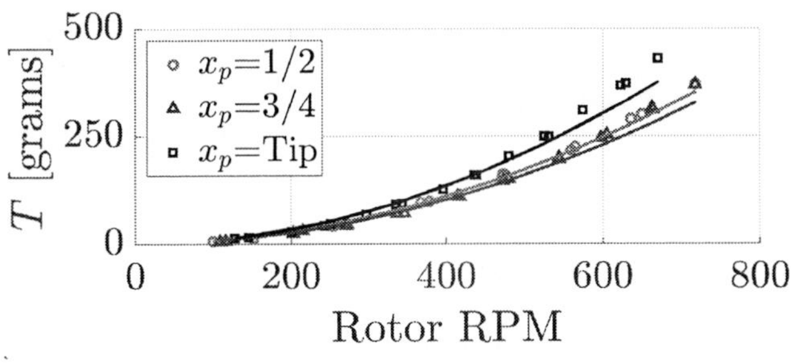

(a) Thrust.

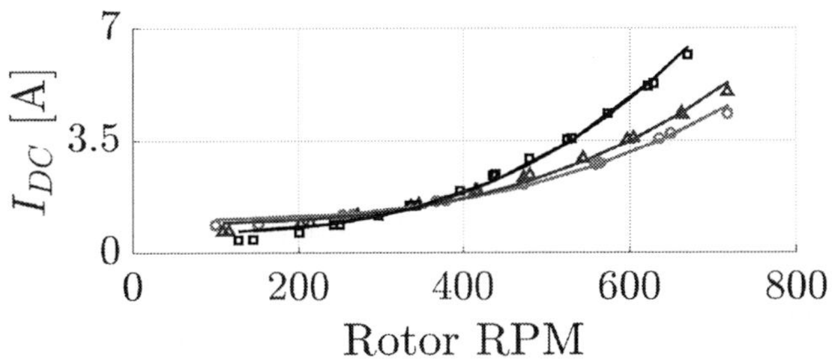

(b) DC Current.

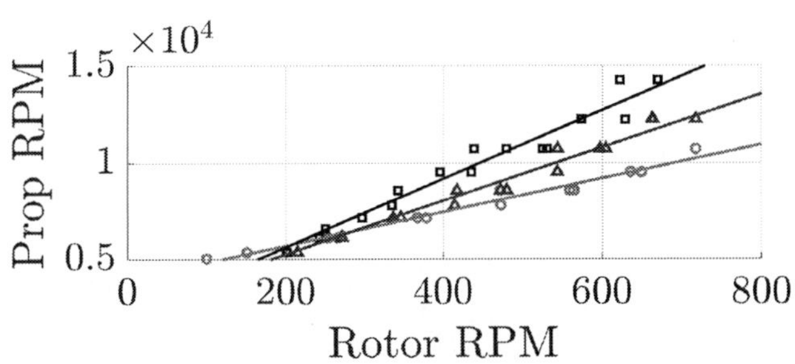

(c) Propeller RPM.

Figure 23: Propeller driven rotor performance results while placing the propeller at various rotor span locations. Dots represent experimental data while lines represent the regression model.

Table 3: Propeller placement variation results.

Propeller Position	$\frac{1}{2}R$	$\frac{3}{4}R$	Tip
C_T/σ	0.0573	0.0533	0.0701
C_{Pe}/σ	0.0224	0.0277	0.0450
I_o[A]	1.03	0.91	0.63
m [Propeller RPM/Rotor RPM]	8.74	13.8	17.2
$\Omega_{p,offset}$ [Propeller RPM]	3950	2490	2075

From Table 3 several deductions can be made. Placing the propeller at the half span location required the propeller offset $\Omega_{p,offset}$ and the DC current offset I_o to nearly double due to the reduced moment arm when compared to placing the propeller at the blade tip. However, placing the propeller at the mid span location resulted in a 20% reduction in rotor thrust loading C_T/σ but a 50% reduction in rotor electrical power loading C_{Pe}/σ. This indicates an initial higher penalty for a mid span propeller placement, but a favorable C_{Pe} for moderate to higher rotor RPMs.

Additionally, the propeller/rotor RPM slope m that relates propeller RPM to rotor RPM was found to linearly vary from $x_p = 1/2$ to 1. The 1/2 span located propeller/rotor RPM slope m is nearly half the tip propeller/rotor RPM slope m value. Lastly, placing the propeller at the 3/4 span location resulted in the lowest C_T/σ value of any rotors, and had a higher C_{Pe}/σ value than the mid span propeller location. This implies that placing the bluff motor body at the rotor 3/4 span location significantly disrupted the rotor's ability to provide thrust, and was found to be the worst performing of any of the propeller locations.

Figures 24 to 27 plot the performance of the different propeller driven rotors versus main rotor thrust. Figure 24 and Figure 25 show two distinct regions when comparing the propeller placements at the midspan and tip locations. For low thrust levels, the tip driven rotor required less electrical power to operate than the midspan located propeller. However, at higher thrust levels, the trend reserves and the midspan located propeller required less power to generate a certain thrust compared to the tip driven propeller. This crossover point occurs at around 125 grams of rotor thrust.

This observation can be explained by examining the values in Table 3. The 1/2 span located propeller had a higher propeller RPM offset of 3950 compared to the tip propeller's value of 2075. However, the electrical power loading coefficient C_{Pe}/σ for the midspan located propeller was 50% the value of the rotor with the tip propeller. This means that there is a higher initial electrical cost for the midspan located propeller but a lower required power once the rotor is spinning due to the lower C_{Pe}/σ.

This creates an interesting design decision for a propeller driven rotorcraft engineer. As indicated in Figure 26, at lower rotor thrust levels the tip located propeller requires less power than the 1/2 span located propeller. However, due to the 1/2 span located propeller's C_{Pe}/σ value, at higher rotor thrust levels the 1/2 span located propeller requires less electrical power than the tip located propeller. The difference in electrical power increases with increasing rotor thrust. At 400 grams of main rotor thrust, the 1/2 span located propeller requires 15% less electrical power compared to the tip located propeller. This crossover point is again evident when considering the T/P_{DC} plot in Figure 25. These propeller placement results indicate that optimum propeller driven rotor performance benefits can be achieved by properly sizing the propeller blade geometry, the propeller location, and the propeller driven rotor configuration.

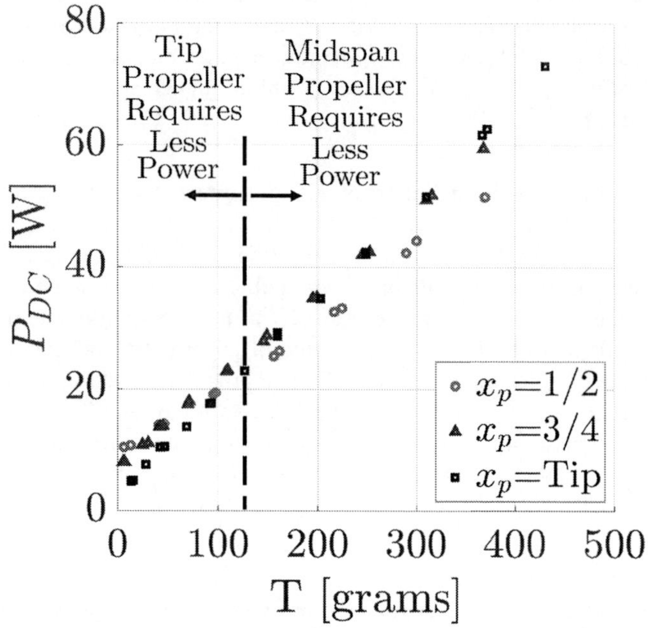

Figure 24: DC electrical power versus main rotor thrust for the different propeller locations.

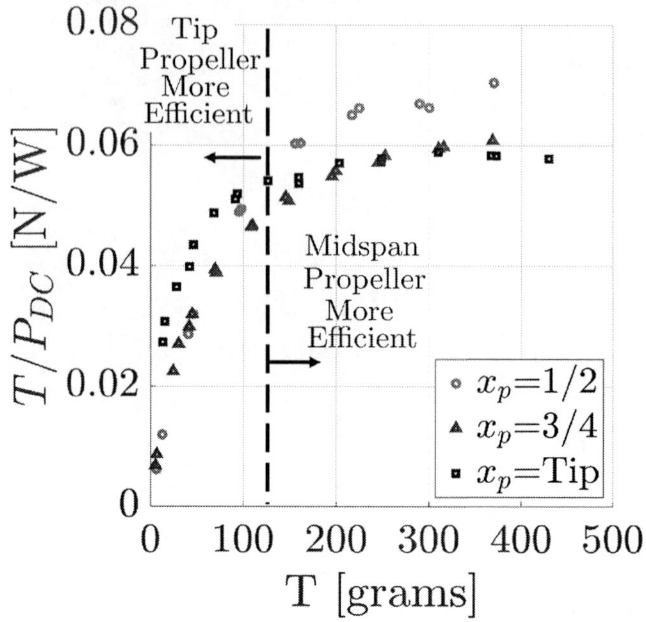

Figure 25: Electrical power loading versus main rotor thrust for the different propeller locations.

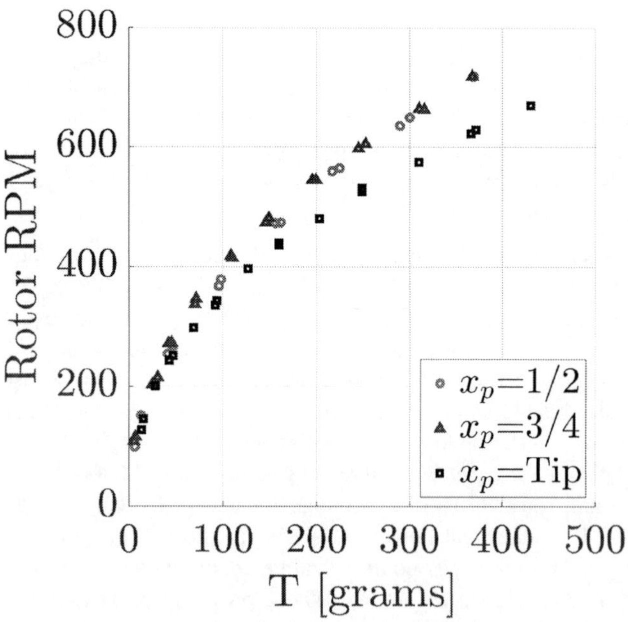

Figure 26: Main rotor RPM versus main rotor thrust for the different propeller locations.

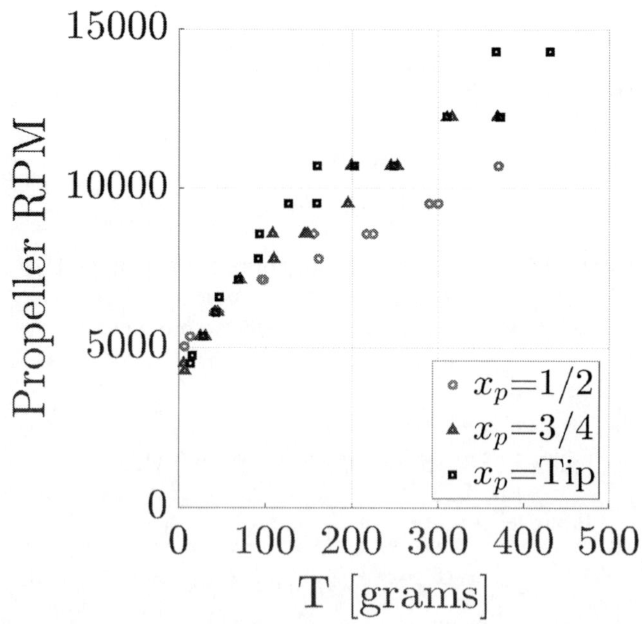

Figure 27: Propeller RPM versus main rotor thrust for the different propeller locations.

Propeller Placement Comparisons

The performance of the propeller driven rotor was also compared by considering three cases: constant main rotor thrust, constant main rotor RPM, and constant propeller RPM.

Case I: Constant Rotor Thrust

This case compares the required electrical power to generate a 300 grams of thrust from the main rotor. The intersection of the 300 gram thrust line and the corresponding electrical power points are shown in Figure 28. The dimensional results are shown in Table 4.

Table 4 shows that placing the propeller at the mid span location resulted in an 18% reduction in electrical power, an 11% increase in rotor RPM, and a 20% reduction in prop RPM compared to placing the propeller at the tip location.

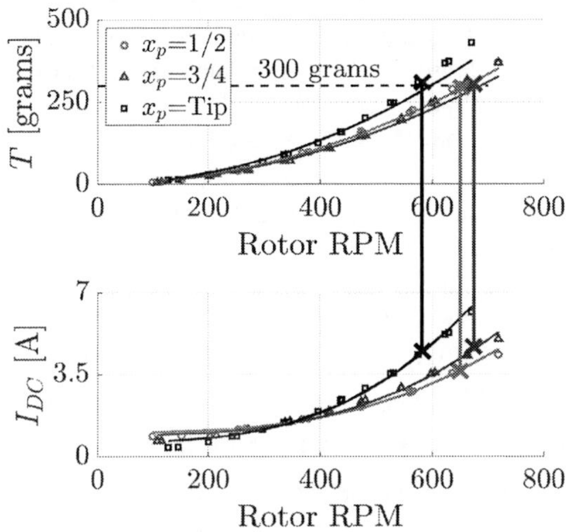

Figure 28: Intersection of 300 grams of thrust for the different propeller driven rotors.

Table 4: Propeller driven rotor performance at 300 grams of thrust.

Propeller Position	$\frac{1}{2}R$	$\frac{3}{4}R$	Tip
T [grams]	300	300	300
Rotor RPM	663	687	600
Propeller RPM	9,800	12,000	12,400
I_{DC} [A]	3.8	4.7	4.7
P_{DC} [W]	45.4	56.6	56.5
T/P_{DC} [N/W]	0.0648	0.0520	0.0521

Case II: Constant Main Rotor RPM

Next, the propeller driven rotor performance is compared at 500 rotor RPM. Figure 29 shows the intersection of the rotor thrust and electrical power, and Table 5. Table 5 shows that placing the propeller at the mid span location results in an 18% reduction in thrust, a 26% reduction in electrical power, and a 22% reduction in propeller RPM compared to placing the propeller at the blade tip.

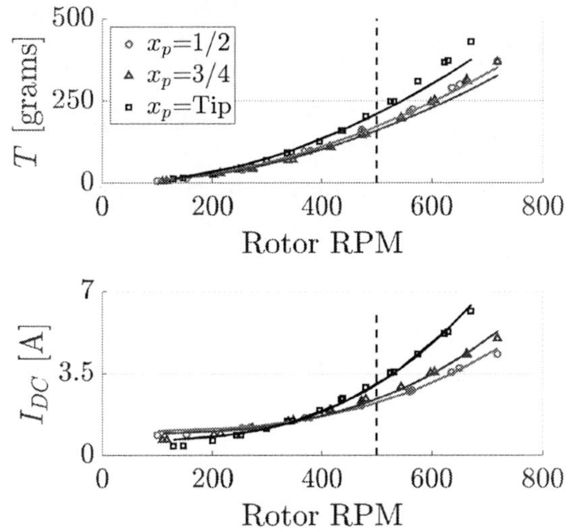

Figure 29: Intersection of 500 RPM of the main rotor for the different propeller driven rotors.

Table 5: Propeller driven rotor performance at 500 RPM of the main rotor.

Propeller Position	$\frac{1}{2}R$	$\frac{3}{4}R$	Tip
T [grams]	171	159	209
Rotor RPM	500	500	500
Propeller RPM	8,300	9,400	10,700
I_{DC} [A]	2.2	2.4	3.0
P_{DC} [W]	26.6	28.6	36.0
T/P_{DC} [N/W]	0.0630	0.0545	0.0569

Case III: Constant Propeller RPM

Finally, the propeller driven rotor performance is compared at 10,000 propeller RPM. Figure 30 shows the intersection of the propeller RPM lines with 10,000 propeller RPM and Table 6 shows the propeller driven rotor performance results. Table 6 indicates that placing the propeller at the mid span location resulted in an 84% increase in thrust, a 68% increase in electrical power. This is due to the 50% increase in rotor RPM compared to the tip propeller placement.

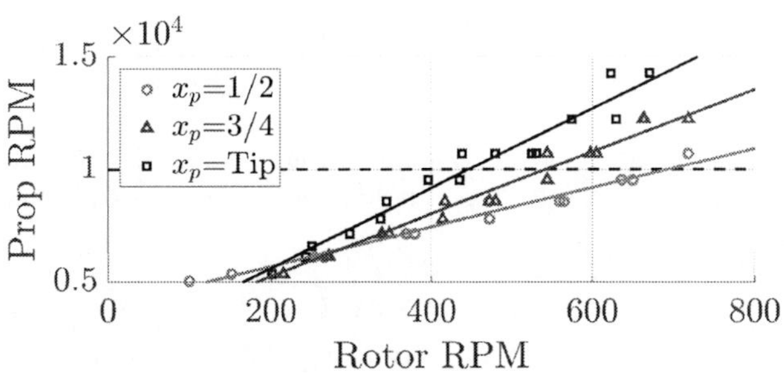

Figure 30: Intersection of 10,000 propeller RPM for the different propeller driven rotors

Table 6: Propeller driven rotor performance at 10,000 propeller RPM.

Propeller Position	$\frac{1}{2}R$	$\frac{3}{4}R$	Tip
T [grams]	327	188	177
Rotor RPM	692	545	460
Propeller RPM	10,000	10,000	10,000
I_{DC} [A]	4.2	2.8	2.5
P_{DC} [W]	50.0	33.6	29.8
T/P_{DC} [N/W]	0.0641	0.0549	0.0582

Next, the propeller driven rotor test stand was modified for shaft driven testing so that mechanical power of the different rotors could be measured.

Shaft Driven Testing

Mechanical power measurements were made by modifying the propeller driven test stand to enable shaft driven testing. The slipring was removed and replaced with an electrical drive motor and a torque cell. These modifications allowed for mechanical rotor power to be measured and compared between the different rotor configurations. The 3D printed hingeless rotors had a rotor collective θ_0 of 8 deg for all shaft driven tests. Figure 31 shows the propeller driven rotor test stand after it was modified for shaft driven testing. The results are compared to an isolated rotor, that is a rotor that does not have a motor nacelle added to it.

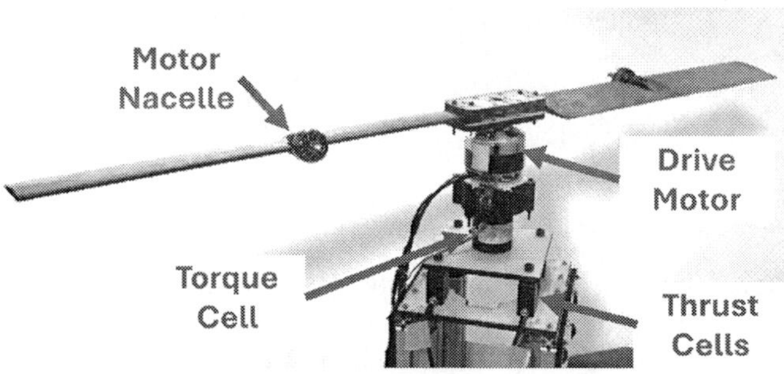

Figure 31: Propeller driven rotor test stand modified for shaft driven testing.

The rotor's mechanical power P was measured from the torque cell and RPM sensor as:

$$P = Q\Omega$$

where Q is the rotor's mechanical torque measured from the torque cell. The rotor's power loading coefficient C_P/σ is defined as:

$$\frac{C_P}{\sigma} = \frac{1}{\sigma} \frac{P}{\rho A (\Omega R)^3}$$

From the measurement of mechanical power P, the rotor's figure of merit (FOM) is defined as:

$$FOM = \frac{C_T^{3/2}/\sqrt{2}}{C_P}$$

and then normalized between 0-100%. Table 7 shows the results of the shaft driven rotor testing for the various rotor blades along with an isolated blade without a motor.

Table 7: Shaft driven rotor results. Rotor collective θ_0 was fixed at 8 deg.

Propeller Position	δ [^{o}]	C_T/σ	C_P/σ	FOM [%]
$\frac{1}{2}R$	0	0.0449	0.0077	27.3
$\frac{3}{4}R$	0	0.0408	0.0079	23.1
R	0	0.0416	0.0127	14.7
R	10	0.0421	0.0167	11.4
R	20	0.0400	0.0181	9.8
Isolated Rotor	-	0.0457	0.0076	28.3

As expected, the isolated rotor performed the best and had the highest figure of merit. Interestingly, the placing the motor at mid span and tip locations resulted in a 1.3% and 67.1% increase in C_P/σ as compared to the isolated rotor. This indicates that placing the motor at the half span locations has significant performance benefits for the rotor and has a rotor figure of merit similar to an isolated rotor.

Table 7 indicates that placing the motor further out created a bluff body was seen to reduce rotor thrust and increase rotor power. Additionally, placing the motor at the 3/4 span was also found to disrupt thrust and increase power, which was also seen during the propeller driven rotor experiments. In the next section, the propeller control scheme from Section II is compared to a swashplate control scheme.

IV. COMPARISONS WITH CONVENTIONAL SWASHPLATE CONTROL

The final set of experiments involved a measuring the rolling and pitching moments of a shaft driven rotor using a swashplate. Afterwards, the shaft driven swashplate based control scheme was compared to the propeller driven rotor's control scheme.

Figure 32 shows the 6 ft diameter, 2 bladed hingeless rotor that was tested. Details on the blade design and construction can be found in (Ref. 2). The 6 ft diameter conventional rotor test stand featured a hub balance, a hydraulic drive motor, and a swashplate. The swashplate enabled a collective sweep of the rotor to be completed, as well as cyclic experiments to quantify control moments.

Figure 32: Conventional rotor test stand using 6 ft diameter composite rotor blades.

The 6 ft diameter, shaft driven rotor blade experiments were conducted using the University of Maryland's Hover Testing Facility, shown in Figure 33. Table 8 contains specifications on the 6 ft diameter rotor. While the 2.7 ft diameter rotor blades and the 6 ft diameter rotor blades are operated at different tip speeds and had different planforms, the presented results have been nondimensional to rotor tip speed and blade area. This ensures a representative physics is captured between the two rotors. Additionally, all control comparisons are conducted at the same rotor collective of 8 degrees for both rotors.

Figures 34 to 37 show the results of a collective sweep using the 6 ft diamater rotor. The collective sweep was between 2 and 14 degrees of rotor root collective while the rotor cyclics were set to 0, and the corresponding rotor thrust and power were recorded. The peak figure of merit for the rotor was found to be 33% for the untwisted rotor. After the collective sweep, the rotor control moments to cyclics were recorded.

Figure 33: 6 ft diameter rotor on the conventional rotor test stand at the University of Maryland.

Table 8: Conventional rotor test stand characteristics.

Rotor Radius R	36 in
Rotor Chord c	3.15 in
Rotor Aspect Ration R/c	11.4
Number of Rotor Blades	2
Rotor RPM	400
Rotor Flap Frequency	1.78/rev
Rotor Airfoil	NACA0012
Rotor Solidity σ	5%
Rotor Collective θ_0	2 to 14 deg
Rotor Twist	None
Rotor Cyclics	-4 to 4 deg

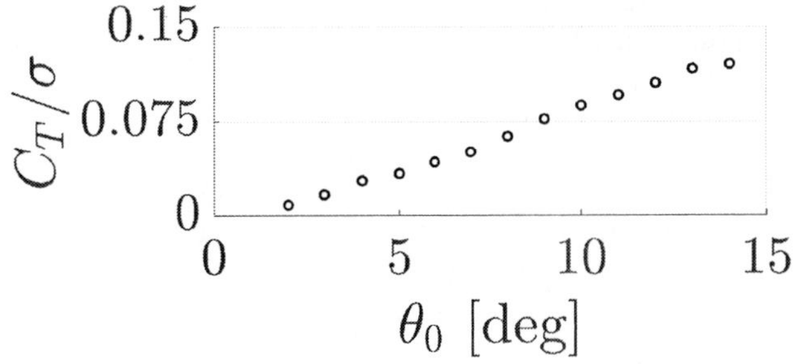

Figure 34: Rotor rotor thrust vs rotor collective.

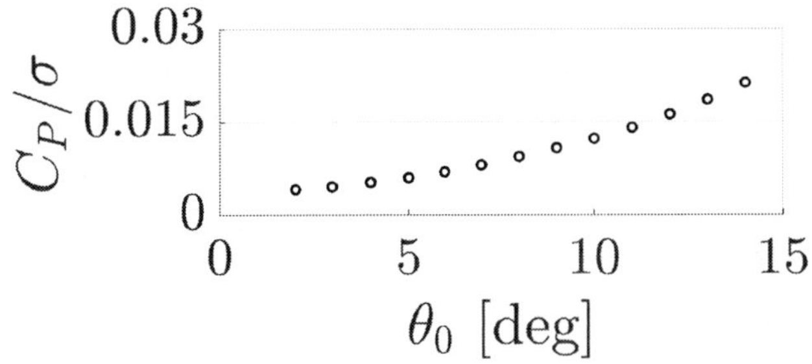

Figure 35: Rotor power vs rotor collective.

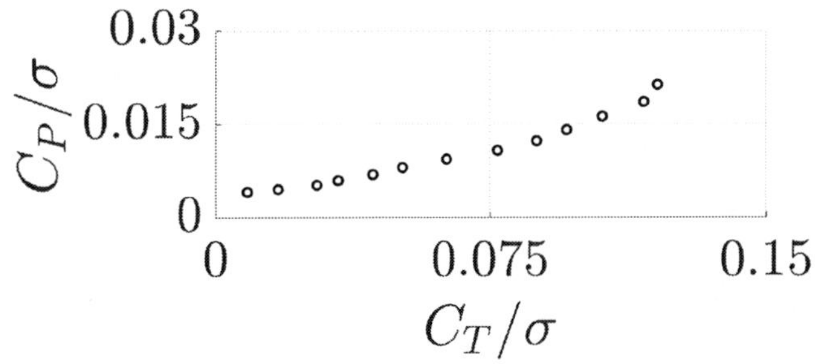

Figure 36: Rotor power vs rotor thrust.

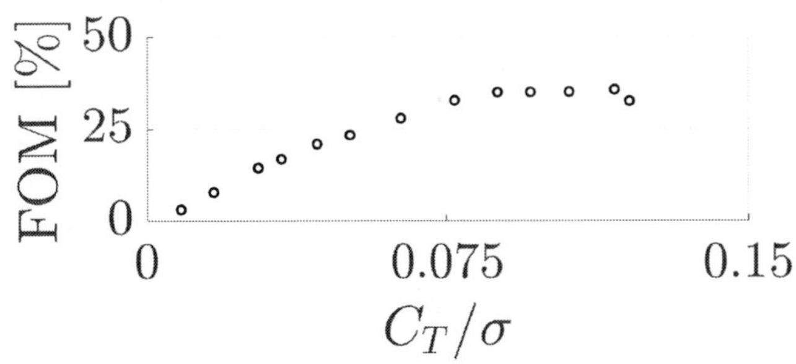

Figure 37: Rotor Figure of Merit vs rotor thrust.

Swashplate Control Sweeps

Next, the rotor collective was fixed at 8 degrees and a swashplate cyclic sweep was performed at 400 rotor RPM. The 8 degrees rotor collective was the same as the 3D printed 2.7 ft diameter rotors from section II and III. The swashplate produced 1/rev input on pitch angle as:

$$\theta(\psi) = \theta_0 + \theta_{1c}\cos\psi + \theta_{1s}\sin\psi \quad (4)$$

where θ_0 is the rotor collective, whereas θ_{1s} and θ_{1c} are the longitudinal and lateral swashplate inputs, respectively. The rotor's rolling C_{mx} and pitching moment C_{my} coefficients are defined as:

$$C_{mx} = \frac{M_x}{\rho A(\Omega R)^2 R}$$

$$C_{my} = \frac{M_y}{\rho A(\Omega R)^2 R}$$

For the experiment, the swashplate sweeps were conducted by varying one cyclic inputs while the other cyclic was kept at 0. Figures 38 and 39 show the results for the longitudinal and lateral sweeps compared to their predicted values, respectively. The control response to a given swashplate input was calculated by solving the rotor blade flapping equation of motion (Ref. 11). The predicted results show good agreement with the experimentally measured values.

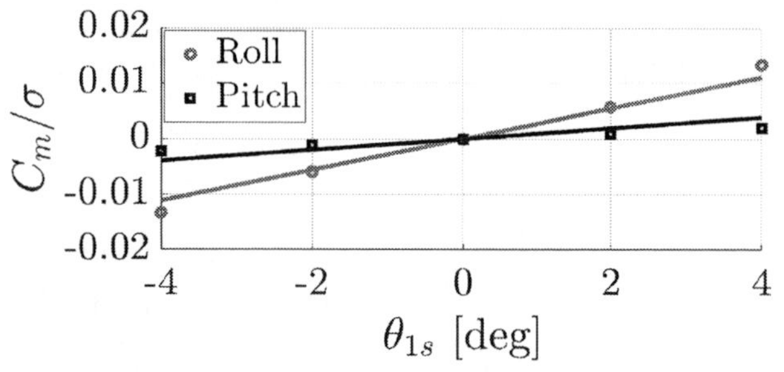

Figure 38: Measured (dots) and predicted (lines) rotor longitudinal cyclic sweeps at 400 rotor RPM.

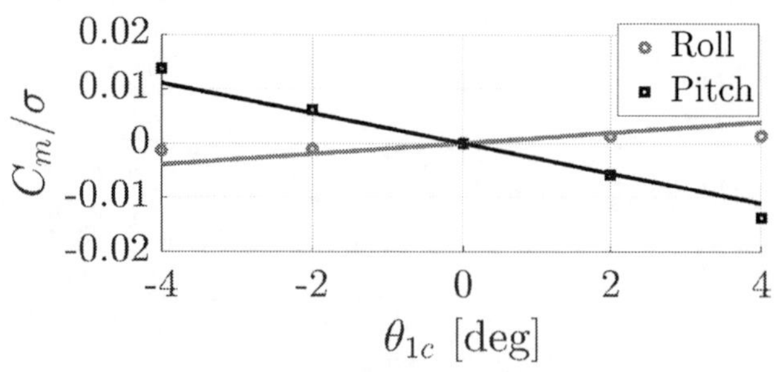

Figure 39: Measured (dots) and predicted (lines) lateral cyclic sweeps at 400 rotor RPM.

Propeller Driven vs Swashplate Control Comparison

Figures 40 to 43 on the next page compare the control response for both the swashplate and propeller driven rotor control schemes. It can be seen that the $\delta = 20^o$ cases produces nearly the same magnitude as the swashplate cyclic controls.

The control response can be compared by determining the control derivative for each input. The control derivative relates the control input u to the control moment C_m, and can be written as:

$$\frac{C_m}{\sigma} = \frac{d(C_m/\sigma)}{du}u$$

where $d(C_m/\sigma)/du$ was determined by examining the slope of the lines in Figures 40 to 43. For the swashplate control, the control moment derivatives were found to be:

$$\frac{C_{mx}}{\sigma} = \frac{d(C_{mx}/\sigma)}{d\theta_{1s}}\theta_{1s} = 0.0034\frac{1}{deg}\theta_{1s}$$

$$\frac{C_{my}}{\sigma} = \frac{d(C_{my}/\sigma)}{d\theta_{1c}}\theta_{1c} = -0.0034\frac{1}{deg}\theta_{1c}$$

A similar argument was made for relating the propeller driven rotor control to the propeller driven rotor inputs u_{1s} and u_{1c}. Consider the propeller driven rotor $\delta = 0^o$ longitudinal control derivative relative to C_{mx}/σ:

$$\frac{d(C_{mx}/\sigma)}{du_{1s}} = 0.0004\frac{1}{\%}$$

Comparing the value of $\frac{d(C_{mx}/\sigma)}{du_{1s}}$ to $\frac{d(C_{mx}/\sigma)}{d\theta_{1s}}$ yields:

$$\frac{d(C_{mx}/\sigma)}{d\theta_{1s}}\frac{du_{1s}}{d(C_{mx}/\sigma)} = \frac{0.0034}{0.0004} = 8.75\frac{\%}{deg}$$

Indicating that for the $\delta = 0$ case, 8.75% motor variation creates the same control moment as 1 deg of swashplate input. Tables 9 and 10 compare the propeller driven rotor control derivative value and their value relative to the swashplate control scheme. These results show that for the $\delta = 20^o$ case, a motor input variation of about 2.3% is equivalent to 1 degree of cyclic variation.

Table 9: Comparison of longitudinal control derivatives.

δ [deg]	$\frac{d(C_{mx}/\sigma)}{du_{1s}}$ [1/%]	$\frac{d(C_{mx}/\sigma)}{d\theta_{1s}}\frac{du_{1s}}{d(C_{mx}/\sigma)}$ [%/deg]
0	0.0004	8.75
10	0.0009	3.89
20	0.0015	2.25

Table 10: Comparison of lateral control derivatives.

δ [deg]	$\frac{d(C_{my}/\sigma)}{du_{1c}}$ [1/%]	$\frac{d(C_{my}/\sigma)}{d\theta_{1c}}\frac{du_{1c}}{d(C_{my}/\sigma)}$ [%/deg]
0	-0.0005	7.12
10	-0.0010	3.50
20	-0.0015	2.31

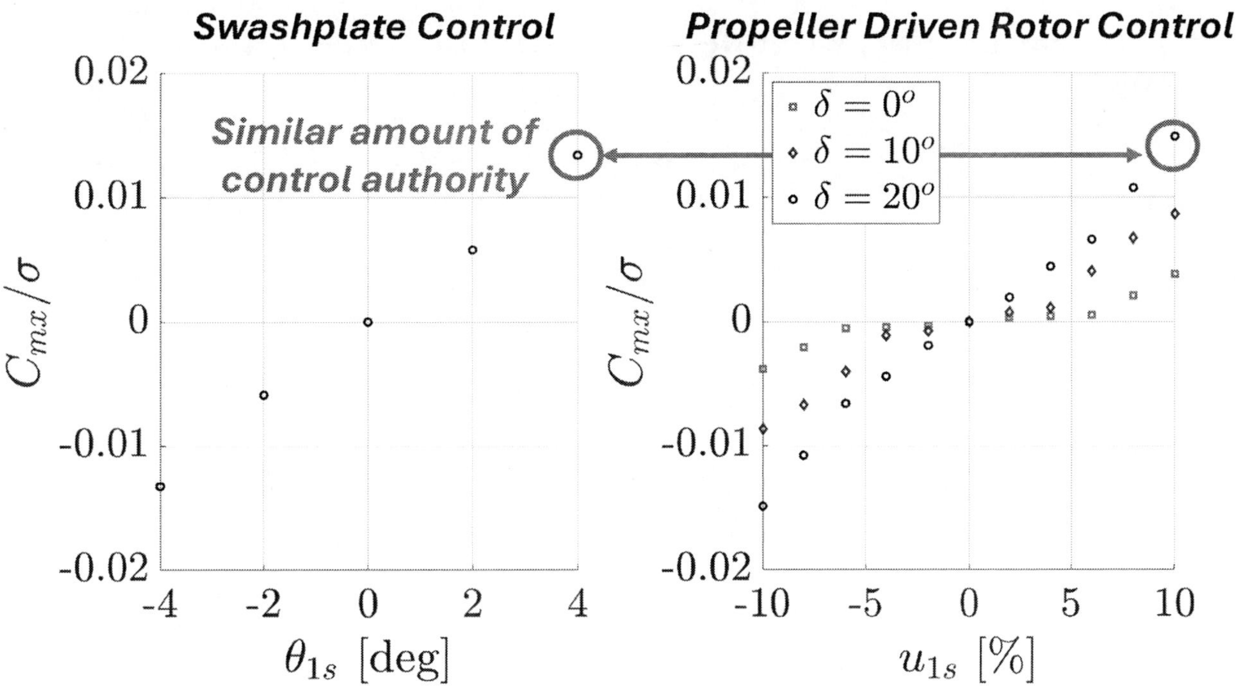

Figure 40: Comparisons of rotor rolling moment to longitudinal input.

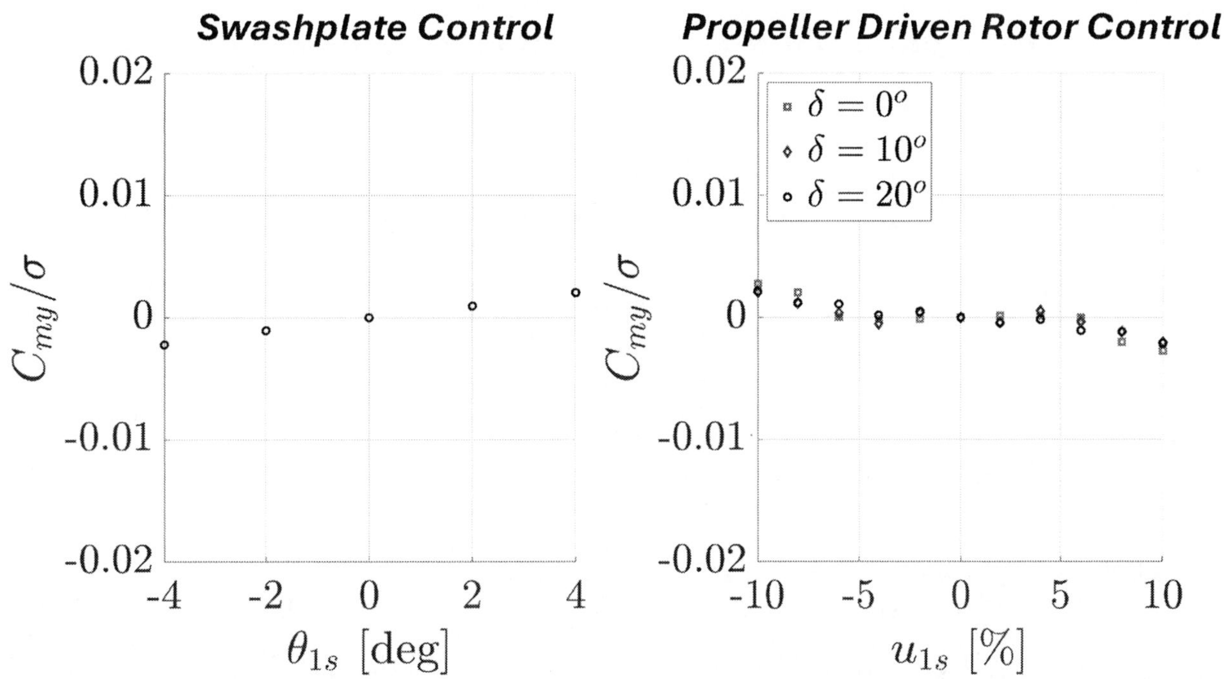

Figure 41: Comparisons of rotor pitching moment to longitudinal input.

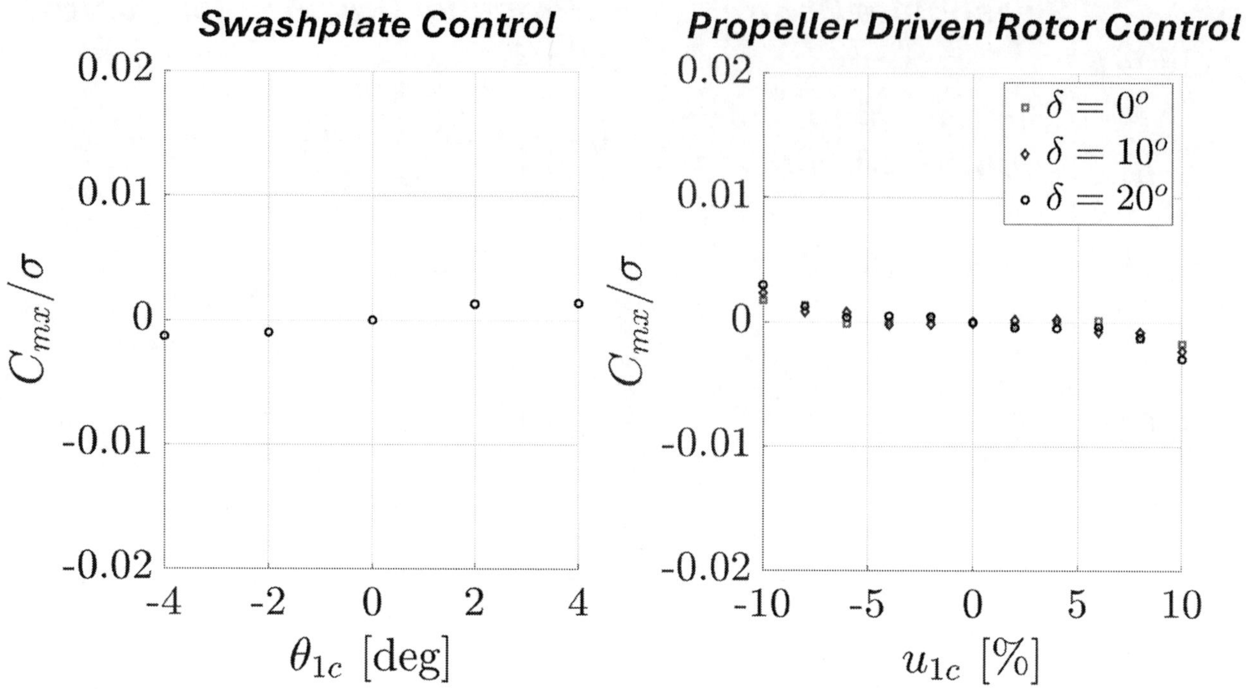

Figure 42: Comparisons of rotor rolling moment to lateral input.

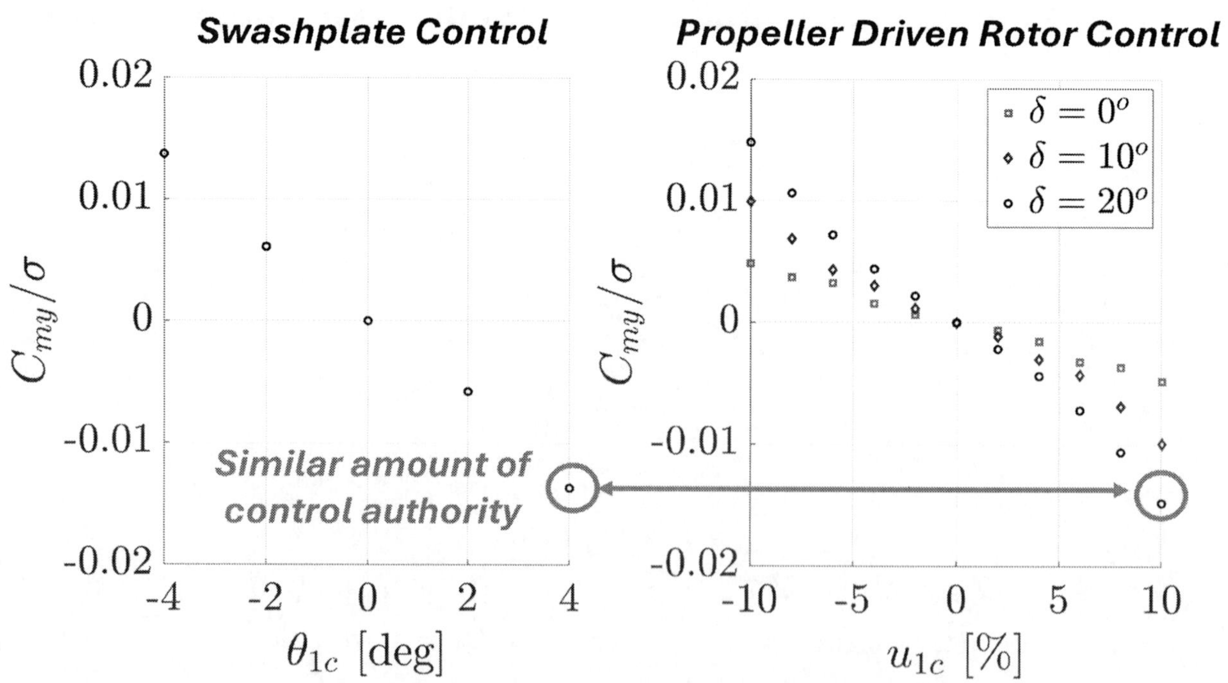

Figure 43: Comparisons of rotor pitching moment to lateral input.

V. SUMMARY, CONCLUSIONS, AND FUTURE WORK

This study documents a novel control methodology for a propeller driven rotor. A 2.7 ft diameter propeller driven rotor test stand was fabricated to experimentally develop and evaluate the novel swashplateless control scheme. The 3D printed rotors enabled rotor blade parameters, such as propeller tip angle δ and motor location, to be varied in a systematic manner. After characterizing the motor throttle variation to control moment performance, a series of experiments were conducted to investigate placing the propeller at alternative locations on the rotor blade span. Finally, the performance of a conventional swashplate was characterized using a 6 ft diameter rotor. This enabled the propeller driven rotor control scheme to be directly compared to a conventional shaft driven rotor with a swashplate.

From this study, the following conclusions were made:

- The propeller driven rotor control scheme was found to produce similar control moments as a conventional swashplate system. For the tip propeller angle $\delta = 20^o$ case, 2.3% of motor throttle variation was equivalent to 1 degree of swashplate input.
- For a tip propeller angle δ of 20°, increasing u_{1s} by 10% required 22.9% additional electrical power.
- Varying the motor throttle with rotor azimuth produced satisfactory on-axis control moments and had negligible cross coupling with the off-axis.
- Placing the propeller at the 1/2 span location reduced the electrical power loading coefficient C_{Pe}/σ by 50% compared to the tip propeller placement.
- Placing the propeller at the 3/4 span location resulted in reduced thrust and increased power compared to the tip and 1/2 span locations.
- For the shaft driven rotor testing, placing the motor at the half span location had similar performance to the isolated rotor, whereas placing the motor at the tip resulted in a 67% increase in rotor power.
- 3D printing rotors was found to be an effective and rapid method for systematically varying and testing rotor designs.

Some areas of future work are:

- Repeat the propeller driven rotor control experiments for mid span propeller location with various δ tip angles.
- Implement the propeller driven rotor control scheme on the 6 ft diameter rotor system.
- Add mechanical power measurements to the test stand to directly measure propeller power.
- Implement closed loop propeller RPM control to adjust and measure propeller RPM with rotor azimuth.
- Create a free-flying technology demonstrator that utilizes the propeller driven rotor control scheme.
- Significantly reduce propeller driven rotor power by optimizing propeller location and propeller geometry.

ACKNOWLEDGMENTS

The authors would like to thank Ismail Shah at the University of Maryland for his assistance with 3D printing and testing the propeller driven rotors.

Author contact:
Robert Brown, robert.h.brown427.civ@us.navy.mil
Inderjit Chopra, chopra@umd.edu

NOTATION

A	Rotor area, πR^2
c	Rotor chord
C_T	Rotor thrust coefficient
C_P	Rotor mechanical power coefficient
C_{Pe}	Rotor electrical power coefficient
C_{Mx}	Rotor roll moment coefficient
C_{My}	Rotor pitching moment coefficient
$d(C_m/\sigma)/du$	Slope of C_m/σ for given input u
FOM	Rotor figure of merit, $(C_T^{3/2}/\sqrt{2})/C_P$
I_{DC}	DC electrical current
I_o	DC electrical current offset
m	Slope of propeller RPM to rotor RPM
M_x	Rotor rolling moment, $C_{Mx}\rho A(\Omega R)^2 R$
M_y	Rotor pitching moment, $C_{My}\rho A(\Omega R)^2 R$
P	Rotor power, $C_P\rho A(\Omega R)^3$
P_{DC}	DC electrical power
$P_{DC,o}$	DC electrical power offset
Q	Rotor torque, $C_Q\rho A(\Omega R)^2 R$
R	Rotor radius
T	Rotor thrust, $C_T\rho A(\Omega R)^2$
T_p	Propeller thrust
u	Motor input command
u_0	Motor nominal input command
u_{1c}	Motor lateral input command
u_{1s}	Motor longitudinal input command
v_c	Propeller Climb speed
V_{DC}	DC voltage
x_p	Nondimensional propeller location on rotor blade
δ	Propeller tip angle w.r.t rotor chord
η_{PT}	Powertrain efficiency
θ_0	Rotor collective angle
θ_{1c}	Rotor lateral cyclic
θ_{1s}	Rotor longitudinal cyclic
ρ	Air density

σ	Rotor solidity
ψ	Rotor Azimuth
Ω	Rotor rotational speed
Ω_p	Propeller rotational speed
$\Omega_{p,0}$	Propeller nominal speed
$\Omega_{p,1c}$	Propeller lateral variation speed
$\Omega_{p,1s}$	Propeller longitudinal variation speed
$\Omega_{p,offset}$	Speed of propeller at 0 rotor RPM

REFERENCES

1. Gibbings, D. "The Fairey Rotordyne - Technology Before Its Time," *The American Journal*, Vol. 108.1089, 2004, pp 565-574. DOI:10.1017/S0001924000000397.

2. Brown, R. and Chopra, I., "Characterization of a Blade-Tip-Propeller Driven Rotor," Vertical Flight Society's 79th Annual Forum and Technology Display, West Palm Beach, FL, May 2023.

3. Brown, R. and Chopra, I., "Interactional Aerodynamics Between a Rotor and a Blade-Tip-Propeller through Wind Tunnel and Hover Stand Testing," Vertical Flight Society's 80th Annual Forum and Technology Display, Montreal, Canada, May 2024.

4. Ulrich, E., Humbert, S., and Pines, D., "Pitch and Heave Control of Robotic Samara Micro Air Vehicles", *Journal of Aircraft*, Vol. 47 (4), July-Aug. 2010, pp. 1290-1299. DOI: 10.2514/1.47197.

5. Ulrich, E., "Design, Development, Analysis, and Control of a Bio-Inspired Robotic Samara Rotorcraft", PhD Dissertation, University of Maryland, College Park, MD, 2012.

6. Bhardwaj, H., Hla Win, S. K., Thura Win, L. S., Sufiyan, D., and Shaohui, F., "P.I.D. Based Sliding Mode Control of Asynchronous Multi-actuator Monocopter", IEEE/ASME International Conference on Advanced Intelligent Mechatronics, Delft, Netherlands, July 2021.

7. Bhardwaj, H., Cai, X., Thura Win, L. S., and Shaohui, F., "Nature-Inspired in-Flight Foldable Rotorcraft", *Bioinspiration & Biomimetics*, Vol. 18, May 2023. pp. 1-15. DOI: 10.1088/1748-3190/acd739.

8. Saxena, A., and Chopra, I., "Design, Development, and Hover Testing of a Mach-Scale Swashplateless Rotor Using Brushless DC Motor-Actuated Trailing Edge Flaps for Rotor Primary Control", *Journal of the American Helicopter Society*, Vol. 62 (2), April 2017, pp. 1-12. DOI: 10.4050/JAHS.62.022002.

9. Saxena, A., and Chopra, I., "Wind Tunnel Testing of a Swashplateless Rotor with Compact Brushless Motor Actuated Flaps for Primary Control", *Journal of the American Helicopter Society*, Vol. 65 (6), January 2020, pp. 1-16. DOI: 10.4050/JAHS.65.012010.

10. Brown, R. and Chopra, I., "Characterization and Modeling of Brushless DC Motors and Electronic Speed Controllers with a Dynamometer," Vertical Flight Society's International Powered Lift Confrence 2020, San Jose, CA, January 2020.

11. Johnson, W., *Rotorcraft Aeromechanics*, Cambridge University Press, New York, NY, 2013, Chapter 6.

Improvements to the Da Vinci Aerial Screw as a VTOL Thrusting Alternative

Ron Berlin†
rberlin8@terpmail.umd.edu
Graduate Student

Koushik Marepally†
koushikm@umd.edu
Postdoctoral Research Associate

James Baeder†
baeder@umd.edu
Igor Sikorsky Distinguished Professor

†Alfred Gessow Rotorcraft Center, University of Maryland, College Park, MD, USA

ABSTRACT

This study explores dominant geometric parameters of aerial screws, drawing inspiration from Leonardo da Vinci's visionary designs dating back to the 1500s. Through Computational Fluid Dynamics (CFD) analyses, a shape-conforming vortex at the tip of the screw functions as a thrusting mechanism, similar to that observed in swept and delta wings. The "da Vinci" vortex, a helical structure at the screw's tip, serves as a focal point for assessing the impact of design variables on aerodynamic efficiency. Detailed analyses of screw pitch and taper reveal nuanced effects on thrust generation, offering valuable insights for optimizing aerial screw performance. This work provides a comprehensive understanding of historical roots and contemporary advancements in aerial screw technology, paving the way for efficient Vertical Take-Off and Landing (VTOL) alternatives.

INTRODUCTION

The concept of aerial screws, designed for vertical flight, traces back to the inventive Leonardo da Vinci in the 1500s, as depicted in Figure 1. Later attempts to realize this visionary concept resulted in the construction of a high-solidity helicopter in the late 1800s (Ref. 1). Several early experimental attempts to understand the physics of screw type propellers were inconclusive (Ref. 2) due to the difficulties in fabrication and experimental measurement. The aerial screw was eventually overshadowed by conventional rotors with large aspect ratios, low solidity, and airfoil blades, which proved more effective for vertical flight. These designs were subsequently left unstudied as vertical flight technology developed.

Much later, these designs were revisited in the context of the VFS annual design competition in 2020 (Ref. 3). A series of contemporary vehicle designs (Ref. 4), inspired by the classical aerial screw, were developed and studied for a simple hovering mission profile. This exploration marks a significant leap forward, utilizing modern analysis tools to research and understand the intricate physics behind aerial screws. The culmination of this effort is showcased through the development of a working prototype for a small-scale UAV employing aerial screws (Refs. 5, 6), offering a compelling proof-of-concept for the thrust generation capabilities inherent in this innovative design. Preliminary designs, estimating figures of merit up to 0.52 (Ref. 4), highlight the significant potential of these rotors in VTOL and hovering applications. Recent research has concentrated on the aerodynamics and underlying physics of aerial screws, with the aim of improving comprehension of the rotor's thrusting mechanism. This research previously identified figures of merit up to .56 (Ref. 7).

Figure 1. Leonardo's Aerial Screw depicted in the page of Paris Manuscript B, folio 83v

Imperative to the thrust generation mechanism of aerial screws is the "da Vinci" vortex (Ref. 4), a shape-adhering helical vortical structure at the screw's tip. The strength of this vortex determines the pressure distribution on the screw's surface and, consequently, the rotor's efficiency. Leveraging the use of modern modeling and simulation tools, the current

Presented at the Vertical Flight Society's 11th Biennial Autonomous Technical Meeting, Phoenix, AZ, USA, Feb. 4–6, 2025.

study aims to unravel the effects of different design variables on the "da Vinci" vortex and the overall efficiency of the aerial screw.

"DA VINCI CODE": THE THRUSTING MECHANISM OF AERIAL SCREW

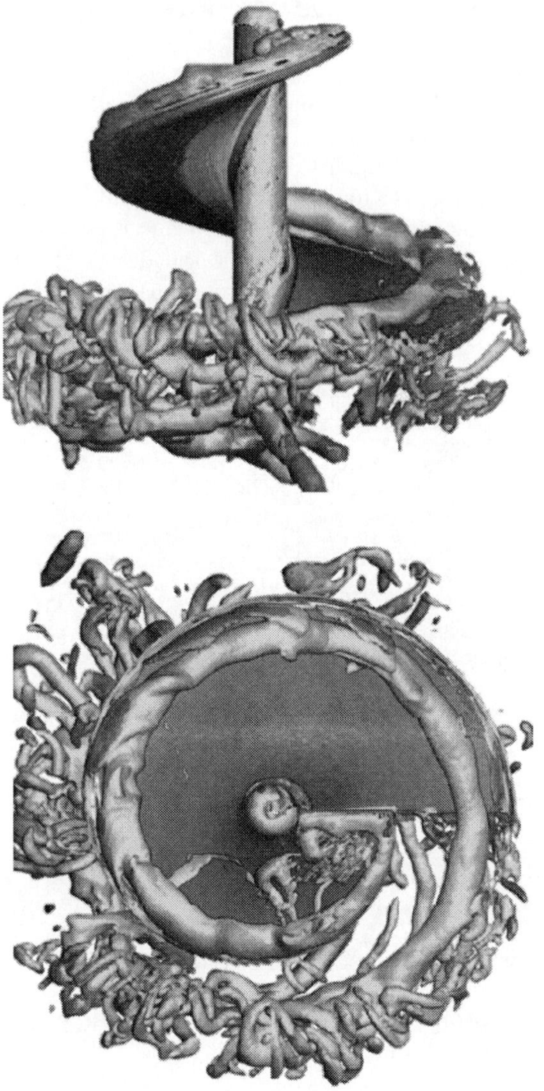

Figure 2. Iso-surfaces of vorticity Q-criterion around a notional hovering aerial screw (top: side view, bottom: top view)

A detailed understanding of the thrusting mechanism within aerial screws is achieved through Computational Fluid Dynamics (CFD)-driven discovery of the flow physics. The thrusting mechanism bears analogies to vortex lift (Ref. 8), an aerodynamic phenomenon of non-linear lift, observed in highly swept and delta wings. Illustrated in Figure 2, the vorticity iso-surface in the flow field show a shape-conforming helical vortical structure on the top surface of a hovering aerial screw. This vortex induces suction which generates a significant pressure disparity across the screw surface and consequently contributes to to the rotor's thrust.

The initiation of the vortical structure's development occurs due to a circulation along the leading edge, where the screw edge meets the freestream. Subsequently, the span-wise vortex moves downstream and rolls up along the screw surface, resulting in the formation of a tip vortex. Further, the presence of taper on the screw surface creates a delta wing-like aerodynamic environment, continuously adding to the strength of the vortex along the blade tip, resulting in a shape conforming helical vortical structure viz. the "da Vinci" vortex. This increases the proportion of the screw surface active in thrust generation. However, it must be noted that the inboard sections of the screw surface exhibit a relatively insignificant contribution to the rotor's thrust generation.

GEOMETRIC DEFINITION OF AERIAL SCREW: DESIGN VARIABLES

The geometry of a typical screw, including an aerial screw, is primarily defined by its radius, pitch, and the number of turns. Pitch of the screw is the length the screw surface advances in the axial direction for one full rotation of the screw. Radius of the screw is the radius of the screw surface projected perpendicular to the axial direction and the number of turns is the total number of full rotations of the screw surface in the azimuthal direction of the screw. For simplicity and due to the impracticality of long, towering screws, the number of turns of the screw is chosen to be 1 throughout this study.

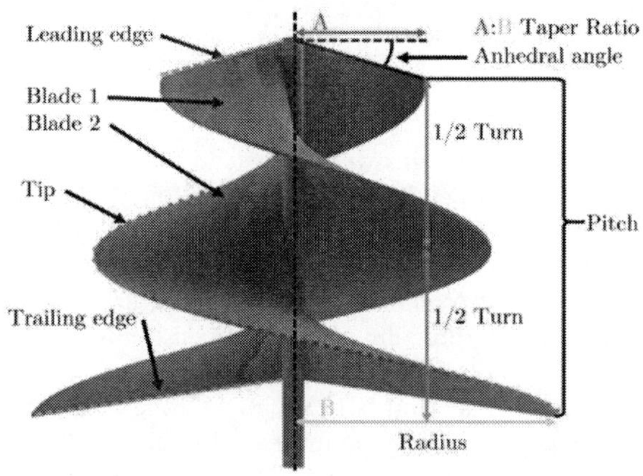

Figure 3. Design variables defining the geometry of a screw

Modern variations of the screw designs include addition of a taper to the screw as well as a dihedral angle between the screw surface and the axial plane. The taper ratio is defined as the ratio between the local screw radii across one full turn of the screw surface. Figure 3 shows different design variables that define the aerial screw geometry. This study excludes the investigation of dihedral angle effects on screw aerodynamics.

Mathematically, the aerial screw designs are defined using the non-dimensional variables and appropriate scaling parameters corresponding to each of these design variables.

Radius of the Screw

The radius of the screw is defined on a projection perpendicular to the screw's axial direction. Since the local radius of the screw can vary due to taper, the radius of the screw is defined as the radius of the projection corresponding to the maximum radius location (R_{max}), i.e., trailing edge. This radius is used as the length scale for non-dimensionalization of other variables and physical quantities. The disc area corresponding to this radius is used as the reference surface area for load coefficients.

A local non-dimensionalization of the radial coordinate is also applied using R_{local}, which is the maximum radius at a particular azimuthal section.

$$r' = \frac{r}{R_{local}} \tag{1}$$

Pitch Rate

The screw pitch rate is defined as the axial length that the screw advance for one full rotation of the screw. For a linearly pitching screw, the advancement per unit change in the screw rotational coordinate is constant throughout the screw surface, i.e., the pitch rate is constant. However, the designs can also have a non-uniform pitch rate, i.e., the screw advances at different rates as the screw rotates at different azimuthal orientations. Within the scope of this study, two different variations of pitch rates are studied, viz., uniform (linear pitch) and bilinear pitch. The bilinearly pitched screw has a different uniform pitch rate for the first and second halves of the screw.

This quantity is non-dimensionalized using the maximum radius of the screw. Thus, if the screw has a pitch rate of ζ, then the screw advances $R_{max}\zeta\Delta N$ when the screw turns by ΔN rotations. It must be noted that the bilinear taper rate is also non-dimensionalized using the same maximum radius of the screw.

$$\zeta = \frac{1}{R_{max}}\frac{\Delta Z}{\Delta N} = \frac{2\pi}{R_{max}}\frac{\Delta Z}{\Delta \theta} \tag{2}$$

Due to a uniform pitch rate, the inboard and outboard section advance the same axial distance per unit change in azimuthal coordinate. Thus, the screw surface has a steeper gradient in the inboard sections compared to the outboard sections, which affect the contributions of the blade normal forces to the axial thrust component. Similarly, as the pitch rate increases, the surface gradient also becomes steeper, resulting in a decreased contribution of normal force to axial loads and an increased contribution to the torque-wise component.

Taper Rate

The screw taper rate is defined as the reduction of the local tip radius (R_{local}) of the screw for one screw rotation. For a linearly tapered screw, the reduction in local tip radius per unit change in the screw rotational coordinate is constant throughout the screw surface, i.e., the taper rate is constant. However, the designs can also have a non-uniform taper rate, i.e., the screw tapers at different rates as the screw rotates at different azimuthal orientations. Within the scope of this study, two different variations of taper rates are studied, viz., uniform (linear taper) and bilinear taper. The bilinearly tapered screw has a different uniform taper rate for the first and second halves of the screw.

This quantity is non-dimensionalized using the maximum radius of the screw. Thus, if the screw has a taper rate of η, then the local tip radius increases by $\eta R_{max}\Delta N$ when the screw turns by ΔN rotations. It must be noted that the bilinear taper rate is also non-dimensionalized using the same maximum radius of the screw.

$$\eta = \frac{1}{R_{max}}\frac{\Delta R_{local}}{\Delta N} = \frac{2\pi}{R_{max}}\frac{\Delta R_{local}}{\Delta \theta} \tag{3}$$

Due to a uniform taper rate, the azimuthal sections closer to the leading edge have a smaller local radius compared to the sections towards the trailing edge. To maintain a same pitch rate, the screw surface has a steeper gradient in the upstream sections compared to the downstream sections of the blade, which affect the contributions of the blade normal forces to the axial thrust component.

INFLUENCE OF DESIGN VARIABLES ON AERIAL SCREW AERODYNAMICS AND PERFORMANCE

The main aim of this study is to identify the influence of each of the design variables that define the aerial screw geometry on the performance, and understand the variations in flow physics attributing to the changes in performance. The performance comparison is performed for a constant thrust requirement for different screws, with figure of merit as a metric of the performance.

Computational Fluid Dynamics (CFD) is used for the aerodynamic analysis of the flow around these hovering screws using HAMSTR (Ref. 9), UMD's in-house CFD solution framework. HAMSTR employs Hamiltonian paths and strand grids based solution methodology, coupled with overset meshes for farfield computations. DDES (Delayed Detached Eddy Simulation) is the mathematical model used to describe the flow, with SA (Spalart-Allmaras) turbulence closure model, and fully turbulent flow assumptions.

Effect of Linear Pitch Rate (ζ)

To understand the effect of pitch rate, several screws with a constant taper rate $\eta = 0.5$ are studied, with each screw having a different uniform pitch rate. All cases in these studies have a maximum radius of $R_{max} = 1$ ft and the tip Mach number at the trailing edge is maintained at 0.25 for all these screw designs, corresponding to a tip Reynolds number of 1.7×10^6.

Figure 4 shows the variation of figure of merit for these screws with the pitch rate. It can be seen that the performance of the screw improves initially when the pitch rate ζ increases from 1 to 1.2, and then decreases continuously with increasing pitch. A maximum figure of merit of .563 is achieved at $\zeta = 1.2$. A comparison of the most efficient design ($\zeta = 1.2$) and the most aggressively pitched design ($\zeta = 2.0$) is discussed to assess the effects of increasing the pitch rate of the rotor.

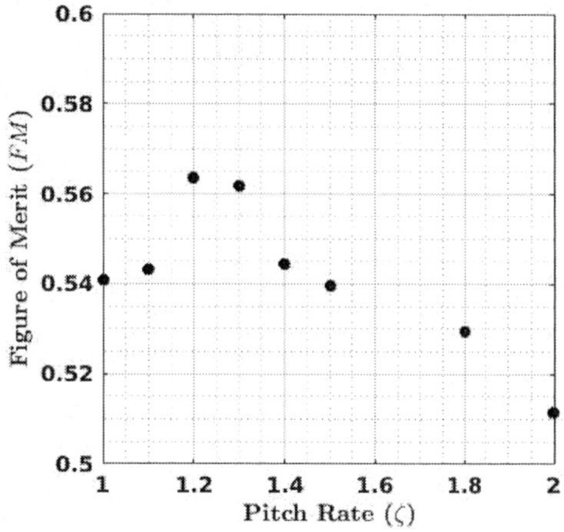

Figure 4. Variation of figure of merit of aerial screws with varying pitch rate

Figure 5 shows the vorticity iso-contour of the $\zeta = 1.2$ linearly pitched design. The visualization depicts an attached "da Vinci" vortex along the blade which remains relatively undisturbed after leaving the trailing edge of the rotor.

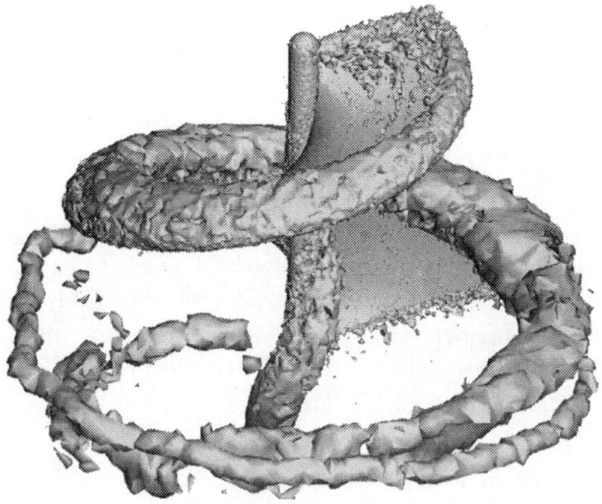

Figure 5. Vorticity contour of the $\zeta = 1.2$ linearly pitched design

Figure 6 shows the vorticity iso-contour of the $\zeta = 2.0$ linearly pitched design. A stronger "da Vinci" vortex is developed, attributed by the aggressive pitch rate. However, as pitch rate increases, the normal vector orientation of the blade points further torque-wise. This results in more power being required to achieve the same amount of thrust, decreasing the figure of merit.

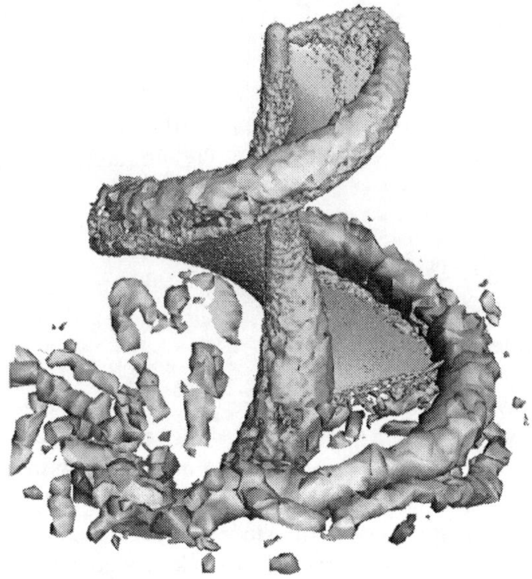

Figure 6. Vorticity contour of the $\zeta = 2.0$ linearly pitched design

Pressure distributions along four azimuthal locations along the blade ($\psi = 0°$ being defined at the leading edge) were extracted to determine the primary locations of lift along the rotor.

Figure 7 shows the pressure distribution of the $\zeta = 1.2$ linearly pitched design.

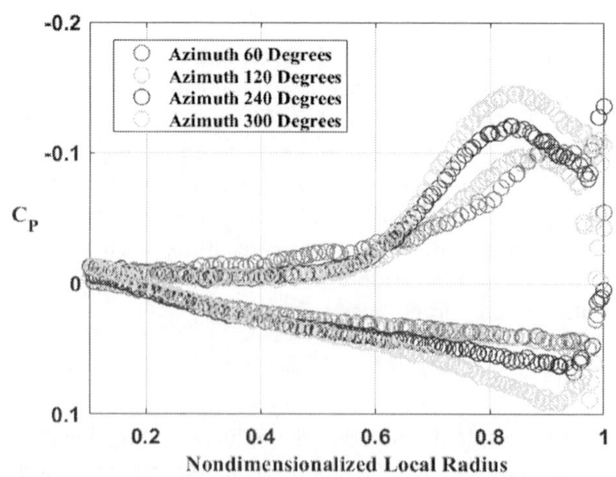

Figure 7. Pressure distribution of the $\zeta = 1.2$ linear pitch design

From this figure, it is seen that the majority of the lift generated by the screw is near the tip of the blade, increasing in strength as the angle of azimuth increases. This lift can be attributed to the "da Vinci" vortex which was visualized in Figure 5. As the vortex advances down the blade, its strength increases and more lift is being generated.

Figure 8 shows the pressure distribution of the $\zeta = 2.0$ linearly pitched design.

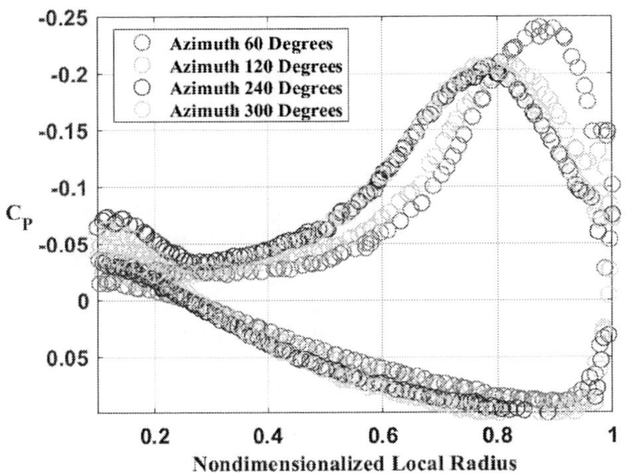

Figure 8. Pressure distribution of the $\zeta = 2.0$ linear pitch design

The pressure distributions change significantly between the $\zeta = 1.2$ and $\zeta = 2.0$ designs. The pressure differences between the top and bottom surface of the blade increase for all azimuthal angles for the $\zeta = 2.0$ design. Along with this, the rotor's vortex appears to be strongest near $\psi = 60°$ for the $\zeta = 2.0$ design. This differs from the $\zeta = 1.2$ design where the vortex continued to grow in strength as angle of azimuth increased. This is contributed to the vortex detaching from the blade surface as it advances, building on the concept of an optimal pitch rate having an upper limit. The increase in pressure differential towards the tip of the blade is visible in Figure 6 as the vortex is visibly stronger than that of the $\zeta = 1.2$ design.

Despite a stronger vortex, the natural shape of an aggressively pitched design results in a significantly larger power requirement. An optimal pitch rate value for a linearly pitched design is found by achieving a balance of vortex strength and a minimal torque component of the blade's surface normal vectors.

Effect of Bilinear Pitch Rate (ζ)

To understand the effect of a nonlinear pitch, a bilinear variation of pitch is developed where the first half of the screw has a differing pitch rate than the second half of the screw. Both pitch rates combine to form an "effective pitch rate" of $\zeta = 1.2$, the optimal linear pitch rate. (A first half pitch rate of $\zeta = 0.8$ with a second half pitch rate of $\zeta = 1.6$ achieves the effective $\zeta = 1.2$) All cases have a linear taper rate of $\eta = 0.5$.

Figure 9 shows the figure of merit plotted against the pitch rate of the first half of each of these screws. The datapoint in blue shows the most efficient linear pitch design with $\zeta = 1.2$ for comparison. An initial pitch rate which is less than $\zeta = 1.0$ is results in slight under-performance compared to the linear pitch rate design. The design with an initial pitch rate of $\zeta = 1.1$ performs nearly identical to the linearly pitched design as the pitch rate change becomes close to negligible. A design where the pitch rate transitions from $\zeta = 1.0$ to $\zeta = 1.4$ raises the figure of merit from .563 to .567, nearly a 1% improvement. The most efficient case with an initial pitch rate of $\zeta = 1.0$ is compared to the initial pitch rate $\zeta = 0.8$ design to discuss the physics.

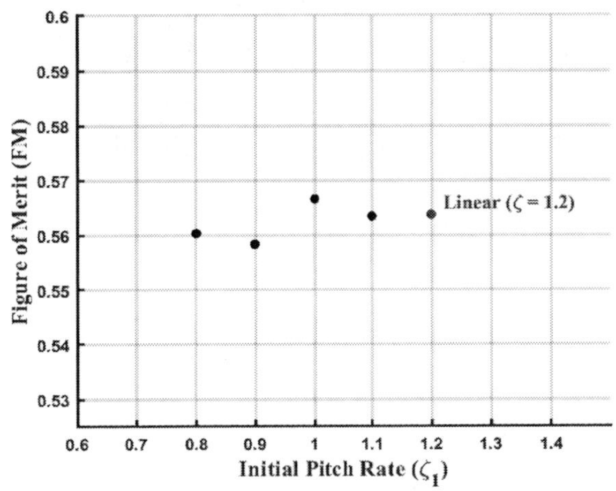

Figure 9. Variation of figure of merit of aerial screws with varying bilinear pitch rates

Figure 10 shows the vorticity iso-contour of the $\zeta = 1.0$ initial pitch design. The vortex remains attached to the blade, similar to the linear pitch design.

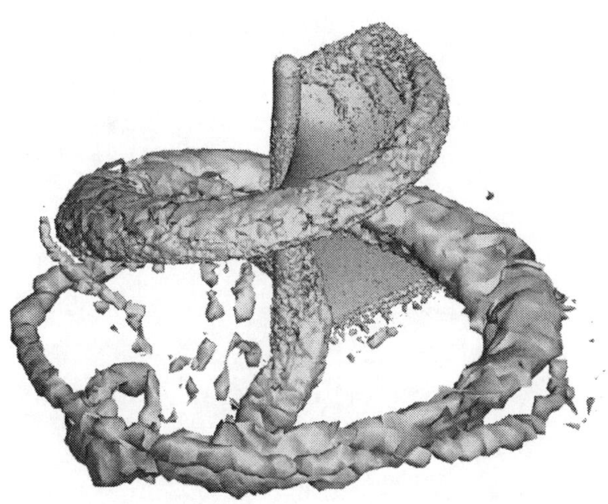

Figure 10. Vorticity contour of the $\zeta = 1.0$ to $\zeta = 1.4$ bilinear pitch design

Figure 11 shows the vorticity iso-contour of the $\zeta = 0.8$ initial pitch design. The "da Vinci" vortex is evidently weak in the first half of the screw. The sharp increase in pitch rate to $\zeta = 1.6$ significantly increases the vortex strength in the second half of the blade.

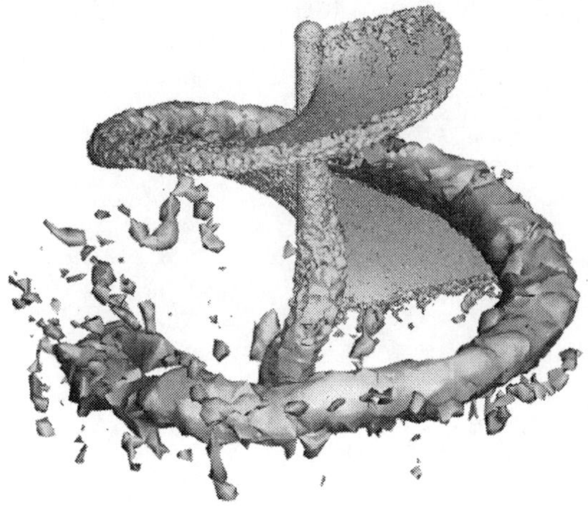

Figure 11. Vorticity contour of the $\zeta = 0.8$ to $\zeta = 1.6$ bilinear pitch design

Assessing the pressure distributions of both of these bilinearly pitched designs provides further insights on the effect of pitch on the aerodynamics and performance of aerial screw rotors.

Figure 12 shows the pressure distribution of the $\zeta = 1.0$ to $\zeta = 1.4$ bilinear pitch design. The pressure differential on the blade's surface continues to grow as the angle of azimuth increases, suggesting an attached vortex on the rotor blade's tip.

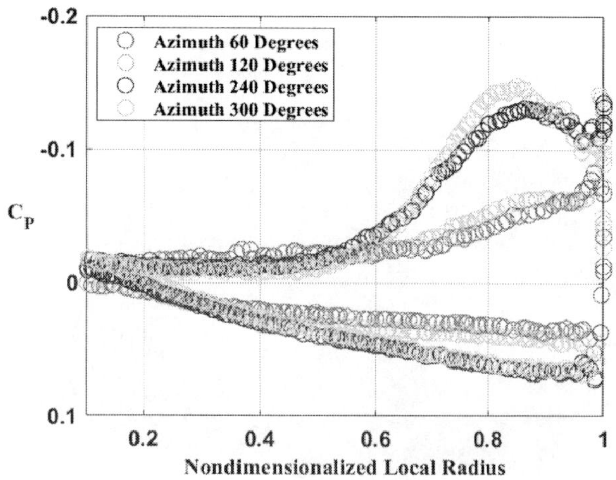

Figure 12. Pressure distribution of the $\zeta = 1.0$ to $\zeta = 1.4$ bilinear pitch design

Azimuthal angles of $\psi = 60°$ and $\psi = 120°$ produce almost identical lift in the first half of the screw. Similarly, angles of $\psi = 240°$ and $\psi = 300°$ produce similar lift in the second half of the screw.

Figure 13 shows the pressure distribution of the $\zeta = 0.8$ to $\zeta = 1.6$ bilinear pitch design.

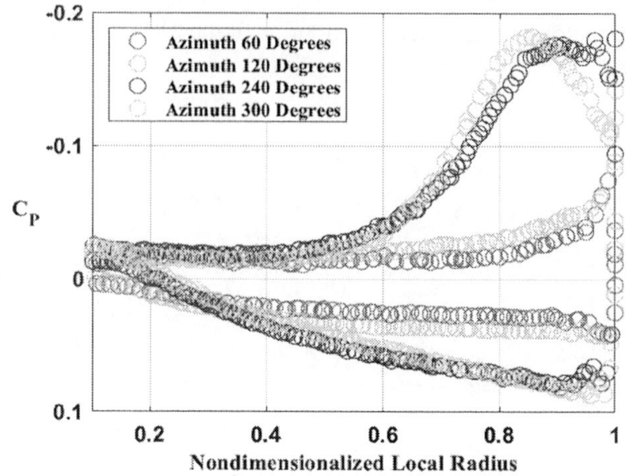

Figure 13. Pressure distribution of the $\zeta = 0.8$ to $\zeta = 1.6$ bilinear pitch design

Comparing the $\zeta = 0.8$ to $\zeta = 1.6$ bilinear pitch design to the most efficient, $\zeta = 1.0$ to $\zeta = 1.4$ bilinear pitch design provides some significant insights on the effect of pitch on aerial screw aerodynamics and performance.

As pitch rate decreases, the "da Vinci" vortex becomes weaker, decreasing the effective lift of the rotor. This is supported by the visible pressure differential decrease near the tip of the blade as pitch rate decreases. This is specifically reflected in the red and green curves in Figures 12 and 13.

Similarly, it is seen that an increasing pitch rate increases the strength of the "da Vinci" vortex. The pressure differential significantly increases in the second half of the blade of where the pitch rate is $\zeta = 1.6$ compared to the most efficient design where the pitch rate is $\zeta = 1.4$.

Despite this increase in lift with a higher pitch rate, the evident loss of lift in the first half of the blade decreases the rotor's overall performance comparing to the $\zeta = 1.2$ linear pitch design. The normal vector orientation points further torque-wise as the pitch rate increases, hindering the rotor's performance as seen in the $\zeta = 0.8$ to $\zeta = 1.6$ bilinear pitch design.

Finding the ideal pitch rate variation requires identifying the optimal trade off between an aggressive pitch rate improving the vortex lift with a minimized power requirement of the rotor.

For a bilinearly pitched design, it is found that this pitch rate change from $\zeta = 1.0$ in the first half of the blade to $\zeta = 1.4$ in the second half of the blade provides the most efficient hover performance with a figure of merit of .567. Differing the location of where the pitch rate changes as well as including more than two pitch rates has potential to further improve aerial screw performance in hover. For this study, only a single pitch rate change at half the span was assessed.

Effect of Linear Taper Rate (η)

Similar to the effect of pitch rate, the effect of taper rate on the performance of the screw is studied by comparing the screws with all other design variables unaltered. Several screws with a constant pitch rate $\zeta = 1.3$ are studied, with each screw having a different uniform taper rate. The tip Mach number and the tip Reynolds number are maintained the same, as well as the maximum radius of the rotors being $R_{\max} = 1$ ft.

Figure 14 shows the variation of figure of merit for these screws with the taper rate. It can be seen that the performance of the screw initially increases from $\eta = 0.4$ to $\eta = 0.5$ then continuously decreases as the taper rate increases. A comparison of the best performing linear taper rate design ($\eta = 0.5$) and the most aggressive linear taper rate ($\eta = 0.66$) is discussed to assess the affects of an increasing taper rate on rotor performance.

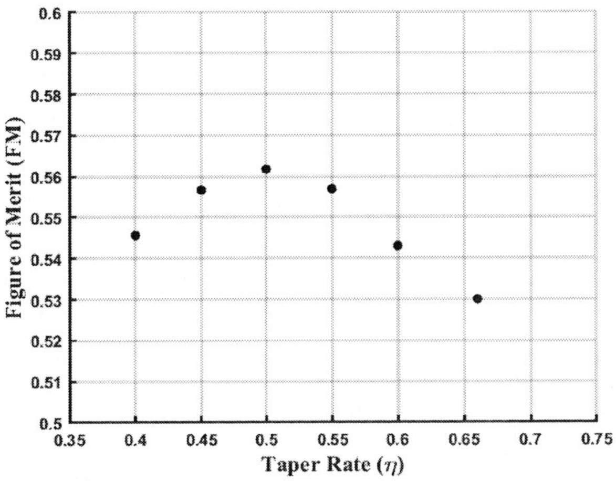

Figure 14. Variation of figure of merit of aerial screws with varying linear taper rate

Figure 15 shows the vorticity iso-contour of the $\eta = 0.5$ linear taper rate design and Figure 16 shows the same for the $\eta = 0.66$ linear taper rate design. Both vorticity contours show an attached "da Vinci" vortex. The distinguishing feature between the differing designs with a linear taper rate is the radius of the leading edge. A larger taper rate corresponds to a smaller leading edge radius. This relationship means that as taper rate is increased, there is an expected loss of lifting area for the rotor.

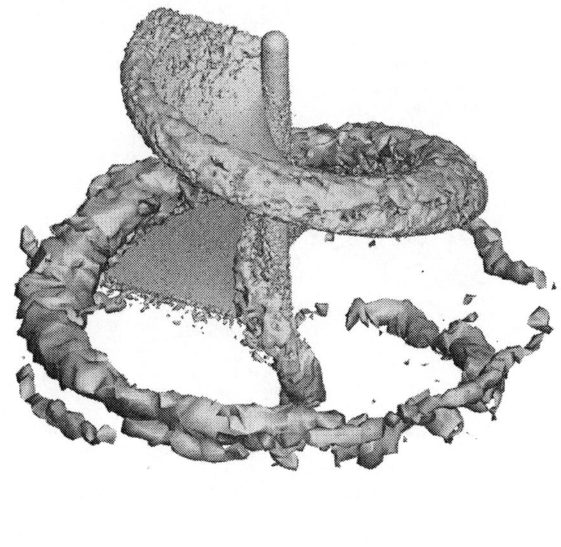

Figure 15. Vorticity contour of the $\eta = 0.5$ linear taper rate design

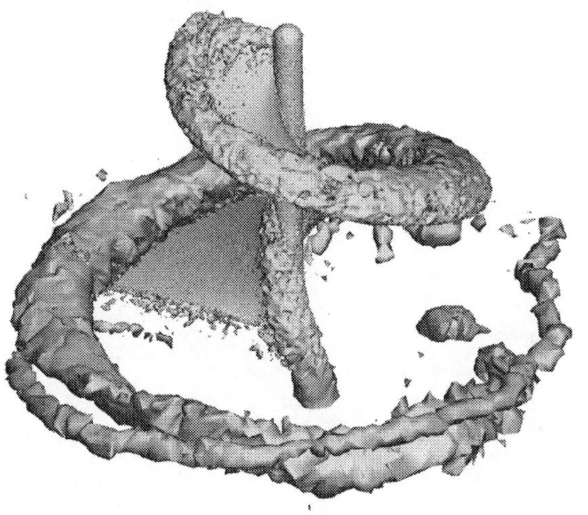

Figure 16. Vorticity contour of the $\eta = 0.66$ linear taper rate design

To understand the benefits of including a taper with an aerial screw rotor, pressure distributions were extracted for the $\eta = 0.5$ design and the $\eta = 0.66$ design. This data provides further insights on the physics of the "da Vinci" vortex and how to further improve rotor performance.

Figure 17 shows the pressure distribution of the $\eta = 0.5$ linear taper design.

This pressure distribution shows that with a tapered rotor, the "da Vinci" vortex moves inboard as it continues to develop along the blade of the rotor. The peaks from the pressure distribution plot can be seen to move from $r = .9R$ to $r = .8R$ as angle of azimuth increases. Since this is a commonality between the linear pitch and bilinear pitch cases mentioned, this phenomenon can be attributed to the taper of the rotor.

Figure 18 shows the pressure distribution of the $\eta = 0.66$ linear taper design.

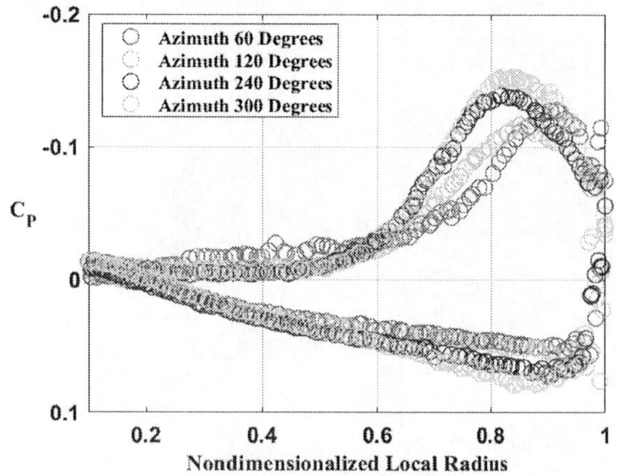

Figure 17. Pressure distribution of the $\eta = 0.5$ linear taper design

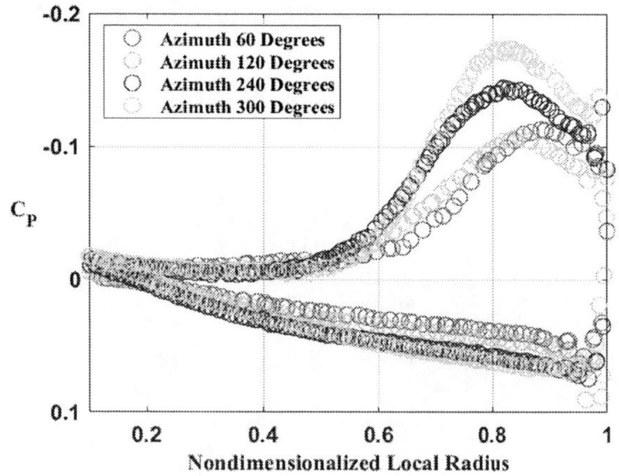

Figure 18. Pressure distribution of the $\eta = 0.66$ linear taper design

Comparing Figures 17 and 18, it is found that increasing the taper rate of the rotor from $\eta = 0.5$ to $\eta = 0.66$ increases the vortex strength as it moves down the span of the blade. The pressure differential increases as angle of azimuth increases in the $\eta = 0.66$ case, displaying the vortex's increase in strength.

Once again, comparing the location of the peaks with respect to azimuthal locations along the rotor blade, the vortex continues to moves inboard as it develops down the rotor blade.

Despite a stronger vortex developed when taper rate is increased, there is a maximum benefit to an aggressive taper rate. As taper rate increases, lifting surface is lost. This study identified that for a linearly tapered rotor at a fixed pitch rate of $\zeta = 1.3$, a taper rate of $\eta = 0.5$ provides the most efficient design with a figure of merit of .562.

Effect of Bilinear Taper Rate (η)

To understand the effect of nonlinear taper, a bilinear variation of taper is developed. The first half of the screw has a uniform taper rate and the second half is untapered. The taper rate must be aggressive for these screws to achieve the maximum radius within half rotation. Figure 19 shows the variation of figure of merit for these rotors, indicating better hover performance when the second half is left untapered. A linear pitch rate of $\zeta = 1.2$ was held constant.

Figure of merit increases as the initial taper rate increases from $\eta = .4$ to $\eta = .8$. The figure of merit then begins to decrease, showing there exists a maximum benefit to an aggressive taper in the first half of the blade.

The design with an initial taper rate of $\eta = 0.8$ proves to be the most efficient out of the dataset, achieving a figure of merit of .59. This design change provides a significant improvement of 5% between the linearly and bilinearly tapered rotors.

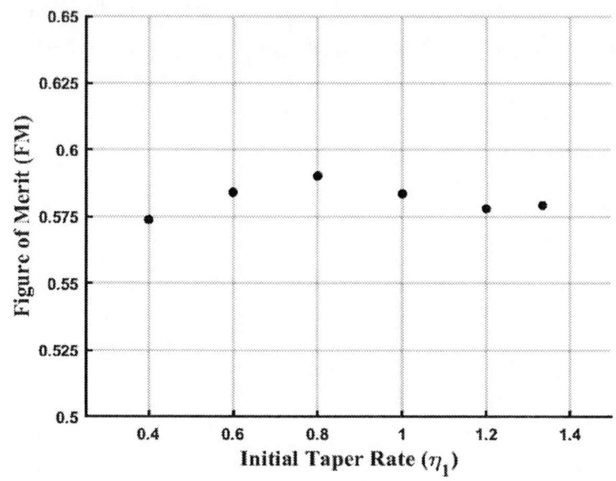

Figure 19. Variation of figure of merit of bilinearly tapered aerial screws with varying initial taper rate

A comparison of the best performing rotor and the most aggressively tapered rotor was done to expand on the effects of taper rate on the aerial screw design. It also provides insights on how the "da Vinci" vortex is affected by an untapered rotor after the maximum radius is reached at half span.

Figure 20 shows the vorticity iso-contour of the $\eta = 0.8$ initial taper rate design and Figure 21 shows the vorticity iso-contour of the $\eta = 1.33$ initial taper rate design.

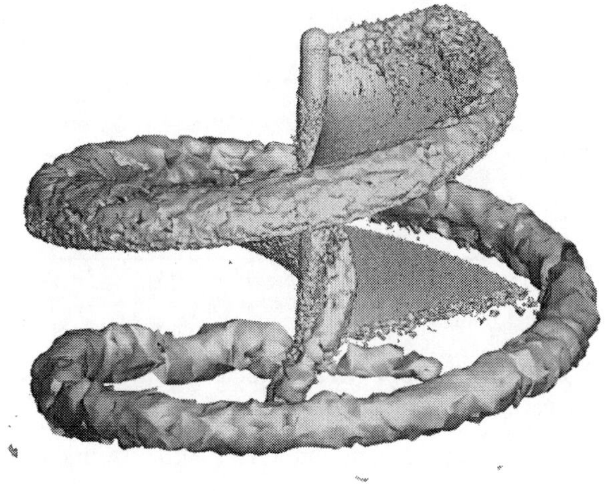

Figure 20. Vorticity contour of the $\eta = 0.8$ initial taper rate design

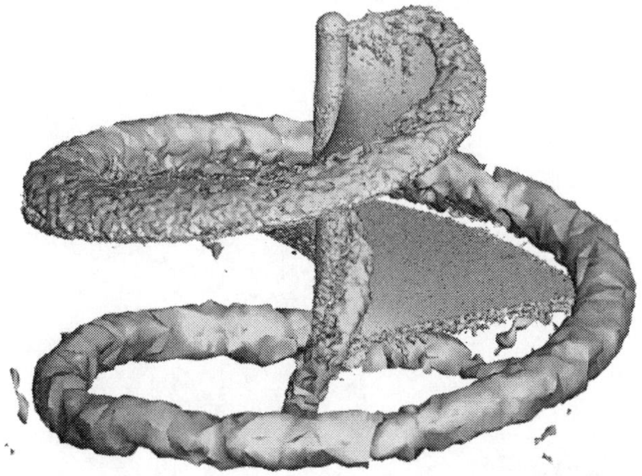

Figure 21. Vorticity contour of the $\eta = 1.33$ initial taper rate design

From these visualizations, it is evident that a taper rate is key in keeping the vortex attached. Once the vortex reaches half way down the rotor blade, it detaches for these designs due to the untapered geometry.

Despite this, the vortex is still able to effectively develop in the first half of the blade. All rotor designs with an untapered second half outperformed the best linearly tapered design. This suggests that using as much lifting surface area as possible for a given maximum radius design is crucial to improving rotor performance.

By assessing the pressure distributions of these two designs, the effects of aggressive tapers as well as untapered segments on the rotor's performance are further understood.

Figure 22 shows the pressure distribution of the $\eta = 0.8$ initial taper design.

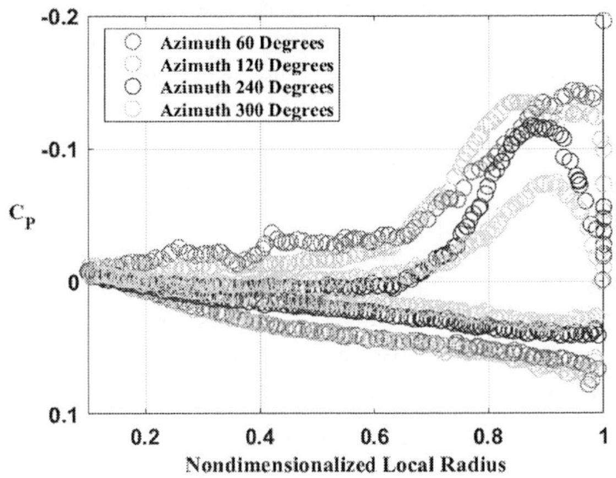

Figure 22. Pressure distribution of the $\eta = 0.8$ initial taper design

The aggressive taper in the first half of the blade generates a significantly larger pressure differential compared to the linearly tapered case in the first half of the blade. It is also confirmed that the "da Vinci" vortex detaches as there is less lift being generated in the untapered section of the blade than the tapered section.

Figure 23 shows the pressure distribution of the $\eta = 1.33$ initial taper design.

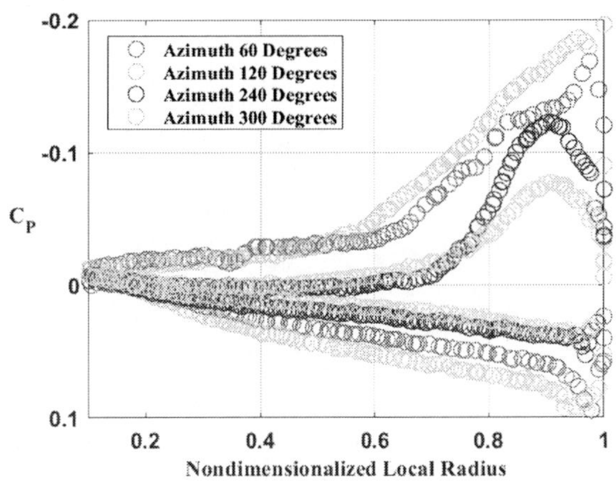

Figure 23. Pressure distribution of the $\eta = 1.33$ initial taper design

The pressure distribution from the aggressively tapered case furthers the idea that an increasing taper rate increases the strength of the "da Vinci" vortex.

Comparing Figures 22 and 23, there is a significant jump in the pressure differential in the first half of the blade for the aggressively tapered case. Regardless of the strength of the vortex in the first half of the blade, transitioning to an untapered design in the second half of the blade detaches the vortex, resulting in similar lift between both of these designs.

Although the pressure distribution seems to suggest more lift is produced for the aggressively tapered case, the loss in lifting area reduces the performance of the rotor. The aggressively tapered design with an initial taper rate of $\eta = 1.33$ achieves a figure of merit of .58, nearly a 2% decrease in performance from the most efficient design.

Combining Bilinear Pitch and Bilinear Taper Rates

The bilinearly pitched design where $\zeta = 1.0$ transitioned to $\zeta = 1.4$ outperformed all linearly pitched designs as previously mentioned. Similarly, the bilinearly tapered design where the initial taper rate was $\eta = 0.8$ achieved a figure of merit of 0.59, the highest out of all datasets.

These results suggest that combining a bilinear pitch rate with a bilinear taper rate would have the potential to further improve the performance the current aerial screw design.

For this study, the most efficient bilinearly pitched and bilinearly tapered designs were combined in hopes to improve the rotor's overall hover performance. This design entails an initial pitch rate of $\zeta = 1.0$ and an initial taper rate of $\eta = 0.8$ in the first half of the screw. The second half of the screw has a pitch rate of $\zeta = 1.4$ and is untapered.

Figures 24 and 25 show that the bilinear design significantly outperforms previous test cases in hover. The red data point shows the bilinearly pitched and tapered design. The blue data point in Figure 24 refers to the $\zeta = 1.2$ linearly pitched design.

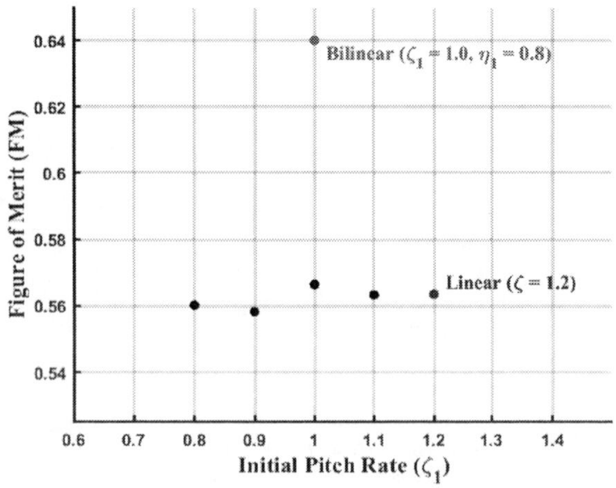

Figure 24. Variation of figure of merit of bilinearly tapered aerial screws with varying initial pitch rate

The bilinearly pitched and tapered design achieved a figure of merit of 0.64.

Combining a bilinear pitch with a bilinear taper resulted in a 12.9% improvement compared to the best bilinearly pitched

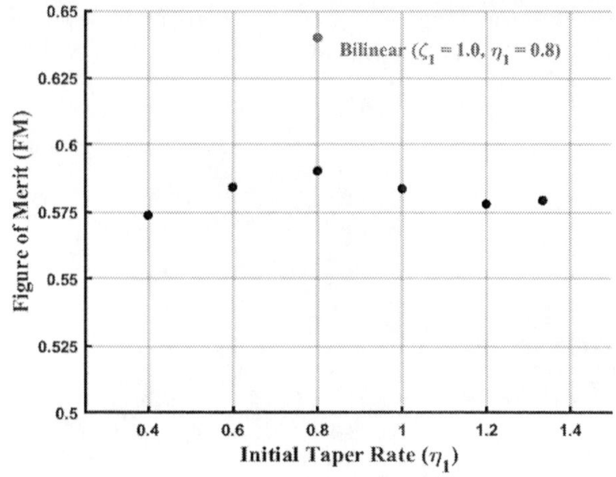

Figure 25. Variation of figure of merit of bilinearly tapered aerial screws with varying initial taper rate

and linearly tapered design and a 13.7% improvement over the best linearly pitched design.

Once again, a significant improvement is seen with the bilinearly pitched and tapered design when comparing to the solely bilinearly tapered rotors. There is an 8.5% improvement in figure of merit when comparing the bilinearly pitched and tapered rotor to the best bilinearly tapered rotor.

Figure 26 shows the vorticity iso-contour of the bilinearly pitched and tapered design.

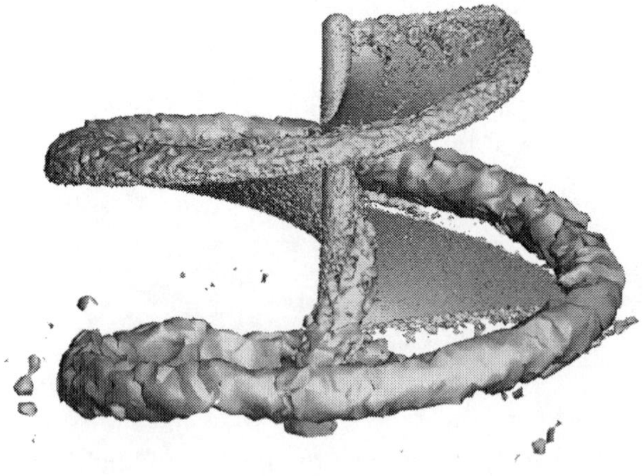

Figure 26. Vorticity contour of the bilinearly pitched and bilinearly tapered design

Figure 27 shows the pressure distribution of the bilinearly pitched and tapered design. A key feature to the improvement in performance of the rotor is the uniformity in the pressure distribution plot across different angles of azimuth. The locations of where the majority of the lift is being generated are more uniform in the bilinearly pitched and tapered design. Improving the lift generation across differing angles of azimuth while maintaining the uniform location of where the vortex lies is key to further improving these aerial screw rotors in the future.

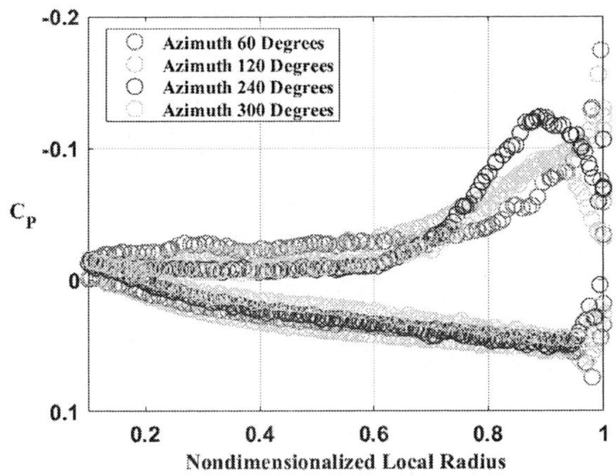

Figure 27. Pressure distribution of the bilinearly pitched and bilinearly tapered design

CONCLUSIONS

This study presents an in-depth exploration analysis of the dominant geometric parameters governing the performance of aerial screws, inspired by Leonardo da Vinci's early concepts. By leveraging modern Computational Fluid Dynamics (CFD) tools, the research identifies the effect of key geometric design variables on the thrust generation mechanism and overall aerodynamic efficiency of the aerial screws.

The analysis reveals that the "da Vinci" vortex, a shape-conforming helical vortical structure at the screw's tip, is central to the thrust production in aerial screws. This mechanism is similar to vortex lift phenomena observed in swept and delta wings, where it contributes significantly to pressure differentials across the screw surface, thereby enhancing thrust.

The influence of each design variable on the aerodynamic performance of aerial screws is studied. The key findings include:

- Effect of Linear Pitch Rate: The pitch rate critically affects the screw's performance. The "da Vinci" vortex increases in strength with increasing pitch rate as a result of stronger separation at the leading edge. The figure of merit is observed to increase from low to moderate pitch rates, reaches a peak at $\zeta = 1.2$, then decreases.

- Effect of Bilinear Pitch Rate: A decreased pitch rate in the first half of the rotor results in a weaker starting vortex. Increasing the pitch rate in the second half of the rotor causes a growth in vortex strength. The figure of merit nears .57 with an initial pitch rate of $\zeta = 1.0$.

- Effect of Linear Taper Rate: The taper rate is a dominant parameter in increasing the strength of the "da Vinci" vortex along the span of the rotor blade. Vortex strength shows improvement in growth with an increasing taper rate, yet a loss of lifting area shows a trade-off. The figure of merit increases from low taper rates and peaks at .56 with a taper rate of $\eta = 0.5$, then drops off.

- Effect of Bilinear Taper Rate: The aggressive taper rate results in an improvement in vortex strength. A decrease in taper rate further along the blade results in a separation of the vortex from the blade's surface. A figure of merit of .59 was achieved at an initial taper rate of $\eta = 0.8$.

Achieving figures of merit as high as 0.55 with variations in just one design variable, 0.59 with nonlinear design variables, and .64 with a combination of nonlinear design variables, this study highlights the considerable potential of these rotors for efficient design. The aerial screws, when meticulously designed, show promise as a suitable alternative for Vertical Take-Off and Landing (VTOL) applications, particularly at smaller scales, offering an innovative alternative to conventional rotor designs.

Future work could focus on further refinement of these design parameters, including the effects of dihedral angles, multiple non-linear tapering and pitching variations, and multi-bladed configurations, to fully unlock the potential of aerial screws in modern VTOL applications. Additionally, the wake region of the Da Vinci Aerial Screw will be investigated and compared to the wake region of a conventional rotor. Moreover, experimental validation of the CFD predictions would provide further confidence in the applicability of these findings to real-world aerial screw designs.

Author contact:
Ron Berlin *rberlin8@terpmail.umd.edu*
Koushik Marepally *koushikm@umd.edu*

ACKNOWLEDGMENTS

The authors would like to thank the University of Maryland's high performance computing resource Zaratan, which was used for the CFD simulations in this study.

REFERENCES

1. Lambermont, P., and Pirie, A., *Helicopters and Autogyros of the World*, Barnes, 1970.

2. Baden-Powell, B., "Experiments with Aerial Screw Propellers," *Aeronautical journal (London, England: 1897)*, Vol. 8, (30), 1904, pp. 28–32. DOI: 10.1017/S2398187300142380.

3. Leonardo Helicopters, "37th Annual Student Design Competition - 2019-2020 Request for Proposal (RFP) - Leonardo's Aerial Screw: 500 Years Later," .

4. Sutherland, J., Brown, R., Ebewele, E., Fisler, E., Krohmaly, K., Linkins, J., Marepally, K., Prete, A., Semenov, I., Nagaraj, V., Chopra, I., and Datta, A., "Leonardo's Aerial Screw: 500 Years Later," University of Maryland, 2020.

5. Prete, A., Nagaraj, V., and Chopra, I., “Revolutionary Flight Vehicle Based on Leonardo da Vinci Aerial Screw: A Paradigm Shift in VTOL Technology,” VFS Aeromechanics for Advanced Vertical Flight Technical Meeting, San Jose, CA, Jan 25-27, 2022.

6. Prete, A. C., *Revolutionary flight vehicle based on Leonardo Da Vinci Aerial Screw: a paradigm shift in VTOL Technology*, Master’s thesis, University of Maryland, College Park, 2022.

7. Marepally, K., Berlin, R., and Baeder, J., “Details of Da Vinci Aerial Screw Physics, Aerodynamics, and Performance,” VFS 50th European Rotorcraft Forum, Marseille, France, September 10-12, 2024.

8. Polhamus, E. C., “A concept of the vortex lift of sharp-edge delta wings based on a leading-edge-suction analogy,” NASA TN D-3767, 1966.

9. Jung, Y. S., Govindarajan, B., and Baeder, J., “A Hamiltonian-Strand Approach for Aerodynamic Flows Using Overset and Hybrid Meshes,” VFS 72nd Annual Forum, West Palm Beach, Florida, 05 2016.

Considerations for both Silicon and Wide-Bandgap Semiconductors in Electric Aircraft

Wiliam Dillard
Sr. Bus. Dev. Manager
Microchip Technology
Chandler, AZ, USA

Jason Chiang
Product Marketing Manager
Microchip Technology
Chandler, AZ, USA

Amit Gole
Product Marketing Manager
Microchip Technology
Garden Grove, CA, USA

Steven Chenetz
Sr Tech. Staff Eng.
Microchip Technology
Chandler, AZ

ABSTRACT

With continued progress towards aircraft electrification, power electronic devices and converter topologies have become increasingly critical to successful designs and to OEM competitiveness. This work treats both but focuses on the former, semiconductor devices in both actuation and propulsion applications. A comparative examination of silicon and silicon carbide is provided within the scope of energy loss mechanisms, EMI/EMC impacts at the system level, as well as non-technical considerations.

INTRODUCTION

Ongoing technical advancements in battery and hydrogen powered aircraft propulsion have significantly advanced the development and deployment of More Electric Aircraft (EMA). Aviation has always prized reduced size, weight and power budgets. These pressures, along with robustness against environmental factors, are severely elevated in Advanced Air Mobility (AAM) aircraft. A reasoned approach is optimizing power efficiencies in power electronics subsystems (batteries, battery management, power conversion and power distribution) while maintaining acceptable thermal conditions. Power devices fabricated in wide-bandgap (WBG) materials like silicon carbide (SiC) and gallium nitride (GaN) are strong technologies for these subsystems and are finding wide acceptance. However, there are motivations for well-designed solutions in silicon as well.

TOPOLOGY CONSIDERATIONS

Since eVTOL primary bus voltages are dc, power converter topologies of interest fall into two classes, dc/ac and dc/dc, with the former more popular in high power multi-phase ac motors for propulsion and the latter in actuation and power distribution to avionics subsystems, nominally at 28 VDC.

Much of the foundational work on these converters can be found in the works of predecessors like Rafin (Ref. 1), Dorn-Gomba (Ref. 2), Langmaack (Ref. 3), and Ebersberger (Ref. 4) which address the growing electric power requirements for traditional aircraft. Consistent conclusions include the need for: higher bus voltages; improved thermal management technologies; reducing the weight and volume of inductors and capacitors by increasing energy storage densities, and reduced losses (increased converter efficiency). A concept schematic for their examinations is shown in Figure 1.

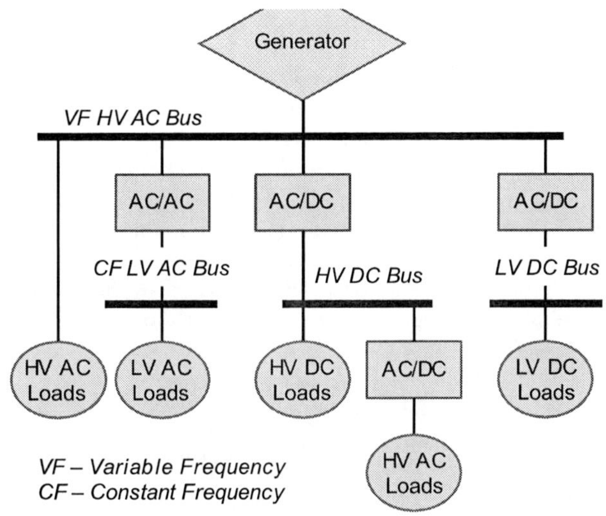

Figure 1. Aircraft power distribution architecture.

Rahn (Ref. 5) and Valente (Ref. 6) have published similar examinations and design advances for eVTOL propulsion motors. In addition to traditional design specifications such as torque, rpm, and power output, the eVTOL application stresses reduced size and weight, increased efficiency, and consideration for the transient limitations of battery stack and its associated power electronics.

Regarding publications on converter topologies specific to eVTOL, Menzi (Ref. 7). Han (Ref. 8), and Liu (Ref. 9) are notable for their contributions to the state of the art as well as their extensive reviews of prior literature. Menzi and Han examine dc-dc converters while Liu focuses on dc-ac

Presented at the Vertical Flight Society's 11th Biennial Autonomous VTOL Technical Meeting, Phoenix, AZ, USA, Feb. 4-6, 2025.

inverters. The applicability of these two converter types to eVTOL propulsion can be seen in the equation below

$$P_R = \frac{m}{N_R n_D n_{hover}}$$

where P_D is power demand per rotor, m is total aircraft mass, N_R is the number of rotors, n_D is the efficiency of the motor and associated drive electronics, and n_{hover} is the aircraft hover efficiency (lift per unit of rotor power).

As the number of rotors increases, PMDC motors become more attractive due to their lower size and weight and the simplicity of their drive topologies. Multi-phase ac motors are preferred in eVTOL with fewer rotors.

SEMICONDUCTOR AND DEVICE CONSIDERATIONS

Referring to Figure 2, the designer has access to several semiconductor and switch types. High voltage/high power is dominated by the silicon IGBT and SiC MOSFET. While the IGBT is limited to lower switching frequencies, as of 2024 it has price advantages over SiC. Hybrid SiC solutions that combine IGBT with SiC diodes are thus a viable option for actuation applications. SiC is the dominant choice for high power and high switching frequencies. The material properties of gallium nitride (GaN), another WBG semiconductor, are also attractive for power electronics. However, fabrication methods, production capacity/rates, and in-field reliability evidence are well behind that of SiC.

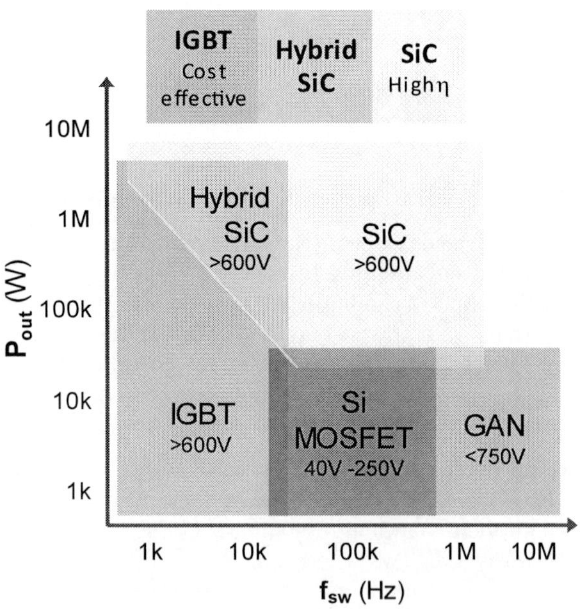

Figure 2. Applicability of various power switches across switching frequency and power level.

LOSS MECHANISMS

While Ren (Ref. 10) and Li (Ref. 11) provide complete examinations of silicon power MOSFET loss mechanisms using MOSFET non-linear terminal characteristics, the simple diagram in Figure 3 is adequate to treat the dominate sources of energy loss.

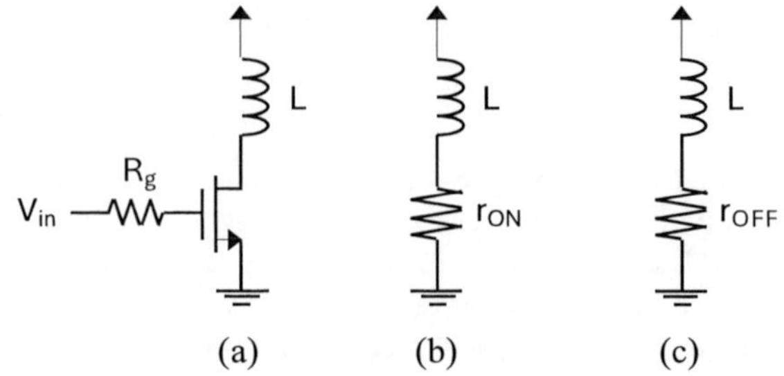

Figure 3. Simple schematics for (a) power MOSFET and inductor, (b) ON-state model, and (c) OFF-state model.

The average switching loss can be simply expressed as linearly related to switching frequency

$$P_{SW} = (E_{ON} + E_{OFF} + E_{T\text{-}ON} + E_{T\text{-}OFF}) \times f_S$$

where f_S is the switching frequency E_{ON} and E_{OFF} are the energy losses in the ON and OFF states, respectively, and $E_{T\text{-}ON}$ and $E_{T\text{-}OFF}$ are the transient losses during turn-on and turn-off. switch turn-on and turn-off, respectively. An alternative equation based on duty cycle is also of value.

$$P_{SW} = P_{ON}\delta_{ON} + P_{OFF}\delta_{OFF} + P_{T\text{-}ON}\delta_{T\text{-}ON} + P_{T\text{-}OFF}\delta_{T\_OFF}$$

where P_i and δ_i are the average power and duty cycle associated with each of the four switch states.

Static On State

This condition, modelled in Figure 3b, is the dominate loss mechanism and is linear with r_{ON}. On-resistance is determined by the nature of the MOSFET channel inversion layer when the MOSFET in the Static On state. It is specified on datasheets at stated V_{GS}, drain current, and temperature conditions. Power loss in this state is expressed below where I_{dss} is the steady state drain current in the static on state.

$$P_{ON} = I_{dss}^2 \times r_{ON}$$

Static Off State

Modelled in Figure 3c, the value of R_{OFF} is sufficiently high, i.e. drain current is limited to leakage mechanisms, that losses are negligible and can be dropped from the analysis.

Turn-On State

Turn-on losses are caused by parasitic device capacitances being charged as the gate voltage rises and the drain voltage drops to a value determined by product of $r_{ON} \times I_d$. This can be visualized through Figures 4 and 5.

Referring to Figure 5, upon turn-on (V_{in} rises) the device goes through four specific charging profiles starting with t_{th} is the time to charge the gate through time constant $R_g C_{iss}$, where C_{iss} is the total input capacitance at the gate. At the end of t_{th} the gate voltage exceeds the threshold voltage, the turn-on process enters time segment t_r, and the drain current starts to rise. Throughout t_r the gate voltage and drain current continue to rise until the drain current reaches its steady state value. In the time segment t_{gd}, gate current discharges C_{ds}, driving V_{DS} down. In the final stage, t_{ov}, the gate voltage rises to V_{in} and the drain voltage stabilizes at r_{ON} x I_d. From Figure 5 it is clear that the bulk of turn-on energy loss (appreciable values for the V_{ds} x I_d product) occurs during t_r and t_{gd}. Losses in the remaining charging segments are negligible.

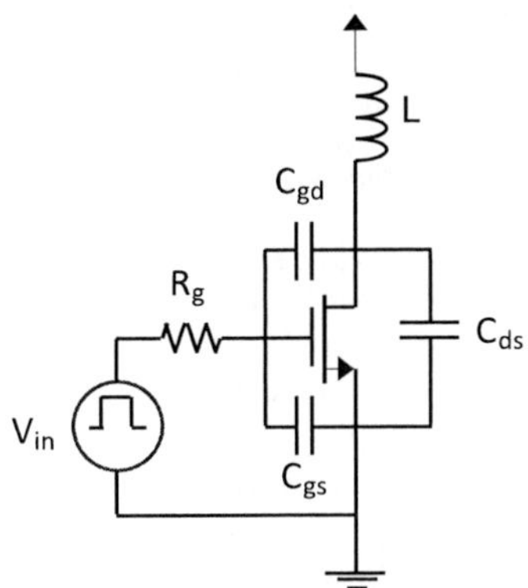

Figure 4. Simple schematics for (a) power MOSFET and inductor, (b) ON-state model, and (c) OFF-state model.

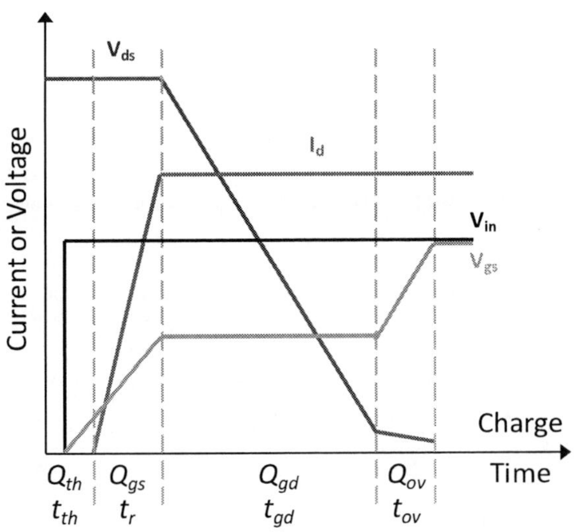

Figure 5. Simple schematics for (a) power MOSFET and inductor, (b) ON-state model, and (c) OFF-state model.

Power loss during turn-on has two components, the first due to the non-zero $V_{ds}I_d$ product and the second to due input capacitance charging. can be estimated as shown below, linearly scaling with switching speed. For higher power applications such as eVTOL propulsion, the $V_{ds}I_d$ term dominates.

$$P_{T-ON} = \left[\frac{V_{dsmax} I_{dss}}{2}\left(t_r + t_{gd}\right) + \frac{V_{in}^2 C_{iss}}{2}\right] f_{sw}$$

Turn-Off State

An analysis of the turn-off losses is essentially the reverse of turn-on and is captured in Figure 6 with an identical equation.

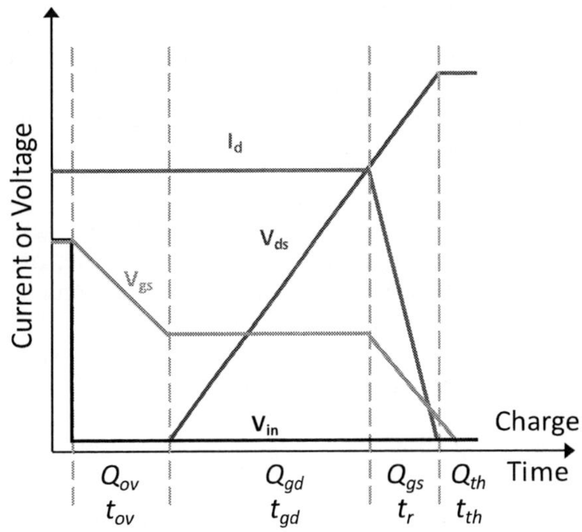

Figure 6. Simple schematics for (a) power MOSFET and inductor, (b) ON-state model, and (c) OFF-state model.

$$P_{T-OFF} = \left[\frac{V_{dsmax} I_{dss}}{2}\left(t_r + t_{gd}\right) + \frac{V_{in}^2 C_{iss}}{2}\right] f_{sw}$$

Inductance-Driven Over-Voltage at Turn-Off

There is another loss mechanism that is caused by the combination of switch and inductor. The voltage across any inductor can be expressed as

$$V_L = L\left[\frac{di}{dt}\right]$$

During turn-off in Figure 6, this equates to

$$V_L = L\left[\frac{I_{dss}}{t_r}\right]$$

which can be significant when t_r is brief, i.e. when switching frequency is high. Referring to Figure 3a, this inductive voltage transient is additive to V_{ds}. In addition to the energy losses, these transients elevate stress on the switch which degrades Failure in Time performance.

Loss Mechanism Comparison – Silicon vs SiC

Compared to silicon, SiC MOSFETs generally experience significantly lower losses, particularly in switching losses, making SiC the preferred semiconductor for high switching speeds. There are several contributing factors including: lower parasitic capacitances, faster reverse recovery in the body diode, and higher electron velocity. Conduction losses advantages of SiC are less dramatic.

There are two additional device-level characteristics on note. Being a wide bandgap semiconductor, SiC can operate safely at higher temperatures. Furthermore, SiC reliability at higher temperatures is superior. However, as of 2024, silicon has the unit cost advantage.

These tradeoffs between switching and conduction losses are captured graphically by Han (Ref. 9) and are provided here in Figure 7. A set of 600V, 20kW half bridges were constructed in silicon, hybrid, and SiC devices and operated at a switching frequency of 20k, 50k, 100k, 150k, and 200kHz. As expected, the "All SiC" implementations showed the lowest switching losses at the same switching frequency. However, conduction losses for the "Hybrid' and "All SiC" are rather comparable.

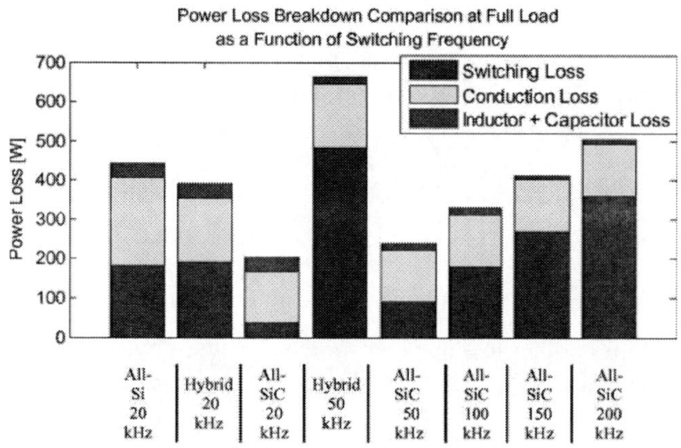

Figure 7. From Han (Ref 9) a comparison of losses based on semiconductor selection.

SWITCHING SPEED AND EMI IMPACTS

Unless mitigated by external filters, the fast switching speed of SiC will generate more EMI energy at higher frequencies. This can be detrimental at the aircraft level where EMI/EMC limitations per standards such as DO-160 requirements for both radiated and conducted emissions can be exceeded. Leuchter (Ref. 12) provides prescriptive mitigation techniques.

NON-TECHNICAL CONSIDERATIONS

There are non-technical considerations related to power device selection that are critical to success at the enterprise level. Among these are cost, supply chain robustness and technology runway. Cost is obvious but must include price of not only the switch, but all secondary elements such as gate drivers and protection devices. Supply chain robustness is the most critical of the three in that it is the least predictable with its negative impacts occurring after all certifications are in place and aircraft are in production.

The top five concerns include the following.

1. Financial stability of the supplier.
2. Supplier's commitment to production lifetimes that align to aviation industry expectations - typically 20 years. A propensity to obsolete part numbers is a risk that can trigger more frequent revision of certification approvals. Seek suppliers that commit to longevity of supply.
3. The supplier's quality management system, preferrable AS9100, reveals their commitment to industry standards and OEM flow-downs related to daily quality across all departments. Component traceability and counterfeit controls are two examples.
4. Maximum supply chain ownership and minimum risk.
5. Geo-political stability of the supplier's sources of materials and components is a growing concern. All suppliers have a plan. Preferred suppliers are taking actions to reduce their risk exposure.

Technology runway is more about opportunity than risk and refers to the possibility of future improvements. For this study, silicon MOSFETs and IGBTs are fully mature with reasonable expectation of incremental advancements. Conversely, SiC MOSFETs are less mature in that they are vetted for use in harsh applications but not yet optimized. The qualitative radar chart in Figure 8 provides a visual interpretation.

Total Cost

The scope of a total cost of ownership analysis is expansive and includes components costs, impacts on cooling systems, impacts on weight, correlations to available payloads and more.

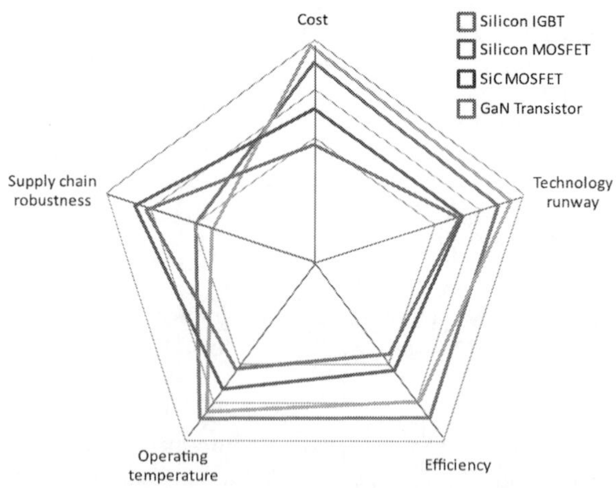

Figure 8. Comparing four switch technologies across technical and non-technical dimensions.

CONCLUSIONS

A comparison for silicon and SiC switches for eVTOL applications has been provided, introducing considerations at the power converter, power switch, and system level. A detailed examination of loss mechanisms has been provided as well as a foundation for the advantage of SiC in switching losses and thereby its preference for high switching frequency implementations. To offer insights into component selection processes, silicon and SiC switches were examined across non-technical characteristics such as supply chain robustness, cost, and technology runway.

Author contact: Bill Dillard william.dillard@microchip.com
Author contact: Amit Gole amit.gole@microchip.com
Author contact: Jason Chiang jason.chiang@microchip.com

REFERENCES

1. Rafin, Sajjad & Haque, Md & Islam, Rejaul & Mohammed, Osama. (2023). *A Review of Power Electronic Converters for Electric Aircrafts.* 1-8. 10.1109/3D-PEIM55914.2023.10052535.

2. L. Dorn-Gomba, J. Ramoul, J. Reimers and A. Emadi, "Power Electronic Converters in Electric Aircraft: Current Status, Challenges, and Emerging Technologies," *IEEE Transactions on Transportation Electrification*, vol. 6, no. 4, pp. 1648-1664, Dec. 2020, doi: 10.1109/TTE.2020.3006045.

3. N. Langmaack, G. Tareilus and M. Henke, "Modular Power Electronics for Electric Vehicles and Aviation Applications," 2019 IEEE 13th International Conference on Power Electronics and Drive Systems (PEDS), Toulouse, France, 2019, pp. 1-5, doi: 10.1109/PEDS44367.2019.8998936.

4. J. Ebersberger, M. Hagedorn, M. Lorenz, and A. Mertens, "Potentials and comparison of inverter topologies for future all-electric aircraft propulsion," *IEEE Journal of Emerging and Selected Topics in Power Electronics*, vol. 10, no. 5, pp. 5264–5279, Oct. 2022.

5. Valente, Wigmore, Saysell, Scott, Gottardo, Johnston, Haise, "Maximizing Power Density in the Design of a 100 kW Geared e-Propulsion Unit,", Vertical Flight Society 80th Annual Forum & Technology Display, Monreal, Quebec, Canada, May 7-9, 2024.

6. Rahn, McDonald, Myers, Paden, Ricci, "Motor and Drive Design for eVTOL Aircraft," Vertical Flight Society 80th Annual Forum & Technology Display, Monreal, Quebec, Canada, May 7-9, 2024.

7. D. Menzi, L. Imperiali, E. Bürgisser, M. Ulmer, J. Huber and J. W. Kolar, "Ultra-Lightweight High-Efficiency Buck-Boost DC-DC Converters for Future eVTOL Aircraft with Hybrid Power Supply," *IEEE Transactions on Transportation Electrification*, doi: 10.1109/TTE.2024.3375026.

8. X. Liu, M. Wei, M. Qiu, K. Hobbs; S. Yang, and A. Dahneem, "FPGA-Based Forced Air-Cooled SiC High-Power-Density Inverter for Electrical Aircraft Applications," *2023 IEEE Applied Power Electronics Conference and Exposition (APEC),* DOI: 10.1109/APEC43580.2023.10131187.

9. D. Han, J. Noppakunkajorn, and B. Sarlioglu, "Comprehensive efficiency, weight, and volume comparison of SiC- and Si-based bidirectional DC–DC converters for hybrid electric vehicles," IEEE Transactions on Vehicular Technology, vol. 63, no. 7, pp. 3001–3010, Sep. 2014.

10. Y. Ren, M. Xu, J. Zhou and F. C. Lee, "Analytical loss model of power MOSFET," in *IEEE Transactions on Power Electronics*, vol. 21, no. 2, pp. 310-319, March 2006, doi: 10.1109/TPEL.2005.869743.

11. X. Li, J, Jiang, A. Huang, S. Guo X. Deng, B. Zhang, X. She, "A SiC Power MOSFET Loss Model Suitable for High-Frequency Applications," in *IEEE Transactions on Industrial Electronics*, vol. 64, no. 10, pp. 8268-8276, Oct. 2017, doi: 10.1109/TIE.2017.2703910.

12. J. Leuchter, V. Stekly and E. Blasch, "Investigation of Avionics Power Switch Loading Versus Aircraft Electromagnetic Compatibility," *IEEE Aerospace and Electronic Systems Magazine*. Vol. 30, No. 9, pp. 24-34, ISSN 0885-8985, 2015.

Assessment of Retrofitting an Existing Helicopter With Electric Propulsion Technologies

Dounya Jarrad
Undergraduate Student
Georgia Institute of Technology
Atlanta, GA, USA

Lakshmi N. Sankar
Professor
Georgia Institute of Technology
Atlanta, GA, USA

Auraluck Pichitkul
Professor
Suranaree University of Technology
Nakhon Ratchasima, Thailand

ABSTRACT

Conceptual design studies are presented for a notional helicopter based on an existing system (Robinson R-44), retrofitted with eVTOL technology. The feasibility of using Lithium-Ion batteries or liquid hydrogen fuel cells has been examined. The impact of the retrofit on the system weight has been examined. Aerodynamic analyses have also been performed with a rotor that uses improved low-drag airfoils to assess the feasibility of the eVTOL systems for air-taxi and inter-city operations. The calculations indicate that Li-Ion and liquid hydrogen-based systems may be able to handle inter-city operations with a 200 km range with improvements to blade aerodynamics, and structural weight reduction strategies.

INTRODUCTION

Vertical lift vehicles are well suited for emergency response in natural disasters or medical emergencies where traditional ground transportation may not be available or effective. Megacities around the world with rapidly growing urban populations, traffic congestion, and transportation challenges would benefit from vertical lift vehicles. In third-world nations without a well-developed transportation infrastructure, vertical lift systems that do not require significant ground support would be an excellent choice for rescue operations and delivery services.

Some major roadblocks to the adaptation of helicopters in these missions are accidents attributable to single-engine failure, community noise, and pollution. Researchers around the world have been looking at advanced VTOL systems that depend on electrical power as a way of eliminating or mitigating these difficulties. In the United States, the rotorcraft research group at NASA Ames Research Center has explored various urban transportation concepts including eVTOL systems [e.g., 1-6]. These studies point to the low energy density of Lithium-Ion batteries as a major limiter in long-range eVTOL missions. Datta has examined the feasibility of using hydrogen fuel cells as a way of increasing the range [7]. It would take several years for these systems to evolve, mature, and be certified for civilian use in an urban environment. Investment would be of the order of several billions, driving up the cost of operations (in dollars per passenger-mile), making them unaffordable for most countries and cities.

Retrofitting existing legacy helicopters with eVTOL capability may be an attractive alternative. The airframe, cockpit, tail boom, main rotor, and tail rotor could be retained from the original configuration with careful redesign. Avionics, guidance/control systems would also be retained. This would reduce the need for training pilots on a new system, and potentially reduce the efforts needed to certify the system for civilian operations. Existing MRO (Maintenance, Repair, and Overhaul) infrastructure may be used to service most of the helicopter components and the avionics package. Only the engine(s), main rotor, and possibly the gearbox and shafts would be replaced with eVTOL options. Operational safety will be enhanced with multiple electrical motors for redundancy, each powered by independent inverter/transformer systems. Given these advantages, helicopter companies, including Tier 1 Engineering, have begun examining electrical propulsion retrofits. (Tier 1 Engineering Pioneers Electric e-R44).

A few feasibility studies have been done on retrofitting existing systems with electric propulsion technology. For example, Datta and Johnson have examined three propulsion strategies — a battery-only, a fuel-cell-only, and a battery + fuel cell hybrid powerplant — for a manned Robinson R-22 ultralight utility helicopter [8]. The present researchers have also examined the impact of retrofitting fixed wing and rotary wing systems with electric propulsion [9-15].

Research Objectives

The primary research objective of this work is to assess the impact of replacing the conventional piston engine system with an eVTOL system powered by a battery pack or hydrogen fuel cell technology, using the approach proposed in Ref. 7 and 8.

Given the low energy density of conventional Lithium-Ion batteries and the larger weight of hydrogen fuel cell systems compared to traditional combustion-based systems, it is important that the total energy required to perform a mission be as low as possible. Towards this end, a secondary objective of this work is to replace the NACA 63015 airfoil section used in the R-44 helicopter rotor with an advanced

Presented at the Vertical Flight Society's 11th Biennial Autonomous VTOL Technical Meeting, Phoenix, AZ, USA, Feb. 4-6, 2025.

low-drag airfoil designed using an inverse design methodology. These efforts build on preliminary studies reported by the present authors in Ref. 13.

A case based on economics and cost of ownership considerations may be made to justify this retrofit. A new R-44 helicopter with the piston engine option costs ~$367,000. A used helicopter may be purchased for less depending on the condition of the vehicle and engine hours. The average insurance cost for a Robinson R-44 is $9,690 for a qualified pilot with $200,000 in hull coverage. Qualified Robinson Pilots will have at least a Private license with IFR rating, 500 total hours and at least 100 in make/model and completed a Robinson training course. Operating costs include fuel, insurance, tie-down or hangar storage fees, regular maintenance inspections, and landing fees. Typical R-44 operating costs average about $190 per hour depending on a wide range of variables. These estimates are based on the data found at the website Robinson R44 Operating Cost | BWI and other links found under this website.

NOTIONAL VEHICLE

In this paper, we focus on a notional single main rotor helicopter, closely patterned after the Robinson R-44 helicopter. This system was chosen because of its relatively low cost of ownership, including financing, maintenance, and crew costs. Figure 1 shows the configuration.

Preliminary Performance Estimates

The power and thrust coefficients, C_P, and C_T, are defined in terms of the rotor power P, and thrust T (approximately equal to the gross weight of the vehicle) in terms of density ρ (at a specified altitude), advance ratio μ (defined as the forward speed V by rotor tip speed), rotor disk area A, rotor angular velocity Ω, and the rotor radius R as follows:

$$C_T = \frac{T}{\rho A(\Omega R)^2} \qquad C_P = \frac{P}{\rho A(\Omega R)^3} \tag{1}$$

The power coefficient for each of the four rotors may be estimated as,

$$C_P = \kappa\lambda_i + \sigma\frac{C_d}{8}(1 + 4.6\mu^2) \tag{2}$$

Here κ is the induced velocity correction factor (set to 1.15) and σ is the solidity of the main rotor (defined as the total projected area of the rotor blades divided by the rotor disk area). A nominal drag coefficient C_d of 0.01 has been used in this preliminary performance estimate.

The induced velocity ratio λ_i is iteratively computed from Glauert's inflow model given below:

$$C_T = 2\lambda_i\sqrt{(\mu^2 + (\mu\, tan(\alpha_{TPP}) + \lambda_i)^2)} \tag{3}$$

Here α_{TPP} is the forward tilt of the tip path plane for the rotors to provide propulsive force in the flight direction. This angle may be estimated from the vehicle drag D based on a specified flight velocity V, the vehicle weight W, and the equivalent flat plate area f as follows:

$$D = \frac{1}{2}\rho V^2 f = W\, tan(\alpha_{TPP}) \tag{4}$$

An equivalent flat plate area of 5 square feet (~0.5 square meters) has been used based on historical data for conventional single main rotor helicopters of comparable weight.

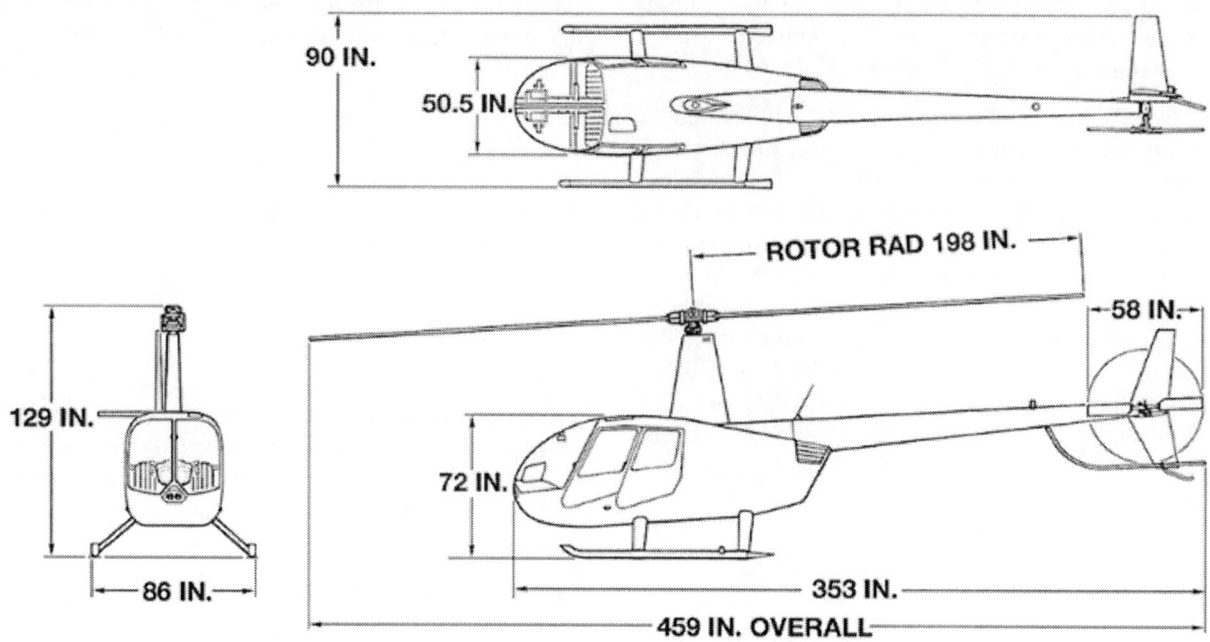

Helicopter	2019 Robinson R44 Raven II
Engine	Lycoming IO-540, six cylinder, fuel injected
Horsepower	Derated to 245 for takeoff and 205 continuous
Maximum Gross Weight	2500 lb
Approximate Empty Weight (including oil & standard avionics)	1505 lb
Standard Fuel	177 lb (80 kg)
Auxiliary Fuel	102 lb (46 kg)
Pilot, Passengers, and Baggage (with standard fuel)	818 lb (371 kg)
Cruise Speed at Maximum Gross Weight	up to 109 kts (202 km/h)
Maximum Range (no reserve)	approx 300 nm (550 km)
Hover Ceiling IGE	8950 ft @ 2500 lb
Hover Ceiling OGE	7500 ft @ 2300 lb
Rate of Climb	over 1000 fpm
Maximum Operating Altitude	14,000 ft

(https://www.aerinova.com/aircraft/helicopters/2019-robinson-r44-raven-ii.html)

Figure 1. Overview of the Notional Helicopter Based on Robinson R-44.

Hover Performance

In hover, the same equations may be used, with the advance ratio set to zero. The tail rotor power may also be computed in an analogous manner. At sea level, under ISA + 20°C conditions, this estimate leads to a total power of 203 hp, ~152 kW for a gross weight of 2500 lb climbing at 100 feet per minute. The engine can supply 245 hp as seen in Figure 1 above, and can meet this requirement.

Forward Flight Performance

Figure 2 shows the forward flight performance at an operating altitude of 4000 feet under ISA + 20°C conditions. The forward speed for best range of 292 nautical miles is 95 knots. The total power consumption is 112 hp (~84 kW), and the piston engine is able to supply this power.

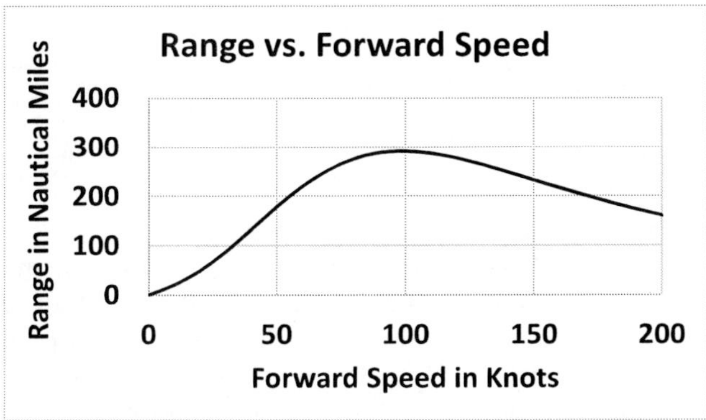

Figure 2. Forward Flight Performance of the Notional Vehicle (2500 lb Gross Weight, ISA + 20°C).

Mission Profile

Next-generation eVTOL systems would operate in a congested urban heliport environment, where the helicopter may be required to pause its vertical climb or descent and hold its position, hover in position, or loiter to maintain minimum distance between nearby helicopters. For this reason, the following mission profile under ISA + 20°C conditions is assumed.

1. It is assumed that the vehicle will be required to vertically climb at 100 feet per minute for ~7 minutes. The preliminary performance estimates for our notional helicopter indicate that 203 hp (~152 kW) would be needed for this initial phase of flight. The energy required for this segment of the flight is thus ~18 kW-hr.
2. A short-haul 36 minute air-taxi cruise segment covering 54 nautical miles (~100 km) in range at an altitude of 4000 feet is considered. As seen earlier, our performance estimates indicate that 112 hp (~84 kW) would be needed in cruise at a cruise speed of 90 knots. The energy need for this segment of the air-taxi mission is 50 kW-hr. A longer hour-long inter-city mission (110 nautical miles, ~200 km) would require 84 kW-hr of stored energy. These estimates include the climb phase and loitering at the landing site prior to landing.
3. A vertical descent of 2 minute duration at 145 kW requiring 5 kW-hr of energy is assumed.
4. Finally, an additional energy of 10 kW-hr to meet the avionics and hydraulic operations has been assumed.

The total energy need for this profile is 82 kW-hr for the air-taxi mission and 115 kW-hr for the inter-city mission, as shown in Table 1 below.

Table 1. Energy Needs of the Air-Taxi and Inter-City Missions for the Notional Helicopter.

Mission Segment	Air-Taxi Mission (54 nautical miles) ~100 km	Inter-City Mission (110 nautical miles) ~200 km	Units
Take-Off Phase, 203 hp (152 kW) including vertical climb at 100 feet per min, for 7 minutes	18	18	kW-hr
Climb to cruise altitude, cruise, descent, 112 hp (84 kW)	50	84	kW-hr
Vertical Descent, 196 hp (145 kW), 2 minutes	5	5	kW-hr
Auxiliary Power, 10 kW per hour	10	10	kW-hr
Total Energy	82	115	kW-hr

The above mission profile was used to estimate the eVTOL gross weight and the battery weight as shown in the next section.

Notional Vehicle Gross Weight Estimates

The following estimates have been made.

Weight of Piston Engine + Fuel Used in the Notional Vehicle

The conventional system is powered by a 6-cylinder piston engine: a Lycoming IO-540, six-cylinder, fuel-injected system. It is assumed that the engine has been derated to provide 245 hp for take-off and 205 hp for continuous operations. The specific fuel consumption is 0.475 lbs per shp per hour. The Lycoming engine weighs 450 lb (Lycoming O-540 - Wikipedia). The amount of fuel burned over a one-hour operation, with a power consumption of 200 hp (allowing for auxiliary system needs), is 100 lb. Thus, the weight of the propulsion system plus fuel is 550 lb.

Weight of the Transmission and Drive Train

Wayne Johnson gives the following estimate for the weight of the gearbox and the transmission shaft [16].

$$W = 95.7634P^{0.78137}RPM^{0.09899}/RPM_{Rotor}^{0.80686} \tag{5}$$

Here, P is the shaft power of the engine (200 hp), RPM is the engine RPM (2700), and the rotor RPM is 250. The above regression fit gives a total of 75 lb for the notional helicopter.

eVTOL VEHICLE GROSS WEIGHT ESTIMATES

The gross weight of the system increases if the helicopter is modified by replacing the piston engine-based drive train with Lithium-Ion batteries and ***dual electric motors***.

The dual electric motor system, each motor equipped with its own electrical system, with each motor capable of supplying the maximum torque and power, ***is expected to***

improve the safety of the operations compared to the baseline single-piston engine vehicle. As stated earlier, safety is paramount in all civilian helicopter applications.

Effect of Lithium-Ion Batteries on eVTOL Vehicle Gross Weight

The following estimates have been made.

Weight of the Lithium-Ion Battery

In the present studies, the energy density of the battery pack was assumed to be 250 watt-hours per kilogram for commercially available Lithium-Ion battery packs. As seen in Table 1, the air-taxi mission requires 82 kW-hr, with a battery weight of ~330 kg or 726 lbs. The inter-city mission requires 115 kW-hr of stored energy, with a battery weight of ~460 kg or 1012 lbs.

Weight of the Motors for the Li-Ion Systems

The weight of the motors may be estimated from the regression data provided in Ref. 16 and 17.

$$W_{per\ motor} = 0.91411\ Q^{0.75} \tag{6}$$

In the above equation, W is in pounds, and the torque Q is in ft-lb. It is assumed that the eVTOL uses the same transmission system and drive shaft as in the notional helicopter, requiring the motor RPM to match the engine shaft RPM. A technology multiplicative factor of 1.5 is applied to the above weight to account for the additional weight required for the voltage controller, inverter, and thermal management components. For a 200 hp motor operating at an RPM of 2700, which corresponds to a tip speed of 622 ft/sec (~0.6 tip Mach number), the weight of each motor is ~120 lbs, and the weight of the dual motor system is 240 lbs, as shown in Table 2.

Weight of the Electrical Control System for the Li-Ion Systems

Duffy et al. [8] give the following weight estimates for the electrical control system as a function of the horsepower P supplied by the motor.

$$W_{per\ motor} = 0.21\ P^{0.96} \tag{7}$$

It is assumed that each of the motors is powered by its own electrical control system to provide redundancy and safety, in the event one of the motors fails. The total weight of the electrical system for the two motors, each capable of supplying 200 hp, is 68 pounds.

Table 2. Estimate of Dual Motor Weight.

	Value	Units
Power, 200 hp	121000	ft-lb/sec
Angular velocity, 2700 RPM	282.74	rad/sec
Torque	427.95	ft-lb
Weight	86	lb
Technology factor, to account for controller/inverter and thermal management	1.50	—
Motor weight	129	lb

Impact of Electric Propulsion on Gross Weight

Table 3 below shows the impact of electric propulsion on the gross weight of the system for the air-taxi and inter-city missions. Our analyses indicate that the gross weight of the eVTOL system with a 110 nautical mile range is ~3188 lb, considerably heavier than the baseline 2500 lb notional vehicle. The shorter air-taxi mission of 55 nautical miles has a weight impact of ~402 lb, with a gross weight of 2900 lb. Thus, neither of the two missions (inter-city or air-taxi) would be feasible without improved rotor technologies, and reduction in the gross weight with advanced materials and more advanced battery technologies.

Table 3. Estimate of Increase in Gross Weight for Lithium-Ion Battery-Based Propulsion System.

	Air-Taxi Mission (54 nautical miles) ~100 km	Inter-City Mission (110 nautical miles) ~200 km	Units
Total energy	82	115	kW-hr
Battery weight, 250 watt-hr per kg	330	460	kg
Battery weight	726	1012	lb
Motor weight	258	258	lb
Control system weight	68	68	lb
Engine weight savings	450	450	lb
Fuel + tank savings	200	200	lb
Net increase in gross weight	402	688	lb

Effect of Liquid Hydrogen Fuel Cells on eVTOL Vehicle Gross Weight

Following the goal of the battery system weight estimations, the weight of a liquid hydrogen fuel cell system for the vehicle is investigated for the inter-city mission. The sizing method, general assumptions, and experimental fuel cell data used are based on those presented by Datta [7], and Datta and Johnson [8]. More detailed background and examination of hydrogen fuel cell systems can also be found there. The sizing of the system is explored for various stack designs and stack operating conditions in this paper. It should be noted that the goal of the hydrogen fuel cell system sizing is to obtain a general idea of the weight requirements for use in the existing helicopter; volumetric constraints and the logistics of installing the system are not considered.

General Sizing Methodology

Sizing of the hydrogen fuel cell system requires sizing all its constituent elements: the fuel cell stack, the subsystems needed for stack operation, and the hydrogen fuel and tank. The subsystems include electronics/controllers, a cooling system for temperature maintenance of the fuel cell stack, a compressor for controlling the pressure of air entering the stack, a separate cooling system for temperature control of the compressed air, and a water management system for the water produced by the stack. Sizing of the subsystems depends on the power requirements, stack design and performance, and constants specific to each respective subsystem. Similarly, sizing of the stack depends on the power requirements, design options (discussed in the following subsection), component weights within each constituent cell, and constants specific to the stack at various operating pressures. All constants used are taken directly from [7].

The hydrogen fuel is assumed to be stored at ~1 atm at -253°C where it is in liquid form, and the oxygen required for the fuel cell reaction is acquired from the air. Weight and power requirements for maintaining the fuel tank storage conditions are not considered for the purposes of this study. For all stack designs, the stack is assumed to operate at 80°C. Corrections, or decrements to be more specific, for the stack voltage due to altitude are also assumed.

Using the same estimation methods for performance of the notional vehicle, the power required of the vehicle with the same max gross takeoff weight is calculated. Because it is the maximum power required of the vehicle, the power required in hover/axial climb is used as the baseline for sizing the hydrogen fuel system, with the same assumed climb rate of 100 ft/min. Additionally, the power and sizing calculations are done for the 4000 ft flight altitude with ISA + 20°C temperature correction. The gross takeoff weight used in the power requirement estimations is 2500 lb (1136.4 kg), the original max gross takeoff weight of the vehicle. As a factor of safety, a power factor of 1.25 is used with the hover/climb power as the required attainable power of the fuel cell system. The resulting required power is found to be 192.5 kW (261.7 hp) for the 2500 lb gross takeoff weight.

The energy requirements for the inter-city mission are assumed to be the same as that of the notional helicopter, 115 kW-hr. Similarly, the total weight savings from the notional helicopter's engine, fuel, and tank is assumed to be 650 lb (295.45 kg). The hydrogen fuel cell system is assumed to use/include the same electric motors as the battery system, with a total weight of 258 lb (117.27 kg). The net weight gain of the system is found by subtracting the total weight savings from the hydrogen fuel system weight. Then, the new gross takeoff weight for the power requirement is found by adding the respective net weight gain to the original gross takeoff weight of 2500 lb.

System Sizing for Variations in Stack Pressure

The hydrogen fuel cell system is sized for a range of stack pressures from 1 atm to 2.5 atm; the range is limited between the maximum and minimum pressures because the stack performance calculations are done based on existing experimental data of stacks operating between these pressures. For stacks with operating pressures that differ from atmospheric pressure by more than ~30 kPa (0.296 atm), a compressor and turbine must be used to pressurize the air to the stack pressure, and an additional cooling system for the compressed air is added. Otherwise, a lighter blower is used, and no corresponding cooling system is needed. The efficiencies of the blower, compressor, and turbine are assumed to be 70%, 70%, and 80%, respectively. The gravimetric weight fraction of the tank, used to determine the required tank weight from the hydrogen fuel weight, is assumed to be 7.5%. The stack parameters for the design/max power and max efficiency cases for the original 2500 lb gross takeoff weight are shown in Tables 4 (a) and (b), respectively.

Table 4. Stack Parameter Estimates for Variation in Stack Pressure.

(a) Design/Max Power Stack Design.

	Stack pressure	Design power	Max power	SFC at design power	SFC at max power	Cell power density	Cell efficiency at design power	Cell efficiency at max power
Units	atm	kW	kW	kg/kW-hr	kg/kW-hr	W/cm^2	—	—
	1	192.5	192.502	0.077	0.077	0.796	0.33	0.33
	1.25	192.5	192.498	0.076	0.076	0.849	0.336	0.335
	1.5	192.5	192.491	0.076	0.076	0.897	0.335	0.335
	1.75	192.5	192.488	0.076	0.076	0.962	0.335	0.334
	2	192.5	192.484	0.077	0.077	1.027	0.332	0.332
	2.25	192.5	192.449	0.077	0.077	1.072	0.332	0.332
	2.5	192.5	192.465	0.077	0.077	1.117	0.332	0.331

(b) Max Efficiency Stack Design.

	Stack pressure	Design power	Max power	SFC at design power	SFC at max power	Cell power density	Cell efficiency at design power	Cell efficiency at max power
Units	atm	kW	kW	kg/kW-hr	kg/kW-hr	W/cm^2	—	—
	1	120	197.117	0.068	0.077	0.481	0.377	0.3
	1.25	120	200.61	0.066	0.076	0.5	0.389	0.304
	1.5	130	194.81	0.068	0.076	0.589	0.378	0.305
	1.75	135	196.59	0.068	0.076	0.648	0.376	0.305
	2	135	193.393	0.068	0.077	0.701	0.374	0.303
	2.25	140	197.727	0.069	0.077	0.741	0.372	0.302
	2.5	140	193.031	0.069	0.077	0.788	0.37	0.302

As expected, the stack designed for the specified design power has a maximum power equivalent to the design power. The variation in the stack pressure does not affect the design/max power for this stack design, as the required achievable power determined for the GTOW of 2500 lb is constant. The cell efficiencies change slightly, and aside from the 1 atm stack case, the efficiency can be seen to decrease with an increase in stack pressure. The reduction in efficiency may explain why the SFC appears to increase with an increase in stack pressure, as more fuel would be needed by the stack to produce the same power. The cell power density increases with stack pressure, meaning the same amount of total power can be delivered by the stack using fewer cells.

The maximum efficiency stack design exhibits much more variation in its design and maximum powers, and the dependency of stack performance on its operating pressure is more noticeable. Higher pressure stacks appear to require higher design powers to reach the same maximum power. These variations in design and max power are further displayed in Figure 3 below. It should be noted that the maximum power is not exactly the same as the required achievable power, likely due to the previously mentioned 5 kW increment in design power used in the iterative design approach; a more accurate maximum power could possibly be achieved if the increment size was reduced. Because the stack is optimized for the lower design power, the cell efficiency at max power is lower than the efficiency at the lower design power across all stack pressures. Aside from the singular increase in the resulting max power and cell efficiency for the 1.25 atm stack, the cell efficiencies, SFC, and cell power densities at the design power follow the same trend as the design/max power stacks. However, for a given stack pressure, the maximum efficiency stacks have a lower cell power density compared to the alternative stack design option.

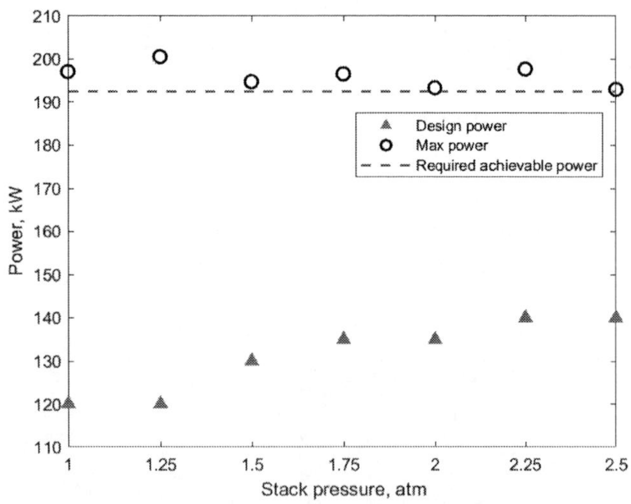

Figure 3. Variations in Design and Max Power for Max Efficiency Stack at Varying Stack Pressures.

The resulting system weights for the design/max power and max efficiency stack design cases for the original 2500 lb gross takeoff weight are shown in Tables 5 (a) and (b). Additionally, the variations in total and individual weights can be seen in Figures 4 (a) through (d).

For both design cases, the stack weight is reduced as stack pressure is increased, owing to the previously mentioned increase in cell power density. The increase in subsystem weight with stack pressure is attributed to the larger air compressor/turbine system and additional cooling system being needed to pressurize the air to stack pressure. The design/max power stack design exhibits lower stack weights and subsystem weights compared to the max efficiency stack design across all stack pressures. Likely due to the fuel weight being estimated based on the SFC at the max power of the stacks, and both stack design options are made to be able to achieve approximately the same max power, the fuel and tank weights are the same.

Table 5. Hydrogen Fuel System Weight Estimates for Variation in Stack Pressure.

(a) Design/Max Power Stack Design.

	Stack pressure	Stack weight	Fuel weight	Tank weight	Subsystems weight	Total weight	Net weight gain	New GTOW
Units	atm	lb	lb	lb	lb	lb	lb	lb
	1	199.83	19.6	241.72	41.03	760.18	110.18	2610.18
	1.25	196.35	19.27	237.62	59.7	770.93	120.93	2620.93
	1.5	189.26	19.29	237.96	79.18	783.7	133.7	2633.7
	1.75	180.32	19.33	238.4	99.03	795.09	145.09	2645.09
	2	172.92	19.45	239.92	118.43	808.72	158.72	2658.72
	2.25	169.51	19.48	240.27	136.94	824.21	174.21	2674.21
	2.5	166.55	19.51	240.61	155	839.66	189.66	2689.66

(b) Max Efficiency Stack Design.

	Stack pressure	Stack weight	Fuel weight	Tank weight	Subsystems weight	Total weight	Net weight gain	New GTOW
Units	atm	lb	lb	lb	lb	lb	lb	lb
	1	204.63	19.6	241.72	47.12	771.06	121.06	2621.06
	1.25	204.65	19.27	237.62	68.16	787.7	137.7	2637.7
	1.5	191.64	19.29	237.96	84.79	791.68	141.68	2641.68
	1.75	184.3	19.33	238.4	104.98	805.02	155.02	2655.02
	2	173.91	19.45	239.92	122.18	813.46	163.46	2663.46
	2.25	174.37	19.48	240.27	143.33	835.45	185.45	2685.45
	2.5	167.27	19.51	240.61	157.46	842.84	192.84	2692.84

As a reminder, the maximum gross takeoff weight of the notional helicopter is 2500 lb (1136.4 kg) including the engine, fuel, and tank weights of 650 lb (295.45 kg). The sizing estimates of the hydrogen fuel cell system are promising in their feasibility for the long-range mission. The net weight gain due to the replacement liquid hydrogen fuel cell system is under 200 lb (~90 kg) for both design options across all stack pressures, considerably lower than the estimated weight gain from the battery system.

System Sizing for Variations in Tank Weight Fraction

Because of its significant impact on the total weight of the system, the effects of reducing the tank weight are also briefly explored for a fixed stack design. The variation in tank weight and net increase in gross weight of the vehicle are estimated for the 1 atm, design/max power stack design, with the gravimetric weight fraction of the tank now varied from 5.5% to 15%. The low-pressure stack is chosen for analysis because it requires the largest amount of fuel. Figure 5 shows the results of the variation.

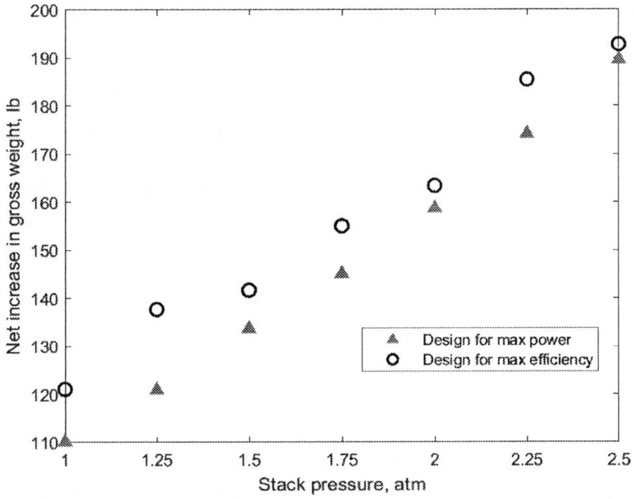

(a) Net Weight Gain.

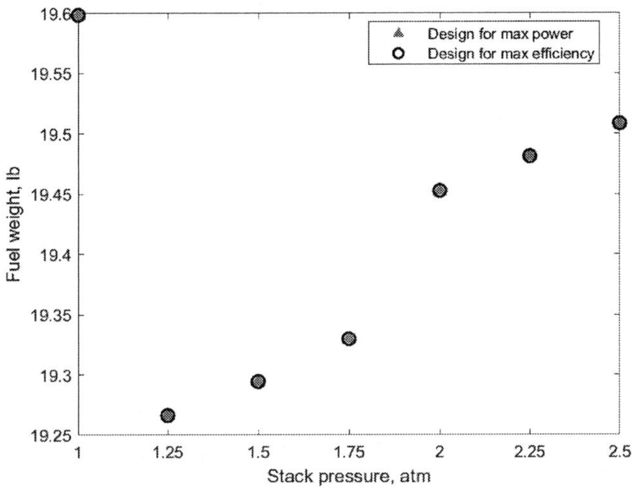

(b) Liquid Hydrogen Fuel Weight.

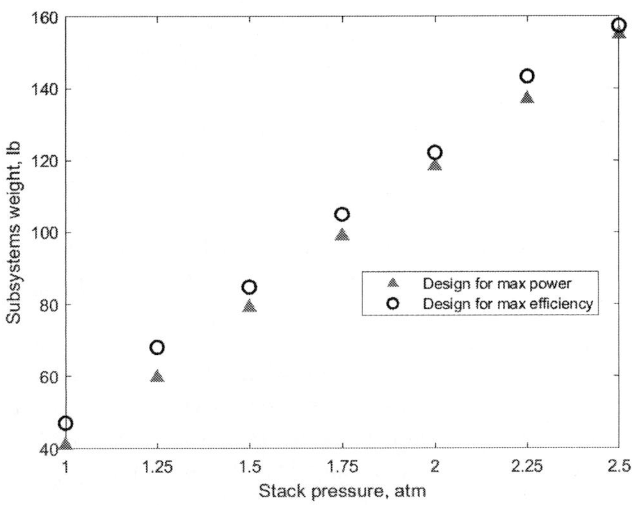

(c) Total Weight of Subsystems.

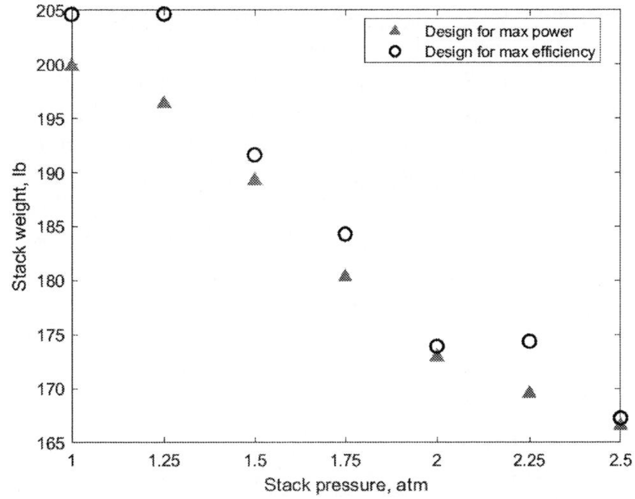

(d) Stack Weight.

Figure 4. Hydrogen Fuel System Weight Estimates for Variations in Stack Pressure.

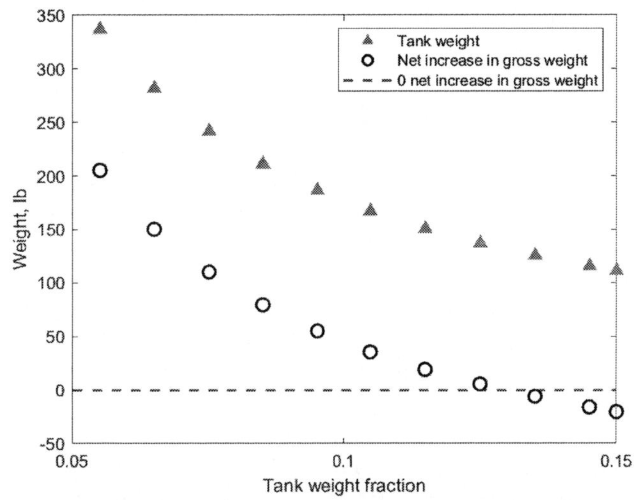

Figure 5. Weight Estimates for Variation in Tank Weight Fraction, 1 atm Max Power Stack Design.

At high enough tank weight fractions, there can be a net decrease in the gross weight of the vehicle, even for the hydrogen fuel system design that uses the largest amount of fuel. The tank weight fraction is dependent on the fuel storage conditions and the structure/composition of the tank. Fuel storage conditions include the pressure and temperature at which the hydrogen is stored, along with considerations of fuel boil-off within the tank over time. Low tank weight fractions limit the amount of boil-off allowed, while high weight fractions allow more room for boil-off. For shorter mission/flight times where fuel will not be stored for an extended period of time, a high weight fraction is feasible. In actual vehicle applications, the aforementioned factors affecting the weight fraction are also constrained by regulations/standards for hydrogen fuel storage.

Factors in Choosing a Hydrogen Fuel Cell System Design

With weight being the driving concern of the new fuel system, stack performance is the main parameter to consider for choosing a stack design. When holding the tank weight fraction constant and only considering the total weight of the hydrogen fuel cell system, the low-pressure stacks result in the lowest total system weight. A lower pressure stack may be the lightest option because of the lighter air management system and respective cooling system. The stacks themselves are lighter for the higher-pressure designs, but unless significant weight reductions can be made in their necessary compressor/turbine and cooling subsystems, the total weight of the entire system is higher.

Choosing between a design/max power and maximum efficiency stack design appears to require a more detailed analysis of energy and power requirements of the prescribed mission, especially for fuel weight estimations. Because the efficiency of the stack varies according to the power it is operating at — dependent on the flight mode of the helicopter — fuel consumption should be calculated based on the SFC at that power and the respective energy needs of the segment. For instance, in the 1 atm stack designed for max efficiency option, the stack operates at its maximum efficiency at its lower design power, 120 kW, where its SFC and fuel weight are therefore reduced. However, the individual power requirements of each mission segment are not the same as this design power, so unless the lower design power is within reasonable range of one of these power requirements, the stack will never operate at maximum efficiency and minimum SFC. For the 1 atm stack designed for design/max power, the efficiency at the lower/max design power is not maximized, so there is a possibility for it to operate at higher efficiencies and lower fuel consumption for a given power requirement of the mission; nevertheless, a more definite analysis would be required for the fuel estimates and comparison between these design options.

Overall, the implementation of a hydrogen fuel cell system for the inter-city mission would result in a new maximum gross takeoff weight of ~2700 lb for the vehicle, which is achievable with the improvements to rotor hover performance presented in the following section. Whether a chosen hydrogen fuel cell stack and system could be designed/implemented in actuality with similar performance characteristics would likely require experimental validation.

IMPROVEMENTS TO ROTOR HOVER PERFORMANCE

To reduce the energy needs of the helicopter, particularly in hover, improvements to the rotor performance under ISA + 20°C operating conditions, particularly in hover, have been considered. The following approach was used.

The NACA 63015 airfoil section used in the notional helicopter has been replaced in this study with a custom-designed low-drag airfoil shown in Figure 5, for optimum performance at a Reynolds number of 3 Million. The airfoil design methodology is discussed in detail in Ref. 18-21.

This airfoil is considerably thinner than the baseline 15% thick airfoil. The rotor structure must be redesigned as part of a detailed design to withstand the loads on the rotor. For single-engine helicopters, the resultant decrease in the polar moment of inertia of the rotor and the autorotative index is also of concern, although the proposed dual motor system would reduce the probability of both motors failing at the same time.

Figure 6 shows the 2-D Drag Polar of the custom airfoil compared to the baseline NACA 63015 airfoil at a Reynolds number of 3 Million at a Mach number of 0.45, computed using XFOIL. It is seen that the new airfoil has a pronounced drag bucket for a broad range of lift coefficient C_l values. The drag coefficient C_d is uniformly lower than the baseline airfoil. Classical blade element theory indicates that the average value of quantity $C_l^{3/2}/C_d$ determines the figure of merit of the rotor.

This improved performance is attributable to the target pressure distribution specified in the inverse design analysis as shown in Figure 7, allowing a favorable or near zero pressure gradient to occur over much of the upper surface. Based on Rayleigh's stability criterion, the reduction and postponement of adverse pressure gradients over the airfoil surface delays boundary transition to turbulent flow, reducing the profile drag.

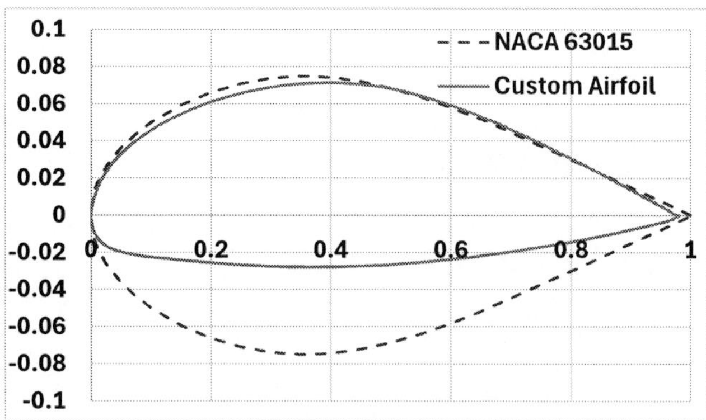

Figure 5. Laminar Airfoil Designed for Optimum Performance at a Reynolds Number of 3 Million.

The new two-bladed rotor has a rectangular planform identical to the baseline rotor. A linear twist of -6 degrees has been used. The tip Mach number has been set at 0.6. This rotor has been modeled in hover using the in-house 3-D CFD solver GT-Hybrid [22] at a collective pitch setting of 6.75 degrees.

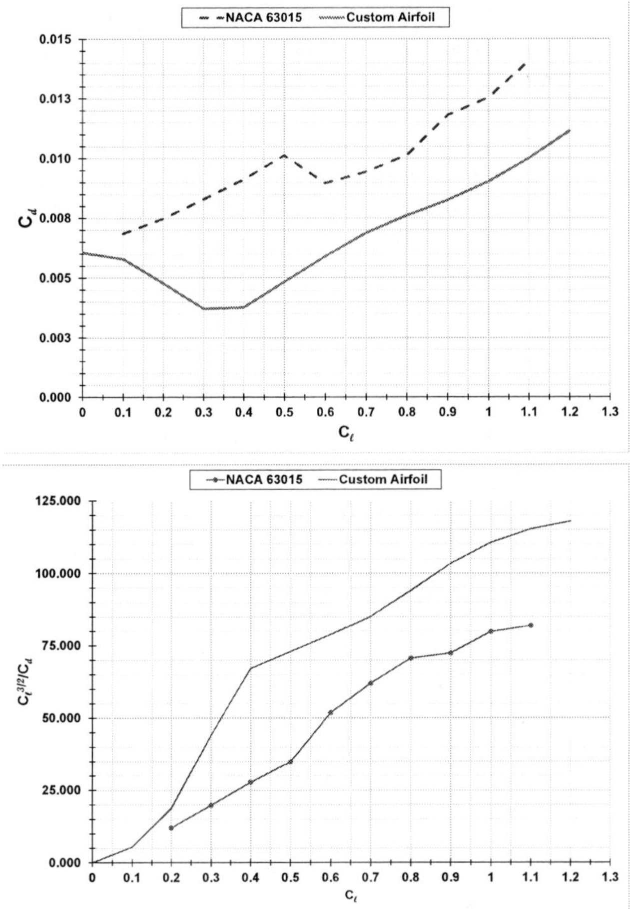

Figure 6. Comparison of the Drag Polars for NACA 63015 and the Custom Airfoil at 3 Million Reynolds Number.

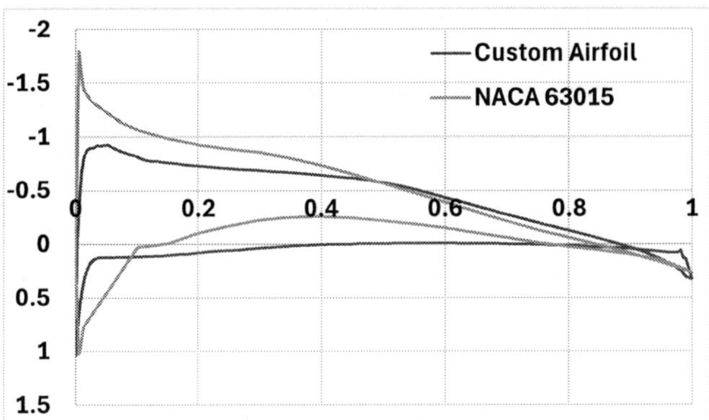

Figure 7. Comparison of Surface Pressure Coefficients for the Baseline Airfoil and the Custom Airfoil (Mach 0.45, C_l of 0.5, 3 Million Reynolds Number).

Table 6 shows the new rotor performance. If a reduction of ~100 lb in structural weight could be made with improved lightweight composite structural components or with higher energy density batteries, this rotor would be able to handle the air-taxi mission with a previously shown gross weight estimate of 2900 lbs.

Table 6. Aerodynamic Performance of the Rotor with Custom Airfoils.

New Rotor, 6.75 degree collective	Value	Units
Thrust Coefficient	0.00302	—
Power Coefficient	0.000172	—
Density, ISA + 20°C	0.002223	slugs/cubic ft
Figure of Merit	0.682287	—
Speed of Sound at ISA + 20°C	1137.553	ft/sec
Tip Mach number	0.615	—
Tip Speed	699.5954	ft/sec
Rotor radius	16.5	ft
Thrust	2809.891	lb
Power	203.5613	hp

CONCLUDING REMARKS

Results are presented for a notional helicopter that is based on an existing system (Robinson R-44), retrofitted with eVTOL technology. The feasibility of using Lithium-Ion batteries or a complete liquid hydrogen fuel cell system has been examined. The impact of the retrofit on the system weight has been examined. Aerodynamic analyses have also been performed with a rotor that uses low-drag airfoils to assess the feasibility of the eVTOL systems for inter-city and inter-city operations. The calculations indicate that Li-Ion and liquid hydrogen fuel cell-based systems may both be able to handle inter-city operations with a 200 km range, with improvements to aerodynamic design of the rotor. The range may be improved further, with a reduction in the vehicle drag (equivalent flat plate area), and a reduction in the airframe structural weight.

The simulations presented in this work did not take into consideration the volume within the fuselage needed to house the batteries or fuel cell system and the electric motors, and the impact of their placement on the vehicle center of gravity and static stability. It would also be necessary to perform a structural redesign of the rotor and the transmission to handle the increase in gross weight, and the changes to the rotor cross-section.

Author contact:

Dounya Jarrad djarrad3@gatech.edu

Lakshmi N. Sankar lsankar@ae.gatech.edu

Auraluck Pichitkul Auraluck.pichitkul@gmail.com

REFERENCES

1. Johnson ,W. ,Elmore ,J. ,Keen , E., Gallaher, A., and Nunez ,G. ,"Coaxial Compound Helicopter for Confined Urban Operations," Presented at the AHS Specialists' Conference on Aeromechanics Design for Vertical Lift, San Francisco, CA, January 20-22, 2016.

2. Johnson, W., and Silva, C., Solis, E., "Concept Vehicles for VTOL Air Taxi Operations," Proceedings of the AHS Technical Conference on Aeromechanics Design for Transformative Vertical Flight, San Francisco, CA, January 16-19, 2018.
3. Johnson, W., "A Quiet Helicopter for Air Taxi Operations" Presented at the VFS Aeromechanics for Advanced Vertical Flight Technical Meeting, San Jose, CA, January 21-23, 2020.
4. Silva, C., Johnson, W., "Practical Conceptual Design of Quieter Urban VTOL Aircraft," Presented at the Vertical Flight Society's 77th Annual Forum & Technology Display, Virtual, May 10-14, 2021.
5. Johnson, W., and Silva, C., "NASA concept vehicles and the engineering of advanced air mobility aircraft," The Aeronautical Journal, 126(1295), 59-91, October 2021.
6. Johnson, W. and Silva, C., "NASA Concept Vehicles and the Engineering of Advanced Air Mobility Aircraft," The Aeronautical Journal (2022), 126, pp. 59-91.
7. Datta, A., "PEM Fuel Cell MODEL for Conceptual Design of Hydrogen eVTOL Aircraft," NASA CR 20210000284, January 2021.
8. Datta, A. And Johnson, W., "Powerplant Design and Performance Analysis of a Manned All-Electric Helicopter, " Journal of Propulsion and Power, Vol. 30, No. 2, March–April 2014.
9. Pichitkul, A., Sankar, L. N., and Ariyarit, A., "Aeroacoustic Modeling and Design of Efficient Low Noise Rotors for eVTOL," The 12th TSME International Conference on Mechanical Engineering, December 2022, Phuket, Thailand.
10. Sankar, L. N., Gahlot, A., Costello, M., and Collins, K. B., "Retrofitting an Existing Helicopter with eVTOL Capabilities: Challenges and Opportunities," Proceedings of the European Rotorcraft Forum, June 2022.
11. Sankar, L. N., Costello, M., and Collins, K. B., "Low and High-Fidelity Approaches for Low Noise eVTOL Modeling and Design," Proceedings of the 56th 3AF International Conference on Applied Aerodynamics, Toulouse, France, March 28-30, 2022.
12. Sankar, L. N. and Tung, C., "A Combined Low and High-Fidelity Approach for an eSTOL Rotor Modeling and Design," Presented at the VFS 2022 Transformative Vertical Flight technical meeting, San Jose, CA, January 24-27, 2022.
13. Pichitkul, A., Tiandum, J., Sankar, L. N., and Tantrairatn, S., "An Engineering Study of Retrofitting a Piston Engine Powered Helicopter with Hydrogen Fuel cells," Proceedings of the Autonomous VTOL Technical Meeting and Electric VTOL Symposium, Mesa, Arizona, January 2023.
14. Pichitkul, A., Tiandum, J., and Sankar, L. N., "Design and Analysis of Efficient Low Noise eVTOL Rotors," The First International Conference on Advanced Air Mobility Systems, Bengaluru, India December 4-6, 2023.
15. Pichitkul, A. and Sankar, L. N., "Aeroacoustic Design and Modeling of eVTOL Rotors for Urban Air Mobility Application," Proceedings of the 10th Asian Australian Rotorcraft Forum, Nov 7-8, 2024.
16. Johnson, W., "NDARC Theory," NASA TP–20220000355, Vol 1, page 314, Table 31.1.
17. Duffy, M., Sevier, A., Hupp, R.. Perdomo, E.. and Wakayama, S.. "Propulsion Scaling Methods in the Era of Electric Flight," AIAA Paper No. 2018-4978, July 2018.
18. Ku, Michelle, Gahlot, Aishwerya, and Sankar, Lakshmi, N., "Application of an Inverse design methodology to Rotor Tip Design," Proceedings of the European Rotorcraft Forum, September 2024.
19. Malone, J., Narramore, J. and Sankar, L. N., "An Efficient Airfoil Design Method Using the Navier-Stokes Equations," Proceedings of the AGARD Specialists Meeting on Computational Methods for Aerodynamic Design (inverse) and Optimization, Loen, Norway, May 22-23, 1989.
20. Malone ,J.B., Vadyak, J. and Sankar, L.N., "Inverse Aerodynamic Design Method for Aircraft Components," Journal of Aircraft, Vol. 24, No. 1, pp. 8-9, 1987.
21. Santos, L. and Sankar, L. N.,"A Hybrid Inverse Optimization method for the Aerodynamic Design of Lifting Surfaces," AIAA Paper 94 1895.
22. Marpu, R., Sankar, L. N., Egolf, A., Hariharan, N., "Analysis of a Rotor in Hover using Hybrid Methodology," AIAA Paper 2014-0210.

Testing Battery Sensitivities of Electric Aircraft Systems Using a Battery Pack Surrogate

Grant Gothing
Chief Technology Officer
Bloomy Controls, Inc
South Windsor, CT, USA

ABSTRACT

Many eVTOL and electric aircraft systems are highly sensitive to battery performance, states, algorithms and behavior, which necessitates thorough testing. However, lab testing with real battery packs is not practical or desirable in most cases due to potential issues with safety, availability, cost, energy, and test coverage. A battery pack surrogate hardware-in-the-loop (HIL) test system may be used in place of the battery in order to safely and efficiently test many battery-sensitive aircraft systems across an extended range of battery conditions. This paper describes an 800V battery pack surrogate comprised of commercially available components including battery cell simulators providing over 200 cells of simulation, as well as signal and power IO, communications, cell models, a real-time controller, and battery management system (BMS). Important design considerations including safety, isolation, topologies, and interconnections are addressed, and applications for systems integration lab (SIL) and iron bird testing are presented.

INTRODUCTION

Electric and eVTOL aircraft are rapidly evolving, with many airframers pursuing certification within the next 1-3 years. For many of these aircraft, lithium-ion (Li-ion) battery packs are the primary energy source that provide power to the entire aircraft. Because Li-ion battery packs contain complex control electronics, they must be tested in a systems integration lab (SIL) to ensure seamless integration with other aircraft systems, including energy management, propulsion, flight controls, avionics, and charging. However, due to the volatile nature and high energy density of these packs, testing with real battery packs is difficult, time consuming, expensive, power intensive, and dangerous. An alternative test stimulus capable of exercising and/or "flying" electric aircraft systems in a manner that is equivalent to the real battery is strongly desired.

This paper introduces the battery pack surrogate as a means for safe and efficient validation testing of an aircraft's battery pack and control electronics as well as for testing how they interact with the rest of the aircraft's battery-sensitive systems. Because each aircraft, battery pack design, and test application is unique, each battery surrogate must be customized to meet the corresponding requirements.

Bloomy Controls, Inc. (BLOOMY®) provides hardware in-the-loop (HIL) and SIL test systems for Li-ion battery management system (BMS), line replaceable units (LRUs) and other units under test (UUTs). Based on our experience developing BMS HIL and battery pack surrogates for a wide range of industries, including electric vehicles as well as electric aircraft, we present the driving factors for requiring battery pack surrogates during SIL and iron bird test. Key considerations for battery pack signal simulation into the BMS from both hardware and software/modeling perspectives will be discussed as will the overall topology and design of a battery pack surrogate. Finally, several example test scenarios are presented as part of this paper.

BATTERY INTERACTIONS WITH ELECTRIC AIRCRAFT SYSTEMS

The battery pack, including the cell modules, cell monitoring units (CMUs), and battery management unit (BMU), is the energy storage system that all of an electric aircraft's electrical and propulsion systems rely on. As such, the battery pack plays a critical role in the performance and safety of the aircraft. The battery pack delivers all required power to the propulsion system, environmental system, vehicle management system, and in-aircraft electronics. The BMS provides the necessary monitoring and control circuitry and logic to ensure that the battery pack is operating safely and at top performance, and communicates the battery's states to many other aircraft systems.

The BMS monitors and controls the operating conditions of the overall pack as well as those of the individual cells. It is comprised of the BMU, CMUs, and disconnect circuitry. The CMUs are connected to the individual battery modules, monitoring cell voltages and temperatures, and controlling balancing. The BMU monitors the overall pack voltage, current, and

Presented at the Vertical Flight Society's 11th Biennial Autonomous VTOL Technical Meeting, Phoenix, AZ, USA, Feb. 4-6, 2025.

temperature, and communicates with the CMUs. It uses this information to estimate the pack's state of charge (SoC) and state of health (SoH), control safety contactors, charge and discharge enable relays, pack environmental factors such as heating or cooling, and output diagnostic trouble codes.

Because every electronic system relies upon the battery pack, its behavior is critical to the operation of the aircraft as a whole. Figure 1 shows the common aircraft systems which interact with the battery pack. Some systems, like the propulsion, charging, and DC/DC conversion are very tightly coupled with the battery pack, often receiving direct power, safety, and control signals from the BMS, as indicated by the red arrows. Other systems like avionics, flight controls, and navigation shown in blue may be less directly coupled, using the aircraft system power derived from the battery pack. However, these systems make decisions and manage aircraft controls based on battery pack information, including SoC, available capacity, or diagnostic trouble codes.

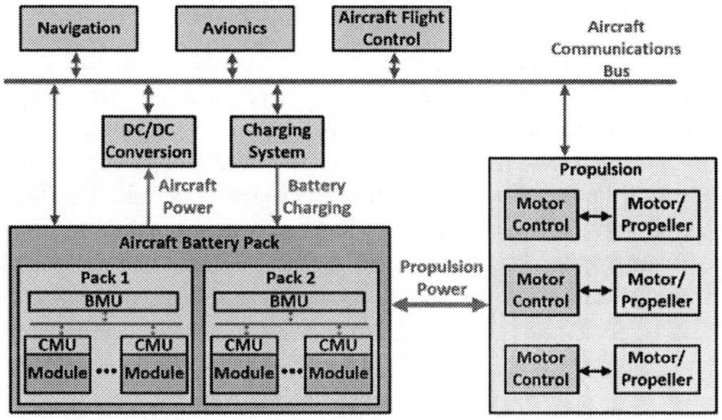

Figure 1. Battery-sensitive electric aircraft systems

AIRCRAFT INTEGRATION TESTING

Overview of Hardware in-the-Loop Test

Hardware-in-the-Loop (HIL) test is a method of validating the performance and functionality of complex control systems by simulating all of the conditions that the system is designed to monitor and control using real hardware. In HIL test, the control system is connected to a simulation system which simulates sensor signals under the control of a software-based model of the plant. The outputs of the control system, as well as other external systems, are measured by the HIL test system and fed back into the plant model, closing the loop of the stimulus and response cycle around the unit under test. HIL testing is often used to characterize and validate how the control system handles situations both within and outside of its normal operating envelope. Once an aircraft enters service, HIL test is further utilized to comprehensively test any revisions to the control systems' software in a process called regression testing.

For example, consider the BMS inside of a battery pack. HIL simulation is a particularly effective means of testing real BMSs and algorithms with simulated battery cells. Real cells can be hazardous, and because their characteristics change slightly with each charge/discharge cycle, they are not a repeatable test stimulus. Using HIL test for BMS development allows simulation of the behavior of multiple battery chemistries and simulates charge and discharge cycles based on the simulated or measured pack currents, balancing currents, and control signals from the BMS. As the mathematical model of the battery discharges and charges, the modeled SoC, cell voltages, and pack temperatures are calculated and simulated as analog signals back into the BMS, effectively making the BMS perform as if it were connected to a real battery pack. The inherent risk of real chemical cells is eliminated, and the test and the test environment is repeatable.

Overview of SILs and Iron Birds

A system integration lab, or SIL, is a test lab where multiple systems are connected together to perform integration testing, validating the individual behaviors and complex interactions between the systems. In a SIL, each control system is typically connected to its own HIL simulator running plant models and simulating signals corresponding to that control system's functionality. These control systems, with their simulated IO interact with each other as if they were in a real aircraft, allowing engineers to validate the overall system as a whole, across a wide range of operating conditions in a repeatable and highly configurable environment. Whereas SILs usually only include the electronic controls connected to simulated sensors, actuators and effectors, an "iron bird" (so-called for their tendencies to be heavily mechanical structures made from steel and other structural metals thereby being completely incapable of flight) is a specialized type of SIL which includes the full complement of an aircraft's flight control systems combined with real sensors, actuators and effectors. The real hardware may include aircraft actuators, interface hardware, flight control computers, avionics, propulsion systems, etc., along with whichever sensors are required for successful operation of the systems. Additional simulated sensors and loads close the loop around the control systems in the iron bird.

Airframers use both SILs and iron birds to validate system integration in different phases of development. SIL testing enables iterative test and debug of the embedded system software starting early in the aircraft development. In addition, SIL testing is more often focused on validating individual control systems and their interactions. Iron bird testing incorporates electrical controls, mechanical systems, sensors and actuators to validate overall integration at a more mature state of the aircraft. Iron birds and SILs rely on test systems designed to test one or more of the control systems within the airframe. With proper design, SILs and iron birds are modular in nature such that the individual control systems may be tested independently, then connected together with other control systems in any appropriate combination. This affords the control development teams the ability to develop their control systems independently of one another and at different paces. After suitable confidence is gained in the independent development environments, the systems are usually connected to each other, both between the simulations as well between the control systems themselves, for system integration testing. If designed in an appropriately modular fashion, the development stations for a single-channel of a multi-channel control may also be connected together to test the control system in its duplex or triplex configuration.

CHALLENGES OF TESTING WITH REAL BATTERIES

Battery packs for electric aircraft contain hundreds of kilowatt-hours of energy, with the ability to instantaneously provide over 400 amperes at 800 volts. This immense power and capacity, while critical to aircraft performance, poses safety risks when testing aircraft systems with real battery cells. Packs are configured in complex topologies of series and parallel cells, making it difficult to set up very specific test scenarios. Furthermore, this energy is stored in a chemical form which does not return to its exact starting state after a discharge/charge cycle, thus adding repeatability of test to the list of challenges of testing with real battery cells.

Dangers of real battery testing

Using real battery packs involves significant risks because of the high amount of chemical energy contained within them, most notably the risks of electrical hazards and thermal runaway. These risks are heightened in validation testing which involves exercising the pack outside of normal operating conditions and which is often experimental in nature. The high voltage and current of the packs present risk of serious electrical hazards, such as electrical shock, short circuits, and arc flash, which can arise from improper handling, incorrect connectivity, faulty equipment, or unvalidated safety hardware and algorithms. The thermal energy available in the pack presents the possibility of thermal runaway, another critical safety concern for test with real batteries. Thermal runaway occurs when a cell overheats and triggers a positive feedback reaction, releasing energy which in turn further heats the cell and may spread to adjacent cells. This results in a rapidly increasing temperature and cascading effects resulting in fire or explosions. Without proper safety controls, the edge conditions tested during validation, including imbalances, fast charging, over charging, high power, short circuit or environmental tests, can all contribute to the risk of thermal runaway. For these reasons, many test labs restrict the use of real chemical cells for any electrical or algorithmic testing. When allowed, test labs that utilize real battery cells must be properly equipped for chemical fire suppression.

Configurability challenges of real-battery testing

Aircraft battery packs contain hundreds of cells in series, and dozens of cells in parallel, providing high voltage, high current, and high capacity. Preparing for a particular test case based on initial capacity and overall pack SoC can take hours to prepare on a real pack, requiring precise charging and discharging of the pack which takes significant time and energy. Setting parameters such as cell, module, or string imbalances can take hours or even days, if possible at all. Consider the imbalanced pack of 9 modules shown in Table 1 where Module 8 is stronger, and Module 9 is weaker than Modules 1-7.

Table 1. Imbalanced pack configuration parameters.

Parameter	Module 1-7	Module 8	Module 9
Initial Capacity	25kWh	28kWh	20kWh
Initial SoC	90%	85%	98%
Initial Temp.	25°C	22°C	38°C

Setting this exact configuration could entail days of setup, requiring specific capacity setup (including adding or removing cells), SoC setup (requiring charging each module individually and maintaining that state until the test is ready), and thermal setup (heating or cooling each module individually). Take this a step further to individual cell imbalances, or to dangerous edge conditions such as over-charged or over-temperature cells and modules, and the configuration is highly dangerous and virtually impossible to accurately create.

Repeatability challenges of real-battery testing

In considering the challenges associated with configuring a particular test scenario, one can imagine the additional difficulty in precisely repeating the same scenario multiple times. The chemical properties of real batteries change slightly with every charge and discharge cycle, and these small perturbations in initial conditions could have a large impact on final test results. Creating the same test conditions multiple times would require replacement of older cells with newer cells and could add significant cost and days to weeks to overall test time. All the while, the chemical properties of real batteries change slightly with every charge and discharge cycle. Real batteries are simply not a repeatable, consistent test stimulus for aircraft systems testing.

THE BATTERY PACK SURROGATE

Because of the hazards and inefficiencies of testing with the real battery pack, a battery pack surrogate should be used in its place. The battery pack surrogate is a type of HIL test system which mimics the complete functionality of the real battery pack by integrating the real BMS with simulated sensors and cells into a closed-loop simulation. It may also contain a high-power supply to provide substantial power needed to drive other aircraft systems, depending on the application of the battery pack surrogate. The battery pack surrogate allows design and validation engineers for aircraft systems to interface with a simulated battery pack that behaves just like the real battery pack, but without the aforementioned issues. This replacement for testing with a real battery pack can dramatically improve safety, accelerate testing time, and help reduce and optimize precious flight test hours. Engineers can test the effect of hardware, firmware, and algorithm changes on the state of charge, state of health, and hazard conditions of the battery. Conversely, engineers can test the effects of initial conditions, such as cell, module, and pack capacity, SoC imbalances, thermal conditions, and states, on the BMS and other aircraft systems.

The battery pack surrogate provides the ability to instantaneously change states and initial conditions for complex tests that would otherwise be time consuming, dangerous, or difficult to test on a real battery pack. Fault conditions, such as a single weak cell, unbalanced cells or modules, broken wires, insulation resistance breakdowns, welded contactors, sensor faults, or extreme temperatures, can readily be implemented in a matter of seconds. Test scenarios, including actual pre-recorded flight data or engineered missions, can be played back into the battery pack surrogate to tune algorithms and system responses to BMS signals, including pack state of charge, diagnostic trouble codes, contactor states, and health alerts. Furthermore, while this simulation is taking place, all channel and communications data can be logged in real-time for post-processing and correlation of system events with battery behaviors.

In the next sections, an overview of a typical electric aircraft battery pack will be presented and then the major components of a battery pack surrogate will be discussed.

Overview of an electric aircraft battery pack

To understand the battery pack surrogate, we must first summarize the details of a typical electric aircraft battery pack. The aircraft battery pack is comprised of thousands of cells, each providing a typical nominal voltage between 3.0V and 4.0V and a capacity between 10Wh to 100Wh. In the battery pack, these cells are organized into series and parallel configurations to achieve the desired voltage, current, and capacity. A typical aircraft battery pack connects between 192 and 250 macrocells in series to generate the 800V pack voltage, with each series-connected macrocell being comprised of dozens of individual battery cells connected in parallel to store hundreds of amp-hours of capacity. Furthermore, where electric traction vehicles will typically disconnect all battery outputs in a fault scenario, most electric aircraft batteries incorporate redundant strings or packs. In a traction EV, disconnecting the pack means that the vehicle loses propulsion and eventually stops. In an aircraft, however, the propulsion loss from disconnecting the entire pack could be catastrophic. Redundant packs are used to ensure that if one pack fails, the aircraft can safely land under reduced power.

Due to the high nominal pack voltage, typically 800VDC, electric aircraft battery management systems employ a distributed BMS topology. This topology is optimized for large, complex, high power applications. While many variations of overall pack topology are used throughout the industry, this discussion will be confined to a basic, high-level pack topology. Figure 2 shows one such topology. This distributed BMS topology groups the entire stack of battery cells into smaller modules. Generally, a module is a convenient subdivision of the overall battery pack which satisfies the distributed topology of the BMS. For instance, a pack of 216 cells in series may be comprised of eighteen 12-cell modules.

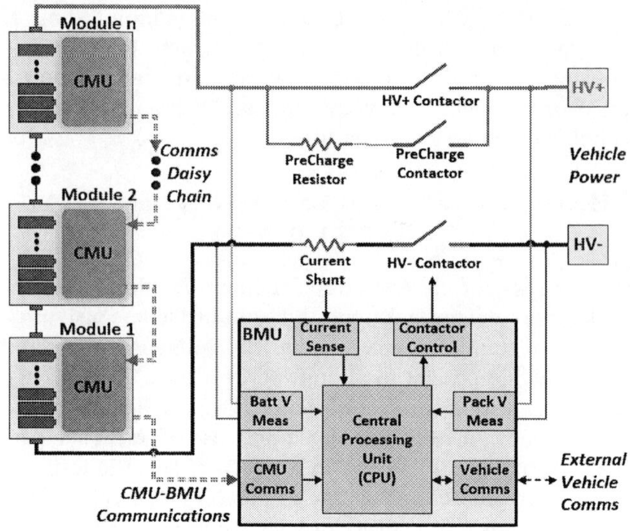

Figure 2. Battery pack block diagram

Because of the high-power density, high volatility, and relatively flat charge curves of lithium-based batteries, advanced monitoring and control electronics are critical to their performance and safety. Each module is equipped with a cell monitoring unit (CMU). The CMU is an electronic assembly which accurately measures cell voltages and temperatures, detects cell-level faults, and controls individual cell balancing, as shown in Figure 3. The CMU only monitors and balances the series-connected cells within the module, as parallel cells only add capacity and cannot be individually measured. Data from each module is reported back to the battery management unit (BMU), over an isolated digital communications interface such as isolated SPI, UART, or CAN. This communications interface is typically daisy-chained from one module to the next, and back to the BMU.

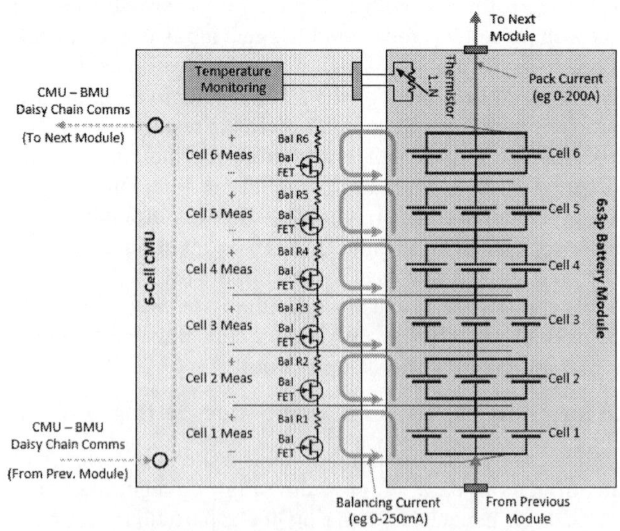

Figure 3. Example module and CMU

The BMU, shown in grey in Figure 2, is responsible for pack-level monitoring and control. The BMU aggregates all cell data reported by CMUs, measures battery and pack voltage (before and after the pack contactors, respectively), pack current, pack temperatures, discrete control and communications, and other pack signals. Because of the high-voltage, high-current nature of the BMU voltage and current input signals, the BMU often employs an analog front end (AFE), shown as red blocks in Figure 2, to perform signal conditioning before feeding measurements to the analog-to-digital converters (ADCs) on the board. The AFE scales the high-voltage signals on the *Batt V Meas* and *Pack V Meas* inputs from 0-1000V down to the signal level required by the ADC, such as 0-10VDC. If a shunt is used for current measurement, the output is typically in the range of ±150mV. The *Current Sense* AFE scales this up to the a higher voltage range for the ADC, such as ±150mV, improving measurement resolution. The BMU uses this information from the cell and pack voltages, currents, and temperatures to calculate overall SoC, SoH, and remaining capacity using mathematical models of the battery. It also executes safety algorithms to control disconnects (either solid-state transistors or mechanical contactors), controls heating and cooling, and reports all battery information back to the rest of the aircraft via digital communications. In redundant electric aircraft systems, the entire pack is often duplicated or triplicated and, with a separate DC power aggregator unit, provides the multi-pack connectivity to the aircraft.

Anatomy of the Battery Pack Surrogate

A complete battery pack surrogate for integration testing employs the real LRUs and control electronics within the pack while using instrumentation to simulate the analog signals that are monitored and controlled by these components. An example battery pack surrogate test system is shown in Figure 4. In this system, the BMS components are located within the system's enclosure with commercial, off-the-shelf instrumentation providing the simulated signals into the BMS.

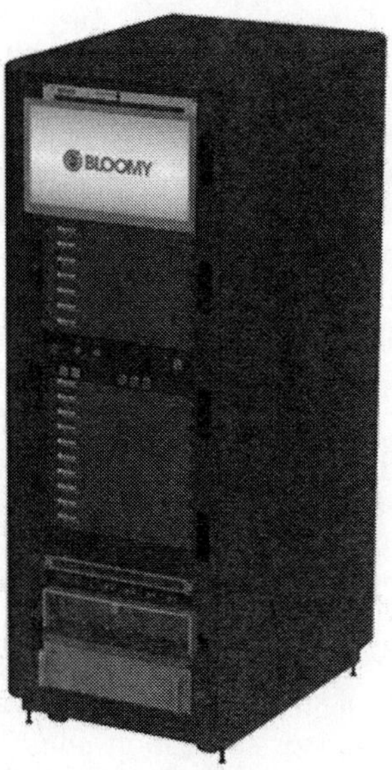

Figure 4. Example 800V battery pack surrogate

The following sections will walk through building up the battery pack surrogate physical hardware from the individual cell simulators to the external aircraft-accessible signals.

CMU/MODULE SIGNAL SIMULATION

CMU signal simulation entails replacing the cell, thermistor, and other external sources monitored by the CMU with programmable simulated signals as shown in Figure 5. A typical CMU monitors and reports cell voltages and temperatures of a single module back to the BMU via a daisy-chained communications interface. All CMUs must be present and communicating to the BMU so that an alarm or fault is not triggered by the BMU safety algorithms. The CMU also controls the balancing of cells as shown in the 6-cell example in Figure 3. While a module will often have many cells in parallel to add capacity, the CMU only monitors the series-connected cell voltages. Furthermore, full module current does not typically pass through the CMU itself, only between the cells and load. Therefore, simulating the cells to a CMU requires only simulating the individual series-connected cell voltages with enough power to satisfy the passive- or active-balancing circuitry. Full pack power is also not required in the CMU cell simulation, as capacity will be simulated with modeling, as discussed later. Temperature monitoring is typically performed by the CMU via thermistors, which can be simulated using either analog voltages or programmable resistors.

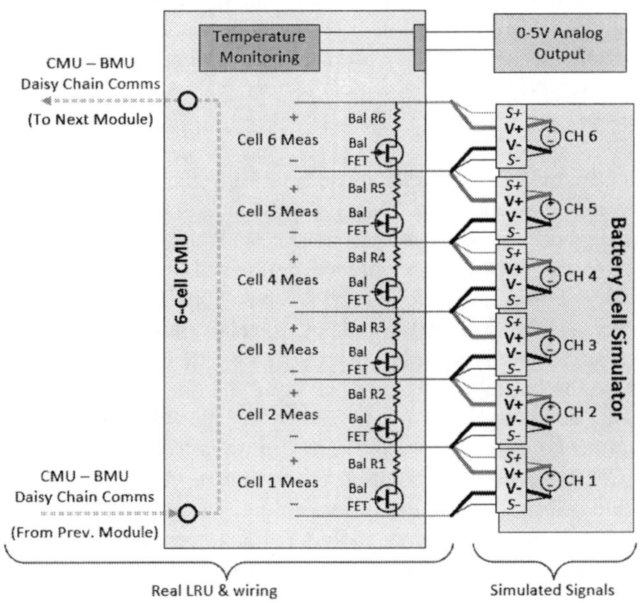

Figure 5. Signal simulation into the CMU

CMU – Cell Voltage Simulation

Just as cells are the fundamental elements of the battery pack, cell simulators are the fundamental building blocks of the battery surrogate. Each series-connected cell that is monitored by the CMU must be simulated using an isolated, high-accuracy voltage source. For instance, comparing Figure 3 to Figure 5, the six series-connected cells would be replaced with individually programmable simulated cells. The three cells in parallel for each series cell only add current capacity and do not need to be individually simulated. Most lithium-based battery chemistries provide an individual cell voltage between 2.5VDC and 4.25VDC, with under-voltage and over-voltage conditions extending to 0V and 5V, respectively. Therefore, the cell simulators should be able to programmatically control individual cell voltages from 0-5VDC.

Output current from each cell must be high enough to satisfy the monitoring and balancing circuitry of the CMU. Nominal CMU current draw is typically on the order of 5-20mA while individual balancing currents may range from 20mA-500mA for passive balancing applications, and higher for active balancing. Cell simulators must be able to source or sink enough current in each individual cell to satisfy CMU nominal current draw plus individual balancing currents.

Accuracy and resolution of the simulated voltage output and current readback are critical. The relatively flat SoC/voltage curves associated with lithium-based batteries means that small variances in voltage can result in large inaccuracies in the SoC estimation algorithms. Most commercially available cell monitoring integrated circuits have a total measurement error of around 2-3mV. Therefore, cell simulators should have a voltage output accuracy of ±1mV or better, with resolutions of 500μV or less. Furthermore, high accuracy current readback is critical when testing balancing algorithms or for feeding balancing current back to mathematical models for overall SoC simulation. Current readback accuracy of ±1mA is generally sufficient for most battery pack surrogate applications.

Finally, because the cells are series-connected at the module, and modules and their CMUs are further stacked in series to create a pack, simulated cells must be galvanically isolated to the highest pack voltage potential connected to a CMU, which could be the full pack voltage of over 800VDC to ground. When selecting a cell simulator, special attention must be given to the type of isolation specified by the instrument. Because the battery pack surrogate could be operating continuously at voltages of up to 1000VDC, the cell simulators should be rated for 1000VDC reinforced insulation per IEC61010 safety for laboratory instruments. Basic isolation, while offering protection of live voltages under normal operating conditions, requires additional safety precautions to ensure user and safety for continuous operation. Reinforced insulation offers a higher level of protection such that even if a single failure occurs, all non-isolated equipment and operators will be protected.

Figure 6 shows the Battery Simulator 1201, a commercial battery cell simulator which provides 12 channels of 0-5V, ±500mA individually programmable cells. Each cell channel of the simulator is individually programmable and has 1000VDC of reinforced isolation to ground. Cell channels across many cell simulators may be stacked in series to simulate all the cells in a module or pack.

Figure 6. Commercial battery cell simulator

CMU – Thermistor Simulation

The majority of CMUs monitor module temperatures using thermistors, either onboard or externally via connector interfaces. Most thermistor monitoring circuits are simple voltage dividers, with the variable resistance thermistor placed in series with a fixed resistor to pull-up to 3.3V or 5V or pull-down to board ground. The CMU will then monitor the voltage at the junction between the thermistor and fixed resistance. Therefore, a thermistor may be simulated by using a variable resistor in place of the thermistor or by providing an analog voltage corresponding to the voltage divider output associated with a particular temperature. Selecting the method of thermistor simulation depends on the type of test fidelity required. Generally, analog voltage-based simulation reduces cost and is sufficient for any algorithmic or behavioral testing associated with a battery pack surrogate. If full hardware testing, including validating the CMU pullups is required, then programmable resistance-based simulation should be selected. In either case, the battery pack surrogate software environment must have a scaling mechanism to match the nonlinearity of thermistors. In most cases, a simple interpolation table from temperature to voltage or temperature to resistance is sufficient and can be derived from the real thermistors' datasheets.

Cell Stack Simulation Topologies

The stack of cells of high-voltage battery packs may be simulated using one of three common methodologies, discussed in the following sections. For illustrative purposes, a 216-cell stack with eighteen 12-cell CMUs will be used, though the concepts can be adapted to other common high-voltage distributed BMS topologies.

The first methodology is *full stack simulation*. In this configuration, shown in Figure 7, all 216 cell voltages of the stack are simulated using eighteen cell simulator instruments, each containing 12 individually programmable cells connected to all 18 CMUs of the BMS. All simulated battery cells are connected in series on each CMU, and all CMUs are connected in series, as they would be in a real pack. In this configuration, full stack voltage may be provided directly into the BMU "Batt V+" measurement terminal. Because CMUs are raised to high voltage, any additional signal simulations or measurements, such as thermistor simulation, must be fully isolated from ground, since they are typically referenced to the bottom-most "Cell 0" signal on the CMU.

The advantage of full stack simulation is that it most closely resembles the actual battery pack configuration by providing the highest-fidelity environment for validation testing at full voltage. In addition, it allows individual cell voltage and balancing current control and monitoring. The main disadvantage is that it requires a large number of high-voltage connections over 100VDC and up to 1000VDC, which may be restricted or heavily regulated in some laboratory environments, as per the NFPA 70E standard (2024) (Ref 1), Table 130.4(D)(b). However, increased safety may be provided by implementing a number of safety precautions within the HIL simulator and supporting equipment, as outlined by NFPA 70E and BLOOMY's application note on using cell simulators in high-voltage applications (Ref 2). These precautions include setting the simulator cell current limits to the minimum required for powering the CMU and exercising its balancing circuitry, restricting access to all high-voltage connections when power is applied using safety enclosures, and implementing electrical hardware interlocks and lockout/tagout systems to ensure the control systems and instrumentation are de-energized when operators have access to them.

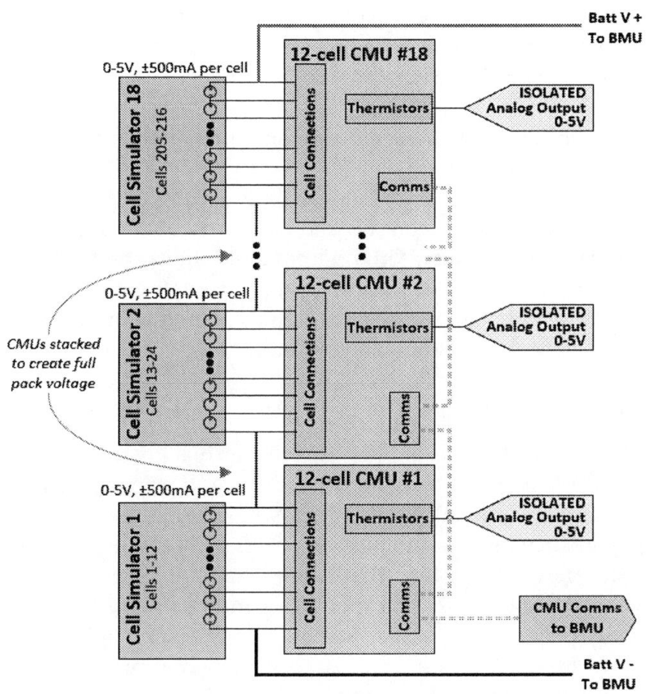

Figure 7. Full stack cell simulation

The second methodology, *low-voltage full cell simulation*, shown in Figure 8, also involves individually simulating all 216 cells into all 18 CMUs. However, rather than stacking all CMUs in series as in a real battery pack, each CMU's "Cell 0" (lowest voltage cell input) is referenced to system ground. In this configuration, the simulated voltages will never exceed a single CMU stack (5V/cell x *n* cells per CMU). Because the CMUs are not stacked to the full pack voltage, a separate high-voltage source must be used to provide the pack voltage to the BMU's "Batt V±" measurement terminals. Note that the high voltage source may be set to a very low current (i.e. milliamps) limit in order to minimize power. This configuration maximizes safety by reducing the number of high voltage connections while still maintaining independent control and monitoring of all cells in the pack. Furthermore, additional signal simulators, such as thermistors can remain non-isolated or single-ended which often saves equipment cost.

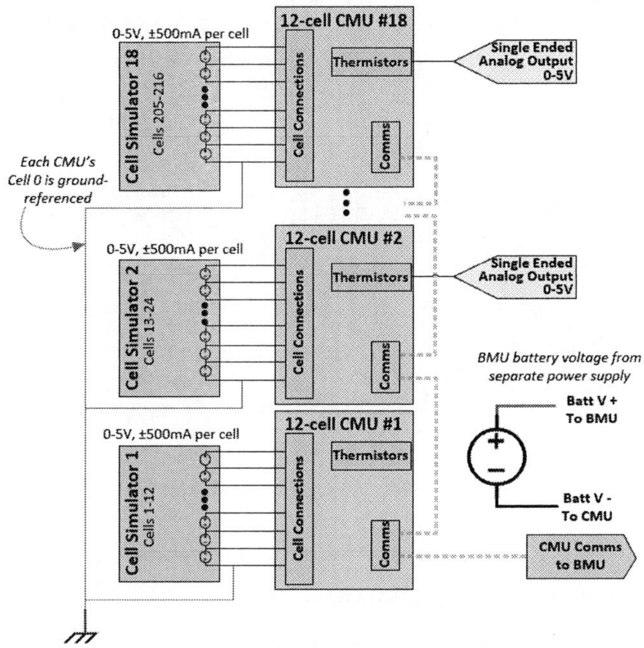

Figure 8. Low-voltage full cell simulation

Shown in Figure 9, *low-voltage shared cell simulation* minimizes the number of cell simulators by sharing simulated cells across multiple CMUs. Like the low-voltage full cell simulation, this configuration maintains a low voltage cell simulation system by ground-referencing the CMU's lowest-potential cells. However, in this configuration a single cell simulator may be shared across two or more CMUs connected in parallel. The primary limitation is that the shared cell currents must not exceed the cell simulator instrument's maximum current per cell. For example, if one CMU balances at 200mA per cell, then three CMUs in parallel may require up to 600mA, potentially exceeding the current limit of the cell simulator instrument. The advantage of this configuration is that it reduces overall system cost by minimizing the number of cell simulators, though at the expense of slightly more complex control software. The major disadvantage is that sharing simulated cells across multiple CMUs limits the individual voltage simulation and balancing measurements associated with full cell simulation.

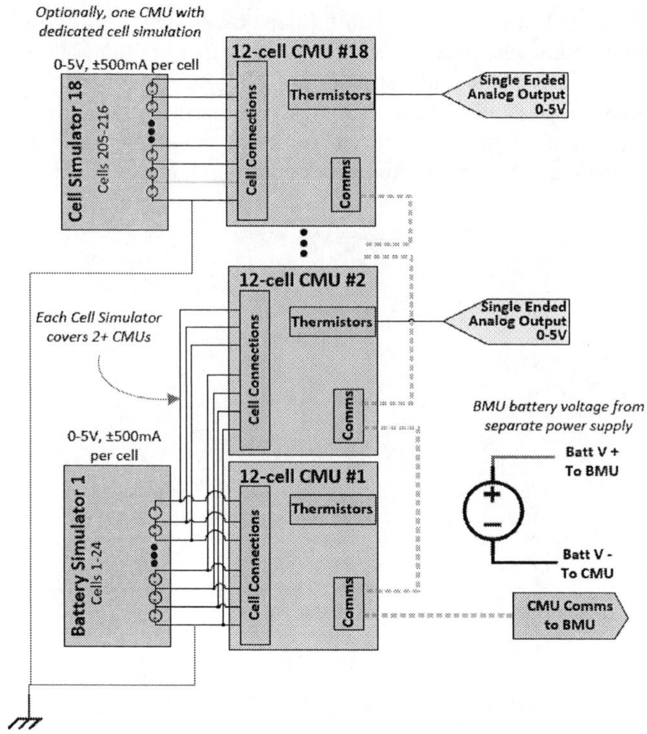

Figure 9. Low-voltage shared cell simulation

CMU to BMU Communications

The CMU to BMU communications is critical to BMU functionality. Without continuous communications to the CMUs, the BMU will fault. When a battery pack surrogate application uses the full complement of real CMUs, no direct simulation of the CMU to BMU communications is required since the CMUs' communications can simply be connected to the BMU as it is in a real battery pack. This provides the only authentic implementation of the BMU to CMU communications interface, matching that of the real battery pack.

One other approach to simulating the CMUs involves removing physical CMUs entirely and using digital communications simulation to replicate the communications from the CMUs to the BMU, only. This approach has a benefit of reduced hardware associated with CMUs, including cell and thermistor simulators and connectivity. However, this is at the expense of highly customized CMU to BMU communications simulation software, as well as loss of authenticity. Communications between the CMUs and BMU is most often a proprietary version of isolated SPI, though isolated UART, ethernet, or CANbus may also be used. To simulate this communications interface in a battery pack surrogate requires custom FPGA-based software to emulate the state machines, clock signals, and tight query/response timings associated with the proprietary communications and cell monitoring integrated circuits on the CMU. While this method does reduce instrumentation costs and space, it is not an authentic representation of the control systems inside the BMS, and may inject unintended faults, or could result in masking real-world faults. For that reason, if simulated CMU to BMU communications is to be employed, it is recommended that this methodology be reserved only for a subset of the battery pack surrogate, in tandem with real CMU-based simulation methodologies. The preferred approach is to implement at least one of the redundant packs of the battery as real CMUs with simulated cells and thermistors, while the other pack's CMUs are simulated via communications.

BMU/PACK SIGNAL SIMULATION

Where the CMU monitors module and cell-level signals, the BMU monitors pack-level signals. While the CMU is generally only monitoring and reporting cell and module data and controlling simple cell balancing, the BMU is aggregating cell- and pack-level sensor data and is running complex algorithms to make decisions about safety and performance. The total number of signals monitored and controlled by the BMU is fewer than the CMUs, but the variety is larger. A BMU will monitor multiple high-voltage measurement points, overall pack current (either via a current shunt or Hall effect sensor), environmental and control signals, as well as communications to the CMUs and to the aircraft. The BMU also has a larger variety of output signals than the CMU, including control of multiple contactors or disconnects, heating and/or cooling, and aircraft/charger interactions.

At a high level, BMU signal simulation entails replacing all pack-level interactions, such as those shown in Figure 2 with simulated signals. Many of these signals, such as thermistors, discrete input/output control, and environmental control can be simulated and monitored with simple digital and analog inputs and outputs. The most critical and complex simulations to the BMU are the overall pack voltage, current, and contactor/disconnect simulation and monitoring.

BMU – Pack Voltage/Current/Contactor Dependencies

A typical BMU will monitor two to four pack voltage-level (i.e., 800VDC) touchpoints, depending on the contactor/disconnect arrangement. Figure 2 shows the most common arrangement, with pre- and post-contactor measurement points, often referred to as the "Batt" and "Pack" voltages, respectively, though many other topologies may be used. In a real battery pack, "Batt" is the internal battery stack voltage and should match the sum of CMU cell voltage measurements. "Pack" measures the external voltage out to the aircraft. "Batt" and "Pack" may differ as a function of contactor state, pack current, and the resistance associated with the current shunt, contactors, precharge resistors, and cabling. When contactors are open, the "Pack" will be close to 0V, whereas when the contactors are closed, "Pack" should be close to "Batt." As the aircraft discharges the battery, "Pack" will drop with respect to "Batt" based on the current flow and series resistance. Conversely, when charging, "Pack" will be higher than "Batt." The BMU monitors pack current and adjusts the expected voltages for "Batt" and "Pack" as a function of the expected system resistances. If there is a discrepancy between the expected values and measured values, the BMU should flag a fault, such as excessive contactor resistance. Due to the tightly coupled nature of these variables, simulation of the voltages, current, and contactor conditions must be implemented as a unified and cohesive simulation.

Because of the high-voltage, high-current nature of the BMU input signals, the BMU employs the AFE to perform signal conditioning that scales the high-voltage and high-current measurements at the connectors to signal levels compatible with the analog-to-digital converters on the board. Therefore, test system engineers must make a critical decision at the outset of battery pack surrogate design: test at high voltage and high current using the complete AFE at full power; test using a mix of high-power and low-power simulations which also use the complete AFE; or create a low-voltage system that simulates high voltage and current signals by tapping into the BMU behind the AFE, bypassing the AFE entirely. The three most common implementations are discussed in the following sections.

BMU Pack Simulation – Full Power, Pre-AFE

In this implementation, shown in Figure 10, a high-power bidirectional power supply is used to emulate the power associated with the real battery pack. The power supply can be specified for the full power (voltage and current) of the real pack, or for derated power based on the limits of the overall system integration test requirements.

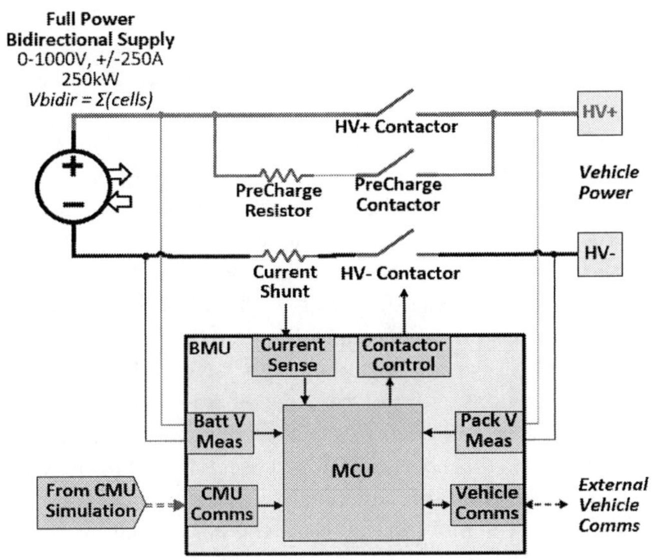

Figure 10. Full power, pre-AFE simulation

The BMU monitors the battery voltage as the output of the bidirectional supply. The full power from the bidirectional supply is routed through the real current shunt and contactors that would be included in the pack and which are controlled directly by the BMU. The post-contactor pack voltage is monitored by the BMU as in the real pack, then output to the outside world via the pack connections, denoted HV+ and HV- in Figure 10. In this way, the BMU monitors all high voltage and current through the onboard AFE signal conditioning, and manages the real contactors for control of the full power outputs. This power may be connected to the external aircraft systems to provide full battery power to propulsion and other high-power consumption systems.

To satisfy the BMU safety algorithms, the bidirectional power supply voltage must match the sum of the cell voltages simulated into the CMUs and reported back to the BMU. This implementation, paired with the full stack cell simulation methodology, most closely mimics a real-world battery pack. It includes all of the real CMUs, the BMU, and high-power control and monitoring electronics, such as the shunt and contactors.

BMU Pack Simulation – Reduced Power, Pre-AFE

While the prior implementation is the most representative of a real battery pack, it has limitations in the types of faults that may be simulated into the BMU because of the inherently coupled nature of the single power supply's voltage and current. It requires hard-wired contactor and power connections, which limits the ability to change contactor configurations, or simulate stuck contactors, or resistance buildup faults on the fly. This limitation can be remedied using the low-power (but still high voltage) pre-AFE simulation implementation shown in Figure 11.

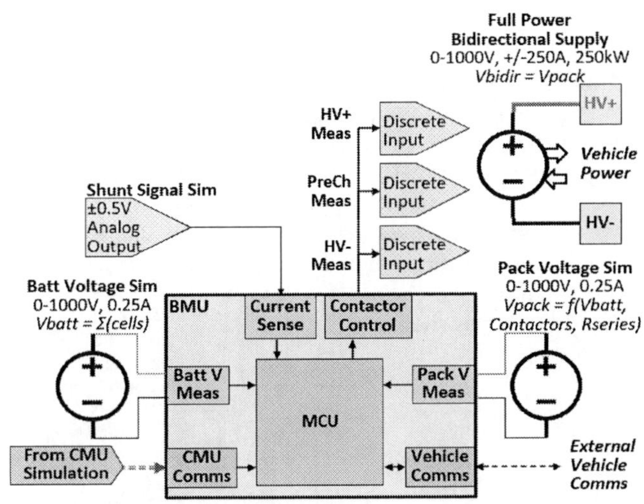

Figure 11. Reduced power, pre-AFE simulation

In this simulation methodology, the signals measured by the BMU are simulated using lower power instrumentation connected to the standard BMU connectors, thus using the BMU's AFE. Battery and pack voltages are provided to the BMU using high-voltage, low-current sources with only enough current to satisfy the measurement signal conditioning in the BMU's AFE. Pack current is simulated using an analog output voltage from the test equipment which is scaled to match the corresponding shunt or Hall effect sensor's output. Contactor control signals from the BMU are monitored using discrete input measurement circuits in the test instrumentation. By monitoring the contactor control signals from the BMU, the HIL test system can use software models to control the "Pack" voltage simulation as a function of the simulated battery voltage, pack current, and simulated characteristics of the current shunt, wiring, and contactors.

Just as with the full power simulation, "Batt" voltage must match the sum of CMU cells to satisfy the BMU. However, this configuration has the advantage of enabling simulation of edge conditions, such as contactor short faults or contact resistance buildup, through individual control of the "Batt," "Pack," and current simulations as a function of the simulated pack voltage and current. For example, if multiple different contactor topologies are to be considered, this methodology also enables the various setups to be modeled in software. In this way, the software can monitor the contactor control signals from the BMU and use contactor topology software models to control the simulated high current and high voltage signals in a more granular fashion. Furthermore, this configuration has the advantage of simulating all BMU I/O at reduced power, though not necessarily at low voltage. If desired, a separate full power bidirectional supply may optionally be used to provide battery power to propulsion, charging, and other aircraft systems. If utilized, the voltage of the bidirectional supply must be commanded by the battery pack surrogate HIL system to match the pack voltage simulation signal, and the current supplied must be monitored and used to control the pack current simulation.

BMU Pack Simulation – Low Voltage, Post-AFE

The pre-AFE simulation solves issues associated with simulated edge conditions pertaining to the contactor, shunts, or wiring. However, the pre-AFE simulation still requires the battery pack surrogate HIL system to make use of high-voltage signals, which may be restricted in some lab environments. Satisfying the requirement for a low-voltage system often requires modification of the BMU circuitry by connecting the simulated signals into the BMU "behind" the AFE signal conditioning as shown in Figure 12.

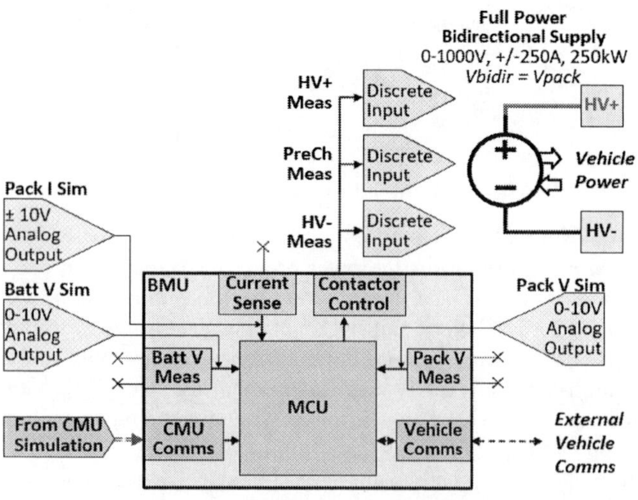

Figure 12. Low Voltage, Post-AFE simulation

Post-AFE simulation entails disconnecting portions of the analog front end signal conditioning on the BMU and replacing it with low-voltage signals as they would be fed into the BMU's analog-to-digital converters. This is typically a case of removing several passive components or lifting pins on the signal conditioning ICs. Once the BMU is modified, "Batt" and "Pack" voltages and pack current can then be simulated by ±10V analog signals, with values scaled to match the BMU AFE gains and offsets. The benefit of this method is that all signals into and out of the BMU may be monitored using common low-cost instrumentation with no high voltages or power present. However, like the pre-AFE methods of simulation, a separate full power bidirectional supply may be optionally added for providing battery power to the rest of the aircraft.

CMU and BMU Integration

Once the desired CMU and BMU simulation methods are selected, the overall battery pack surrogate hardware may be integrated. Figure 13 shows an integration of the CMU using full stack simulation with the BMU using full pack power, pre-AFE simulation. The battery pack surrogate HIL system represented by this block diagram corresponds to the actual system shown in Figure 4. As shown here, eighteen real CMUs are connected to 216 simulated cells and isolated thermistor simulations. All simulated cells are stacked in series to create the full 800V stack that would be seen in a real battery pack. The isolated communications from the CMUs are then routed to the real BMU. Because the simulated cells cannot provide full battery power, a full power bidirectional power supply is employed to deliver full battery power through the real current shunt and contactors. The BMU monitors communications from the CMUs, Batt and Pack voltage, and pack current, while controlling the real contactors. All BMS boards, high power control and monitoring, as well as simulation instrumentation are enclosed within the battery pack surrogate system. The output of the battery pack surrogate system includes the exact connections that would be provided by the real battery pack, including high power (HV+ and HV-), communications to the aircraft, and other I/O (such as interlocks, switches, and environmental feedback). This configuration provides the most accurate representation of the battery pack, and therefore, highest fidelity test capability. For this reason, most iron bird test setups use this configuration of battery pack surrogate.

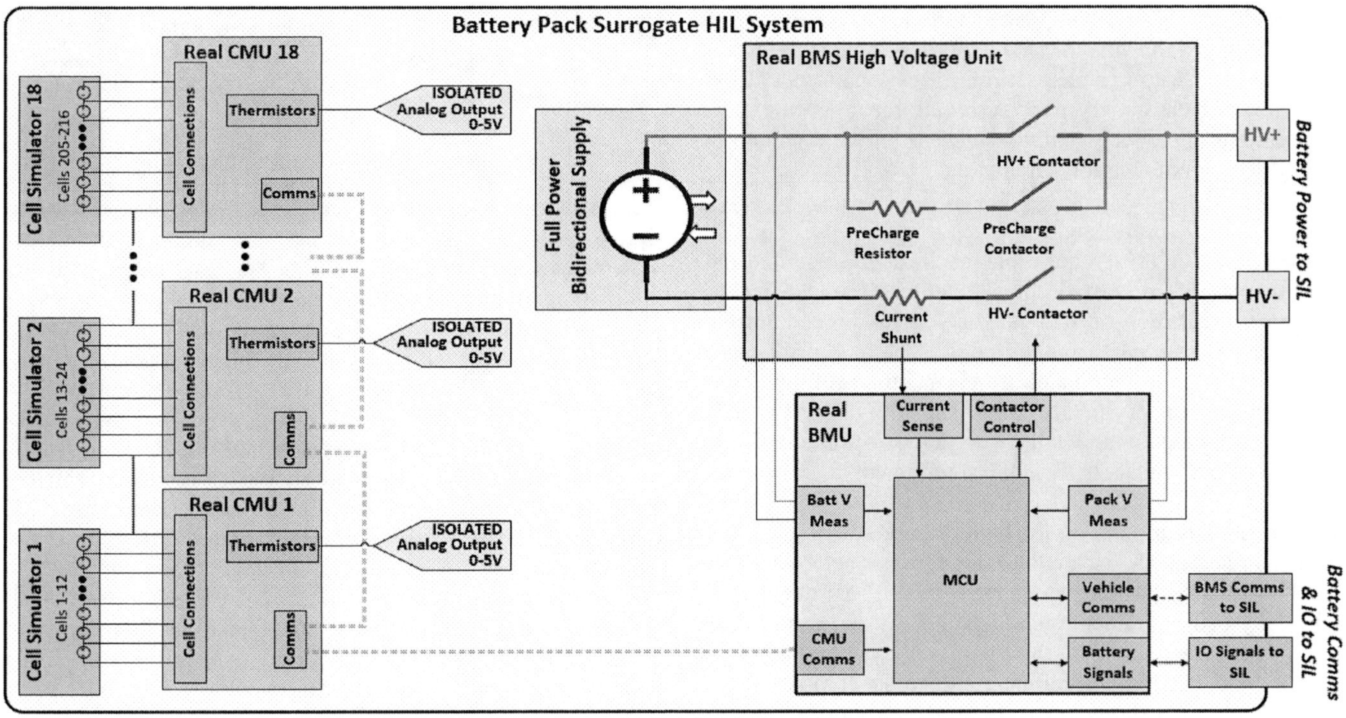

Figure 13. Battery pack surrogate CMU & BMU integration

BATTERY PACK SIMULATION SOFTWARE AND MODELING

While simulated signals enable the battery pack surrogate to affect and monitor the internal measurement and control circuitry of the BMS, software is required to make these signals behave like a real battery to the electric aircraft systems. Central to the software is a HIL simulation environment. Commercial software toolchains including, but not limited to NI VeriStand, The MathWorks Speedgoat, and ADI ADEPT provide this environment. Furthermore, the battery pack surrogate may be integrated into airframers' proprietary software tools using direct control or via reflective memory. The HIL environment, shown in Figure 14, provides a real-time control, model execution, and mapping engine that drives the battery pack surrogate. This real-time HIL engine executes cell chemistry, pack control, and other plant models, then maps the inputs and outputs of the models to physical instrumentation signals and calculated channels. It controls the instrumentation and communicates with the BMS, all within a deterministic control loop which executes at a rate of between 200Hz and 2,000Hz, depending upon computing resources, model complexity, and number of IO. External to the HIL engine, applications for test automation, graphical user interfaces, data playback, and model development may be used to augment or interact with the real-time control systems.

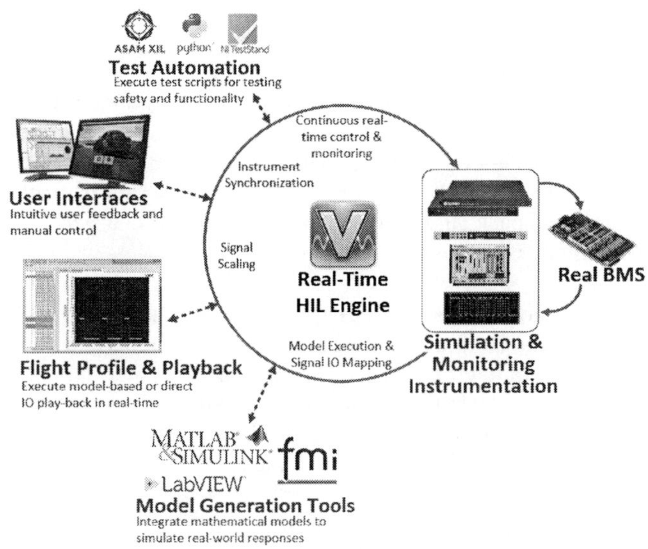

Figure 14. HIL Engine block diagram

The real-time HIL engine is the heart of the battery pack surrogate software. Once configured with modeling, scaling, mapping, and signal control, the real-time engine controls the previously-described simulation hardware to behave like a real battery. Python, NI TestStand, or some other test scripting environments are used to automate common setup and test scenarios, such as pack startup and shutdown, setting initial test conditions, or causing fault or edge conditions.

A graphical user interface is an important tool for providing engineers with easily-manipulated and viewed access to individual signals and model parameters for setup, monitoring, and control. Customizable user interfaces can provide different views of the battery pack surrogate for tasks like direct signal control, model preparation and execution, and monitoring BMS-communication parameters.

In addition to controlling the simulation with user controls, some test cases may require playback of pre-recorded data into the various aircraft systems under test. Using a HIL environment that supports data playback, with direct control of both all I/O as well as model inputs such as a pack power flight profiles, further enhances the capabilities of the battery pack surrogate. Typically, this capability allows basic comma-separated value tables to define millisecond-level control of any signals in the HIL environment.

Lastly, the key to making the battery pack surrogate behave like the real battery pack comes down to the battery cell and pack models that are executed in real-time. These may be provided by cell manufacturers, developed by in-house battery scientists, or generated by third party test system integrators. While MATLAB and Simulink are the most common model generation applications, open standards such as FMI are becoming increasingly popular, and can be executed by most commercial HIL environments.

Cell and Pack Models

Just as the cell is the basic building block of the battery pack and the cell simulator is the foundation of the simulated hardware, the cell model is one of the most critical components of the battery pack surrogate software. The cell model enables the entire system of signals to behave like a battery of a particular chemistry, adjusting cell voltages, temperatures, pack voltage, and pack current as the simulated pack is charged and discharged.

A typical model for a battery pack surrogate HIL system will receive initial conditions, including number of series and parallel cells, initial per-cell SoC and capacity, thermal and chemical characteristics. It uses these initial conditions as the starting point for the simulation. The simulated pack's state is updated by the model as it receives pack current and air/coolant temperature information, either from a real or simulated propulsion system, through the system's bidirectional power supply, via a manual user interface, or from an engineered flight profile or playback. Additionally, the state is updated by inputs to the model from instrumentation monitoring the BMS and other external system components. The model's outputs are mapped to the physical instrumentation which provides inputs to the BMS as shown in Figure 15.

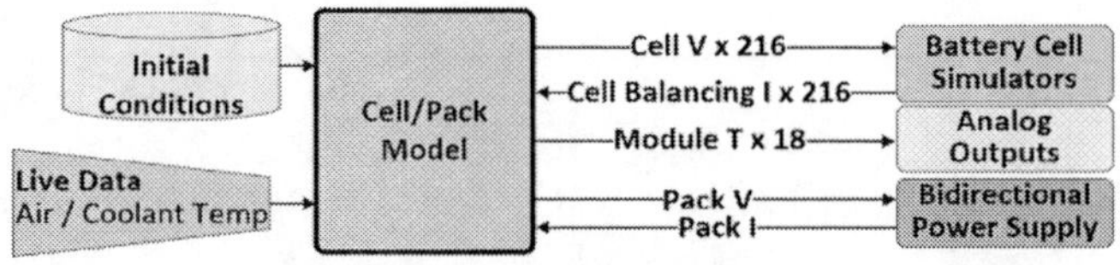

Figure 15. Cell/Pack model overview

The HIL real-time engine executes the model in discrete time steps. As shown in Figure 16, at each time step, the model calculates each cell's total current, state of charge, voltage, and temperature. The cell total current is calculated as a sum of the overall pack current plus the individual cell's balancing current. This is then used to calculate the cell's SoC as a function of the model's execution time step, cell current, and the total cell capacity (individual cell capacity times the number of cells in parallel). Cell voltage is then calculated using the current, SoC, and specific cell chemistry model parameters which are dependent on the cell chemistry model in use. Cell temperature is then calculated as a function of the time step, cell current, cell resistance, cell thermal parameters, and the air/coolant temperature. Finally, pack-level parameters, such as pack voltage, adjusted pack current, and pack temperature may be calculated based on the individual cell values. These calculated parameters are then output from the model to be mapped to physical channels, such as cell simulator instruments for cell voltage, analog output signals for temperatures, and power supplies for battery output voltages. The bidirectional supply reads output current to the aircraft and maps it back as a model input for the next iteration to drain or charge the cell models. Cell simulators read individual cell balancing and quiescent currents and feed these back into the model, allowing the model to drain or charge each cell individually. This repeats at the HIL real-time execution's control loop rate, typically on the order of every 0.5ms to 5ms.

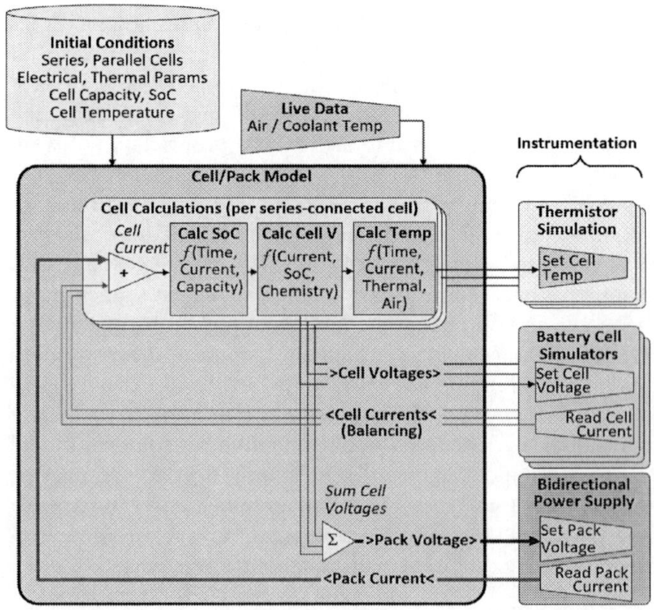

Figure 16. Cell/Pack model block diagram

Cell and pack models come in many different flavors, though there are two major categories of model: physics-based and equivalent circuit. Physics-based models simulate the internal electrochemical and physical properties of a lithium-ion cell. While an excellent model for predicting cell behavior, including voltage, thermals, degredation, aging, and mechanical stresses, they tend to require high levels of computational power and are typically not used in high channel count, real-time HIL test systems. An equivalent circuit model (ECM), like that shown in Figure 17, is a simplified mathematical model that uses basic electrical circuit components to represent the behavior of the cell.

In an ECM, the open circuit voltage (OCV) and circuit parameters are each a function of the cell's state of charge and temperature. This is usually represented by 2D interpolation tables for each parameter. The OCV dictates the resting voltage of the cell as if it were to sit with zero current for several hours. R_0 represents the effective series resistance (ESR) of the cell, while R_1, C_1, R_2, and C_2, represent R-C circuits that model the dynamic characteristics of the cell (Ref 3). The cell voltage, V_{cell}, can be calculated using standard circuit analysis methods. Many additional variants of the ECM exist that provide different features such as ageing, hysteresis, or additional R-C pairs.

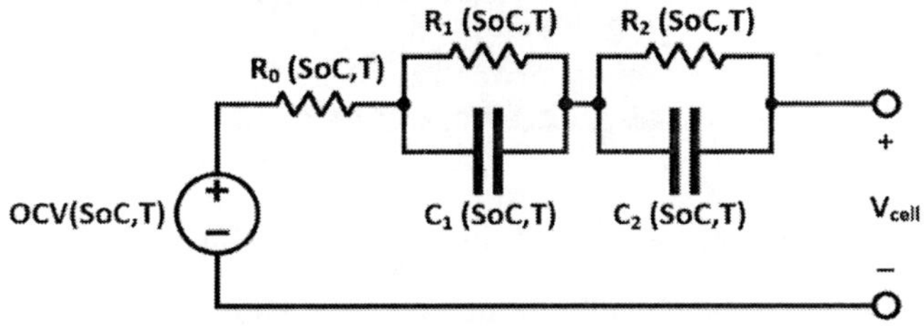

Figure 17. Equivalent circuit model

While the ECM is not a perfect representation of a cell, it provides a "good enough" approximation for most electric aircraft systems integration testing purposes. Where Figure 17 shows a single cell, the full pack model must replicate this once for each series-connected cell in the pack. Because the validation team will likely reconfigure the battery pack or cell initial conditions frequently, the model should be implemented such that the coefficients for each of the circuit parameters can easily be modified and reloaded for testing. Therefore, it is recommended that the model code utilize some external configuration file for storing of coefficients. A graphical user interface (GUI), like that shown in Figure 18, makes cell model configuration especially straightforward for test operators looking to develop specific test cases.

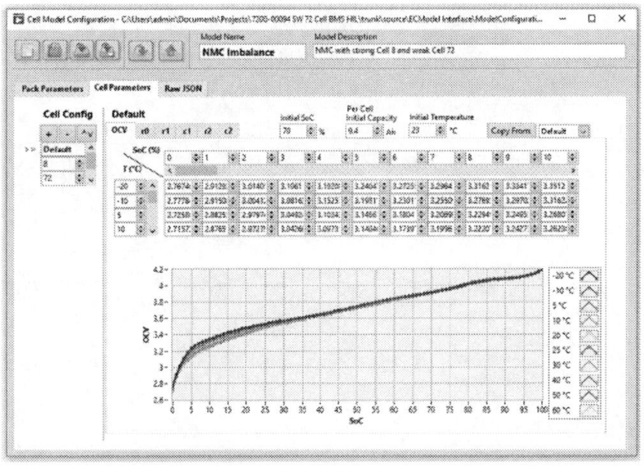

Figure 18. Cell model configuration GUI

PUTTING IT ALL TOGETHER

Once the entire battery pack surrogate's connectivity to the BMS and software control with modeling have been integrated with simulated signals, the system may be used for testing with other aircraft systems. Referring back to Figure 13 and Figure 16, the physical outputs, including pack power and vehicle communications from the battery pack surrogate HIL system are controlled by the real BMS, while all signals into the BMS are simulated by instrumentation controlled by the cell and pack models. In this way, the entire battery pack surrogate HIL system acts as a real battery. Each battery pack surrogate must be configured to the requirements of the aircraft, battery, and validation team.

The example system shown in Figure 4 is a battery pack surrogate which includes the BMS components housed inside the rack alongside the simulation instrumentation. This system contains eighteen 12-cell simulators for a total of 216 cells. All cells are stacked in series to each CMU, and all CMUs are stacked in series to provide full pack voltage to the BMU, as shown in Figure 19. A bidirectional power supply provides full power out to the external aircraft systems.

Figure 19: Full stack battery surrogate CMU connections

Figure 20 shows a different battery pack surrogate configuration for a low-voltage lab. In this system, multiple BMS prototypes were to be tested. Therefore, all simulated signals are routed through commercial industry-standard connectors on the front of the system. The system contains 192 simulated cells, 96 simulated thermistors, and various analog and digital IO for CMU and BMU interfaces. Because the simulated cells are routed through front-panel D-subminiature connectors, with a maximum voltage of 250VDC to ground, and because all thermistor simulation signals are non-isolated, it is intended to be configured for the low-voltage full cell CMU simulation method. The system also contains 10x 0-1000VDC, 1mA outputs for simulating the high voltage measurement points into the BMU. This particular system uses contactor simulation models to control the 10 high voltage outputs as a function of the simulated pack current and BMU contactor control signals, allowing the validation team to simulate multiple different contactor arrangements, without rewiring the contactor interfaces.

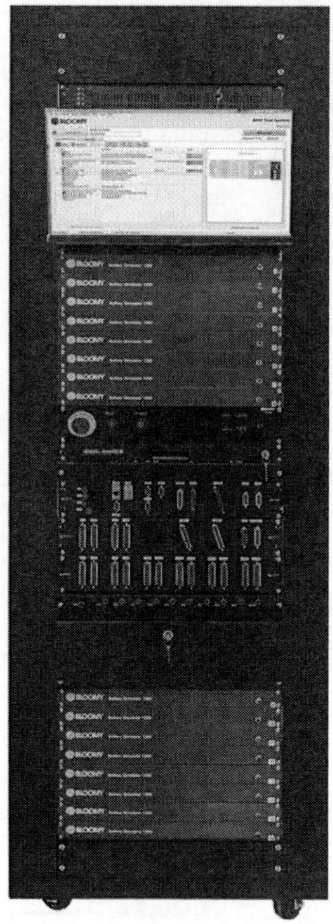

Figure 20. Example low-voltage battery pack surrogate

An operator may configure the initial conditions of the cell and pack models, such as pack topology, per-cell capacity, initial cell states of charge, and initial temperatures, and then initiate the model execution. At this point, the HIL system is running independently as if it were a real battery. Any current drawn by the external systems will be read by the HIL instrumentation and fed back to the model. Meanwhile, the real BMU is managing communications and interactions with the external aircraft systems and is controlling disconnects, safety, and performance algorithms. For all intents and purposes, any aircraft system connected to the battery pack surrogate will perform as if it were interacting with a real battery. The model will drain and charge the pack based on the bidirectional power supply readback, and the corresponding signals to the BMS LRUs will be adjusted such that the BMS responds as if it were connected to and controlling real chemical cells.

The following sections outline several examples of applications of the battery pack surrogate HIL system and the resulting acceleration of electric aircraft integration and validation.

Example 1. Iron bird systems integration lab

An iron bird is used to ensure that all major systems within the aircraft are functioning in a fully-integrated environment. In an iron bird, many aircraft systems and LRUs will annunciate faults without receiving battery data. Systems such as mission planning, aircraft management, and propulsion require information such as available capacity, state of charge, fault states and pack voltages to be supplied by the BMS. Other systems such as environmental controls, actuators, DC/DC converters, and auxiliary electrical systems require power from the battery. Furthermore, the interactions of the various aircraft systems will have complex behaviors dependent on the battery pack's state, including fault and edge conditions. The battery pack surrogate allows these systems to operate normally during basic integration testing, as well as testing of error conditions caused by the battery pack.

Using a battery pack surrogate also ensures that validation teams have nearly infinite control over the battery pack state from the high-level initial pack state of charge down to individual cell chemistry, capacity, SoC, and temperature parameters. At any point during integration testing, the battery pack surrogate allows any pack and cell parameters to be reconfigured. This empowers the validation team to repeatedly run different full aircraft integration tests without waiting for a real battery pack to charge or to be reset to a particular set of initial conditions.

Example 2. Propulsion system dynamometer testing

A dynamometer test system for the electric aircraft propulsion system is shown in Figure 21. The validation team would use this system to understand how different initial states and edge conditions of the battery pack may affect the motor controller across various flight profiles. Their end goals include tuning the motor controller's parameters to optimize performance and flight time, as well as to configure the battery fault response of the motor controller. In this setup, the battery pack surrogate is providing real power to the motor controller by way of the bidirectional power supply. The BMS is communicating with the motor controller as if it were connected to a real battery, although all I/O of the BMS are simulated by the battery pack surrogate. In this scenario, the battery pack surrogate is configured with initial conditions for the desired cell chemistry and the initial pack state of charge, capacity, and temperatures. A flight profile is used with the real aircraft motor controller which powers the real motors that then provide feedback to the controller using real speed sensors. The flight profile also controls a dynamometer to load the motors for simulating takeoff, cruise, and landing.

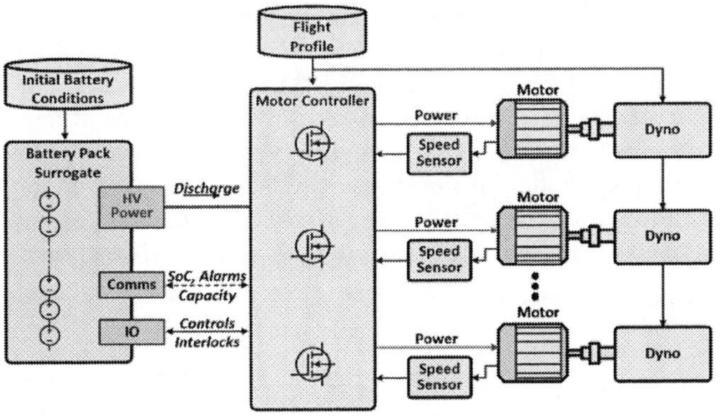

Figure 21. Example propulsion dyno test

Because the motor controller draws real power from the battery pack surrogate during the flight profile execution, the surrogate feeds the current readback from the bidirectional supply into the cell and pack models. The models then drain the pack SoC, causing cell and pack voltages to drop as if they were chemical cells of a real pack. These voltages are then fed to the cell simulators, bidirectional power supply, and other instrumentation. In addition, because the actual BMS is included in the surrogate, the validation engineers may easily test potentially dangerous edge conditions, such as a weak module or cell, over-temperature, or low initial SoC. In these edge cases, the validation engineers may even test scenarios where part of the redundant pack faults and disconnects – ensuring that the remaining pack provides sufficient power to safely land the aircraft. At the end of testing, the HIL simulator provides a data log of the pack current draw, calculated cell voltages, temperatures, pack voltage, and SoC, as well as BMS reported values like cell and pack voltage, current, SoC, and alarm information.

Example 3. Charger and ground support integration

Ground support equipment, such as chargers, are usually a separate system external to the aircraft and are often developed separately from the aircraft itself. However, the charger and ground support equipment are tightly coupled with the battery pack functionality. Safe and efficient charging is critical to the success of any electric aircraft program. Overcharging due to improper charge control algorithms poses a high risk of battery damage and thermal runaway. This is especially true for quick flight industries, such as electric air taxis, which may endure several fast-charge cycles per day. Validation of the interaction between the battery, the BMS's safety, control, and balancing systems, and the external charger is essential.

A block diagram of charger integration testing is shown in Figure 22. During charger integration testing, the battery pack surrogate is configured to an initial pack SoC, capacity, and temperature. Once model execution is started, the battery pack surrogate runs stand-alone, behaving like the real battery pack. Validation engineers can then validate and monitor the constant current/constant voltage charge profile, protection circuits, charge controller, and AC/DC conversion of the charger. Charger to BMS communications can be monitored, interlocks and protection circuitry can be validated, and the overall charging algorithm can be tuned. Furthermore, individual cells may be set to varying SoCs, capacity, and temperatures to simulate cell, module, or pack imbalances. Because a real battery is not in use, validation engineers do not need to worry about inadvertently overcharging and causing damage to the pack while validating the control algorithms.

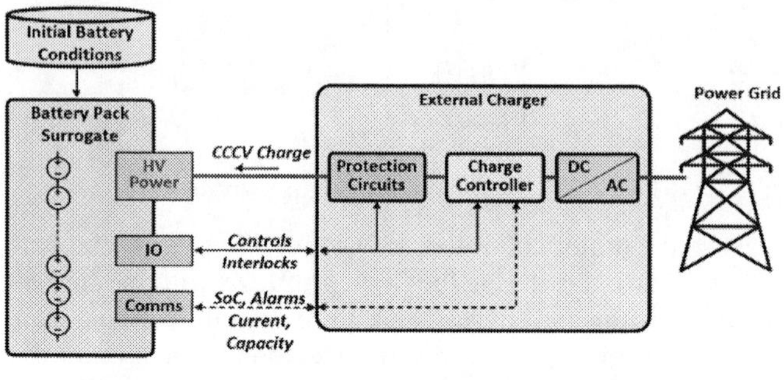

Figure 22. Example charger integration test

Test Data Collection

One of the keys of successful validation testing is data collection. Because the battery pack surrogate is a HIL test system, the real-time HIL engine allows continuous and deterministic data collection on all physical I/O channels, model inputs and outputs, and intermediate data channels. This collection can be helpful for troubleshooting integration issues between systems, when the culprit of an unintended behavior is not obvious. Beyond troubleshooting, proper data collection, management, and validation is critical to certifying flight readiness. Figure 23 shows an example of data collected for the model-calculated SoC, the real BMU-reported SoC, and overall pack voltage as a function of a charge/discharge pack current profile.

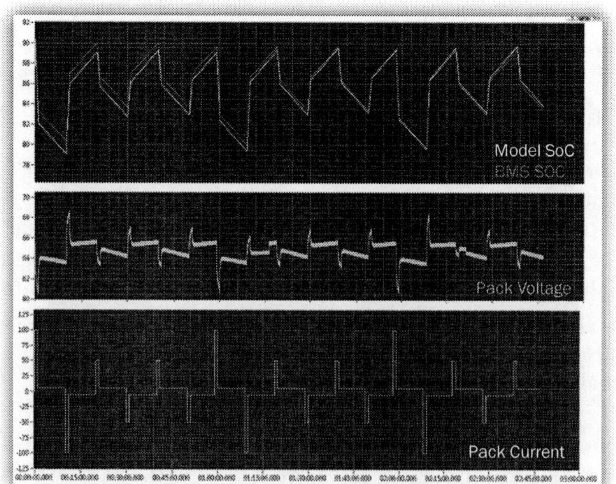

Figure 23. Example charge/discharge profile data

CONCLUDING REMARKS

A battery pack surrogate is a hardware in-the-loop simulation system which includes real BMS components, such as the cell monitoring units (CMUs) and battery management unit (BMU), connected to instrumentation which simulates the physical signals of the battery pack. The simulated signals of the battery pack surrogate are controlled by mathematical models of the cell chemistry, disconnects, and thermal behaviors of the battery, effectively making the BMS perform as if it were connected to real physical cells, sensors, contactors and other pack infrastructure. Combining these components with quickly-reconfigurable models into a single HIL system provides marked improvements in validating electric aircraft systems' interaction with the battery pack.

Author contact: Grant Gothing grant.gothing@bloomy.com

ACKNOWLEDGMENTS

Peter Blume: President of Bloomy, provided R&D funding for cell simulation, BMS test, and battery pack surrogate development. Edited final copy. William N Eccles: technical support, SIL/iron bird and aerospace expertise. Edited final copy.

REFERENCES

1. *Standard for Electrical Safety in the Workplace*, NFPA 70E, NFPA, 2024

2. Bloomy, "Using the Battery Simulator 1200 in High-Voltage Applications", 2022, https://www.bloomy.com/support/resources/white-paper/using-battery-simulator-1200-high-voltage-applications

3. Tran, M.-K.; DaCosta, A.; Mevawalla, A.; Panchal, S.; Fowler, M. Comparative Study of Equivalent Circuit Models Performance in Four Common Lithium-Ion Batteries: LFP, NMC, LMO, NCA. *Batteries* **2021**, *7*, 51. https://doi.org/10.3390/batteries7030051

Silicon Batteries with High Energy and Power Density for Electric Mobility

Ionel Stefan
CTO
Amprius Technologies, Inc.
Fremont, CA, USA

ABSTRACT

Amprius is continuously enhancing its silicon anode technologies, which have resulted in lithium-ion batteries with energy density and specific energy performance surpassing current graphite cells by 50-80%, depending on the cell size and form factor. Amprius products have demonstrated that silicon anode-based batteries can achieve up to 1,300 Wh/L and 500 Wh/Kg while maintaining a cycle life suitable for aerospace, military, and other high-end applications. Additionally, the high-capacity electrodes with reduced thickness allow cells to operate at high rates of charge and discharge without overheating, reaching 4000 W/kg power density in cells with over 400 Wh/kg specific energy density. Amprius began large-scale manufacturing and commercialization of its batteries, delivering over 100,000 units to customers in 2024. Furthermore, the range of applications, markets, and cell form factors developed has expanded significantly, with cell capacities ranging from 1Ah to 100Ah, available in both pouch and cylindrical formats. The current products, their performance, and applications, as well as the specific anode features that enable this performance, will be presented.

INTRODUCTION

Amprius is a Li-ion cell manufacturer with silicon anode and related technologies that enables an increase in energy density of over 50% compared to cells containing graphite anodes. Amprius' main technology, a highly engineered silicon structure made of 100% silicon, enables Li-ion cells with specific energies reaching 500 Wh/kg and energy densities of over 1300 Wh/L. The technology thus provides a significant step forward in lithium-ion battery development, roughly equivalent to the entire improvement over the last two decades with graphite anodes.

Amprius' silicon nanowire anode is unique in the industry because it is the only anode that is made of pure silicon (i.e., no binders, conductive carbons, or other inactive materials) and physically rooted in (i.e., grown from) the current collector foil, which confers unprecedented mechanical stability to the anode structure. The growth-rooted low tortuosity structure is optimal for high power performance (fast charge/discharge) due to very good electrical connectivity between the active anode material and the foil and an open, short pore structure. This combination of properties led to the development of cells that are not only energy dense (Wh/kg and Wh/L) but also power dense (W/kg and W/L), capable of fast charge (minutes instead of hours) and high-power delivery. Amprius cells are used in unmanned drone applications, such as high-altitude pseudo satellites, unmanned and manned flying drones and vehicles. Multiple eVTOL OEMs are currently testing and integrating silicon cells into batteries designed for specific aircraft frames.

The silicon nanowire technology, in MWh manufacturing capacity in 2024, will scale to hundreds of MWh by 2026. This capacity will be enough to support many aerospace and defense applications as well as to demonstrate competitive manufacturing costs at scale, further enabling investment for GWh capacities for electric mobility (electric vehicles and electric flight), replacing graphite-based cells with energy dense storage solutions in most applications, with game changing impact on energy storage weight and endurance. The technology relies entirely on a domestic supply chain, with silane, the silicon precursor available in large quantities, made in the USA.

TECHNICAL DESCRIPTION

Structure and Properties

The silicon nanowire structure, shown in Figure 1, includes a metallically conductive nanowire core, metallurgically connected to the current collector. This robust connection is essential in creating and maintaining a stable electrical connection and mechanical structure at the electrode level. Each silicon wire is directly connected and does not need to

Presented at the Vertical Flight Society 11th Biennial Autonomous VTOL Technical Meeting, Phoenix, AZ, USA Feb. 4-6, 2025.

rely on particle-to-particle contacts for conductivity. Since this connection is very robust compared to connections between particles created by electrode calendaring and binders, the silicon material remains in electrical contact, i.e., electrochemically active, over the entire cycle life of the cell. The fast loss of electrical connectivity and changes in mechanical structure that plague most other silicon technologies are virtually non-existent in the rooted silicon nanowire anodes. Moreover, this significant stability is achieved in a structure that is virtually 100% silicon, without any binder or conductive additive to dilute its active material content.

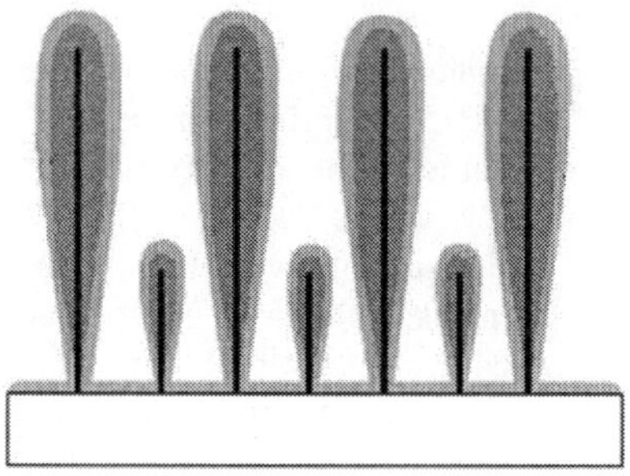

Figure 1. Amprius' growth rooted silicon nanowire structure

The silicon nanowire has a tapered shape, with a much lower thickness toward the root, where it connects to the current collector. This shape is essential in reducing the mechanical stress at the interface during lithiation/delithiation cycles. When silicon films or particles are coated on foils as thin as those used currently in lithium-ion cells, e.g. 6-12 μm in thickness, upon lithiation the foils are stretched and that leads to wrinkles, ripples, and foil breaking. In the case of silicon nanowire electrodes, the minimum contact surface between the root of the silicon wire and foil eliminates mechanical stress in the foil during cell cycling.

A second feature of Amprius' silicon nanowire material is its nanostructure. The internal porosity and amorphous structure accommodate and dissipate some of the mechanical stress induced in the material during lithiation/delithiation processes. Since these processes start from the silicon/electrolyte interface, they occur mostly radially, and the stress and material expansion are of the same geometry. The silicon nanowire structure can accommodate a large amount of lithium without significant radial expansion. Even at full lithiation, the axial expansion, which is responsible for the cell expansion, is less than 25%, resulting in less than 10% cell thickness increase. This allows electrode loadings (mAh/cm^2) that are significantly higher (approaching 100% increase) than conventional graphite electrodes.

Because the nanowires, which work as electron conduction paths, are attached to the current collector, the structure does not rely on particle-to-particle contact for electronic conductivity and is able to achieve high conductivity, energy, and power. This structure difference is shown in Figure 2, where the difference in the ionic and electronic conductive paths are indicated by arrows. Both paths are short and straight in the nanowire anode case, making the structure ideal for high power applications. Combined, these properties result in superior results to particle type graphite, silicon, or lithium metal anodes, exceeding them both in energy density and power capability, including fast charge capability and low temperature operation.

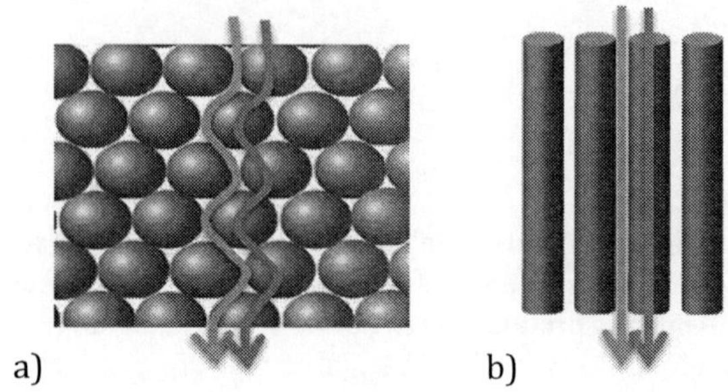

Figure 2. Electronic and ionic conductivity paths in silicon nanowire anode (b) are shorter and straight compared to the tortuous paths in graphite anodes (a

The nanowire material structure also enables a relatively high 94% first coulombic efficiency in half cells and 90% coulombic efficiency in full cells with LCO cathode. These values are competitive with graphite materials and reduce or eliminate the need for prelithiation steps that compensate for the first cycle loss in the cell fabrication. A typical first cycle at a 0.05C rate of a silicon nanowire electrode is shown in Figure 3.

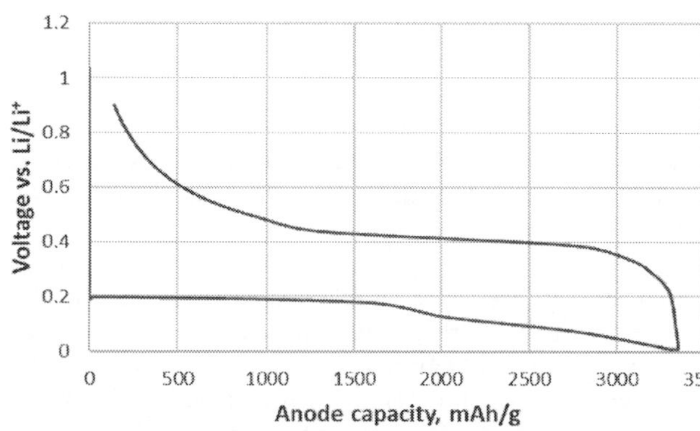

Figure 3. Silicon nanowire voltage profile in half cell, lithiated to 10mV vs. Li/Li^+ electrode and delithiated to 0.9V at a 0.05C rate

Energy and Power Density

As described in the previous section, the silicon nanowire anode has intrinsically high-power capability. The Ragone plot in Figure 4 shows the energy and power application range of silicon nanowire cell types that were developed by matching silicon nanowire anode with different cathode structures. With high loading cathodes, the specific energy reaches over 500 Wh/kg, but has a relatively narrow power range. This platform is used in applications that need operation times of many hours, such as day/night cycles or

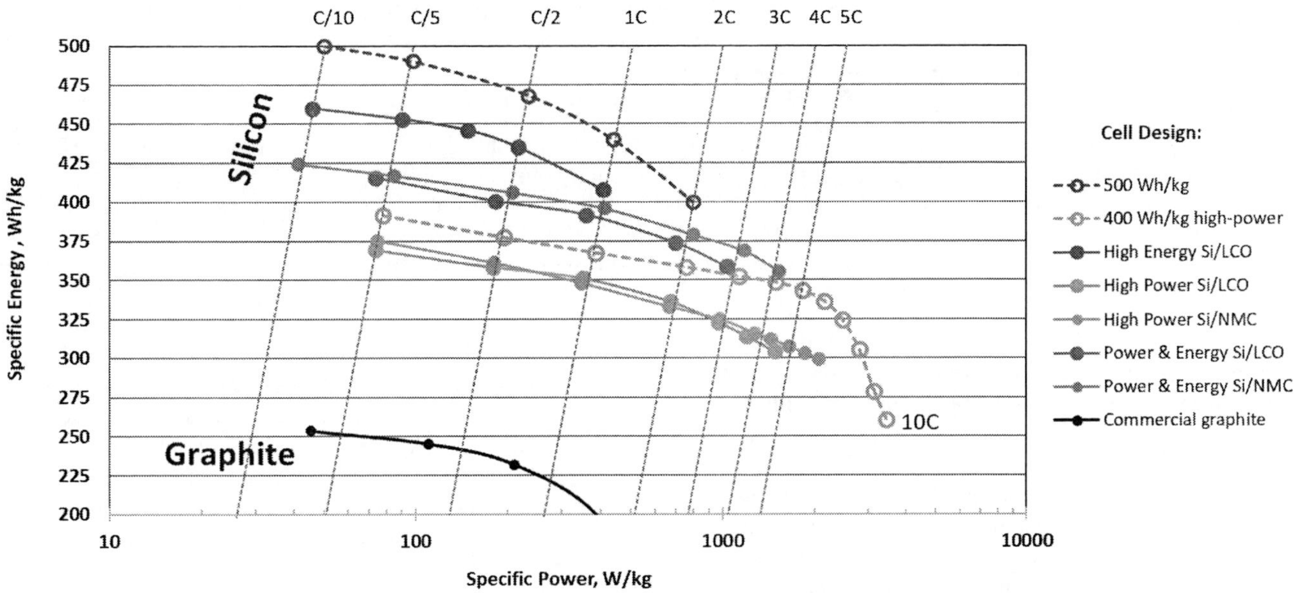

Figure 4. Amprius has developed three power-energy platforms by matching silicon nanowire anode with Lithium Cobalt Oxide (LCO) and Nickel Manganese Cobalt (NMC) cathodes with different loadings.

long mission applications. A porosity optimized high loading structure resulted in cells with a specific energy of 400-450 Wh/kg and specific power of 1000 W/kg. This power-and-energy platform is used in long endurance drones and eVTOL applications where target endurance exceeds one hour. When matched with a high-power, thin cathode, the cells still have over 380 Wh/kg specific energy and can support 3000 W/kg of continuous power load. This power platform is used for applications that require very high specific power, such as load carrying power drones and eVTOL aircraft with designs that require very high power for take-off and landing. In most cases, silicon cells double the running times compared to graphite cells. In high-power applications, typically the limits are cell body temperature and the thermal management of the battery.

Cycle Life

The silicon nanowire anodes can cycle for hundreds of cycles in full cells due to limited swell at the material and cell level, and stable material and electrode structure. The stability of the material and partial utilization of the theoretical capacity of silicon also limits lithium consumption for SEI repair and, thus, extends the cycle life to numbers approaching those of graphite cells. In real applications, when cells are not fully discharged, cycle life exceeds 1000 cycles (Figure 5) and in typical eVTOL flight profiles, cycle life spans thousands of cycles due to lower depth of discharge and energy reserve requirements, both factors that have a significant impact on the silicon lithium loss mechanisms.

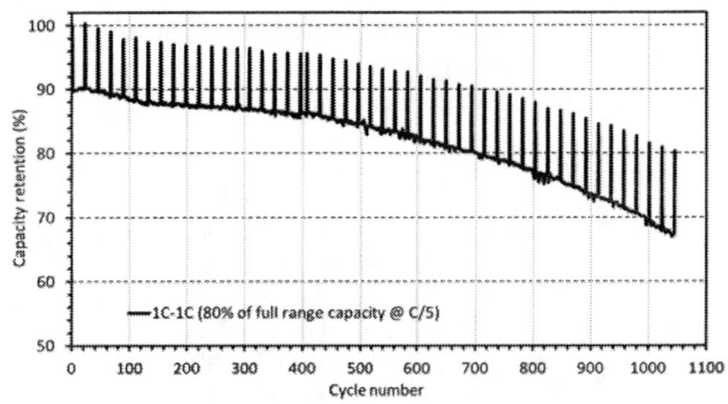

Figure 5. 1C/-1C cycling with 80% depth of discharge and +0.2C/-0.2C capacity check every 20 cycles.

Extreme Fast Charge

As described in Figure 2, the silicon nanowire anode has a low diffusion distance (thin electrode), low tortuosity structure (straight pores), and short electric path to current collector (all silicon is "metallurgically" connected to the foil). These properties have led to cells products that have demonstrated 5-minute charge (Extreme Fast Charge, XFC) capability in cells with 360 Wh/kg specific energy (see Figure 6). Thus, additional capabilities that are only possible with silicon anode are available for deployment**Error! Reference source not found.**.

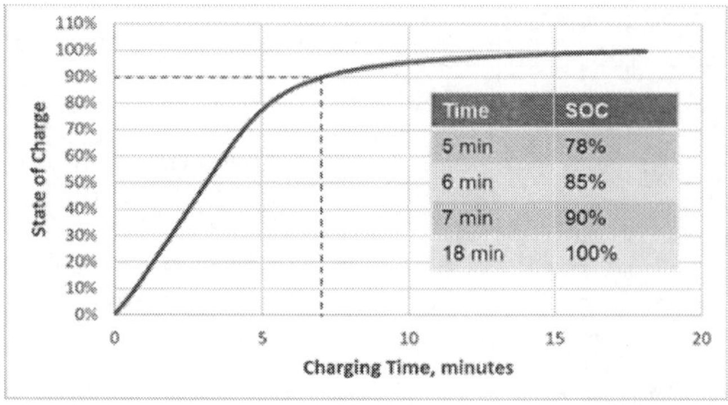

Figure 6. Silicon cells are capable of extremely fast charge rates, charging to 80% in under 6 minutes.

Low Temperature

Silicon has intrinsically better capability at low temperatures, due to its slightly higher operating voltage than graphite. For example, Amprius cells operate down to -30°C without low temperature optimized electrolytes and can operate even at -50°C with low freezing point electrolyte formulations. Similarly, the high temperature limit can be tailored by electrolyte formulation to meet targets. Cell data shown in Figure 7 indicates that about 80% of the energy available at 20°C is available at -20°C, for cells with specific energy over 400 Wh/kg. This capacity retention is about double compared to graphite cells at low temperatures.

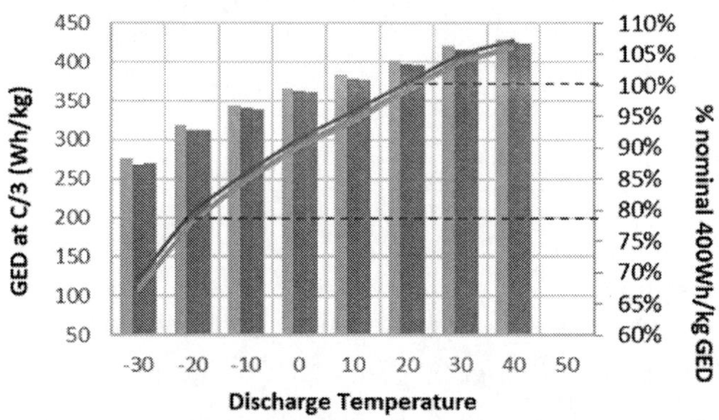

Figure 7. Silicon cells retain 80% of room temperature performance at -20°C. The figure shows the retained specific energy of three cells between -30°C and 50°C.

Safety

Silicon nanowire cells developed in the last few years were successfully tested in multiple abuse and safety evaluation conditions, including electrical, mechanical, and thermal factors. For example, 5.4 Ah pouch cells with silicon nanowire anode and LCO cathode were certified to pass the tests required for air cargo shipping, according to UN 38.3 specifications. Amprius has developed a gel polymer electrolyte formulation that insulates penetrating objects, such as nails, preventing thermal runaway. Separately, silicon nanowire cells passed other abuse and safety tests, including hot box at 130°C and overcharge by 150% charge capacity. These results suggest that silicon materials have similar safety profiles to graphite anodes, when used in similar cells and with similar electrolyte and cathode materials.

Products and Manufacturing

Amprius has developed cells in multiple sizes, from 200mAh capacity for wearable applications to 80Ah for electric vehicles. Pouch cells can be customized in different form factors with relative ease, and Amprius has experience in prototyping and developing such cells. Moreover, the silicon chemistry platform can be used as developed or with small optimization changes in multiple cell form factors. The cell pictures shown in Figure 9 illustrate the variety of designs in different cell form factors and stages of production. Cells of 5.4 Ah and 10 Ah capacity were built into battery packs for drone applications by primary defense contractors, setting records of endurance in multi-rotor flight. In smaller form factors, Amprius developed a battery for the nano-UAS drones and has demonstrated an increase in flight endurance of over 80%.

The complete set of manufacturing processes have been integrated since 2016 into a continuous roll-to-roll pilot scale anode fabrication line. Two pilot lines and associated cell production are currently operated in Fremont, California. A larger, MWh manufacturing line was installed in 2023 in Fremont, CA, and a 500MWh manufacturing scale project also started in 2023 in Brighton, CO, the next step toward the commercialization of the silicon technology for larger markets. Customers in aerospace and defense will be able to use Amprius' high-performance lithium-ion cells to the full requirements of these markets at prices competitive to graphite cells.

SUMMARY AND CONCLUSIONS

Silicon is recognized as one of the most promising materials for next generation lithium-ion battery anode to replace the conventional carbon-based anode due to its high theoretical capacity, similar discharge potential and reliable operation safety. However, in most silicon material structures and compositions, the critical drawback of huge volume expansion upon lithiation causes series of adverse consequences, leading to very poor cyclic stability and preventing its large-scale commercialization.

The rooted silicon nanowire structure developed by Amprius has largely mitigated the drawbacks that afflict other silicon materials and has led to the development of lithium-ion pouch cells with energy and power performance significantly above those of commercial state of the art graphite anode cells, in most cases by 50-100% improvement. Moreover, the cycle life of these cells is closing in on graphite cell performance, already meeting the requirements of a variety of markets. Some of the silicon nanowire cell products have already been qualified for applications, including safety and abuse tolerance validation. In general, the development of the first silicon nanowire products has shown that the safety and abuse tolerance of silicon nanowire cells can be as good or better

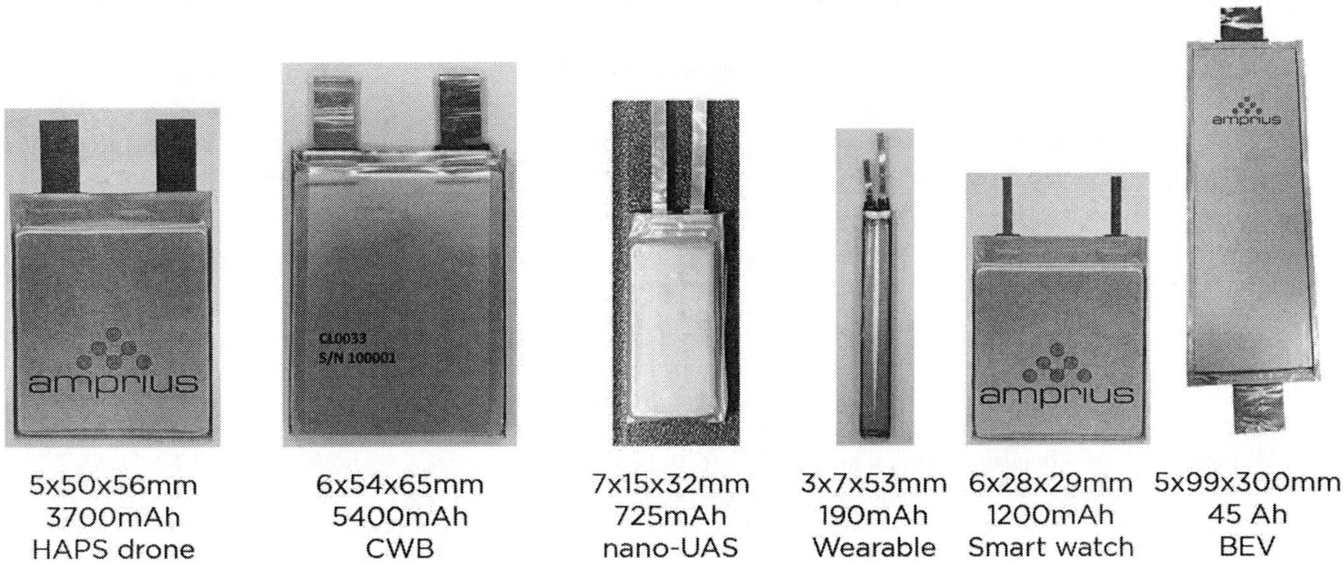

Figure 8. Multiple cell designs were developed with Amprius' silicon anode chemistry.

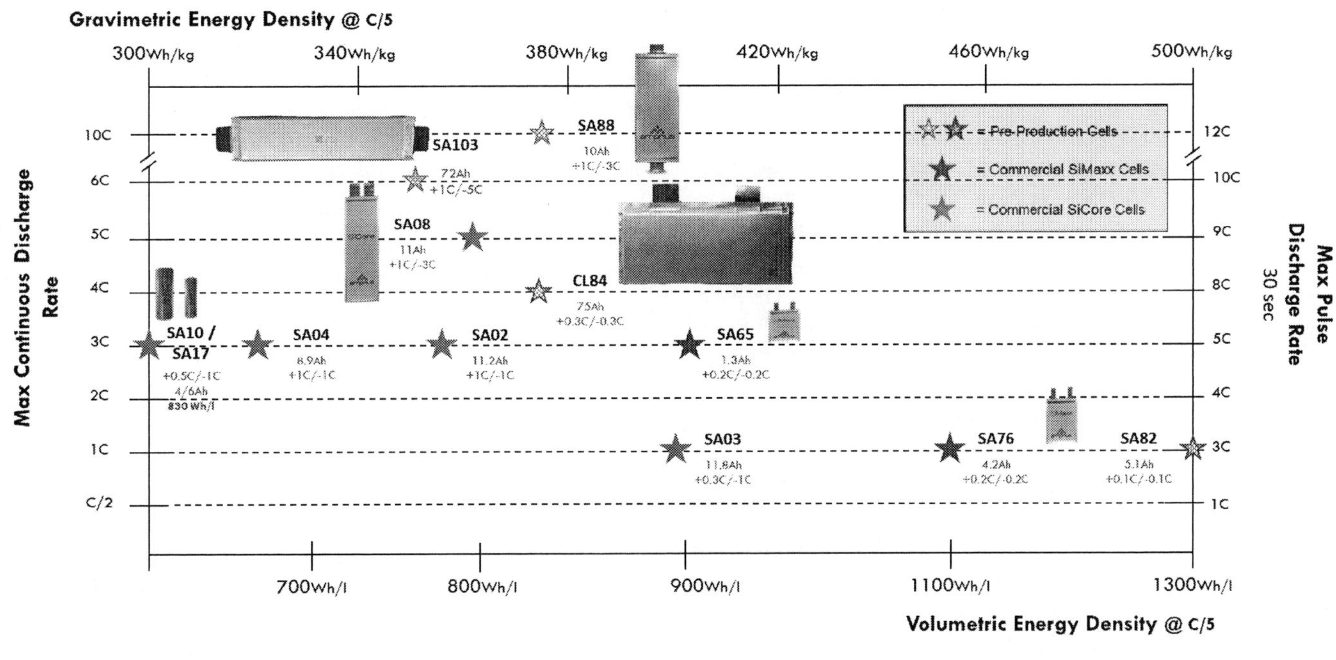

Figure 9. Commercial and pre-commercial Amprius products in a multi-metrics map

than that of similar graphite cells. It is expected that, in parallel with an increase in manufacturing capability and capacity, the application scope of silicon nanowire cells will expand from premium and mission critical markets to consumer and commercial application in the next few years.

Author contact: Ionel Stefan, ionel@amprius.com

ACKNOWLEDGMENTS

Amprius wishes to acknowledge the funding received from US Army (W56KGU-18-C-0025/STEP-EGS), DOE (EE0010224) and USABC (DE-EE0006250).

CFD Analysis of Quadrotor Biplane Tailsitter Hover-to-Cruise Transition

Paulo Arias
Graduate Research Assistant
University of Maryland
College Park, MD, USA

Umberto Saetti
Assistant Professor
University of Maryland
College Park, MD, USA

James Baeder
Professor
University of Maryland
College Park, MD, USA

ABSTRACT

Among various UAV configurations, quadrotor biplane tailsitter designs are particularly promising due to their efficiency in both hover and forward flight modes. This study investigates the hover-to-cruise transition of the CRC-80, a scaled-up, 80-pound, pitch-controlled UAV, based on the CRC-10 developed at the Army Research Laboratory (ARL). A flight dynamics and controls model of the aircraft is developed and coupled with three different levels of aerodynamic fidelity: a dynamic inflow solver, a vortex particle method code, and a Reynolds-Averaged Navier-Stokes based computational fluid dynamics solver. The different aerodynamic models are used to achieve accurate and force-balanced transition flight simulations. Unsteady and quasi-steady simulation approaches combined with multi-fidelity analysis are used to determine the sufficient level of simulation necessary for understanding the aircraft loads and acoustics.

INTRODUCTION

The rising interest in advanced air mobility (AAM) has driven the development of innovative designs that prioritize versatility, efficiency, and specialized capabilities. Among these advancements, numerous unmanned aerial vehicle (UAV) configurations have emerged, varying widely in scale and functionality. UAVs are now being employed in diverse fields, including package delivery, search and rescue, agriculture, surveying, and recreation, among others (Refs. 1–4). One particularly promising configuration is the quadrotor biplane tailsitter, which combines the efficiency of a quadcopter in hover with the forward-flight capabilities of a two-wing biplane. This study focuses on the common reference configuration CRC-80, a scaled-up, pitch-controlled version of the CRC-10, a design developed collaboratively by the Army Research Lab (ARL) and the University of Maryland (UMD).

This aircraft's unique design leads to notable differences in behavior between hover and cruise flight modes. The complexity of the CRC-80's flight in combination with its large number of interacting components spurs the need for in-depth aerodynamic analysis. While prior research has examined its performance under various steady-state flight conditions (Refs. 5–7), the dynamics of its transition flight remain less understood, particularly regarding performance and noise implications. Although previous studies have explored trajectory planning and optimization for both hover-to-cruise and cruise-to-hover flight (Refs. 4, 8), this work seeks to delve deeper into the effects of non-periodicity on its flight. Our in-house flight dynamics code has been successfully coupled and tested in a vortex particle method code (Ref. 9) in previous work. The study also looks to determine the level of fidelity necessary to capture essential phenomena for different types of analysis, such as flight dynamics and acoustics, in a complex design like the CRC-80. The insights gained from these simulations could play a crucial role in refining design strategies to optimize performance and reduce noise during unsteady flight.

Despite advancements in computational power, unsteady maneuver simulations continue to be prohibitively expensive, especially for intricate configurations that demand highly detailed meshes. This study aims to evaluate the trade-offs between simulation accuracy and computational cost across different levels of aerodynamic fidelity coupled with flight dynamics. For this analysis, the lowest aerodynamic fidelity used is dynamic inflow combined with lifting lines. The mid-fidelity simulations consist of a vortex particle and panel method (VPM). Finally the highest fidelity is Reynolds-Averaged Navier-Stokes (RANS) computational fluid dynamics (CFD). In addition to analyzing the effect of aerodynamic fidelity, the results of unsteady flight simulations will be compared to quasi-steady simulations at a set of way points throughout the maneuver. This will determine whether non-periodicity plays a large role in the results. While unsteady maneuver flight implementation in UMD's in-house CFD Framework was not completed in time for this study, quasi-steady results are included and compared to the other aerodynamic models.

METHODOLOGY

This study seeks to identify the most effective method for performing accurate unsteady simulations of a quadrotor biplane tailsitter aircraft during its hover-to-cruise transition. Different levels of aerodynamic fidelity are analyzed to identify the most suitable and computationally efficient approach for flight

Presented at the Vertical Flight Society's 11th Biennial Autonomous VTOL Technical Meeting, Phoenix, AZ, USA, Feb. 4-6, 2025.

dynamics and aeroacoustic studies. Additionally, the study compares quasi-steady and fully unsteady analyses to evaluate their respective advantages and limitations.

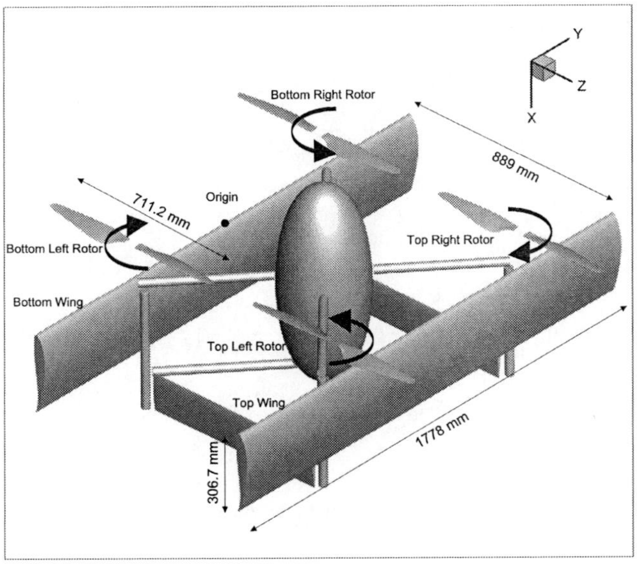

Figure 1: Model Geometry

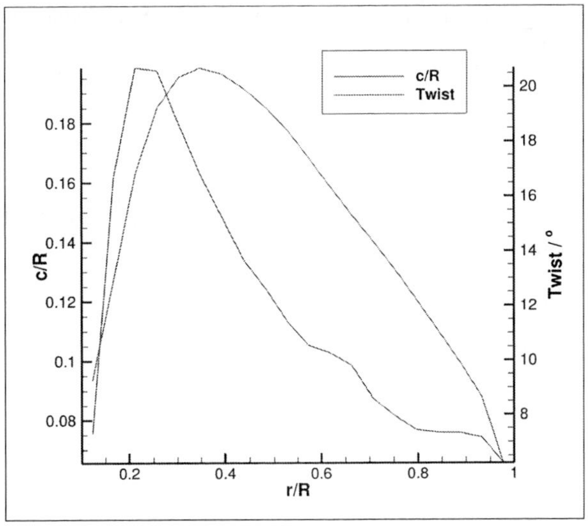

Figure 2: Chord and twist distribution of rotor blade

Aircraft Model Features

The aircraft analyzed in this study, the CRC-80, is a scaled-up version of the CRC-10, a design developed by the Common Research Configuration (CRC) group under the Army Research Lab (ARL). The CRC-80 simplified CFD model is shown in 1. This scaled model assumes a gross takeoff weight of 80 lbs, compared to the original CRC-10's 10 lbs. The change in weight required a scale factor of 3.5, and T-MOTOR G28x9.2 blades were selected to match the new airframe. The blade chord and twist distributions are illustrated in Figure 2. Following these modifications, the rotor radius (R) measures 355.6 mm, the wingspan is 1778 mm, and the wing chord is 306.7 mm. For improved responsiveness and control authority, the aircraft flight is governed by pitch control. The rotational speed still varies, but is modified based on the flight condition. The optimized rotational speeds range from 3900 RPM in hover, to 2300 RPM at a cruise speed of 20 m/s.

Flight Dynamics Model The flight dynamics simulation model is developed using an in-house MATLAB®/Simulink-based generic multi-rotor/wing flight dynamics and control framework (GenMR) at the University of Maryland (UMD). The model has the following features:

- The fuselage is represented as a rigid body. Aerodynamic forces acting on the fuselage are computed using equivalent flat-plate areas for frontal, lateral, and vertical surfaces (Ref. 10).
- The framework supports an arbitrary number of rotors and wings with user-defined orientations, which can vary over time.
- Rotor aeromechanics are modeled at different levels of fidelity, ranging from static inflow with blade flapping to dynamic inflow (Ref. 11) and flexible blade dynamics based on finite element methods (FEM). Rotor interactions, such as rotor-on-rotor effects, are also included (Refs. 12–14).
- Wing aerodynamics can also be modeled with varying complexity, with the most advanced implementation using lifting-line theory. Interactions between rotors and wings are computed using Biot-Savart methods and circulation-based approaches (Ref. 15).
- The model is implemented in MATLAB®, facilitating the design and testing of flight control laws.
- A baseline control law, employing Dynamic Inversion (Refs. 16–18), is integrated into the model. The control architecture includes inner attitude and outer velocity loops.

The rotorcraft flight dynamics are described by a nonlinear time-invariant system:

$$\dot{\boldsymbol{x}} = \boldsymbol{f}(\boldsymbol{x},\boldsymbol{u}) \; \boldsymbol{y} = \boldsymbol{g}(\boldsymbol{x},\boldsymbol{u}) \tag{1a}$$

where $\boldsymbol{x} \in R^n$ represents the state vector, $\boldsymbol{u} \in R^m$ denotes the control input vector, and $\boldsymbol{y} \in R^l$ is the output vector. The nonlinear functions $\boldsymbol{f}$ and $\boldsymbol{g}$ define the system's dynamics and output equations, respectively. The state vector is expressed as:

$$\boldsymbol{x}^T = \left[\boldsymbol{x}\mathrm{F}^T \; \boldsymbol{x}\mathrm{R}^T\right] \tag{2}$$

where $\boldsymbol{x}\mathrm{F}$ represents the fuselage (rigid-body) states and $\boldsymbol{x}\mathrm{R}$ corresponds to the rotor states. The fuselage states are given by:

$$\boldsymbol{x}_{\mathrm{F}}^T = [u \; v \; w \; p \; q \; r \; \phi \; \theta \; \psi \; x \; y \; z] \tag{3}$$

where:

- u, v, w are the longitudinal, lateral, and vertical velocities in the body-fixed frame,
- p, q, r are the roll, pitch, and yaw rates,
- ϕ, θ, ψ are the Euler angles, and
- x, y, z are the positions in the North-East-Down (NED) coordinate frame.

The dynamics of the fuselage states are described by the rigid-body equations of motion. The rotor state vector is defined as:

$$\mathbf{x}\mathrm{R}^T = \left[(\lambda_0\ \lambda 1c\ \lambda_{1s}\ \Omega\ \psi)1\ \cdots\ (\lambda_0\ \lambda 1c\ \lambda_{1s}\ \Omega\ \psi)_4\right] \quad (4)$$

where:

- λ_0, λ_{1c}, λ_{1s} are the rotor induced inflow ratio harmonics,
- Ω is the rotor angular velocity, and
- ψ is the azimuth angle of a reference rotor blade.

The subscript $()_\mathrm{i}$ identifies the ith rotor. The pilot input vector is given by:

$$\boldsymbol{u}^T = \left[\delta_{\mathrm{lat}}\ \delta_{\mathrm{lon}}\ \delta_{\mathrm{col}}\ \delta_{\mathrm{ped}}\right] \quad (5)$$

where:

- δ_{lat} and δ_{lon} are the lateral and longitudinal control stick positions,
- δ_{col} is the collective control input, and
- δ_{ped} is the pedal control input.

The pilot inputs are converted to control effector inputs via the following relationship:

$$\boldsymbol{u}_\mathrm{c} = \boldsymbol{G}(V)\,(\boldsymbol{u}+\boldsymbol{u}_\mathrm{b}) + \boldsymbol{b}(V) \quad (6)$$

where $\boldsymbol{G}$ is the mixing matrix, $\boldsymbol{u}_\mathrm{b}$ is the pilot bias vector, and $\boldsymbol{b}$ is the control effector input bias vector. Additionally, θ_0 θ_{1s} θ_{1c} are the collective, longitudinal cyclic, and lateral cyclic swashplate inputs, Ω is the commanded rotor speed, and $V = \sqrt{u^2+v^2+w^2}$ is the total speed. Note that the mixing matrix and control effector bias vector are scheduled with flight speed, such that $\boldsymbol{G} = \boldsymbol{G}(V)$ and $\boldsymbol{b} = \boldsymbol{b}(V)$. Because during hover-to-cruise transition the tailsitter accelerates from low-speed flight ($V < 5$ kts) to high-speed flight ($V < 38.8$ kts), control mixing is scheduled with flight speed and optimized for each flight condition. Below 5 kts, hover control mixing is used; above 38.8 kts cruise flight mixing is used; between 5 and 38.8 kts a blend between the two is used. These three control mixing strategies are summarized as follows:

$$\boldsymbol{G}(V) = \begin{cases} \boldsymbol{G}_{\mathrm{LS}}, & V < V_{\mathrm{LS}}, \\ \boldsymbol{G}_{\mathrm{LS}}\left(\dfrac{V_{\mathrm{HS}}-V}{V_{\mathrm{HS}}-V_{\mathrm{LS}}}\right) + \boldsymbol{G}_{\mathrm{HS}}\left(\dfrac{V-V_{\mathrm{LS}}}{V_{\mathrm{HS}}-V_{\mathrm{LS}}}\right), & V_{\mathrm{LS}} \le V < V_{\mathrm{HS}}, \\ \boldsymbol{G}_{\mathrm{HS}}, & V \ge V_{\mathrm{HS}}. \end{cases} \quad (7a)$$

$$\boldsymbol{b}(V) = \begin{cases} \boldsymbol{b}_{\mathrm{LS}}, & V < V_{\mathrm{LS}}, \\ \boldsymbol{b}_{\mathrm{LS}}\left(\dfrac{V_{\mathrm{HS}}-V}{V_{\mathrm{HS}}-V_{\mathrm{LS}}}\right) + \boldsymbol{b}_{\mathrm{HS}}\left(\dfrac{V-V_{\mathrm{LS}}}{V_{\mathrm{HS}}-V_{\mathrm{LS}}}\right), & V_{\mathrm{LS}} \le V < V_{\mathrm{HS}}, \\ \boldsymbol{b}_{\mathrm{HS}}, & V \ge V_{\mathrm{HS}}. \end{cases} \quad (7b)$$

where V is the total speed of the aircraft, $V_{\mathrm{LS}} = 5$ kts, and $V_{\mathrm{HS}} = 38.8$ kts. The control effectorvector is:

$$\boldsymbol{u}_\mathrm{C}^T = [(\,\theta_0\ \theta_{1s}\ \theta_{1c}\ \Omega\,)_1\ \cdots\ (\,\theta_0\ \theta_{1s}\ \theta_{1c}\ \Omega\,)_4] \quad (8)$$

whereas the control mixing and control effector bias vector are:

$$\boldsymbol{G}_{\mathrm{LS}} = \boldsymbol{G}_{\mathrm{HS}} = \frac{1}{100}\begin{bmatrix} 1 & 1 & 1 & -1 \\ 0 & 0 & 0 & 0 \\ 0 & 0 & 0 & 0 \\ 0 & 0 & 0 & 0 \\ -1 & 1 & 1 & 1 \\ 0 & 0 & 0 & 0 \\ 0 & 0 & 0 & 0 \\ 0 & 0 & 0 & 0 \\ -1 & -1 & 1 & -1 \\ 0 & 0 & 0 & 0 \\ 0 & 0 & 0 & 0 \\ 0 & 0 & 0 & 0 \\ 1 & -1 & 1 & 1 \\ 0 & 0 & 0 & 0 \\ 0 & 0 & 0 & 0 \\ 0 & 0 & 0 & 0 \end{bmatrix}, \quad (9a)$$

$$\boldsymbol{b}_{\mathrm{LS}} = \begin{bmatrix} 9\pi/180 \\ 0 \\ 0 \\ \Omega_{\mathrm{LS}} \\ 9\pi/180 \\ 0 \\ 0 \\ \Omega_{\mathrm{LS}} \\ 9\pi/180 \\ 0 \\ 0 \\ \Omega_{\mathrm{LS}} \\ 9\pi/180 \\ 0 \\ 0 \\ \Omega_{\mathrm{LS}} \end{bmatrix}, \quad \boldsymbol{b}_{\mathrm{HS}} = \begin{bmatrix} 9\pi/180 \\ 0 \\ 0 \\ \Omega_{\mathrm{HS}} \\ 9\pi/180 \\ 0 \\ 0 \\ \Omega_{\mathrm{HS}} \\ 9\pi/180 \\ 0 \\ 0 \\ \Omega_{\mathrm{HS}} \\ 9\pi/180 \\ 0 \\ 0 \\ \Omega_{\mathrm{HS}} \end{bmatrix} \quad (9b\text{-}c)$$

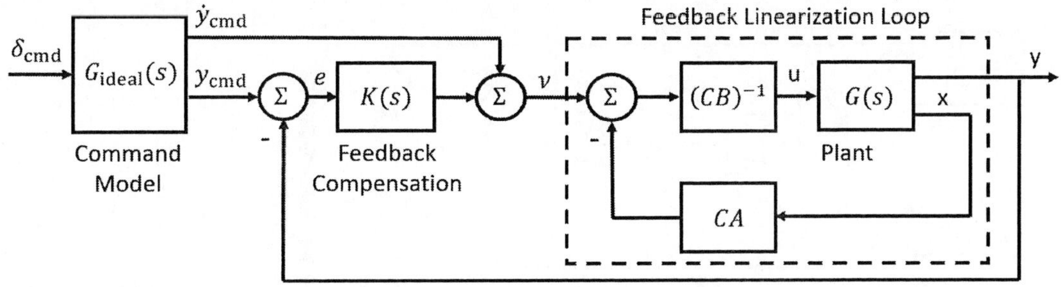

Figure 3: DI controller as applied to a linear system.

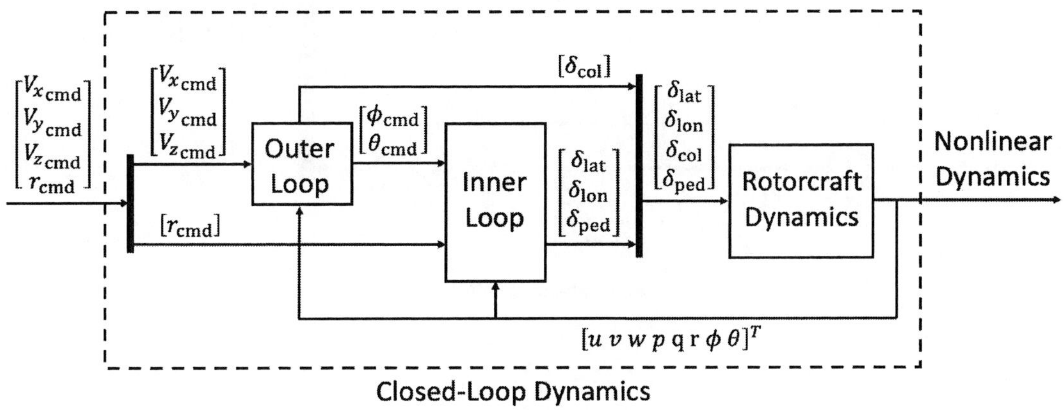

Figure 4: Schematic of the closed-loop tiltrotor dynamics.

Finally, the pilot input bias vector is:

$$u_b^T = \begin{bmatrix} -50 & -50 & -50 & -50 \end{bmatrix} \tag{10}$$

The configuration under consideration is controlled exclusively via collective pitch. Consequently, the longitudinal cyclic and lateral cyclic controls are set to zero, and the rotor speed is prescribed. Due to the fact that the tailsitter may operate at pitch angles of $\theta = 90°$, particularly during hover, the rotational kinematics equations of motion are prone to singularities (ie, gimbal lock). To address this issue, alternative equations for the time rate of change of the Euler angles are employed when $89.99° \leq \theta \leq 90.01°$ (Ref. 19):

$$\dot{\phi} = \frac{p}{2} - \frac{1}{2}\left(\frac{\dot{r}q - \dot{q}r}{q^2 + r^2}\right), \tag{11}$$

$$\dot{\theta} = \begin{cases} \sqrt{q^2 + r^2}, & \theta \text{ is increasing,} \\ -\sqrt{q^2 + r^2}, & \theta \text{ is decreasing,} \end{cases} \tag{12}$$

$$\dot{\psi} = -\frac{p}{2} - \frac{1}{2}\left(\frac{\dot{r}q - \dot{q}r}{q^2 + r^2}\right). \tag{13}$$

Linearized Models Linearized time-invariant (LTI) models are obtained by trimming the rotorcraft flight dynamics at a specified flight condition, followed by linearization using perturbation methods (Ref. 20). To simplify the dynamics and make the models more suitable for practical control design, the order of the linearized dynamics is reduced. This is achieved by eliminating states associated with higher-order dynamics, including rotor and higher harmonic dynamics. This process, known as residualization, is part of singular perturbation theory applicable to LTI systems (Ref. 21). The application of residualization to rotorcraft flight dynamics has been documented in various studies (Refs. 16, 17, 22, 23).

The flight control architecture employed in this study is Nonlinear Dynamic Inversion (NDI). The use of NDI control laws for rotorcraft has been extensively discussed in the literature (Refs. 18, 22–29). A distinguishing feature of NDI is its reliance on model inversion to cancel plant dynamics and achieve tracking of a desired reference model. An advantage of NDI is that it inherently inverts the plant model within its feedback linearization loop. This eliminates the need for gain scheduling, a requirement in other conventional model-following strategies such as explicit model following (EMF). However, the plant model used for feedback linearization still requires scheduling based on the flight condition.

A generic DI controller as applied to a linear system is illustrated in Fig. 3. The main components of the controller include a command model (also referred to as a command filter or reference model) that specifies the desired response to pilot inputs, feedback compensation for tracking errors, and an inner feedback loop that performs model inversion (ie, feedback linearization).

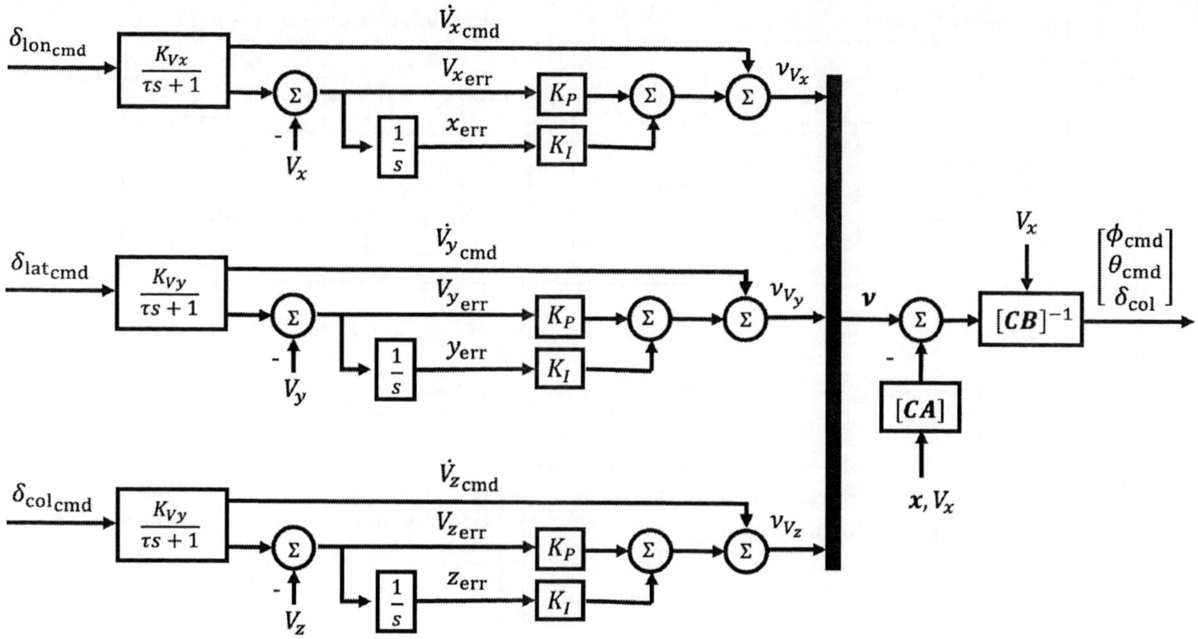

(a) Outer-velocity control loop.

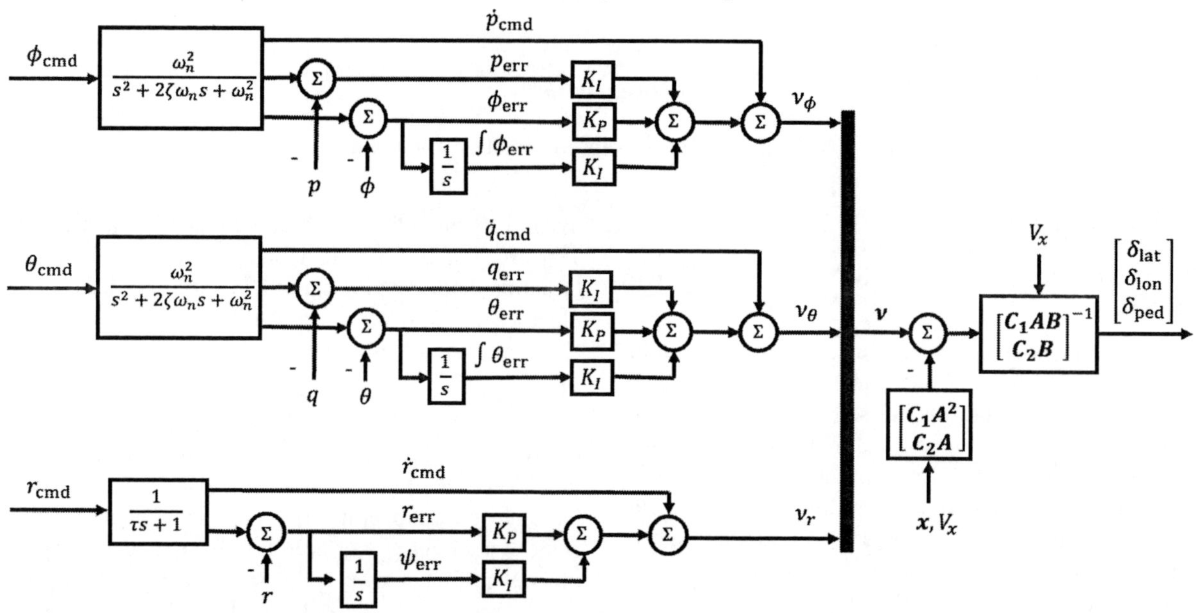

(b) Inner-attitude control loop.

Figure 5: Dynamic inversion inner-attitude and outer-velocity control loops.

Flight Control Design A multi-loop NDI control law based on (Refs. 16, 17, 25, 26) is designed to enable autonomous flight across low-speed flight (helicopter mode), high-speed flight (airplane mode), and transitions between the two. The closed-loop rotorcraft dynamics are represented schematically in Fig. 4.

The outer loop controller, shown in Fig.5a, tracks longitudinal, lateral, and vertical ground velocity commands in the heading frame. It computes the desired pitch and roll attitudes for the inner loop to track, along with the collective control input. The outer loop achieves a Translational Rate Command (TRC) response. The inner loop, depicted in Fig.5b, ensures stability, disturbance rejection, and desired response characteristics around the roll, pitch, yaw, and heave axes. When coupled with the outer loop, the system employs an Attitude Command / Attitude Hold (ACAH) response for roll and pitch axes, and a Rate Command / Attitude Hold (RCAH) response for the yaw axis.

CFD Framework and Model

All high-fidelity simulations of the CRC-80 hover-to-cruise transition presented in this paper are performed using quasi-steady analysis. In the context of this paper, quasi-steady analysis consists of freezing the control inputs and orientation of the aircraft in a point in time of the transition, and then allowing the case to run until convergence (periodic loads). Note that these cases will still vary in time since the blades are still changing position with every time step.

High-fidelity aerodynamic simulations were conducted using the RANS-based in-house CFD framework, Mercury (Ref. 30). The framework's structure, along with its components and capabilities, is illustrated in Figure 6. For this study, the near-body domains were solved using the Hamiltonian/Strand (HAMSTR) solver (Ref. 31), while the off-body Cartesian domains were handled by the Graphics Processing Unit Accelerated Rotor Flow Field (GARFIELD) solver (Ref. 32). The connectivity between the overset meshes was managed by the Topology Independent Overset Grid Assembler (TIOGA) (Ref. 33).

The simulations employed a 2nd-order backward difference (BDF2) implicit time-marching scheme and a 5th-order weighted essentially non-oscillatory (WENO) scheme for spatial discretization. To accelerate convergence, an initial simulation phase was performed with a 1° time step over 4-9 revolutions depending on the free-stream velocity, followed by 2-3 revolutions at a 0.5° time step with 30 sub-iterations per step to obtain the final results. Simulations closer to hover take longer to converge than fast axial flight. Turbulence closure was achieved using the Spalart-Allmaras delayed detached eddy simulation (SA-DDES) model (Ref. 34).

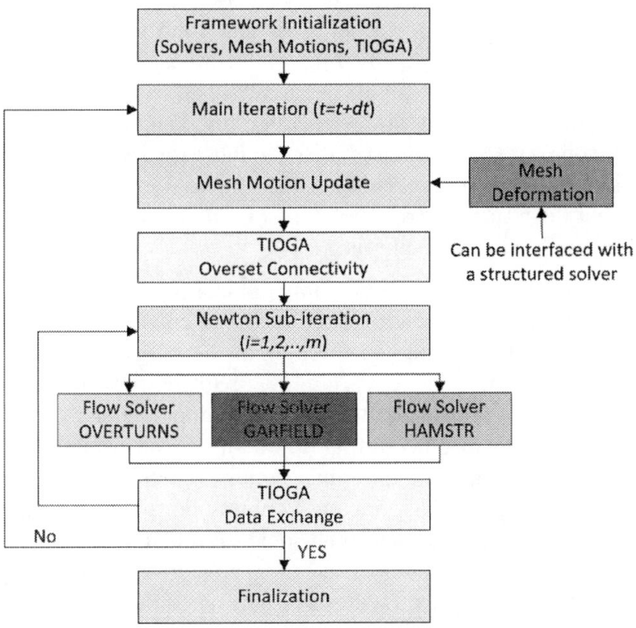

Figure 6: Mercury CFD Framework

The surface mesh for the simulation was generated using a combination of triangular and quadrilateral elements, which

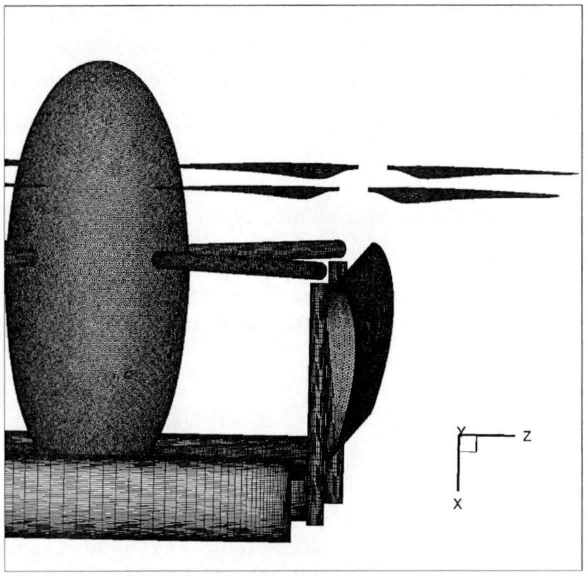

(a) Sub-divided all-quad surface mesh

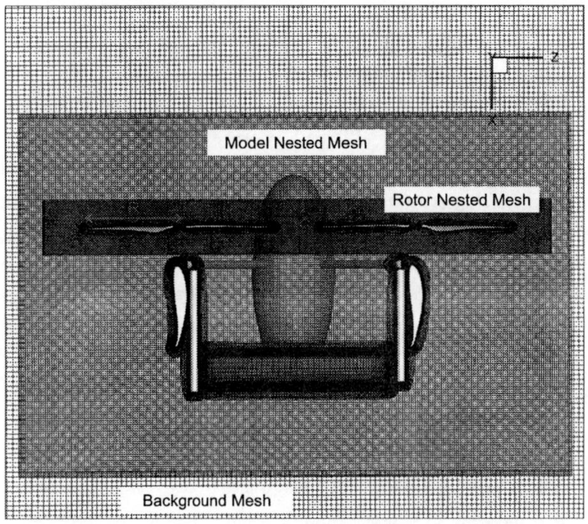

(b) Sectional view of overset volume mesh system

Figure 7: Computational mesh system

were subsequently subdivided into a 434,704-element all-quadrilateral mesh to meet the requirements of the in-house solver HAMSTR for identifying Hamiltonian paths (Ref. 31). From this surface mesh, 43 strand layers were generated to construct the near-body volume meshes. The initial wall spacing, set at $4.5 \times 10^{-6}R$, corresponds to a y^+ value of 1, with a stretching ratio of 1.16 applied between layers.

For the off-body region, six nested Cartesian meshes were used, including four rotor-nested meshes, one model-nested mesh, and one background mesh. These meshes were designed with a graded spacing to ensure finer resolution near the aircraft and coarser resolution towards the domain boundaries. The simulation domain extends to $26r$ in the y-direction and $20r$ in both the x- and z-directions, where r is the rotor radius. The coordinate system is defined as illustrated in Figure 1, and the outer boundaries are assigned a free-stream boundary condition. In order to lower computational cost, the mesh system was cut in half to create a half-span model, for which an inviscid wall was implemented at the symmetry plane.

The near-body strand meshes for the half-span mesh consist of approximately 10 million elements, while the nested Cartesian meshes comprise 30 million elements in total. The surface mesh and the overset volume mesh system are depicted in Figure 7.

Mid-Fidelity Aerodynamic Tool and Model

The in-house flight dynamics simulation will be coupled with the mid-fidelity aerodynamic code, DUST. DUST (Ref. 35) is an open-source mixed boundary element-vortex particle method developed at Politecnico di Milano. It uses the Helmholtz decomposition of the velocity field and is well-suited for analyzing the interactional aerodynamics of complex advanced air mobility aircraft. The code allows users to assign varying fidelity levels to different components as needed.

The mid-fidelity aerodynamic model includes a central pod, two wings, eight rotor blades, and two stabilizers. The mesh created for the mid-fidelity aerodynamic model are shown in Figure 8. For the CRC-80, the rotor blades are represented as lifting lines with 20 spanwise sections. The blades are truncated at the root to 7% radius to allow for blade pitch. The central body and wings are modeled using panels, with the wings discretized into 20 spanwise and 10 chordwise sections. In the mid-fidelity model, the connecting rods between the wings and the central body are excluded. Simulations are conducted using a 7.5° time step, determined based on the average of the maximum and minimum RPM values chosen for the maneuver. Although the physical time step in terms of azimuth degrees varies with RPM, a sensitivity study confirmed that this variation does not significantly affect the aerodynamic loads.

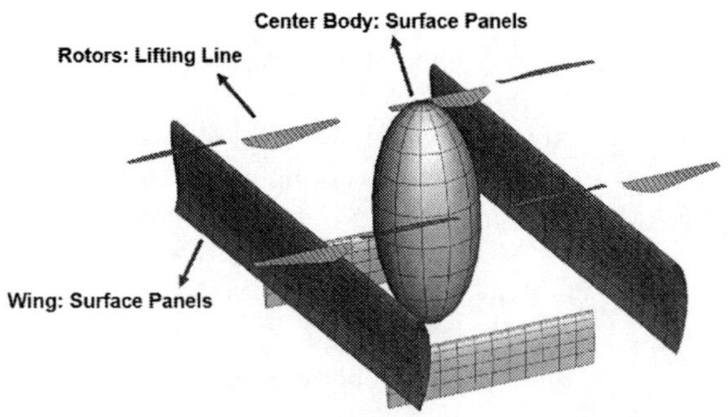

Figure 8: Mid-Fidelity (VPM) Mesh

Perturbations to the DUST simulations were performed in order to inform the control and state derivatives in the flight dynamics model. To expedite this process, a wrapper for DUST was developed to efficiently create trim Jacobians, and use them to trim desired steady states that could then be perturbed. Based on the objective of modeling hover-to-cruise transition, the problem was confined to three degrees of freedom, horizontal motion, vertical motion, and aircraft pitch.

Two types of analysis were used in the hover-to-cruise transition simulations:

Unsteady: The time-varying aircraft velocities, aircraft attitude, rotor angular velocity and rotor pitch describing the maneuver are prescribed to DUST from GenMR. Major discrepancies between the resulting loads were addressed by refining various aspects of the simulations. The capabilities to input a time history for the aircraft motion and controls were added to DUST in collaboration with its developers.

Quasi-steady: Once a successful hover-to-cruise simulation was performed with the coupled VPM and flight dynamics, eight equally spaced instances in time throughout the maneuver were selected for quasi-steady analysis. At each of these stations, the aircraft attitude and controls were frozen in time, and the aircraft wake was converged until it reached a periodic behavior. Note that, due to the nature of rotorcraft simulations, there is almost always time dependence due to blade motion. Therefore, the term "steady" is used to refer to steady trimmed flight, as opposed to true time invariance.

RESULTS

Flight Dynamics and Control

The first step in accurately modeling the hover-to-cruise transition of the CRC-80 starts with defining realistic controls that result in the desired flight trajectory. For this we used GenMR, our in-house flight dynamics code coupled with dynamic inflow and lifting lines for aerodynamics.

The aircraft is trimmed at flight speeds ranging from 0 to 80 kts, with the resulting trim attitudes and control inputs presented in Fig. 9. Specifically, Fig. 9a illustrates the trim Euler angles as a function of flight speed. While all trim conditions maintain zero roll and yaw angles, the pitch angle decreases from approximately 80° to 0° as speed increases. Most of this change in pitch angle occurs between 0 and 45 kts, suggesting that the transition from helicopter mode to airplane mode occurs within this velocity range.

Figure 9b shows the corresponding control inputs as a function of flight speed. Lateral and directional control inputs remain constant and centered throughout the range. The collective control input increases slightly up to approximately 30 kts, after which it increases linearly at a steeper rate. This suggests that flight speed could be limited by power availability or rotor blade stall. Longitudinal control inputs exhibit less significant changes: they increase slightly up to about 5 kts, decrease up to 30 kts, and then increase quadratically with speed. This behavior compensates for the increasing pitching moment of the cambered wings as flight speed increases.

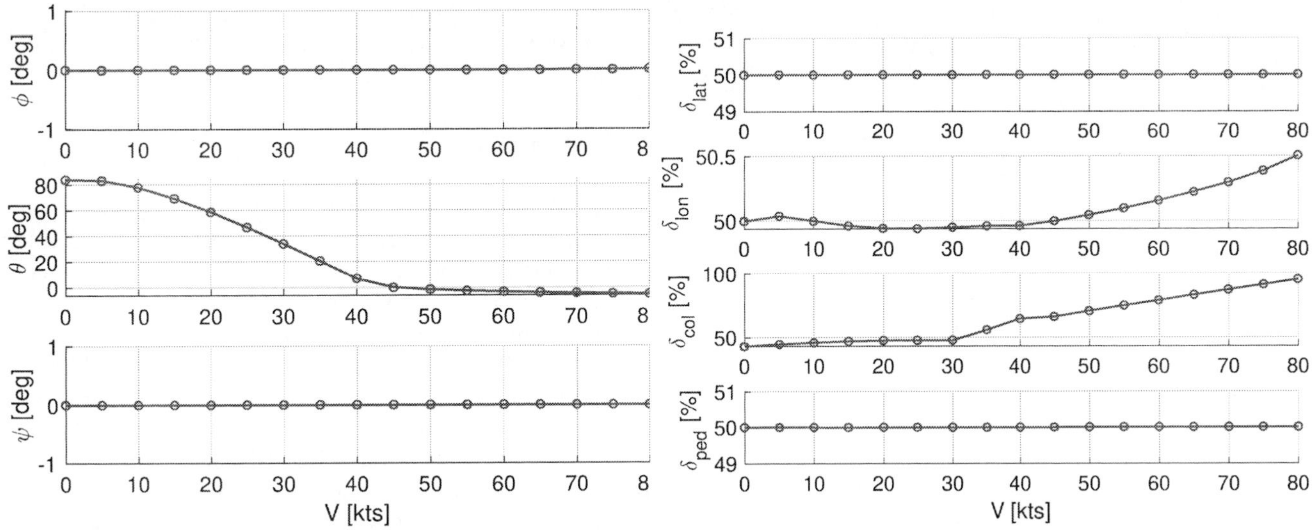

(a) Attitude.

(b) Control inputs.

Figure 9: Trim variables at incremental airspeeds.

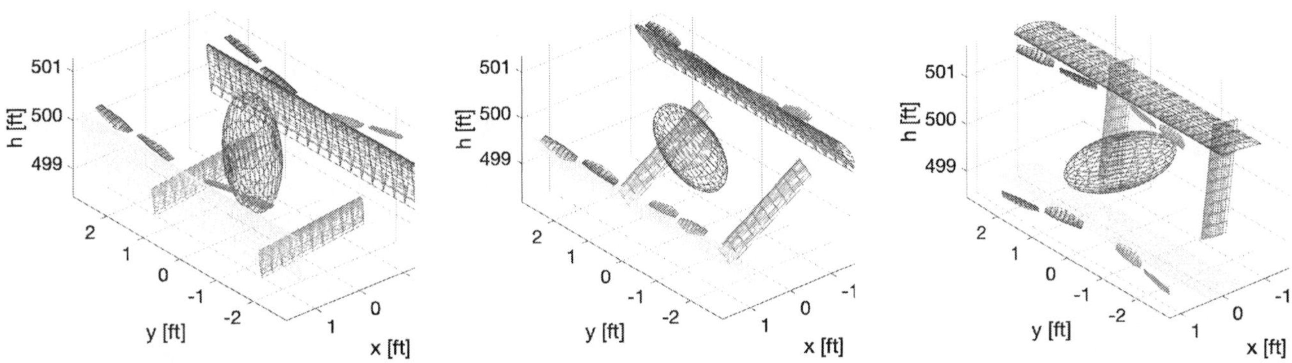

(a) Hover.

(b) 20 kts.

(c) 40 kts.

Figure 10: Rotorcraft geometry in trimmed flight at incremental airspeeds.

Figure 10 illustrates the geometry corresponding to selected trim conditions (ie, 0, 20, and 40 kts). The figure shows that the aircraft pitches forward with increasing speed, transitioning from relying primarily on rotor lift in hover to wing-generated lift in forward flight. At higher speeds, the rotors function primarily as propulsive devices. It is noteworthy that the hover pitch attitude is less than 90° due to rotor downwash acting on the cambered wings, which generates lift. As a result, the aircraft must pitch forward to achieve zero net force along the longitudinal axis.

Dynamic inversion control laws, based on residualized models obtained at the previously discussed trim conditions, are applied in a closed-loop simulation of an autonomous hover-to-cruise transition. The transition is executed over 3.2 seconds, covering a speed range from 5 to 40 kts. This simulation incorporates the nonlinear rotorcraft dynamics and is illustrated in Fig. 11.

Figure 11a presents the heading-frame velocities. The longitudinal velocity closely follows the reference trajectory, while the lateral and vertical velocities exhibit minimal off-axis responses. The commanded velocity profiles are sinusoidal for both longitudinal and vertical velocities. The longitudinal velocity (x-direction) has a sinusoidal profile with a period twice the maneuver time, while the vertical velocity (z-direction) has a period equal to the maneuver time. Additionally, the maximum z velocity is set to half of the maximum x velocity. These velocity profiles are inspired by previous work in (Ref. 36).

The minimum transition time, ensuring the rotorcraft avoids blade stall, was determined by parametrizing the velocity trajectories based on maneuver time and progressively reducing it until blade stall was observed. Blade stall typically occurs at the midpoint of the maneuver. For this configuration, the minimum estimated transition time is 3.2 seconds, significantly influenced by rotorcraft weight, rotor speed, and blade airfoil characteristics.

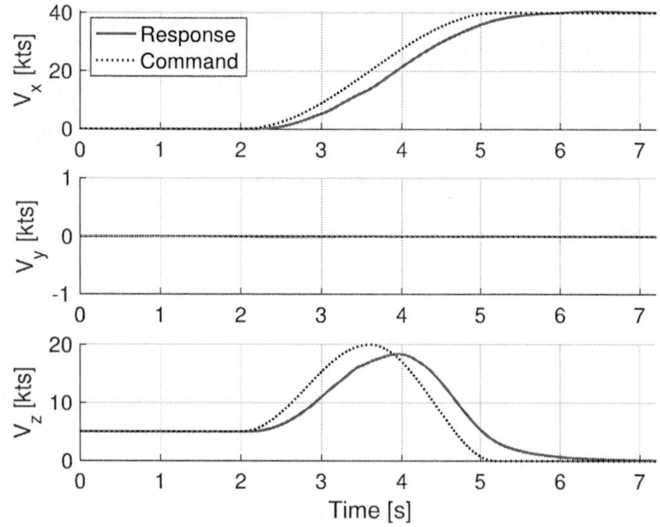

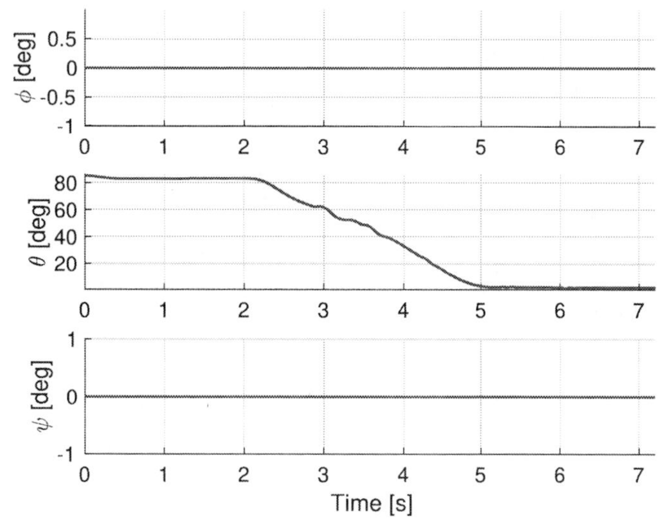

(b) Attitude.

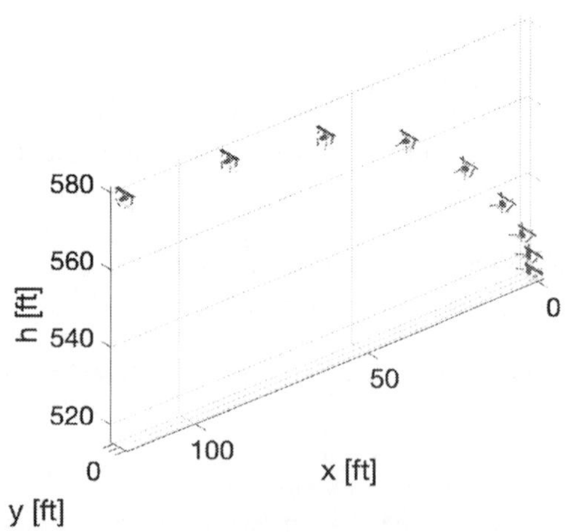

(c) Trajectory snapshots.

Figure 11: Closed-loop hover-to-cruise transition.

Figure 11b demonstrates the performance of the inner attitude loop, showing accurate tracking of the desired pitch attitude while maintaining zero roll and yaw attitudes. Finally, Fig. 11c provides snapshots of the flow trajectory throughout the transition.

Prescribed Maneuver in Mid-Fidelity

The results of the flight dynamics simulation, including rotor collectives, rotor angular velocities, aircraft attitude, and aircraft velocities, are prescribed to DUST, the mid-fidelity solver based on VPM. The time histories used for this prescribed simulation are shown in Fig. 12. The resulting forces are shown in Fig. 13. The most noticeable difference between the two fidelity simulations is the presence of high-frequency fluctuations in the VPM solver, not seen in the low-fidelity. This is due to the rotor-on-wing and wing-on-rotor interactions captured in VPM. While this may not affect the aircraft motion, it would likely have an effect on acoustic predictions.

The rotor forces are shown in Fig. 13a, the plot includes the vertical and horizontal forces separately in the inertial frame for both GenMR (dynamic inflow) and DUST (VPM). At the beginning of the maneuver, the aircraft is mostly climbing, so the vertical forces of the rotor dominate. As the transition continues, the vertical force from the rotors start to decrease since the wings begin to carry the weight of the aircraft, and the rotors' horizontal force dominates as the CRC-80 accelerates to 20 m/s. The loads from the low- and high-fidelity simulations match very well for the rotors. The peaks observed in the dynamic inflow are not present because the collectives where smoothened when prescribing the maneuver from the flight dynamics to VPM.

The airframe results shown in Fig. 13b do not match that well between low- and mid-fidelity. The mid-fidelity airframe shows generally higher lift than the low-fidelity analysis. The mid-fidelity should capture the interactions between the rotors and the wing better, but the wing panels do not include airfoil table corrections. Meanwhile, the low-fidelity analysis has a simpler model for rotor-on-wing interaction, but does include airfoil tables in the wing model. The advantages and shortcomings of each simulation make it tough to decide what the more correct solution is without high-fidelity analysis.

Unsteady Maneuver vs. Quasi-Steady Simulations

Part of the objective of this work is to understand wether a fully unsteady simulation of the CRC-80 transition unveils additional information not attainable with quasi-steady analysis. As a reminder, in this study, quasi-steady simulations refer to freezing the control inputs, aircraft position, and aircraft orientation in time. All simulations are, however, time-variant since the rotor blades continue to change position. While the unsteady maneuver simulation in high-fidelity RANS CFD was not completed for this paper, there are results for the transition quasi-steady analysis.

Quasi-steady simulations were performed at eight equally spaced states throughout the maneuver. The results of the quasi-steady analysis are shown for both the mid- (VPM) and

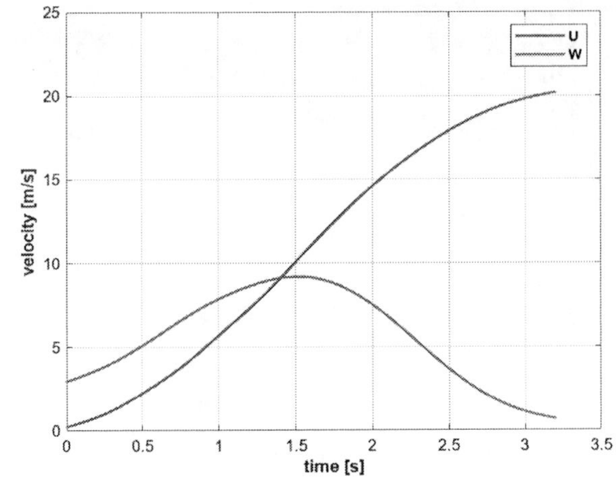

(a) Inertial-frame velocities.

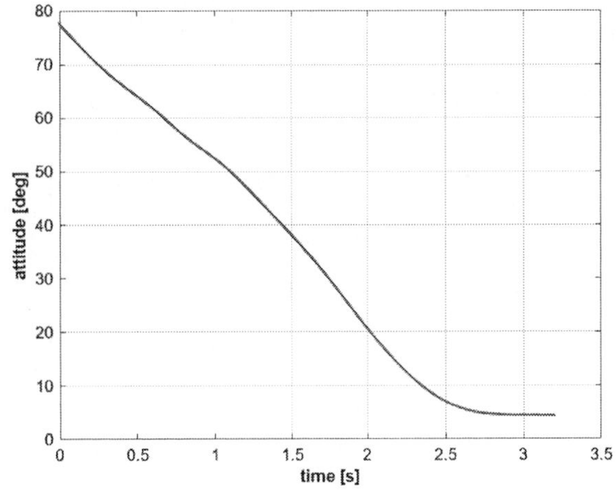

(b) Attitude.

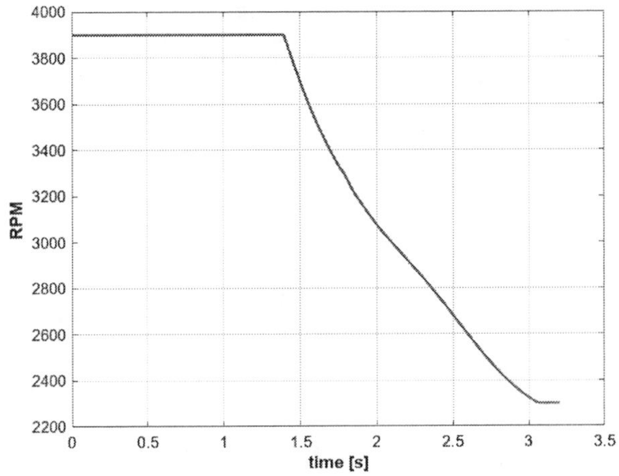

(c) Rotor angular velocity.

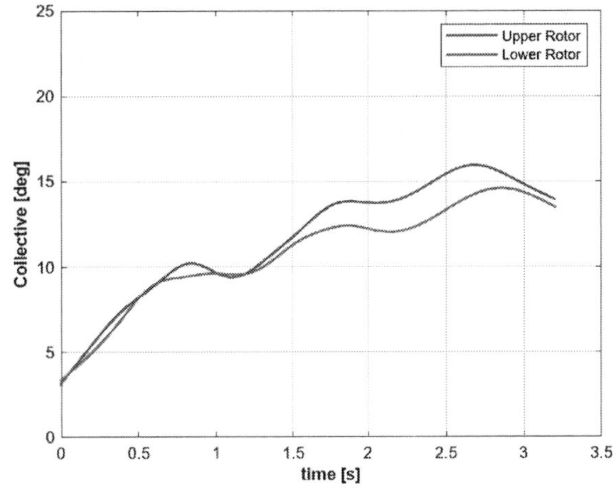

(d) Rotor collective.

Figure 12: Inputs to mid-fidelity solver for prescribed maneuver.

high-fidelity (RANS CFD) simulations in Fig. 14, and plotted next to the unsteady results from the low- (dynamic inflow) and mid-fidelity simulations.

Focus first on the comparison between the unsteady and quasi-steady VPM simulations, that is, the solid lines and diamonds. For both the rotor and airframe, the quasi-steady simulations are almost the same as the mean of the unsteady simulation. This implies that, for this maneuver and aircraft, in the mid-fidelity, there is no large impact on the mean loads from non-periodic effects. The airframe lift loads are only different at the third state, just before the one second mark, which is the state closest to maximum rotor thrust.

Looking at the RANS CFD quasi-steady results, shown as squares, there is good correlation to both unsteady and quasi-steady VPM, as well as to the dynamic inflow results. It is encouraging to see that the rotor controls created in the flight dynamics simulation give the expected loads for all levels of fidelity tested. Interestingly, for the airframe loads, the RANS simulation matches closer to the low-fidelity than to the mid-fidelity results. Although the mid-fidelity should be better at modeling the rotor-on-wing interactions, it appears that the airfoil tables included in the low-fidelity wing model provides a more accurate result in this case than the panels used in the VPM code. Overall, the quasi-steady RANS CFD results match the desired loads from the flight dynamics simulation well.

While we have determined that the loads match well between the different fidelity solvers for both unsteady and quasi-steady simulations, the high-frequency variations in the aircraft loads have a bigger impact in acoustic analysis than the mean loads. It is therefore important to compare the magnitude of the fluctuations of the forces, so that we can make informed predictions on the acoustic results expected from the different simulations. For this purpose, the converged, periodic loads from the quasi-steady simulations are compared to the unsteady mid-fidelity maneuver in Fig. 15. For this plot, the mean was removed from all results to emphasize the high-frequency fluctuations. To better quantify and compare the unsteadiness of the loads, the root-mean-square of the forces,

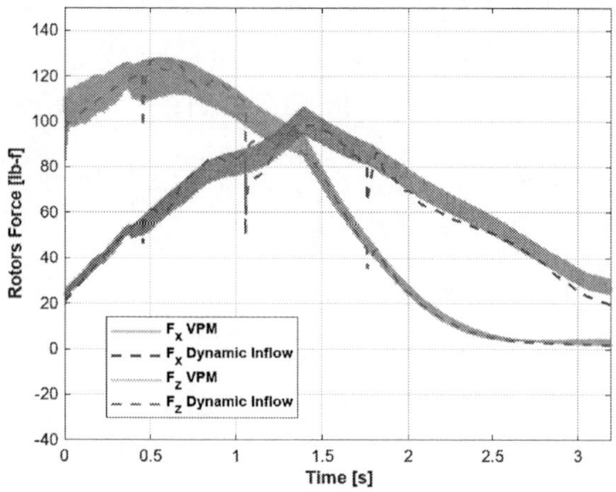

(a) Rotor Forces.

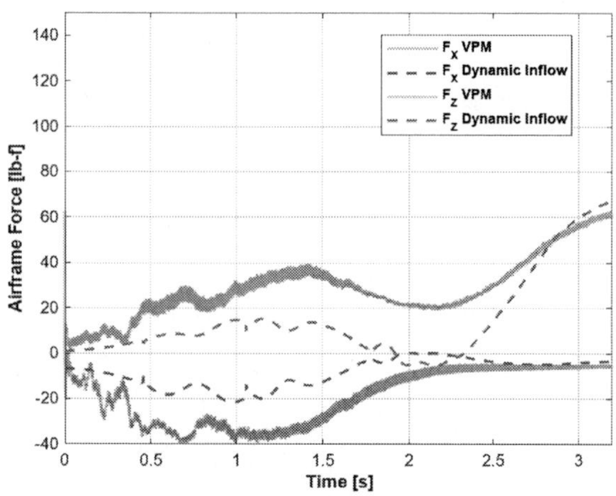

(b) Airframe Forces.

Figure 13: GenMR (Dynamic Inflow) vs. DUST (VPM) unsteady maneuver forces comparison.

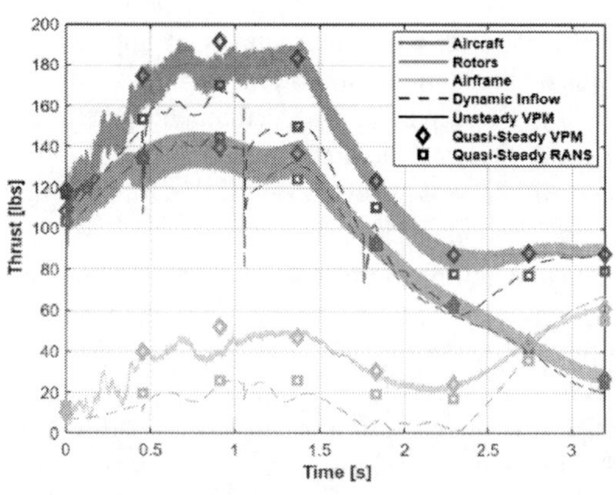

Figure 14: Unsteady maneuver vs. quasi-Steady analysis for three different fidelity simulations.

with the mean subtracted, are shown in Fig. 16. Comparing the unsteady and quasi-steady VPM results shows slightly larger fluctuations from the unsteady simulation at the first

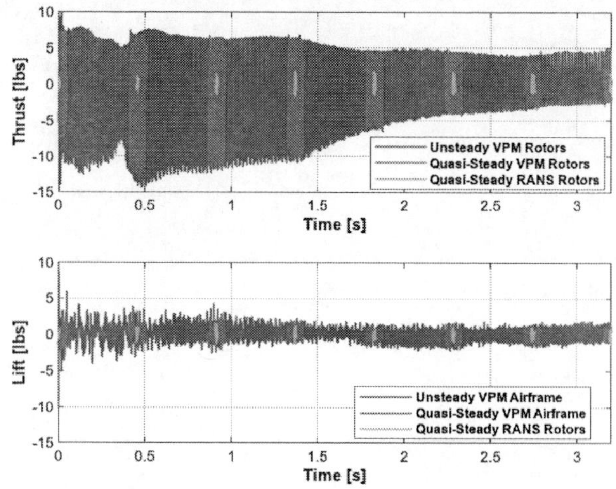

Figure 15: Load variability in unsteady and quasi-steady simulations. (mean subtracted)

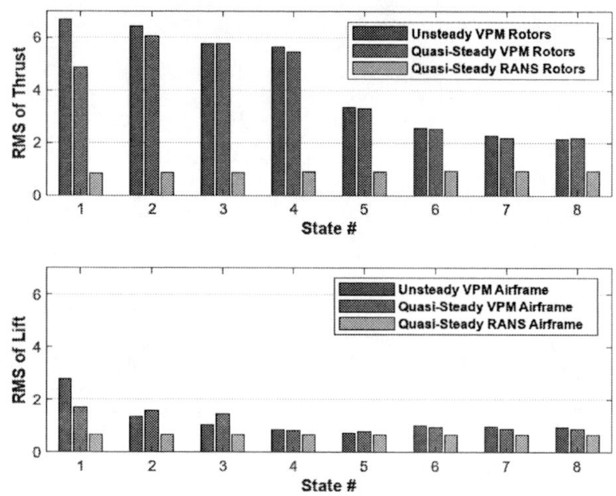

Figure 16: Root-mean-squared of forces with mean subtracted of unsteady and quasi-steady simulations.

few states. The largest difference is seen at the first state, which is most likely a result of wake instabilities encountered in the mid-fidelity solver when close to hover. For the most part however, the unsteady and quasi-steady high-frequency fluctuations are similar, and therefore acoustic analysis using either method is likely to be comparable at any given point of the maneuver. Both the airframe and rotors, but the rotors in particular, show a trend of higher fluctuations at the beginning of the transition, and reduced fluctuations as the aircraft approaches the cruise condition.

While the unsteady and quasi-steady VPM simulations math well with each other, the RANS results have major differences. The most noticeable discrepancy is that the fluctuations of the RANS rotors are several times smaller than those of both VPM simulations, suggesting that the mid-fidelity simulations are considerably overestimating the wing-on-rotor interactions, and would most likely also overestimate the resulting noise. The RANS quasi-steady simulation also shows similar fluctuations between the wing and rotor, as well as close to no changes in fluctuations between the eight states of the maneuver. It will be insightful to compare the quasi-steady RANS to its unsteady counterpart once it is completed.

CONCLUSIONS

A flight dynamics model of the CRC-80 was created using the in-house UMD code GenMR, and successfully coupled to higher fidelity aerodynamics by prescribing the time varying controls and aircraft motions. A hover-to-cruise transition maneuver was simulated using this model. The loads predicted by the flight dynamics code using a dynamic inflow model, were satisfactorily recreated using VPM analysis with a quasi-steady and fully unsteady methodology, as well as quasi-steady RANS CFD simulations. The mean rotor loads were similar between all fidelities and methodologies used. However, airframe loads were consistent only between low- and high-fidelity simulations, as the mid-fidelity VPM simulation significantly overestimated mean airframe loads.

When comparing the unsteady and quasi-steady simulations for the VPM solver, both the mean values of the aircraft loads, as well as the magnitude of the oscillations of the forces were similar. This result suggests that the least expensive of the two methods, the quasi-steady analysis, is sufficient for analyzing both the forces that affect aircraft motion, and acoustic analysis which is impacted by higher-frequency fluctuations. This is true for a 3.2 second hover-to-cruise transition simulated using the VPM based solver DUST.

It was found that the rotor loads fluctuations obtained using RANS CFD quasi-steady analysis are several times smaller than the VPM fluctuations. The RANS simulations also showed similar fluctuations at different points of the maneuver, while the VPM simulations exhibited a pattern of larger fluctuations at the beginning of the transition, which then diminished as the aircraft reached a more steady cruise condition. The large discrepancies between RANS and VPM load unsteadiness suggest that the VPM simulations could significantly overestimate rotor noise. The differences in load variability were smaller for the airframe loads.

The findings suggest that quasi-steady analysis is sufficient to accurately capture the loads impacting the motion of the CRC-80, even for a complex configuration and fast maneuver. A comparison to an unsteady RANS CFD maneuver still has to be performed to confirm that this is also the case for higher-fidelity analysis.

FUTURE WORK

An unsteady hover-to-cruise simulation in RANS CFD is essential to conclusively compare quasi-steady and unsteady analysis of the aircraft transition. Aeroacoustic analysis also will be performed to confirm the predictions inferred by observing the thrust and lift fluctuations. Different and potentially more complicated maneuvers, such as a cruise-to-hover transition, will also be tested.

ACKNOWLEDGEMENTS

This study was funded by the Vertical Lift Research Center of Excellence (VLRCOE) under the technical monitoring of Dr. Mahendra Bhagwat. The computational resources required for the high-fidelity CFD simulations were provided by Zaratan high-performance computing cluster at the University of Maryland. We thank Dr. Alessandro Cocco for his guidance and development assistance for the DUST simulations.

REFERENCES

1. González-Jorge, H., Martínez-Sánchez, J., Bueno, M., and Arias, P., "Unmanned Aerial Systems for Civil Applications: A Review," *Drones*, Vol. 1, (1), 2017, pp. 2.

2. Avera, M., and Singh, R., "Scalability of Hybrid-Electric Propulsion for VTOL UAS," Proceedings of the NATO Research Symposium on Hybrid/Electric Aero-Propulsion Systems for Military Applications, Trondheim, Norway, 2019.

3. Phillips, B., Hrishikeshavan, V., Rand, O., and Chopra, I., "Design and Development of a Scaled Quadrotor Biplane with Variable Pitch Proprotors for Rapid Payload Delivery," Proceedings of the American Helicopter Society 72nd Annual Forum, West Palm Beach, FL, USA, 2016.

4. Reddinger, J.-P., McIntosh, K., Zhao, D., and Mishra, S., "Modeling and Trajectory Control of a Transitioning Quadrotor Biplane Tailsitter," Proceedings of the Vertical Flight Society 75th Annual Forum, Philadelphia, PA, USA, 2019.

5. Jayasundara, D., and Baeder, J., "Aerodynamic and Aeroacoustic Analysis of a Quadrotor Biplane Tailsitter in Forward Flight," Proceedings of the 79th Vertical Flight Society Annual Forum, West Palm Beach, Florida, 2023.

6. Jayasundara, D., and Baeder, J., "Aerodynamic and Aeroacoustic Analysis of a Hovering Quadrotor Biplane Tailsitter," Proceedings of the 10th Biennial Autonomous VTOL Technical Meeting and 10th Annual Electric VTOL Symposium, 2023.

7. Arias, P., Jayasundara, D., and Baeder, J., "Aeroacoustic Analysis of a Quadrotor Biplane Tailsitter in Climb and Synchrophased Hover," Proceedings of the Vertical Flight Society's 6th Decennial Aeromechanics Specialists' Conference, 2024.

8. McIntosh, K., Reddinger, P., Mishra, S., and Zhao, D., "Optimal Trajectory Generation for Transitioning Quadrotor Biplane Tailsitter using Differential Flatness," Proceedings of the 77th Annual Forum of the Vertical Flight Society, Virtual, May 10-14, 2021. DOI: 10.4050/F-0077-2021-16858.

9. Arias, P., Saetti, U., and Baeder, J., "Multi-fidelity Analysis of Quadrotor Biplane Tailsitter Hover-to-Cruise Transition," AIAA SCITECH 2025 Forum, 2025.

10. Leishman, J. G., *Principles of Helicopter Aerodynamics*, Cambridge University Press, 2006.

11. Pitt, D. M., and Peters, D. A., "Theoretical Prediction of Dynamic-Inflow Derivatives," Proceedings of the 6th European Rotorcraft and Powered Lift Aircraft Forum, Bristol, England, September 16-19, 1980.

12. Guner, F., and Prasad, J. V. R., "Combined Momentum Theory and Simple Vortex Theory Inflow Model for Multirotor Configurations," *Journal of the American Helicopter Society*, Vol. 67, (2), 2022, pp. 1–15. DOI: https://doi.org/10.4050/JAHS.67.022007.

13. Guner, F., "A Multirotor Inflow Model Based on Combined Momentum Theory and Simple Vortex Theory (CMTSVT) for Flight Simulations," Proceedings of the 78th Annual Forum of the Vertical Flight Society, Fort Worth, TX, May 10-12, 2022.

14. Saetti, U., and Guner, F., "Interactional Aerodynamics Modeling and Flight Control Design of Multi-Rotor Aircraft," Proceedings of the 80th Decennial VFS Aeromechanics Specialists' Conference, Santa Clara, CA, Feb 6-8, 2024.

15. Hyeson, H. H., "Equations for the Induced Velocities Near a Lifting Rotor with Nonuniform Azimuthwise Vorticity Distribution," Technical report, NASA TN D-394, 1960.

16. Horn, J. F., "Non-Linear Dynamic Inversion Control Design for Rotorcraft," *Aerospace*, Vol. 6, (38), 2019. DOI: https://doi.org/10.3390/aerospace6030038.

17. Saetti, U., and Bugday, B., "Tiltrotor Simulations with Coupled Flight Dynamics, State-Space Aeromechanics, and Aeroacoustics," *Journal of the American Helicopter Society*, Vol. 69, (1), jan 2024. DOI: https://doi.org/10.4050/JAHS.69.012003.

18. Saetti, U., Horn, J. F., Lakhmani, S., Lagoa, C., and Berger, T., "Design of Dynamic Inversion and Explicit Model Following Control Laws for Quadrotor UAS," *Journal of the American Helicopter Society*, **65**, 032006 (2020). DOI: https://doi.org/10.4050/JAHS.65.032006.

19. Regan, F. J., *Re-Entry Vehicle Flight Dynamics*, AIAA Education Series, 1984, Chapter Appendix D.

20. Saetti, U., and Rogers, J. D., "Revisited Harmonic Balance Trim Solution Method for Periodically-Forced Flight Vehicles," *Journal of Guidance, Control, and Dynamics*, Vol. 44, (5), May 2021, pp. 1008–1017. DOI: https://doi.org/10.2514/1.G005553.

21. Kokotovic, P. V., O'Malley, R. E., and Sannuti, P., "Singular Perturbations and Order Reduction in Control Theory, an Overview," *Automatica*, Vol. 12, (2), 1976, pp. 123–132. DOI: 10.1016/0005-1098(76)90076-5.

22. Saetti, E. J., U., , and Horn, J. F., "Flight Dynamics and Control of an eVTOL Concept Aircraft with a Propeller-Driven Rotor," *Journal of the American Helicopter Society*, Vol. 67, (3), 2022, pp. 153–166. DOI: https://doi.org/10.4050/JAHS.67.032012.

23. Saetti, U., Rogers, J. D., Alam, M., and Jump, M., "Tau Theory-Based Flare Control in Autonomous Helicopter Autorotation," *Aerospace*, Vol. 33, (1), 2024. DOI: https://doi.org/10.3390/aerospace11010033.

24. Horn, J. F., "Non-Linear Dynamic Inversion Control Design for Rotorcraft," *Aerospace*, Vol. 6, (38), 2019. DOI: https://doi.org/10.3390/aerospace6030038.

25. Berger, T., Tischler, M. B., and Horn, J. F., "Outer-Loop Control Design and Simulation Handling Qualities Assessment for a Coaxial-Compound Helicopter and Tiltrotor," Proceedings of the 77th Annual Forum of the Vertical Flight Society, Virtual, Oct 5-8, 2020. DOI: 10.4050/F-0076-2020-16392.

26. Berger, T., Tischler, M. B., and Horn, J. F., "High-Speed Rotorcraft Pitch Axis Response Type Investigation," Proceedings of the 77th Annual Forum of the Vertical Flight Society, Virtual, May 10-14, 2021. DOI: 10.4050/F-0077-2021-16793.

27. Saetti, U., Horn, J. F., and Berger, T., "On the Effects of Rotor Induced Vibrational Stability on Helicopter Flight Dynamics," *CEAS Aeronautical Journal,* Article in Advance, 2024. DOI: https://doi.org/10.1007/s13272-024-00718-w.

28. Saetti, U., and Bugday, B., "Tiltrotor Simulations with Coupled Flight Dynamics, State-Space Aeromechanics, and Aeroacoustics," *Journal of the American Helicopter Society*, Vol. 69, (1), 2024. DOI: https://doi.org/10.4050/JAHS.69.012003.

29. Saetti, U., "Dynamic Inversion Flight Control Laws for Autonomous Transition of Tilt-Rotor Aircraft," Proceedings of the Second International Conference on Advanced Air Mobility Systems (ICAAMS-2), Singapore, Dec 4–6, 2024.

30. Jude, D., Lee, B., Jung, Y. S., Petermann, J., Govindarajan, B., and Baeder, J., "Application of a Heterogeneous CFD Framework Towards Simulating Complete Rotorcraft Configurations," American Helicopter Society, 74th Annual Forum Proceedings, Phoenix, Arizona, May 14-17, 2018.

31. Jung, Y. S., Govindarajan, B., and Baeder, J., "Turbulent and Unsteady Flows on Unstructured Line-Based Hamiltonian Paths and Strands Grids," *AIAA Journal*, Vol. 55, (6), 2017, pp. 1986–2001.

32. Jude, D., and Baeder, J. D., "Extending a Three-Dimensional GPU RANS Solver for Unsteady Grid Motion and Free-Wake Coupling," 54th AIAA Aerospace Sciences Meeting, 2016.

33. Brazell, M. J., Sitaraman, J., and Mavriplis, D. J., "An Overset Mesh Approach for 3D Mixed Element High-Order Discretizations," *Journal of Computational Physics*, Vol. 322, 2016, pp. 33–51.

34. Spalart, P. R., Deck, S., Shur, M. L., Squires, K. D., Strelets, M. K., and Travin, A., "A New Version of Detached-Eddy Simulation, Resistant to Ambiguous Grid Densities," *Theoretical and Computational Fluid Dynamics*, Vol. 20, (3), 2006, pp. 181.

35. Montagnani, D., Tugnoli, M., Fonte, F., Zanotti, A., Syal, M., and Droandi, G., "Mid-Fidelity Analysis of Unsteady Interactional Aerodynamics of Complex VTOL Configurations," In 45th European Rotorcraft Forum, Warsaw, Poland, 2019.

36. Reddinger, J. P., "Performance and Control of a Scalable Quadrotor Biplane Tailsitter," Proceedings of the AIAA Scitech Forum, San Diego, CA, Jan 7-11, 2019.

Aeroelastic Simulation of Complex VTOL Configurations

Dr. Thomas Hays
Principal Engineer
Answer Engineering
Englewood, CO 80112

Connor Schultz
Senior Engineer
Answer Engineering
Englewood, CO 80112

ABSTRACT

Answer Engineering has implemented a combination of mid-fidelity aerodynamic and structural tools to simulate the aeroelastic behavior of VTOL transition from hover to forward flight. The transition maneuver is a complex aerodynamic environment that has become a common challenge in recent VTOL designs. Proven methods for aeroelastic analysis exist for fixed surfaces, rotors, prop whirl, control surface buzz, and other singular circumstances, but combinations of these phenomena require new methods. Solutions simultaneously coupling aerodynamics and elasticity of rotors, motor mounts, booms and wings in a multi-physics simulation allows for insight to the aeroelastic stability and aero-structural interaction of the vehicle. Both time and frequency domain outputs are created to study spectral content and general stability of aeroelastic systems. Stiffness requirements for large VTOL aircraft have been shown to dominate structural sizing of many VTOL components when compared to typical static, modal, and vibration load analysis on the same elements. Bend-twist couplings of rotor and vehicle structures can be modeled and optimized through composite ply angle changes to reduce aeroelastic loads and stabilize structures in previously problematic conditions. The aeroelastic simulation is implemented on cloud resources to simultaneously compute a large array of flight states varying edgewise flow velocity, motor torque, RPM, aircraft angle of attack, and sideslip. This solution sequence previously might take a single engineer months to iterate through and can now be simulated and reported within a day. This scalable aeroelastic simulation approach allows prediction and remediation of aeroelastic failure throughout the transition envelope before designs are committed into expensive test aircraft.

INTRODUCTION

Aeroelastic stability of VTOL aircraft is a challenging flight condition to simulate and predict. The complex aerodynamic and vibratory nature of VTOL transition from hover to forward flight further complicates an already challenging aerodynamic and structural analysis task. The workflow described in this paper enables mid-fidelity aeroelastic simulation of multibody representations of VTOL craft through the transition flight envelope from pure hover to forward flight. Simulations including the wing, boom, motor, rotor, wake, and other desired component interactions of a VTOL system can be conducted to discover total system stability and estimate structural margins. Fully detailed finite element models are simplified into representative beams by translating super elements into coupled stiffness matrices. Simulations using coupled DUST And MBDyn, software detailed in References 1 and 2, are batch processed with the generated multibody aeroelastic model. Output deformation data is filtered based on a multidimensional strain limit function to down select to conditions where flutter, aeroelastic divergence, or resonance threaten the structure and warrant a specific high fidelity FEM solution of that deflected state. The simulation of an entire transition map is accelerated by either single or multi-node batch processing of flight states using cloud-based computation resources.

VTOL AEROELASTIC SIMULATION

The developed workflow begins and ends with full-fidelity finite element meshes, while using mid-fidelity aeroelastic simulation in the middle to generate solutions in reasonable time. The full fidelity FEA models are first used to provide multibody structural definitions, and then later map mid-fidelity aeroelastic model outputs back onto a fully meshed finite element representation of aircraft structure. The key differentiation of this approach to other options is the use of viscous vortex particle aerodynamic methods, multibody structural representations, and the ability to solve aero-structural solutions at a speed suitable to calculate an entire VTOL transition flight envelope each day.

Model Development

Detailed CAD and FEM models are simplified to average section properties needed to predict the aeroelastic response of an aircraft. A series of beam and mass elements coupled with vortex particle wake panels yields the aerodynamic and structural dynamic models suitable to predict aeroelastic response when coupled. Typically, the main structural components (wing, empennage, fuselage) are reduced to

---

Presented at the Vertical Flight Society's 11th Biennial Autonomous VTOL Technical Meeting, Phoenix, AZ, USA, Feb. 4-6, 2025.

beams along their respective axes. Each beam's stiffness is defined by 6x6 constitutive and damping matrices, described in Reference 2, determined by the average properties between multiple structural stations. Structural mass and inertial properties are determined in a similar fashion and are applied via mass elements which are mated to structural grid points. Aerodynamic models can be sectioned differently, as the aerodynamic mesh and solver converge separately from the structural solver. The two models are developed independently but are coupled via the software PreCICE to update aerodynamic loads to the structural solution, and structural deflections to the aerodynamic solution for each time step. Time step size is determined by the structure's natural frequencies and the frequency components of system inputs to get stable simulations. The required timestep size will vary with each problem, but for mainstream VTOL multirotor designs a step size of one thousandth of a second has proven suitable.

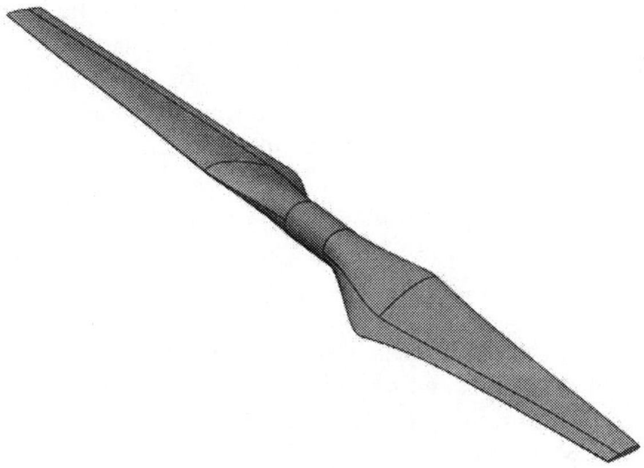

Figure 1. 3D VTOL rotor model for high resolution FEA meshing and translation into simplified MBDyn beam element model

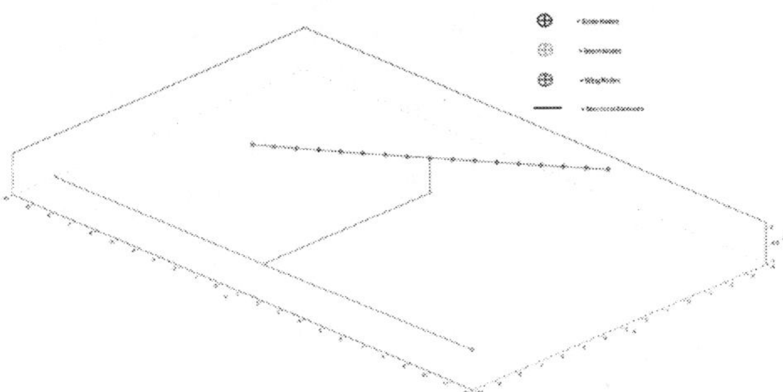

Figure 2: Simplified Structural Skeleton Model

The structural nodes above demonstrate a foundational set of wing, boom, and rotor mass and stiffness elements. Additional connections, joints, and complexity can be added as necessary to represent the VTOL, fixed wing, helicopter, or other rotorcraft problem in consideration. Each skeletal element behaves as a beam element with a 6x6 compliance matrix to describe simple to complex bend-twist behaviors.

Loads and Limitations

There are accepted methods for loads development of conventional rotorcraft and aircraft outlined in Part 25 or 27 and 23 respectively, which can be found in Reference 8. Preliminary sizing load cases are developed through simplified equations to determine loads on each part of the vehicle for a specific flight event. Typically, there are nominal, maneuvering and gust cases that size the structure of an aircraft in the well-defined flight regimes of conventional airborne configurations. Today's VTOL configurations utilize the well understood rigid body loadings created from hovering and forward flight, but programs often delay or neglect the aeroelastic loads of the wing-boom-rotor structural combination. In the most severe of flight circumstances a VTOL aeroelastic system can become unstable which leads to powered structural divergence of the boom-rotor system in a small number of rotor revolutions.

Rotor Divergence, Flutter, and Whirl

Aeroelastic stability boundaries and structural loads arising from excited elastic dynamics quickly grow to magnitudes exceeding rigid body static and resonant loads. Aeroelastic divergence is an especially harmful condition where the advancing blade in an edgewise flow condition exceeds its local stability airspeed, and the combination of twist and deflection past the velocity boundary only leads to higher loads, and higher deflections. The onset of advancing blade divergence in simulation leads to structural failure of the rotor system in less than 10 revolutions.

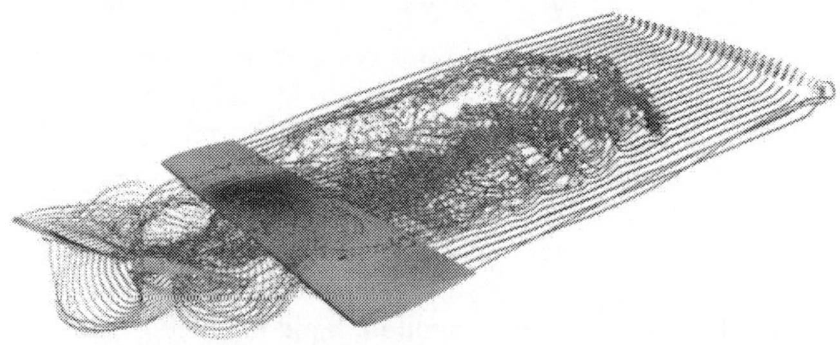

Figure 3. Unstable Rotor Due to Blade Flutter

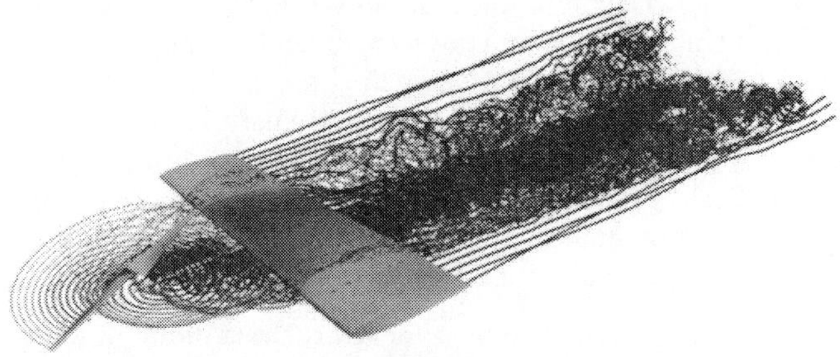

Figure 4: Stable Outcome After Rotor Structural Modification

Blade flutter occurs when the flapping and twisting modes of the blade couple with positive feedback. This negative damping leads to the blade extracting energy from the flow in the classical flutter flapping motion translated into a rotating frame. This problem can be worsened if the blade passing frequency coincides with the bending and/or torsional natural frequencies of the rotor system. Given most VTOL systems must transit from zero speed hover through edgewise forward flight, the likelihood of a VTOL passing through a resonant crossing is high. The structural reaction from flutter is synonymous with classic resonance, so failure occurs as the coupled deflections increase exponentially. Figure 5 shows a blade flutter simulation that accelerates up to time step 700 where the linear solution deflects sufficiently to cause structural failure. In this linear structural solver, the deflection and drag continue to rise past the point of real-world failure, slowing the rotation of the system.

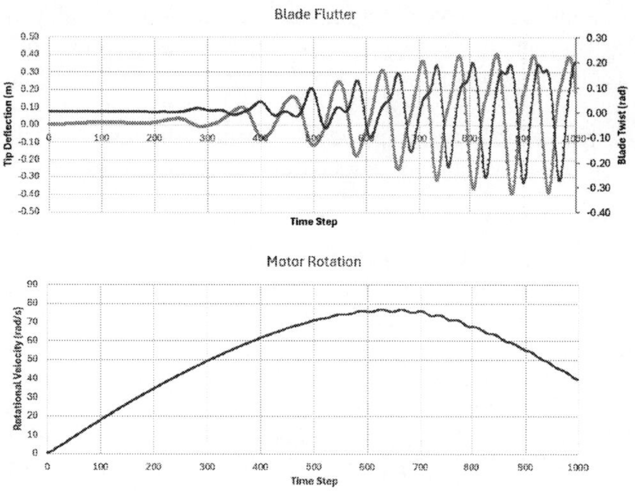

Figure 5: Example of Bend-Twist Coupling and Build During Flutter

Whirl is an instability of the rotor system and not necessarily the rotor or blades alone. In this mode, stability is more dependent on the stiffness of the mounting structure for the rotor. When the axis of the rotating disk changes due to flight maneuver or disturbance, gyroscopic reactions are created about the axis perpendicular to the imposed angular rate. Past critical rotor speeds, the flexure of the rotor mount grows with each cycle leading to large vibration, rapid fatigue, or outright failure of the rotor and motor mounting structure.

VTOL edgewise flight, even at slow speeds, in a direction perpendicular to the booms of a craft can lead to a torsional vibration feedback mechanism in the booms not predicted by rigid aerodynamic, or modal structural analysis alone. As peak hub moments occur 90 degrees after peak lifting loads, lateral flight produces twisting around forward boom axis' and can set up reinforcing aeroelastic feedback leading to torsional vibrational failure at slow lateral speed.

The benefit of multi-body aeroelastic simulation of these phenomena is that an appropriately detailed structural model can check for each type of aeroelastic failure within the same simulation setup. Flight velocities and operating RPMs across the transition flight envelope are run as an array of cases, and the deflections resulting from each type of aeroelastic failure mechanism can be seen if they occur.

Critical Frequencies

Many VTOL vehicles use a fixed pitch, fixed position vertical lift system to avoid the added mass and complexity of articulated hub systems. This decision has some considerable consequences when it comes to rotor design for transition flight loads.

Fixed pitch vertical rotor systems must change RPM for control, the vertical lift systems also slow as the VTOL transitions to forward flight because of the increased lift production efficiency brought by edgewise flight. The rotor's RPM operating range must also expand to cover a wider range of thrust capability for maneuverability and motor out operations. With such a wide band of input frequencies required because of the direct relationship between thrust production and RPM, it can be difficult to impossible to avoid all cases of coinciding motor and rotor forcing and natural frequencies.

The number of blades per rotor sets the "nP" frequency which is commonly referred to as the blade passing frequency. The blade passing frequency is typically the highest amplitude forcing function because of advancing-retreating blade effects. The advancing blade, spinning into the free-stream airflow, experiences an increased effective airspeed; generating higher lift than the retreating blade which experiences the opposite effect; for greater detail on the aerodynamic and performance ramifications of this see Reference 6. This phenomenon results in a generally large force imbalance that is coupled out by the rotor mounting hardware and propagated as an internal moment in the structure. As the number of blades increase, the magnitude of this imbalance decreases, but the increased inertia of additional blades is commonly in direct opposition with VTOL maneuverability requirements because the motor must change RPM to change lift production. Because of this strong relationship between vehicle control response and rotor inertia, and the desire for low drag when rotors are stowed during forward flight, multi-blade rotors are often undesirable.

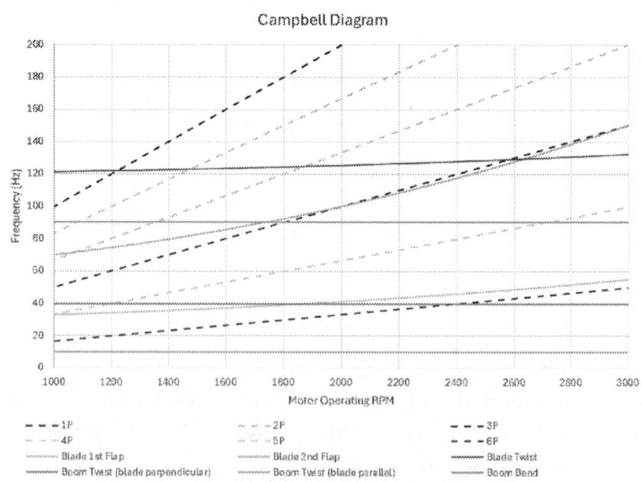

Figure 6: Example of Campbell Diagram

The simple approach to predicting aeroelastic interactions is by use of the Campbell Diagram as shown in Figure 6. This allows for a quick inspection of "modal crossings" which indicate dangerous operating conditions. When the motor operating frequency (MOF) and its harmonics intersect with structural modal frequencies (which increase with operating RPM), there is an increased chance to resonate the structure and cause failure. There are several factors this chart does not consider like damping forces from mass, inertia, structure and aerodynamics. Damping generally widens the band of modal excitation, but also reduces the amplitude at which the structure will react. This reduces the certainty of determining "stay-out zones" for motor operation as some crossings may result in severe reaction, and others nearly none. Another caveat to the Campbell Diagram is that not all harmonics of the MOF are equal in input energy, or critical when coupling with structural modes and determining which are critical is system dependent. It is a good place to start, but designing a VTOL system primarily with a Campbell Diagram is both incomplete and misleading.

Spectral Loads

Figure 7 shows an example postprocessing output similar to a traditional Campbell Diagram, but uses aggregated simulation data to display actual interaction peaks. This identifies which crossings create dangerous interactions, and which do not. The fan of blade passing frequency and its harmonics are readily apparent, with increased response near crossings with blade, boom, and other structural natural frequencies.

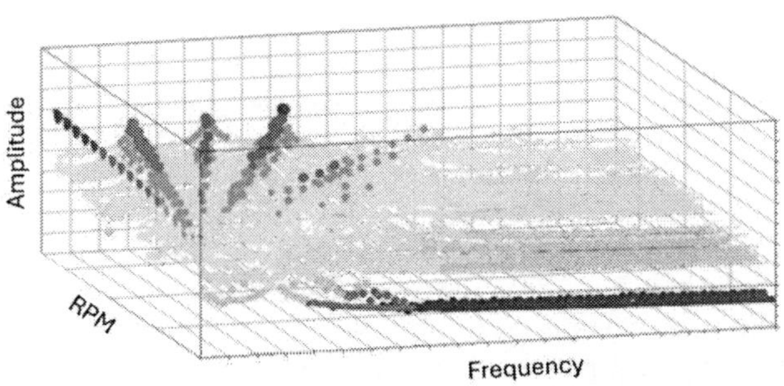

Figure 7: Spectrogram Showing Amplitude and Frequency for a Given Airspeed

Spectral loads are also a good way to calculate a statistically sufficient life cycle for fatigue critical components. In most cases, it is not efficient to size entire aircraft to infinite fatigue life stress allowables. Time and frequency output from the aeroelastic simulation can be extracted from each run as shown in Figure 8. This spectral and temporal data provides the frequency, reversal, and peak stress necessary for data quick identification of fatigue problem areas. Having a predicted spectrum for hover, transition, and forward flight is a good start at doing life cycle analysis for components that are fatigue critical. The compounding risk and importance of vibration prediction, fatigue analysis, flight after after rotor loss, and likelihood of rotor cascade failures in VTOL flight makes early and fully informed aeroelastic analysis a key safety measure.

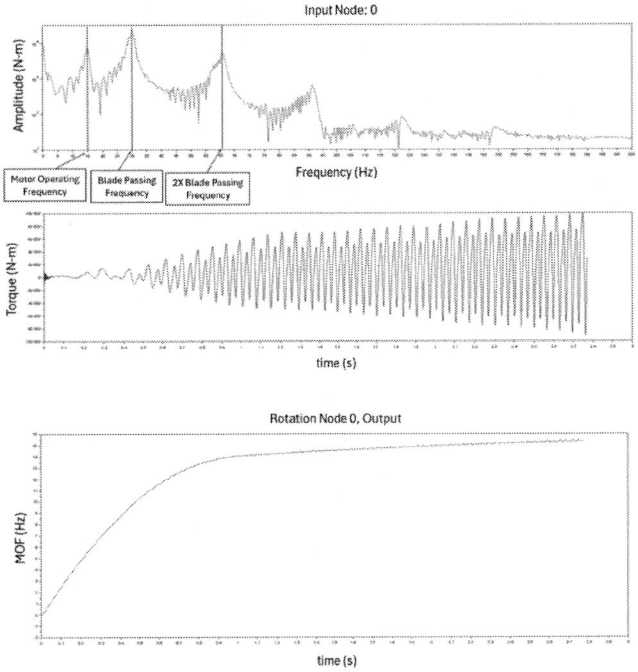

Figure 8: General Loads/Spectral Analysis of Stable Operating Condition

Implementation Into Detailed FEA Models

After reduced order strength and deflection checks are done for the entire dataset, a subset can be down selected to analyze in detail using full fidelity finite element models. The generic anisotropic beam section equation, described in Reference 2, can be used to relate internal force and moment values to section strain and curvature for each time step.

$$\begin{Bmatrix} f \\ m \end{Bmatrix} = \begin{bmatrix} K_{f\nu} & K_{f\kappa} \\ K_{m\nu} & K_{m\kappa} \end{bmatrix} \begin{Bmatrix} \nu \\ \kappa \end{Bmatrix}$$

For a complex composite part, failure can come from many different mechanisms that are difficult to predict with reduced order models. During targeted studies of the identified high-risk set of cases, simulation outputs are transformed into load decks for higher fidelity analysis tools like NASTRAN. This allows us to calculate specific and local margins of safety, using material specific classical failure criteria, to further reduce risk or improve structural design. This process is iterated until the VTOL structure passes in the necessary transition flight envelope and the design is closed.

GUST LOADS

To investigate the effects of gust loads on VTOL systems, a nominal frontal cosine gust profile was assigned as a far field boundary condition. The front had a vertical perturbation with a maximum velocity of 15 m/s. The front encounters the rotor at time step 300, where each time increment is 0.001 seconds. As expected, the output thrust of the rotor grew in proportion to its effective angle of attack with the passing of the gust front through the rotor disk which can be seen in Figure 9. The rotational rate of the constant torque rotor was also affected by the gust front, but not directly proportional to the change in effective flow angle.

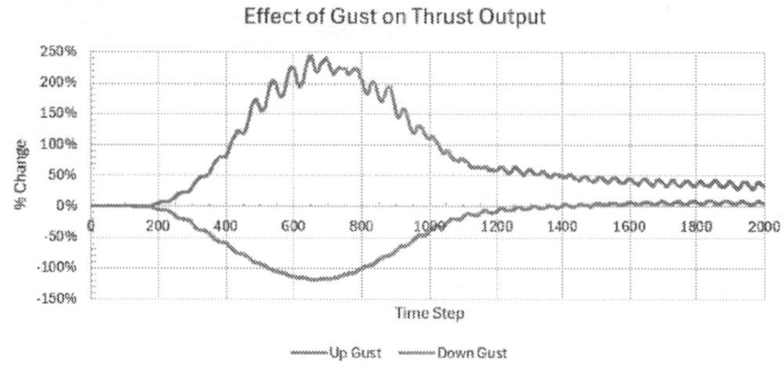

Figure 9: Gust Front Changes Output Thrust at Constant Torque

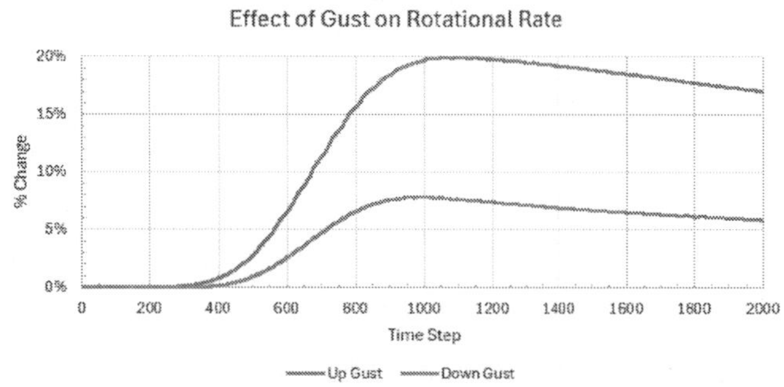

Figure 10: Gust Front Increases Rotational Rate at Constant Torque

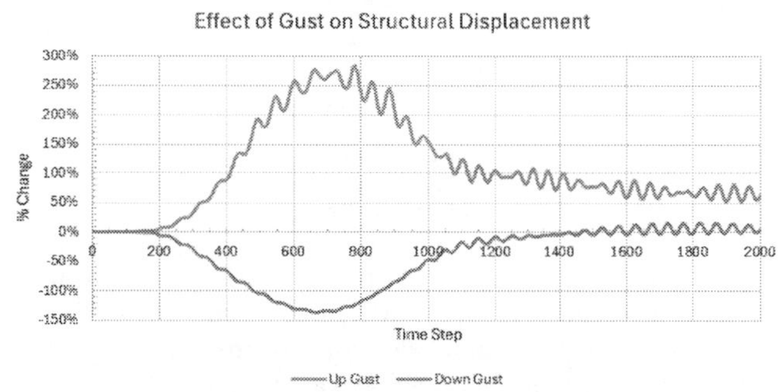

Figure 11: Gust Front Changes Structural Displacement at Constant Torque

During both transient simulations, rotor speed increased, but opposite effects on thrust output were seen, driven by change in rotor disk angle of attack. Significant displacement of the rotor and boom were seen during gust reactions which further decrease structural margins of safety at critical operating points of the transition and edgewise flight envelope.

In a gust scenario, the rotational rate of the simulated rotor under constant torque increased up to 20% over its intended rate in less than 0.5 seconds. This acceleration of the rotor could quickly exceed aeroelastic limits if already flying in a critical state. Vehicle control systems can help to limit rotor overspeed, but outcomes will vary with implementation. Aeroelastic simulation at critical aeroelastic conditions in combination with required gust front survivability is essential for proper structural margin and aeroelastic stability determination. VTOL operation in urban environments is likely to generate sudden freestream velocity change, so the ability to simulate vehicle aeroelastic response to urban air environments is critical.

BATCH SIMULATION OF VTOL TRANSITION ENVELOPE

Batch simulation utilizes cloud compute resources to conduct multiple simulations simultaneously. Often, there will be a need to analyze a matrix of airspeed, sideslip, and motor operating conditions. This results in many test points which can be computationally expensive. Analyzing each point in parallel reduces run time from potentially months if run in series, to finishing in hours when cases are solved in parallel.

To illustrate the power and necessity of batch simulation, we consider the following configuration. A notional VTOL aircraft requires a maximum rotor velocity of 2400 RPM for take-off and stable hover. It's minimum wing-borne flight velocity is 60 m/s, and each wing holds 2 vertical lift 3 bladed rotors. Part 25, in Reference 8, dictates that a "sufficient number of gust gradient distances in the range 30 ft to 350 feet must be investigated to find the critical response". A specific transition flight window has not been determined, so the full envelope of 0 to 2400 RPM and 0 m/s to 60 m/s is considered. We choose a resolution of 200 RPM and 10 m/s evenly within those bounds resulting in a matrix of 72 simulations. According to Federal Regulations, we must pick some resolution for gust gradient distances between 30 and 350 feet, so 40 feet is our baseline increment. That dimension brings the number of simulations up to 576. If each simulation takes 4 hours, the total computational time is 2,304 hours (over 3 months). That does not include postprocessing to identify and study critical cases, which can take several days. Another layer of complexity is maneuvering and further grid refinement for problem areas. Custom pre and post processing scripts assist in moving efficiently through the hundreds of simulations and large datasets generated from the batch runs. Serial processing of multibody aeroelastic simulations is time prohibitive for most aerospace programs.

With the above example of 2,304 hours of computational time, a pool of 576 virtual machines could divide the full simulation matrix described above in 4 hours. It is better to divide the total CPU's available into many simulations using few CPU's instead of highly parallelizing each simulation to solve the set in minimum time.

As important as efficient simulation and computation is, all the data produced is useless without effective processing. Multiple gigabytes of output structural and aero data need to be parsed through to draw meaningful conclusions from the flight scenarios being simulated. Answer Engineering has developed a suite of postprocessing tools to automatically distill the large dataset into informative charts and outputs.

Early Design Phase Integration

Ideally, aeroelastic trade studies can be performed in the early program phases where there is flexibility to adjust major design decisions. If critical aeroelastic interactions are discovered early, the most effective way to correct them is by changing the structural section to engineer corrective natural frequency separation and stabilizing mode shapes into the system.

This aeroelastic simulation workflow permits inverse aeroelastic design early in the aircraft design process. Stiffness studies can be conducted to determine component requirements before components are drawn. This method significantly reduces program risk as the primary structure can be designed in early stages with informed stiffness requirements.

Late Design Phase Integration

In the circumstance that the aircraft has been designed before checking aeroelastic loads, deflection-based failure criteria can be derived for existing structure and checked against all transition envelope simulations to predict whether an aeroelastic problems exist. This yields a contour of predicted failure indices throughout the potential operating conditions as shown in Figure 12.

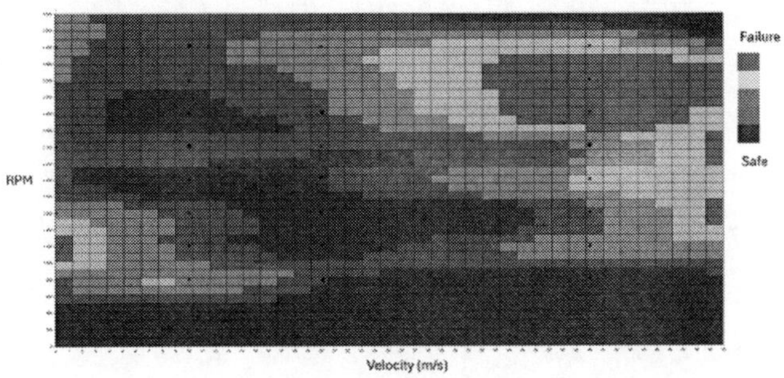

Figure 12: Example Output of Failure Prediction Through Operation Envelope Without Blade Tailoring

The velocity and RPM combinations in the red zone of Figure 12 exceed deflection limits and are likely to lead to structural failure. These peaks in structural deflection indicate regions of problematic aeroelastic interactions. This chart does not label the type of aeroelastic problem, only that a deflection limit has been exceeded and that specific operating point should be viewed, considered, and corrected if that operating point is essential for transition.

PASSIVE AEROELASTIC STABILITY

Design of intentionally "unbalanced" layups naturally couple bending inputs advantageously with twisting deformation. In the case of a blade in edgewise flow, this designed coupling combats aeroelastic divergence by causing the blade to twist nose-down as lifting forces bend the blade up. The basic principle behind this approach can be found within classical lamination theory as described in the following equation:

$$\begin{bmatrix} N_x \\ N_y \\ N_{xy} \\ M_x \\ M_y \\ M_{xy} \end{bmatrix} = \begin{bmatrix} A_{11} & A_{12} & A_{16} & B_{11} & B_{12} & B_{16} \\ A_{12} & A_{22} & A_{26} & B_{12} & B_{22} & B_{26} \\ A_{16} & A_{26} & A_{66} & B_{16} & B_{26} & B_{66} \\ B_{11} & B_{12} & B_{16} & D_{11} & D_{12} & D_{16} \\ B_{12} & B_{22} & B_{26} & D_{12} & D_{22} & D_{26} \\ B_{16} & B_{26} & B_{66} & D_{16} & D_{26} & D_{66} \end{bmatrix} \begin{bmatrix} \varepsilon_x \\ \varepsilon_y \\ \varepsilon_{xy} \\ \kappa_x \\ \kappa_y \\ \kappa_{xy} \end{bmatrix}$$

For clarity, the portion of the 6x6 matrix which couples bending and twisting deformations is the upper right and lower left 3x3 quadrants. These quadrants are otherwise referred to as the B matrix of a laminate which is nonzero when the layup is not symmetric. Assuming the two designs have similar failure strain limits, the decreased overall deflection results in a safer structure. The decrease in angle of attack on the advancing blade naturally reduces the cyclic load on the system.

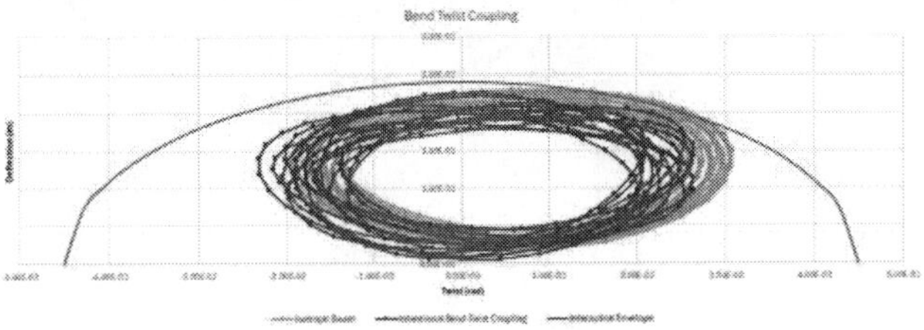

Figure 13: The Benefits of Bend-Twist Coupling in Composite Structure

In Figure 13 above, a study is conducted on the test point where the rotor is spinning at about 4000 RPM with an edgewise flow of 50 m/s. The blade modeled as an isotropic beam without preferential bend-twist coupling has its tip deflection and twist shown in purple. In blue, is an anisotropic beam with designed composite bend-twist coupling, resulting in lower overall deflections and positive margins of safety where the uncoupled beam predicted failure.

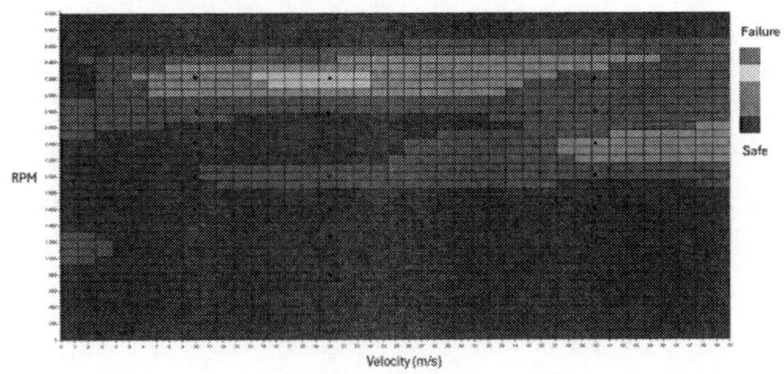

Figure 14: Failure Prediction with Favorably Designed Bend-Twist Coupling

An example of the efficacy of passive bend-twist coupling can be seen in Figure 14, which repeats the transition simulation conducted in Figure 12 but with a rotor featuring favorable bend-twist coupling. All predicted aeroelastic failure points are eliminated in the transition window, with one green peak of interest remaining at 3200 rpm and 20 m/s. This demonstrates the method's capability to predict and model design mitigations for multiple types of aeroelastic structural failures.

CONCLUSIONS

The multibody VTOL aeroelastic analysis workflow created by Answer Engineering around the DUST and MBDyn coupled aeroelastic analysis toolset allows for highly parallelized analysis of the hover to forward flight transition envelope required for safe VTOL operation. This approach allows for programs to reduce aeroelastic configuration risks during all phases of design without waiting for large scale testing to commence.

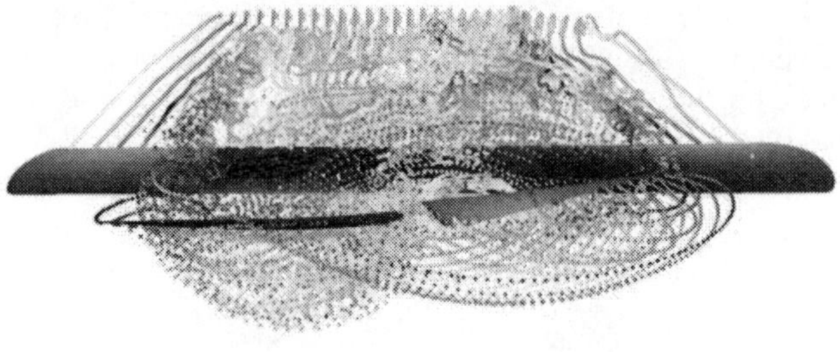

Figure 15: Deflection of Advancing/Retreating Blades

ACKNOWLEDGMENTS

The authors would like to thank Re:Build manufacturing and Answer Engineering for funding this technical paper and associated study. We would also like to thank the authors of DUST and MBDyn from Politecnico di Milano – Polimi whose aerodynamic and structural solvers are critical to this developed workflow.

REFERENCES

1. Montagnani, D., Tugnoli, M., Fonte, F., Savino, A, Cocco, A., and Colli, A., "DUST User Manual", December 14, 2023, public.gitlab.polimi.it/DAER/dust
2. Masarati, P., "MBDyn Input File Format", Version 1.7.3, January 12, 2017, https://www.mbdyn.org/userfiles/documents/mbdyn-input-1.7.3.pdf
3. Savino, A., Cocco, A., Zanotti, A., and V. Muscarello, "A Coupled Multibody-Mid Fidelity Aerodynamic Tool for the Simulation of Tiltrotor Manoeuvres", *47th European Rotorcraft Forum*, Glasgow, UK, September 2021.
4. Piccinini, R., Tugnoli, M., and Zanotti, A., "Numerical investigation of the rotor-rotor aerodynamic interaction for eVTOL aircraft configurations", *Energies*, 13(22), 5995, 2020
5. Tugnoli, M., Montagnani, D., Syal, M., Droandi, G., and Zanotti, A., "Mid-fidelity approach to aerodynamic simulations of unconventional VTOL aircraft configurations", *Aerospace Science and Technology*, 115, 106804, 2021.
6. McCormick, B. W., *Aerodynamics of V/STOL Flight*. Academic Press, New York, New York, 1967, Chapter 4.
7. Ashley, H., and Bisplinghoff, R. L., *Principles of Aeroelasticity*. Dover Publications, Inc., 1962, Chapter 7.
8. *Code of Federal Regulations,* Title 14: Aeronautics and Space, Federal Aviation Administration, Department of Transportation, Chapter 1-C, Part 23 - "Normal Category Aircraft", Part 25 – "Transport Category Aircraft" & Part 27 – "Normal Category Rotorcraft".
9. Leishman, J. Gordon, *Principles of Helicopter Aerodynamics,* Cambridge University Press, 2000, Chapter 2.14.

Emerging CFD Simulation Technologies for eVTOL Aerodynamics, Fluid Management, and Powertrain Optimization

Jun Li
Senior Engineer
AVL Mobility Technologies, Inc.
Plymouth, MI, USA

ABSTRACT

This paper explores advanced CFD simulation technologies tailored to address design and operational challenges in autonomous eVTOL aircraft. Tools such as AVL's PreonLab and FIRE M are highlighted for their contributions to enhancing aerodynamics, fluid management, and powertrain optimization. PreonLab employs a meshless particle-based approach ideal for highly dynamic fluid flow applications, including gearbox oil distribution, electric motor cooling, water landing and soiling simulations. Meanwhile, FIRE M offers aerodynamics simulation solution based on a proprietary immersed boundary method implementation (Embedded Body Method), virtually eliminating the need for CAD clean-up and manual meshing. These technologies support the eVTOL industry's goals for electrification, sustainability, and operational efficiency, accelerating the deployment of safe and effective urban air mobility solutions.

INTRODUCTION

The emergence of electric vertical takeoff and landing (eVTOL) aircraft represents a transformative shift in urban air mobility. Achieving the efficiency and safety required for these vehicles necessitates cutting-edge simulation tools. Computational Fluid Dynamics (CFD) simulations play a pivotal role in enabling the industry to overcome technical barriers in aerodynamics, thermal management, and powertrain optimization. Meshing has been one of most time-consuming tasks for CFD simulations. NASA CFD Vision 2030 [1] proposed that mesh generation should be invisible to the CFD user or engineer.

This paper delves into the innovative approaches provided by AVL's PreonLab and FIRE M. PreonLab's meshless particle-based framework caters to dynamic fluid interactions, offering robust solutions for applications such as gearbox oil flow, electric motor cooling, water landing and soiling management. FIRE M, leveraging its Embedded Body Method (EBM), introduces an efficient approach to aerodynamics modeling, drastically reducing simulation setup times.

This work underscores AVL's role in pioneering simulation technologies that empower the eVTOL industry to meet environmental and operational challenges, facilitating the deployment of autonomous, efficient, and sustainable air mobility systems.

Presented at the Vertical Flight Society's 11th Biennial Autonomous VTOL Technical Meeting, Phoenix, AZ, USA, Feb. 4-6, 2025.

PREONLAB: FLUID MANAGEMENT EXCELLENCE

PreonLab is a cutting-edge simulation tool that leverages Smoothed Particle Hydrodynamics (SPH), a particle-based method that eliminates the need for traditional mesh generation. This approach streamlines the simulation workflow, significantly reducing setup and computation times. This makes it an ideal solution for dynamic and complex fluid interactions, such as those encountered in eVTOL applications.

Key features of PreonLab [2] include:

- Versatility Across Fluid Scenarios: Capable of simulating single-phase and multiphase flows, PreonLab adapts seamlessly to different conditions, including interactions between oil and air in high-speed gearbox scenarios.

- Dynamic Particle Resolutions: PreonLab stands out among other SPH tools primarily due to its advanced particle refinement and coarsening capabilities, particularly its Continuous Particle Size (CPS) feature, which offers the ability to adjust particle sizes dynamically for greater precision where needed, while optimizing computational efficiency.

- Ease of Use: PreonLab's intuitive interface and automated workflows enable non-CFD experts to achieve high-quality results with minimal training.

- Validation and Accuracy: Extensive validation studies, including comparisons with experimental data, have demonstrated PreonLab's ability to predict phenomena such as oil distribution, torque losses, and lamella formation with remarkable accuracy.

- Compatibility: PreonLab can be run on both CPU and GPU, with significant performance improvements on GPU.

The tool's core applications include water management, gearbox oil distribution, and soiling simulations. Its innovative approach empowers engineers to address challenges in fluid management, ensuring enhanced performance and reliability for eVTOL systems. The technology's key applications include, but not limited to:

- Gearbox Oil Distribution: Optimizing oil supply for bearings and gears under various operating conditions.

- Electric motor cooling: Dealing with the jet cooling for e-motor

- Water landing: Accessing vTOL pressure and wave behavior at emergent water landing events

- Soiling: Addressing soiling from rain, and snow to improve sensor functionality.

Gearbox Oil Distribution

PreonLab's simulation capabilities are particularly valuable for gearbox oil management in eVTOL systems. Using a meshless particle-based approach, engineers can evaluate oil distribution patterns under a range of operating conditions, such as varying shaft speeds, oil levels, viscosities, and tilting angles. Key engineering tasks include:

- Oil Distribution Optimization: Ensuring adequate lubrication of bearings and gear contacts to minimize wear and enhance longevity.

- Torque Loss Analysis: Identifying and mitigating sources of energy loss due to oil movement and interaction with rotating components.

- Breather Shielding Design: Assessing the positioning and effectiveness of breathers to prevent oil leakage and contamination.

- Thermal Behavior Assessment: Investigating the heat transfer characteristics of the oil to maintain operational temperatures within safe limits.

PreonLab has been validated against FZG test rig data [3,4] for oil distribution and torque loss. For low-speed scenarios (240-540 rpm), visual comparisons between test rig results [3,4] and PreonLab simulations demonstrated close agreement. The simulation of physical oil flow accurately captured effects such as oil bridging and lamella formation (Figure 1).

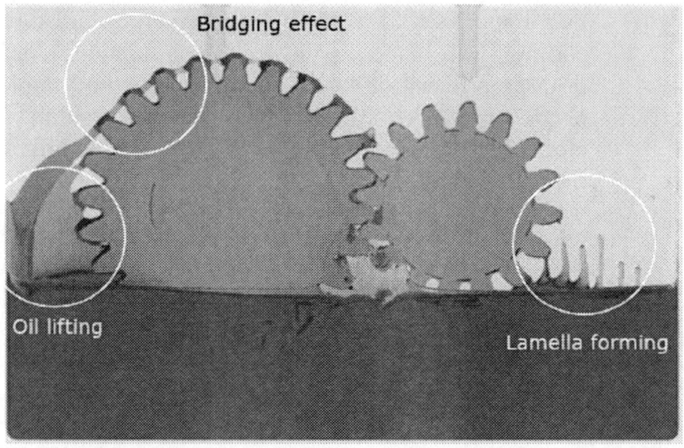

a) Simulation with PreonLab

b) Experiments [4]

Figure 1. ***Visual* comparison *of oil lifting and lamella formation between test rig and PreonLab simulation at 240rpm.***

At higher speeds, such as 2734, 4101, and 5225 rpm, where the air phase in the gearbox becomes significant, a multiphase approach was utilized in PreonLab simulations to account for the interaction between oil and air. It was observed that the oil flow structure appeared less patterned in both simulations and experiments. Power loss predictions using PreonLab were within the error margin of experimental results (Figure 2).

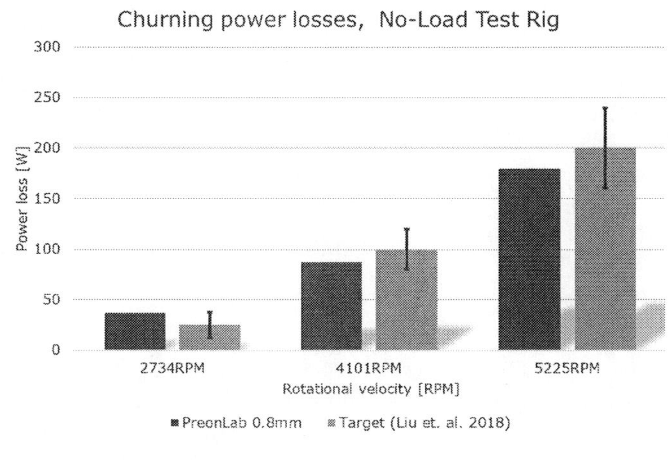

Figure 2. Power *loss at high speeds.*

Compared to conventional finite volume methods (FVM), PreonLab offers several distinct advantages for gearbox oil simulations:

- Elimination of Complex Meshing: FVM requires the creation of detailed, body-fitted meshes, which can be time-intensive and challenging for intricate geometries. PreonLab's meshless approach bypasses this step entirely, enabling faster setup and design iteration.

- Enhanced Handling of Dynamic Interfaces: Gearboxes involve complex interactions between moving components and fluids. PreonLab's particle-based framework naturally adapts to dynamic interfaces, while FVM often struggles with mesh deformation and remeshing requirements.

- Reduced Computational Overhead: Traditional FVM simulations require fine grids to capture flow details, leading to high computational costs. PreonLab optimizes resource usage by distributing particles only where needed, improving efficiency without compromising accuracy.

- User-Friendly Workflow: The automated nature of PreonLab simplifies workflows, allowing non-CFD specialists to achieve high-quality results with minimal training.

E-Motor cooling

eVTOL aircraft demand compact and lightweight motors to minimize weight while delivering high power output for vertical take-off and hovering. This high-power density generates significant heat, requiring effective cooling to prevent thermal stress and maintain operational efficiency. As motors are often integrated with inverter and other powertrain components, effective motor cooling minimizes heat transfer to nearby systems, preserving the overall thermal balance of the aircraft. Innovative cooling solutions, such as impinging jet or liquid cooling systems, help achieve efficient heat dissipation without compromising aerodynamics or adding excessive noise.

Impinging jet simulations are challenging to perform because of the strong non-linearities in the temperature field that require a fine resolution to capture accurately. For this reason, the multi-level refinement feature of PreonLab is important to optimize the runtime.

The cooling oil used has a viscosity that strongly depends on the temperature of the material. To improve the realism of the simulation, the temperature-based viscosity feature is used.

The first benchmark is to validate the PreonLab against experimental data and empirical models for impinging jet cooling. The experimental setup is described in [5] and shown in Figure 3.

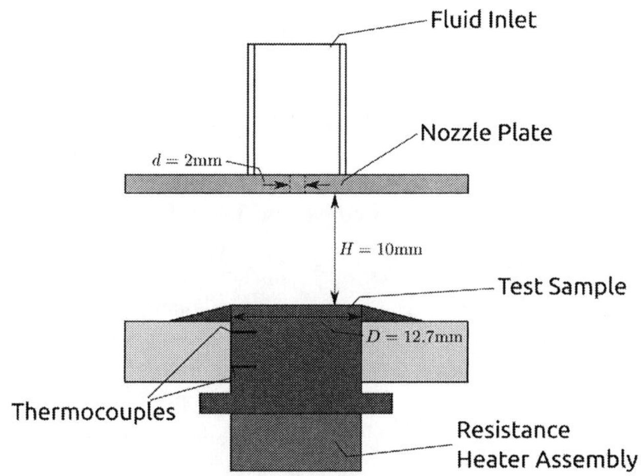

Figure 3. *The* experiment *setup for impinging jet [5].*

The surface of the test sample can either be flat or textured to mimic the copper windings found in electric machines as shown in Figure 4. For the winding targets, 3 wire gauges are considered:18, 22, and 26 American Wire Gauge (AWG) is also shown in Figure 4.

In the experiment, the top surface temperature is maintained at 110 °C with a resistance heater underneath the copper with thermocouples.

For the simulation, adaptive refinement is used in the vicinity of the test sample to resolve the non-linearities near this boundary. Away from it, the particle size can remain coarse to reduce the computational effort. Like in the experiment, the temperature of the target surface is maintained at 110 °C. A developed pipe flow velocity distribution is used at the inlet. The Ford Mecron oil [6] is used as the jet liquid.

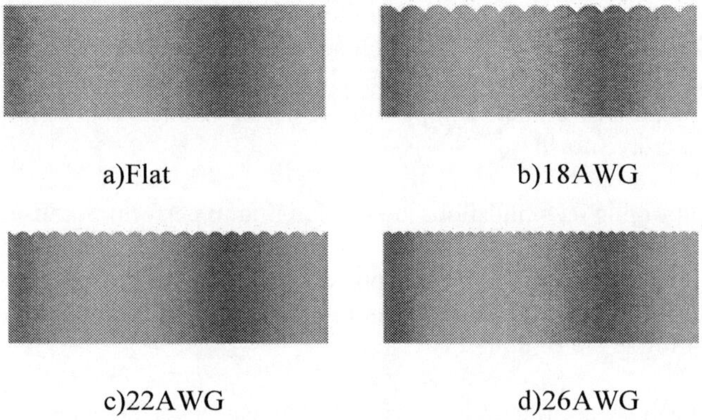

Figure 4. ***The*** **various** ***targets used for impinging jet.***

Once the simulation is converged, we measure the local heat flux on the test sample and derive the average heat flux on the plate and the associated heat transfer coefficient (HTC). For the application, the HTC is defined as:

$$h = \frac{q}{T_{\text{wall}} - T_{\text{jet}}}$$

Where q, Twall, Tjet, are respectively the heat flux, the wall temperature, and the temperature of the jet at the inlet.

The simulated impinging jet on flat surface with PreonLab is shown in Figure 5. After imping on the plate, the oil jet spreads outwards until the entire target is covered. As the oil moves outward, it slows down leading to an increase in the oil film thickness. This can also be seen in Figure 5, where the oil film thickness increases toward the edge of the disc.

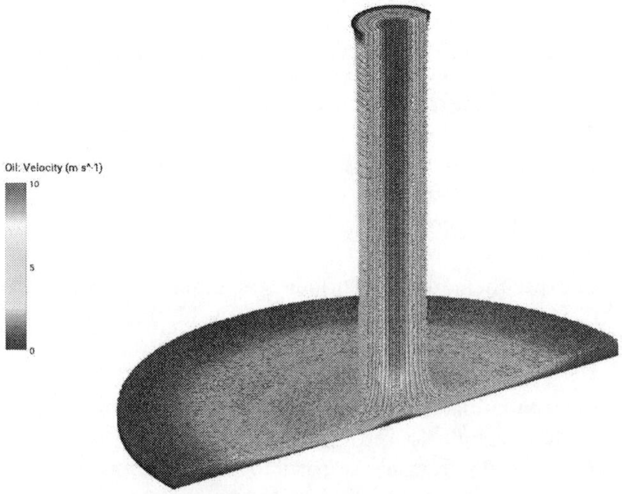

Figure 5. ***A cross-*****section** ***of the result of the impinging jet on a flat surface.***

The average HTC as a function of the inlet velocity is shown in Figure 6. The results from PreonLab follow closely (within 10% of relative error) the experimental average HTC for the whole range of inlet velocities.

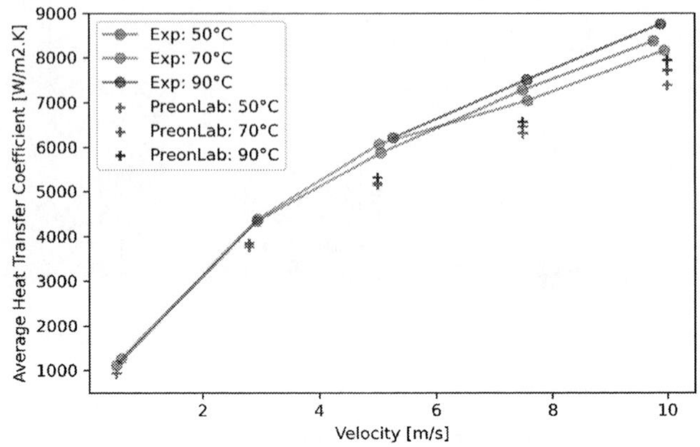

Figure 6. Average heat transfer coefficient as a function of velocity.

Figure 7 shows the comparisons between the simulation result and the experimental data at 50, 70, and 90 C respectively. PreonLab is in good agreement with the experiment for the range of velocities, inlet temperatures, and wire gauges studied in [5].

The benchmark proved that PreonLab can capture the complex heat transfer phenomena generated by an impinging oil jet.

Two different jet cooling from rotating shaft and from injection points along the stator are simulated and shown in Figure 8. For demonstration purposes and computational effort, the cooling of only a portion of the hairpins is shown.

The initial copper hairpins have an initial temperature of 110°C. It is seen that the jet cooling from rotating shaft can reach the hairpin more evenly and therefore the hairpin has more uniform cooling effect. While for the jet from fixed locations along stator, if not optimized, the temperature reaches to around 116°C wherever the oil does not interact with the hairpins and rea hotspots are created because of the heat generated by the heat field(Joule loss).

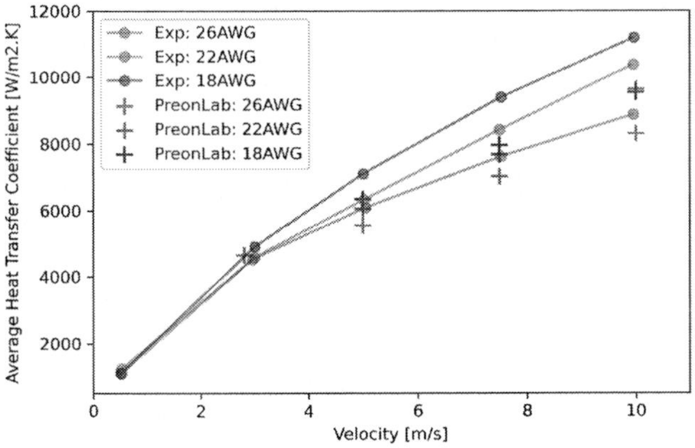

a) 50°C

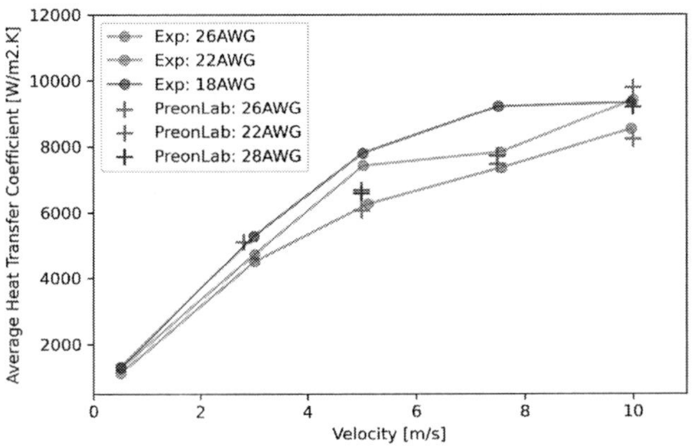

b) 70°C

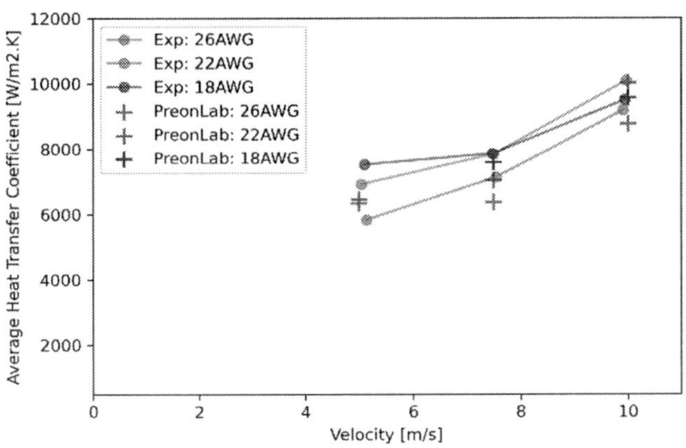

c) 90°C

Figure 7. Average heat transfer coefficient of plate for various inlet velocities and various wire gauges at three different temperatures.

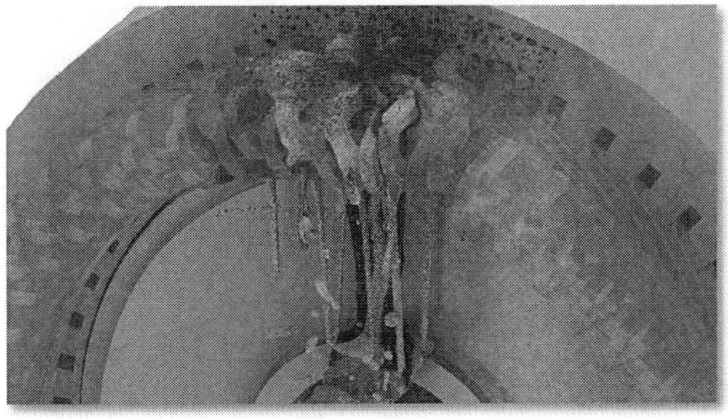

a) Jet cooling from fixed injection points along stator

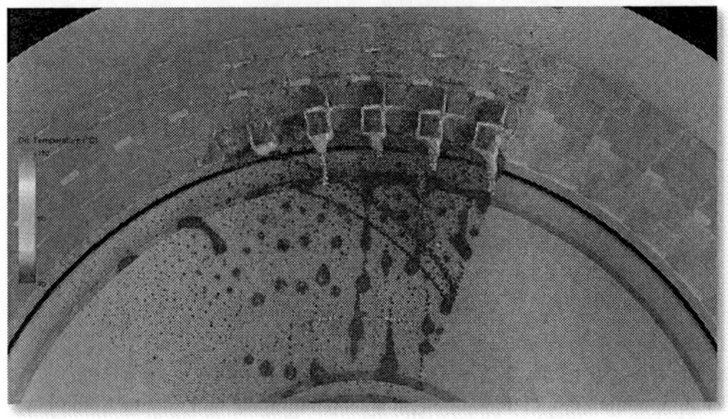

b) Jet cooling from rotating shaft

Figure 8. Jet cooling of the copper hairpins by oil emitted from various locations.

Water Landing

Emergency water landing (or Ditching) in an eVTOL Aircraft is similar to ditching in a helicopter. After impact, it's best to allow the aircraft to settle before releasing harness and escape. Due to current battery technology, eVTOL may pose a higher risk for emergency water landings compared to other aircraft, primarily due to shorter trip times, low altitude and weather. The proposed advisory circular (AC) 21.17-4 published by the FAA has certain certification topic, which propose safety requirements for vTOL aircraft flying over water to have a buoyant and water-tight passenger cabin in case of a water landing. It is important for an eVTOL to remain intact during the impact with water and it should allow evacuation in water waves.

One interesting topic would be the pressure on the aircraft during water landing at various speed. A benchmark case is a wedge dropping into water. Yattou et al. [7] performed large-scale experiments with free falling wedges entering water in a water tank. In the following section, PreonLab will be validated against this experimental data. It is worth noting that the original goal for this study is to facilitate ship hull design.

The wedge is shown in Figure 9 with pressure transducer mounted on the wedge surfaces.

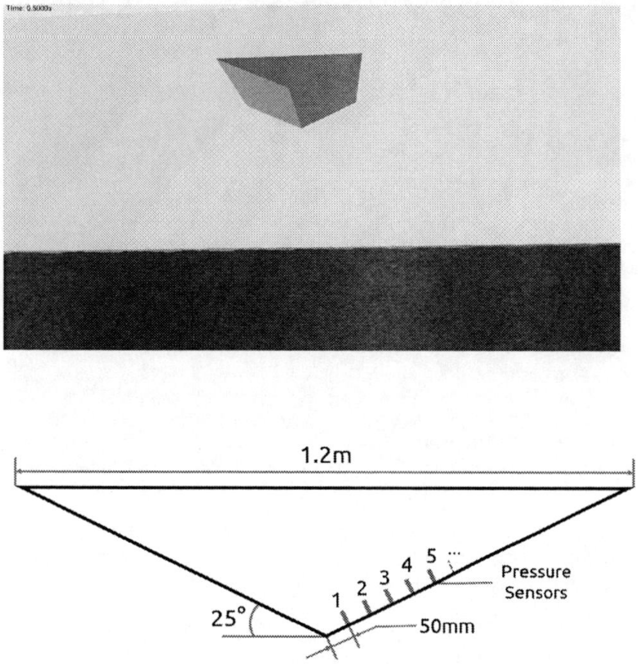

Figure 9. Sensor positions on the wedge which drops into water.

The simulation of such a case is challenging, since around 80% of the kinetic energy of the wedge dissipates within less than 100ms after the initial contact with the water surface. Another difficulty arises regarding the spatial resolution. Fine particles are required near the wedge surface to capture the small-scale hydrodynamic phenomena of the water entry properly. The Continuous Particle Size (CPS) feature in PreonLab is suitable for this task, as fine particles (6.25mm) can be used near the wedge surface and large particles (25mm) are used in the other region of the tank.

By looking at the velocity of the wedge over time shown in Figure 10, the strong correlation between experimental data and the results of the simulation in PreonLab can be seen. After the initial free fall period, the wedge hits the water surface at around 0.5s and decelerates rapidly. The wedge continues to decelerate and eventually floats back to the surface because of its lower density.

The expected surface pressure is an important metric when designing cabin strength. The comparison of the peak surface pressure between PreonLab and experimental data for various deadrise angles are shown in Figure 11. It is seen that the simulation results obtained with PreonLab are in good agreement with the experimental data. Conclusively, PreonLab enables us to gain insight into the structure load that the wedge must withstand during water landing.

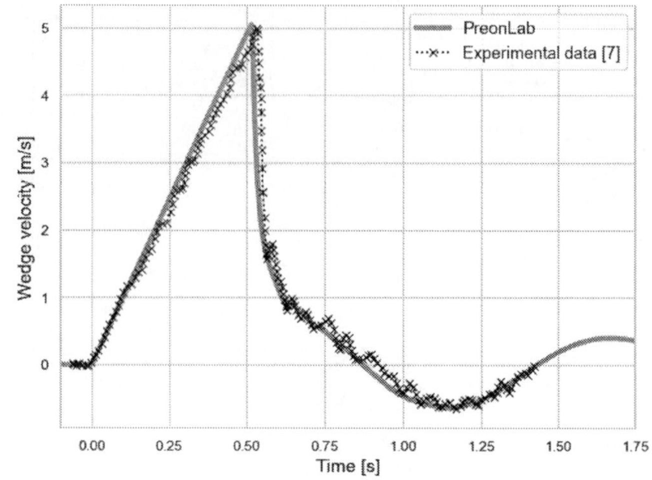

Figure 10. Vertical velocity profile for the free-falling wedge.

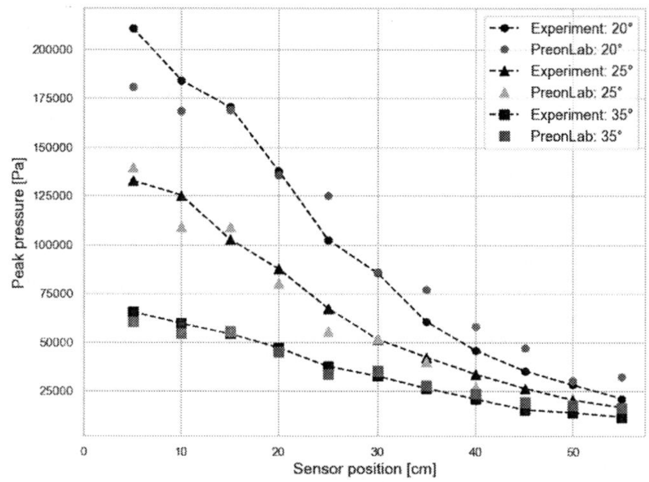

Figure 11. Comparison ***of the peak surface pressure between PreonLab and experimental data [7] for multiple deadrise angles.***

Another scenario PreonLab can be useful is to virtually simulate the motion of aircraft in water, sometimes in wavy conditions. Figure 12 shows the simulation of a ship's motion in a wavy water tank. A ship can right itself due to shift of center of buoyancy. However, a periodic change in righting moment aligned with natural roll period can cause dangerous amplification known as parametric roll. The tool can be used to simulate this kind of phenomena. For vTOL aircrafts, it will be very interesting to see if it can right itself as the buoyancy center will be shifting differently from ships. This will affect how the aircraft settles in emergency water landing and when and how to evacuate.

Figure 12. Ship *motion simulation for a moving ship in a water tank with PreonLab.*

Soiling

Autonomous aircraft rely on sensors such as LiDAR and cameras to stay on their designated path. The accuracy of these sensors can be affected by environmental factors like rain and snow soiling. To mitigate these influences and prevent potential hazards, sensors should be positioned in locations where such effects are minimized. A well-considered design ensures clear detection of the surroundings, reducing the need for additional cleaning devices like sprayers or heaters. By achieving early confidence in sensor placement and design using simulation, costly changes in the later stages of development can be avoided.

Simulating rain or snow soiling requires imported air flow velocities. Snow is computed as an elasto-plastic continuous material, where the elastic part is solved implicitly. Different types of snow can be modelled ranging from chunky to powdery snow, from wet and adhesive to dry. Just like the Preon fluid solver, the snow solver can handle any imported geometries without meshing and offers competitive performance.

Figure 13 shows a validation of snow deposit formation on a wedge. The wind tunnel test is done at Volvo[8]. The transient air flow field at 80 km/h is imported with a frequency of 40 frames per second. On the left of Figure 13, the simulation results show the snow deposit formation on the side and back of the wedge. Overall, the snow deposition locations agree with the experimental results. The simulation is done for 30 seconds physical time with enhanced adhesion to the wedge surface to reduce the simulation effort.

In situations where rain or snow accumulates and obscures sensors, cleaning mechanisms, such as dispensing cleaning fluid onto the sensor surface, can effectively remove debris and restore an unobstructed view for the sensors. The process of water dispensing to clear dirt or snow can also be simulated, as illustrated in Figure 14.

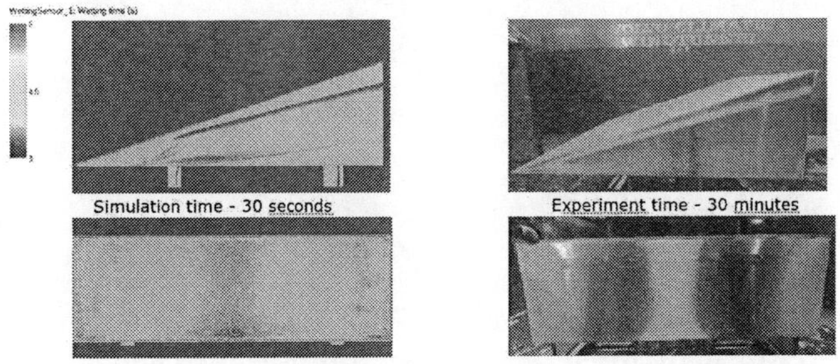

Figure 13. Comparison of snow deposit formation on a wedge model.

Figure 14. *Simulation of water dispensing onto a screen with PreonLab.*

FIRE™ M: MESHINGLESS AERODYNAMICS

FIRE™ M is a finite volume based code which provides state-of-the-art turbulence modeling that accounts for the complexity of real flows. Traditionally, CFD mesh (Figure 15 a) is aligned with the surface/body and suitable configuration of boundary layers is applied. As CFD engineers often encounter complex geometries, especially surface mesh of poor quality, it can take significant amount of time to get from dirty surface mesh to final CFD volume mesh. This is the reason the "Embedded body" is integrated into FIRE™ M's standard finite volume method. This embedded body is proprietary immersed boundary method implementation in AVL FIRE™ M, which simplifies the handling of geometrically complex bodies at the fluid-solid interface in terms of discretization of fluxes, boundary conditions, rate of change, gradient calculation, etc.

A simple two steps are used for this embedded body approach (Figure 15b): first, a simple structure mesh is generated in the domain, then we employ body forcing to introduce wall boundaries. The key point is the fluid-solid interface with respect to the discretization of fluxes, gradient calculation, differencing scheme and boundary conditions. This embedded body method implementation fully leverages advantages of the advanced turbulence modeling combined with smart adaptive mesh refinement strategy.

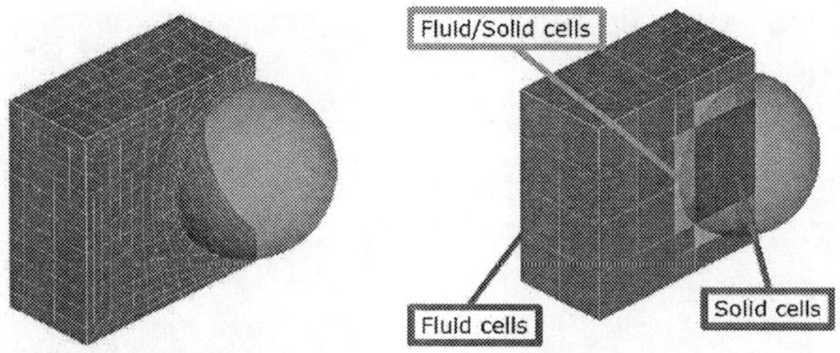

a) Body fitted **b)Embedded body**

Figure 15. Comparison of body fitted mesh and embedded body mesh.

Figure 16 shows an analytical test case [9] for determining solver quality. At a near infinite Reynolds number, it does not consider friction or turbulence. Any dampening impact of turbulence modelling is not active. This way, it is very useful in showing implementation shortcomings of the basic solution procedure.

The problem consists of an infinite band of profiles that is modeled using periodic boundary conditions. The reference shows several analytical solutions for flows with different angle of attack. Two solutions were carried out: a solution based on the boundary fitted mesh and an embedded body simulation based on a hexahedral background mesh. The pressure distribution along the airfoil surface is compared against the analytical solution. In wide parts of the pressure distribution, the simulation yields the exact same results for both embedded body method and body fitted mesh. As adaptive mesh refinement is constricted to a feasible value, small fluctuations at the pointed trailing edge can be observed due to the local resolution.

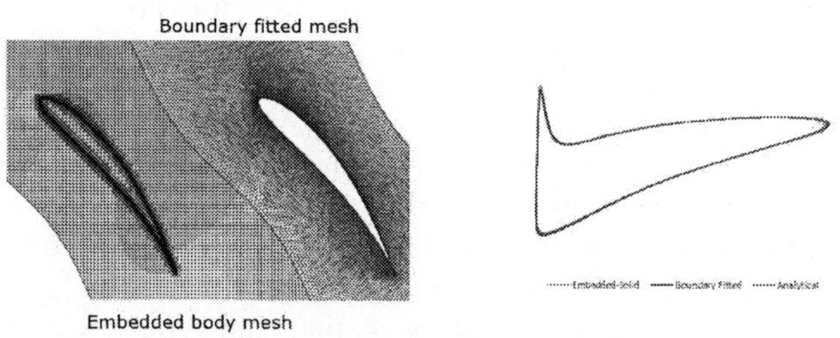

Figure 16. Pressure distribution around the airfoil with body fitted mesh, embedded body mesh and analytical solution.

This approach has been further validated on DrivAer benchmark case for vehicle aerodynamics simulations [**10**]. The accuracy of car lift and drag are in line with body fitted approach.

Embedded body approach in the field of aeronautic simulations is illustrated on the example of helicopter model (Robin mod7). The first case is the take-off from platform in the water as shown in Figure 17. The moving geometry can be easily handled within embedded body framework in combination with adaptive mesh refinement.

a) Iso-surface VF=0.5 colored by pressure

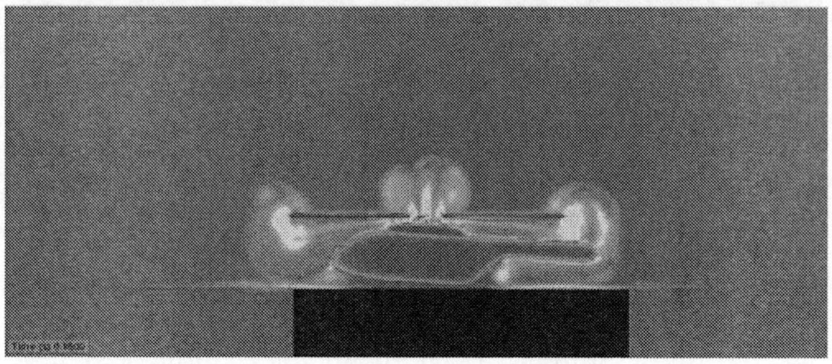

b) Vorticity in Y-cut through rotation axis; black line indicates water surface

Figure 17. Simulation of helicopter model take-off the platform in the water.

The second case is to change rotor pitch angle during the simulation run. Change of angle is performed during one revolution of the rotor as shown in Figure 18. It is predicted that the lift force increases during this change from 0.096s to 0.144 s. After that, the lift force will decrease a little and settle at a higher level. A stronger downwash is also noticed via rendered video once the attack angle is increased.

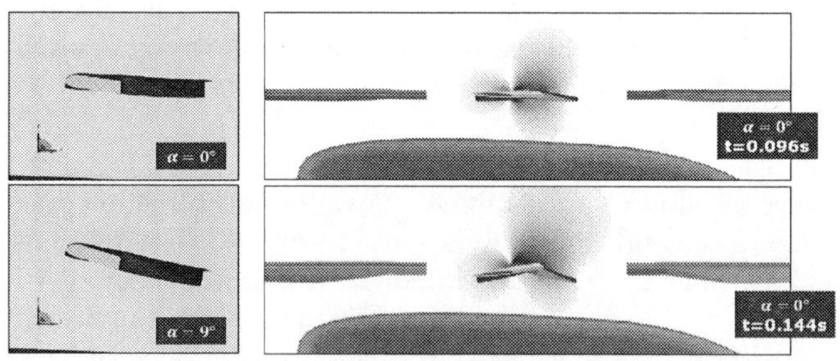

Figure 18. helicopter rotor pitch angle changes during one revolution.

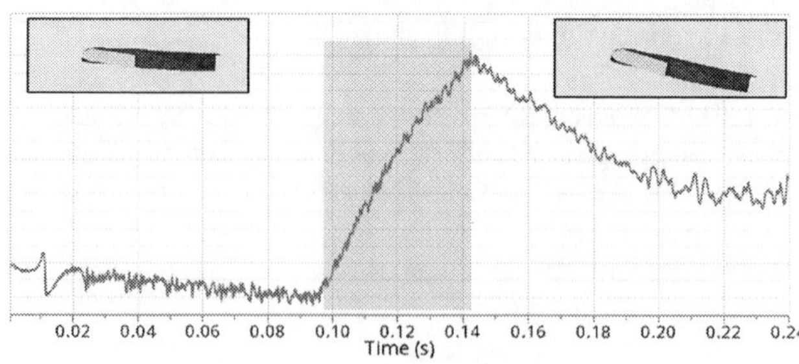

Figure 19. Predicted lift force on the rotor.

CONCLUSIONS

This paper highlights the transformative potential of advanced CFD simulation technologies, such as AVL's PreonLab and FIRE M, in tackling key challenges associated with eVTOL design and operation. PreonLab's meshless particle-based approach enables efficient and precise modeling of dynamic fluid scenarios, including gearbox oil distribution, soiling management, and water ingress mitigation. Meanwhile, FIRE M's Embedded Body Method enhances aerodynamic simulation by eliminating complex meshing processes, allowing faster design iterations and accurate rotor blade modeling.

The integration of these technologies fosters significant advancements in areas such as thermal management, aerodynamic efficiency, and powertrain optimization. By addressing critical aspects like eMotor cooling and water landing safety, these tools ensure compliance with stringent regulatory requirements while promoting sustainability and operational reliability. Moreover, their ability to streamline workflows and provide real-time insights supports engineers in delivering cost-effective and innovative solutions for urban air mobility.

As eVTOL aircraft are shaping the future of air transportation, the role of simulation will remain indispensable. By leveraging cutting-edge tools and methodologies, the industry can achieve safer, more efficient, and environmentally friendly aviation solutions, paving the way for widespread adoption of autonomous air mobility.

Author contact: Jun Li jun.li@avl.com

ACKNOWLEDGMENTS

I would like to express my sincere gratitude to Sanjin Saric, Alexander Oliva, Robert Fridman and Zoran Pavlovic for their valuable comments and suggestions on this paper.

REFERENCES

1. National Aeronautics and Space Administration (NASA). *CFD Vision 2030 Study: A Path to Revolutionary Computational Aerosciences.* 2014. NASA/CR-2014-218178. Available at: https://ntrs.nasa.gov/api/citations/20140003093/downloads/20140003093.pdf.
2. FIFTY2 Technology GmbH. *PreonLab User Manual.* Version 6.2, 2024. FIFTY2 Technology GmbH.
3. Gorla, C.; Concli, F.; Stahl, K.; Höhn, B.-R.; Michaelis, K.; Schultheiß, H.; Stemplinger, J.-P. "CFD Simulations of Splash Losses of a Gearbox." *Adv. Tribol., 2012,* 616923
4. Liu, H.; Jurkschat, T.; Lohner, T.; Stahl, K. "Detailed Investigations on the Oil Flow in Dip-Lubricated Gearboxes by the Finite Volume CFD Method." *Lubricants*, 2018, 6(2), 47
5. Bennion K. and Moreno G., "Convective Heat Transfer Coefficients of Automatic Transmission Fluid Jets with Implications for Electric Machine Thermal Management," In International Electronic Packaging Technical Conference and Exhibition, volume 56901, page V003T04A010. *American Society of Mechanical Engineers*, 2015.
6. Ford Motor Company, Product Data Sheet Mercon LV, **link**
7. Yettou EM., Desrochers A., Champoux Y., Experimental study on the water impact of a symmetrical wedge, *Fluid Dynamics Research*, Vol. 38, Issue 1, p. 47-66, 2006
8. Enmark, M. CFD Modeling of Snow Contamination on Cars: Implementation of a Snow Adhesion Regime Map by User Defined Functions. Master's Thesis, Chalmers University of Technology, 2017.
9. Gostelow, J. P., & Horlock, J. H. (1964). Potential Flow through Cascades – A Comparison between Exact and Approximate Solutions. Ministry of Aviation, Aeronautical Research Council (C.P. No. 807).
10. Saric, S., & Basara, B. (2023). Embedded Body-Based App for External Vehicle Aerodynamics. Presented at the 2023 Academic Annual Conference of Automotive, Shenzhen, China, September 7-8, 2023. AVL List GmbH.

Agricultural Application using Autonomous UAS

Daniel A. Wachspress
Senior Associate
Continuum Dynamics, Inc.
Ewing, NJ USA

Milton E. Teske
Senior Associate
Continuum Dynamics, Inc.
Ewing, NJ USA

Robert M. McKillip Jr.
Senior Associate
Continuum Dynamics, Inc.
Ewing, NJ USA

Michael Piasecki
President
Dragonfly Pictures, Inc.
Essington, PA USA

Joseph Pawelczyk
VP of Operations
Dragonfly Pictures, Inc.
Essington, PA USA

Jeffrey Field
Senior Engineer
Dragonfly Pictures, Inc.
Essington, PA USA

Joseph Kujawski
VP of Space Systems
Brandywine Photonics
College Station, TX, USA

ABSTRACT

Interest in Unmanned Aircraft Systems (UAS) for civilian use has increased greatly in recent years and is expected to grow significantly in the future. This paper describes work performed in a NASA-sponsored effort to develop and demonstrate the capability for a UAS used for agricultural spraying to make autonomous real-time decisions based on sensor data of the aircraft state and prevailing wind conditions. An intelligent controller was developed and demonstrated that can autonomously direct efficient agricultural chemical application of a UAS aircraft. The controller is able to provide good application uniformity while minimizing spray drift for given wind conditions by utilizing predictions obtained with Continuum Dynamics, Inc. (CDI) AGDISP (AGricultural DISPersal) software enhanced with a free wake model from the CDI CHARM (Comprehensive Hierarchical Aeromechanics Rotorcraft Model) software. The effort demonstrated the ability to perform the required onboard sensing and communication to allow inflight decision-making and execute a pre-determined flight path/spraying strategy autonomously using a UAS.

INTRODUCTION

In July 2014, NASA launched a partnership with Dragonfly Pictures, Inc. (DPI) in a landmark study of requirements for civil airworthiness certification for operation of the DP-14 Field Hawk unmanned tandem helicopter for agricultural application [1]. The goal of this NASA/DPI effort was to develop UAS to study airworthiness requirements to support FAA commercial certification. The requirements developed were targeted specifically for precision agricultural application, and it was intended that the airworthiness requirements developed for the DP-14 would help create a roadmap for other UAS developers to build safe systems that can be certified for commercial use in the U.S. In November 2015, NASA came out with a suggested certification basis for unmanned rotorcraft being used for precision agricultural spraying based on the DP-14 Field Hawk [2].

This paper describes a subsequent joint effort performed by Continuum Dynamics, Inc. (CDI) and DPI to support the goals of this NASA/DPI initiative while also addressing NASA's objective to facilitate the process of introducing UAS into the national airspace for commercial purposes by advancing technology in support of UAS platforms performing complex operations autonomously. In terms of safety and commercialization of UAS, advancements in autonomy are essential to expand beyond operations that can be performed through direct control by experienced pilots. The ultimate breakthrough in commercial UAS use will hinge on the ability of UAS to fly autonomously and safely in the civilian airspace alongside manned aircraft, thereby eliminating the need for an experienced pilot to be assigned to each aircraft. The goal of work described in this paper was to develop and demonstrate a specific autonomous, UAS commercial operation: agricultural spraying managed by software designed to guarantee airworthiness according to NASA and FAA standards as well as efficient, cost effective, chemical deposition with minimal spray drift. CDI focused on the development of agricultural application modeling software while DPI focused on autonomous UAS flight control, with both companies collaborating in the development of an interface between the two technologies that supports efficient field operations. Demonstration flight tests were performed with a representative UAS.

Presented at the Vertical Flight Society's 11th Biennial Autonomous VTOL Technical Meeting, Phoenix, AZ, USA, Feb. 4-6, 2025.

The autonomous flight control was demonstrated using a small (10 lb) UAS. The flight test demonstrated an ability to perform advanced onboard sensing on a UAS platform, establish communication between the UAS and flight controller at high enough bandwidth to allow inflight decision-making, and execute a pre-determined flight path/spraying strategy autonomously. A subsequent test was performed with a Yamaha RMAX where data on spray distribution was collected and compared with analysis prediction. This paper focuses on the first flight test.

The prototype system developed can perform high-level decision-making during flight and satisfy NASA's draft certification basis for UAS performing precision agricultural spraying. The prototype system incorporates new software, (called MakeFly), that can provide intelligent real-time flight control commands to a UAS autopilot system during agricultural application. The MakeFly software is "trained" for a particular UAS system using an upgraded and enhanced version of CDI's AGDISP agricultural application deposition and drift analysis. CDI's AGDISP analysis is currently being used by the US EPA for spray drift management of manned aircraft. AGDISP was enhanced during this effort for modeling UAS, primarily by coupling AGDISP with CDI's CHARM rotorcraft aerodynamics module to support time-accurate modeling of rotary-wing UAS wakes operating at low speeds near the ground. The new software is called AGDISP-CHARM. A prototype system was developed by CDI and DPI designed for use by DPI's DP-12 mid-size (500 lb) tandem UAS which can be controlled autonomously using the MakeFly software by sending control commands to DP-12's MicroPilot autopilot system during spraying operations.

BACKGROUND

Since its inception in 1979, CDI has been involved in the development of numerical models of helicopter and fixed-wing aircraft wakes for a variety of applications including aerodynamic design, concept evaluation, flight simulation, acoustics, helicopter brownout, shipboard operations with ship air wakes, multiple aircraft (e.g. airport) operations with wake interactions and wake impact on ground personnel and structures. Within this body of previous research is the ongoing development of mechanistic models for the prediction of the consequences of the aerial and ground application of pesticides, initially for NASA but primarily for the United States Department of Agriculture Forest Service (USFS), through cooperation with the Spray Drift Task Force (SDTF), U.S. Environmental Protection Agency (EPA), and the Agricultural Research Service (ARS). This prior research and development at CDI includes ongoing support for the 3,000 global users of CDI's AGDISP agricultural application software.

CDI has developed several models of chemical release from fixed- and rotary-wing aircraft and subsequent particle trajectories as chemicals are entrained in the aircraft wake and wind. The baseline model used in this effort is CDI's AGDISP model for the accurate prediction of deposition of spray material released from aircraft [3]. This model addresses the behavior of aerially released material from a wide range of aircraft types beginning after the nozzle spray breaks-up into droplets, through interaction with the aircraft wake and atmospheric turbulence, to deposition onto crops and the ground. A physics-based model follows the droplets with a Lagrangian solution technique validated against a series of field studies conducted by the USFS and the SDTF [3-10].

The key motivation for use of these models is that the uniform aerial spray application of pesticides within a clearly defined area must be at levels sufficient to achieve efficacy while limiting unwanted drift of pesticides from the target site to within EPA-standards. Spray drift is a source of environmental concern due to potential human health impacts, downwind damage to crops and livestock, and endangerment of ecological resources, especially streams and water bodies. The AGDISP analysis determines spray deposition patterns of aerial pesticide applications bypassing the need for exhaustive flight testing. The AGDISP code is currently utilized for agricultural application by the USFS, while CDI's AGDRIFT is the SDTF regulatory version of the model. AGDRIFT and AGDISP are PC-based codes, operating in a Windows environment with a highly capable GUI to guide user operation (Figure 1).

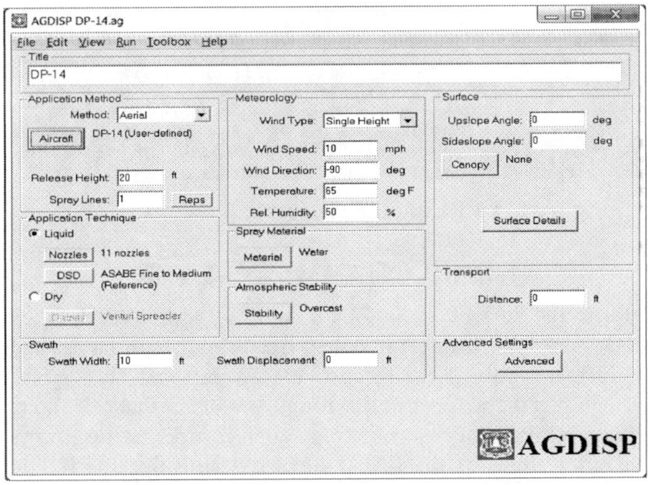

Figure 1. Main Menu – AGDISP GUI.

In the work described here, AGDISP was enhanced for modeling UAS in hover and low-speed flight, where wake characteristics are far more complex and influential on ground deposition and spray drift. This was achieved by coupling CDI's CHARM rotary-wing aircraft wake model as an option for use within a specially enhanced version of AGDISP software. The CHARM wake model can quickly and accurately predict the rotor wake flow field of rotary-wing UAS flying at low speeds near the ground [11-13], a necessary capability required for UAS spray drift and ground deposition prediction which often involves a completely different aircraft wake environment than manned vehicles flying at higher speeds and higher altitudes.

DPI was founded in 1992, and has become an industry leader in the development and implementation of small rotary wing UAS in the U.S. DPI UAS have successfully flown thousands of hours in military applications and DPI is now actively pursuing their deployment in commercial, industrial applications. DPI's fleet of UAS, (some of which are shown in Figure 2), are designed to be easily tailored for specific applications and specialize in areas where stable precision flight, self-launch/land, and high payload capability are critical required platform elements. DPI is recognized for robust low-cost UAS, world-class innovative rotorcraft design, extensive testing, and successful UAS deployment. DPI has extensive experience in developing and implementing autopilots for unmanned air vehicles.

Figure 2. DPI Unmanned aircraft systems (UAS).

AGDISP-CHARM ANALYSIS

Prior to the current work, CDI's AGDISP agricultural spray model was applicable for either fixed-wing or rotary-wing aircraft, but only if the rotary-wing aircraft was flying fast enough that the wake structure roughly matches that of a fixed wing aircraft, with two rolled-up super-vortices trailing from the outer edges of the lifting surface, (wing tips for fixed-wing, outer sides of the rotor disc for rotary-wing), (see Figure 3). In general, when the speed of the apparent wind, (aircraft + wind velocity) is greater than one tenth of the rotor tip speed, the far wake generated by a lifting rotor can be approximated by a super-vortex pair. The ratio of apparent wind speed to rotor tip speed is referred to as the "advance ratio" and is designated by the symbol $\mu = V/(\Omega R)$.

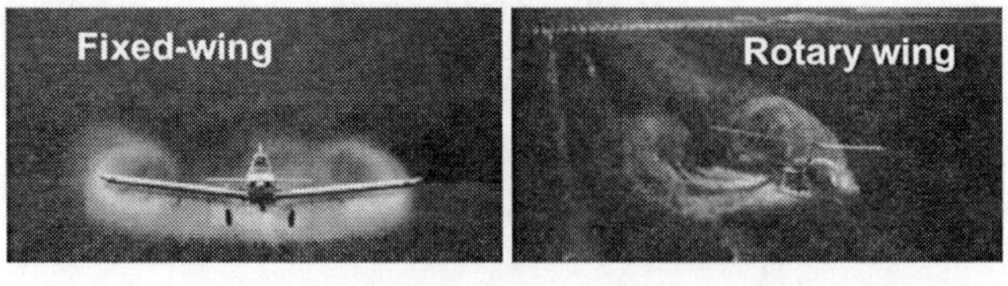

Figure 3. Similarity of fixed- and rotary-wing aircraft wakes in moderate speed cruise.

A primary cost/benefit of using UAS for agricultural application is their ability to hover and fly at very low speeds and altitudes which allows more targeted, precision spraying. To support low speed UAS operation modeling, it was necessary to enhance AGDISP to accurately model the wake dynamics in the hover/low speed flight regime, below μ=0.1, for single main rotor / tail rotor and multirotor aircraft. This flight regime is often referred to as "translational lift" because it is the portion of flight where the aircraft is typically "translating" from take-off to cruise. For agricultural application, the UAS will often be operating entirely within this flight regime. Enhancing the AGDISP software for modeling within this flight regime was accomplished by incorporating the CHARM rotary-wing aircraft wake model as an option within AGDISP. The original AGDISP software assumes a quasi-steady super-vortex pair structure with trajectory and dissipation specified based on fly-by tests performed by the SDTF [14]. The new AGDISP-CHARM software includes all the current AGDISP capabilities plus an ability to use the CHARM wake to provide the wake-induced velocities to the spray model in a time-accurate manner for rotary-wing aircraft flying at low speed.

Figure 4 shows snapshots from animations of a CHARM model of a tandem UAS similar to the DPI DP-14. The colored lines in the top snapshot represent the tip vortices from the blades and are used to determine the wake-induced velocity in the flow field generated by both rotors and their wakes. The bottom snapshot shows chemical being released from a spray boom beneath this aircraft as determined by an AGDISP-CHARM model. In this animation, homogeneous, 50 micron diameter, water-based chemical droplets were released from a spray boom 6 feet beneath the aircraft extending from rotor tip to rotor tip. Two lobes, related to the two super-vortices formed by the rolled-up wake are clearly apparent, as is drift and spray lofting outboard of the primary wake, related to the unsteady interaction of the wakes from each rotor with one another and the ground.

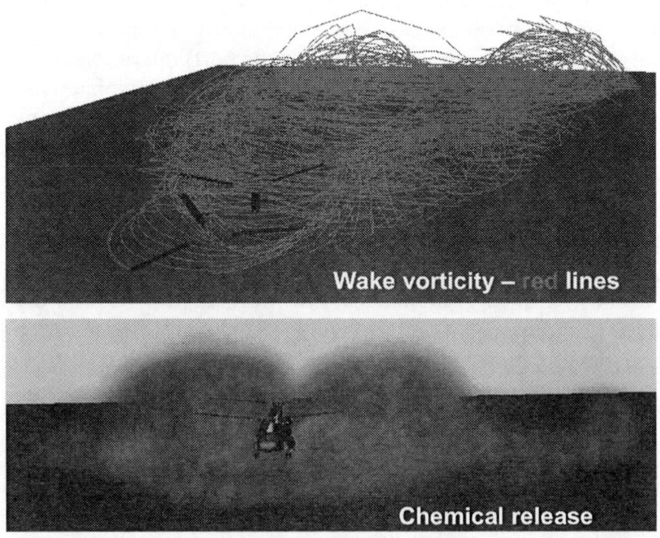

Figure 4. Snapshots from an animation of a CHARM model of a tandem UAS.

The original AGDISP model determines the strength and location of two super-vortices that represent the vehicle wake based on matching computational predictions and test data. The chemical cloud is modeled as particle clusters. After release, the subsequent motion, diffusion and evaporation of the chemical cloud clusters is determined using validated formulas found within the AGDISP software. The cloud evolves based on the existing wind conditions and aircraft flow field. At each time step, the wind + wake flow field velocity at each particle cluster location is determined from the super-vortex model.

In the updated software, the CHARM free wake solution is added as an option to generate flow field information. To avoid having to incorporate a full helicopter/UAS flight simulation within AGDISP-CHARM, the trim characteristics of each UAS modeled within AGDISP-CHARM is determined offline by the CHARM Stand Alone analysis and this information is incorporated directly into the software as part of a library of aircraft supported. This reduces the complexity of the flight simulation software within AGDISP-CHARM and reduces the computation time required by the AGDISP-CHARM calculation because all the aircraft trim (pilot control) settings are pre-determined. Within the AGDISP-CHARM GUI, the user loads a particular aircraft from a library of aircraft, defines the nozzle and chemical configuration from libraries or interactively, executes the calculation, and then selects desired information from a menu of various options, (e.g. ground deposition, amount of spray material still aloft at a critical downwind distance).

The key aspect of the AGDISP solution that needed to be modified to support hover and low speed flight is the modeling of the rotor wake flow field close behind the aircraft. The baseline AGDISP helicopter rotor outwash model is partitioned between a downwash model at small times and a super-vortex pair model at long times. The new implementation, referred to as AGDISP-CHARM, allows an option to replace the "small time" flow field with the CHARM free wake solution. If the advance ratio is large enough that super-vortices will form, ($\mu > 0.1$), then the original super-vortex solution for the long time wake-induced velocity solution is retained. Once the aircraft is a certain distance downstream of the position where the chemical has been released, the AGDISP-CHARM rotor wake solution defaults back to the original AGDISP super-vortex pair solution. However, close behind the air vehicle, the higher fidelity CHARM model is used. Consider a traditional helicopter flying at 40 kts. If the rotor diameter is 50 ft then after roughly 1.5 seconds the super-vortex pair wake solution will dominate the flow field. This is why the AGDISP solution without CHARM is accurate for speeds of 40 kts and more. At much slower speeds, say 5 kts, the transition to the super-vortex flow field never occurs, and the CHARM model replaces the super-vortex model. At intermediate speeds, the AGDISP model transitions from the CHARM free wake solution to the super-vortex solution over the period of time it takes for the super vortices to form and dominate the flow field. By implementing the modified model in this way, the CHARM solution can be used to extend the range of accuracy of the original AGDISP model to low speed and hover while retaining the well-validated AGDISP wake solution for higher aircraft speeds. Furthermore, the extent of the CHARM free-vortex wake model required can be reduced because wake is only needed in the region behind the aircraft before the super-vortex pair is established. This helps support the ability of the model to provide fast turnaround since the wake does not need to be extended for long distances downstream. However, at very low speeds and in hover, a large wake extent is required, which requires a significant amount of calculation time. Therefore, to support field operations, it is necessary to generate a database of information that can be accessed in real-time rather than attempting to run the AGDISP-CHARM calculation directly in the field.

AGDISP-CHARM uses an interpolation function "F" to manage the transition from the CHARM near wake to the AGDISP super-vortex far wake. A good functional form of the interpolation function F was determined through calculations, and is shown in Figure 5. The interpolation function, F, is now set to 1.0 until the aircraft has traveled one rotor radius upstream and thereafter $F = 1/(x/R)^2$ which drops to $F = .01$ once the aircraft has traveled five rotor diameters downstream. This transition is only performed if the aircraft advance ratio is greater than $\mu=0.1$. For lower values of advance ratio, it was found that the CHARM free wake model must be used throughout in order to obtain proper predictions.

Snapshots of an AGDISP-CHARM solution for a sample result using the interpolation function are shown in Figure 6. Side and frontal views of the wake and particle positions are shown. The tip vortices trailing from each of the six blades (three on the front rotor and three on the back rotor) are colored red in these figures while the center of each particle cluster is shown by a single blue dot with an attached blue line showing the CHARM wake-induced velocity at the point. These are identified by green circles surrounding the particles in the figures. The particles are released at the nozzle locations along the boom line with an initial velocity equal to the aircraft velocity. The particles quickly decelerate due to air drag and then are transported by the rotor wake. (This is seen much better in animated visualizations rather than snapshots.) Once the particles are significantly downstream, the CHARM wake is replaced by the AGDISP super-vortex pair wake, and the CHARM wake-induced velocity goes to zero. Note, the blue lines in these figures represent only the CHARM component of the wake-induced velocity. They do not include the super-vortex pair induced velocity.

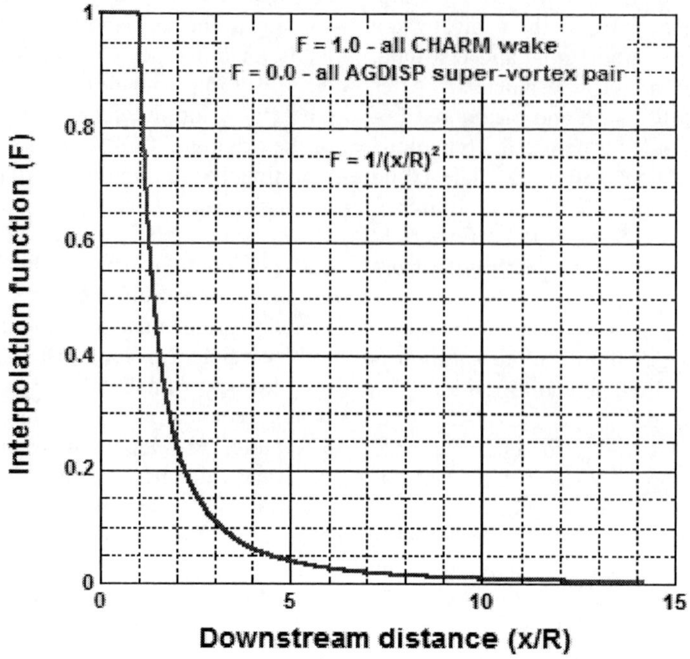

Figure 5. Interpolation function, *F*, used in the new AGDISP-CHARM software.

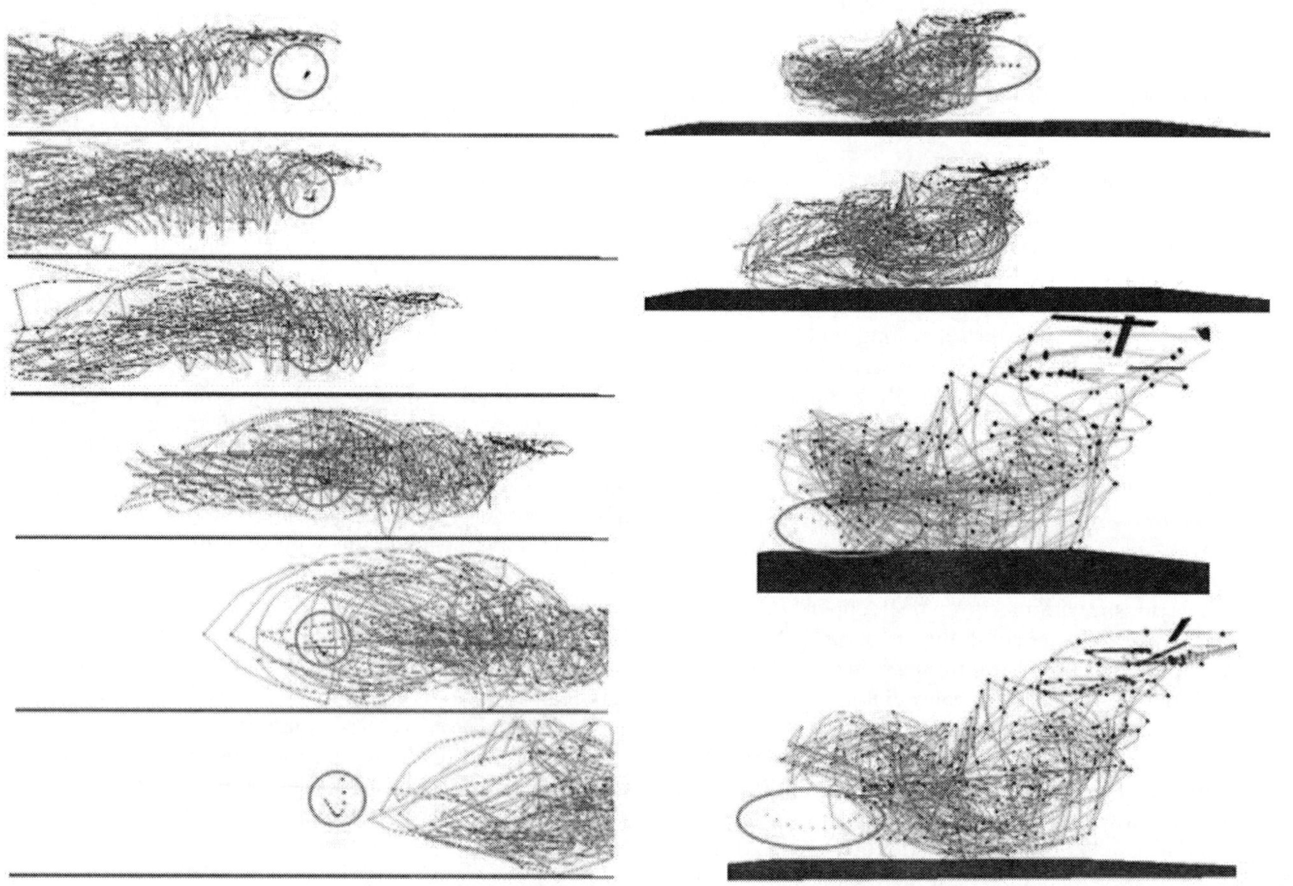

Figure 6 Snapshot sequence of an AGDISP-CHARM solution for a DP-14 flying at 30 mph. A 10 mph cross wind is responsible for the right to left drift.

OPTIMAL TRAJECTORY

A process was developed for determining optimum spraying strategies for a given flight plan, aircraft, wind conditions and user-specified criteria using the AGDISP-CHARM software. For example, given a 5mph wind from the NE, what is the optimum height and speed of flight to minimize spraying time while maintaining a specified level of uniformity of application with spray drift limitations. To perform this optimization in the field would be too time consuming. It is much more efficient to interrogate a database in the field that has been generated through a prior optimization process performed using the stand alone AGDISP-CHARM software. This enables the UAS controller to respond quickly in the field to changing wind conditions rather than needing to perform new AGDISP-CHARM optimization calculations whenever the wind changes. The process is now described using an AGDISP-CHARM model of the DP-12 500 lb tandem helicopter Figure 7.

Characteristic	Dragonfly DP-12 Rhino
Rotor Radius (m)	1.37
Number of Rotors	2
Blades/Rotor	3
Spraying Weight (kg)	220.0
Spraying Speed (m s^{-1})	5.0 to 15.0
Rotation Rate (RPM)	1050
Downwash (m s^{-1})	9.1

Figure 7. DPI DP-12 Rhino tandem rotor UAS.

The DP-12 was modeled with AGDISP-CHARM across a wide range of forward speeds and nominal flight/wind conditions, including several payload levels. Vortical flow fields generated by the DP12 are initiated by counter rotating rotors. It is seen from Figure 8 that the tandem wake of the DP-12 involves a complex wake interaction.

DP-12 calculations placed a spray boom 1m below the main rotor hub, extending the boom out to roughly 1m on either side of the UAS, with 9 evenly spaced nozzles positioned on the boom. The spray release height was set to 1.5m above the ground; a swath width of 3m was assumed while tracking the centroids of released spray material from the nozzles. Looking down at the DP-12, the front rotor rotates counterclockwise and the rear rotor rotates clockwise.

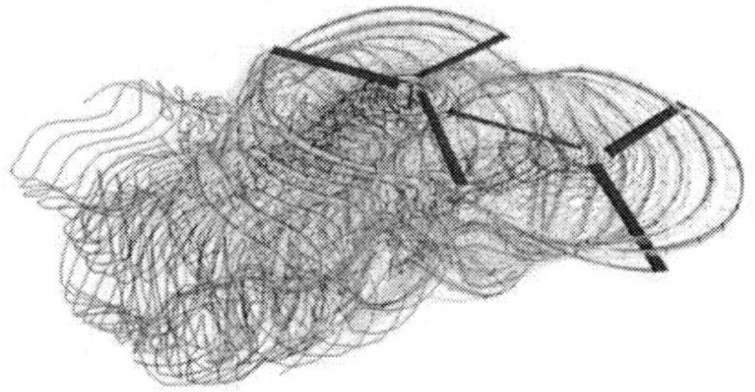

Figure 8. CHARM wake model of the DP-12 tandem.

Calculation results are shown for in Figure 9 and Figure 10 for two crosswinds: 0.22 m/s and 4.47 m/s. The plotted results include four application parameters of interest: (1) coefficient of variation; (2) application efficiency; (3) downwind deposition; and (4) airborne drift. Coefficient of variation (COV) is a measure of how uniformly spray material deposits across the swath; smaller values of COV suggest a more even spray deposition pattern on the ground. Typically, a value of COV equal to 0.3 or lower is considered desirable [15]. Application efficiency is a measure of how much spray material lands within the swath. If the application efficiency is 1.0, all active ingredients in the spray have deposited within the swath; contrarily, if the application efficiency is 0.0, none of the active ingredients has landed within the swath. Downwind deposition is the fraction of active ingredients deposited downwind of the swath, while airborne drift is the fraction of active ingredients still airborne at a specified downwind distance. For these calculations the downwind distance was set to 6m from the downwind swath edge.

Figure 9 shows the predicted application properties for the low crosswind case (0.22 m/s). It is seen that there is no airborne drift and the application efficiency is essentially 1.0. Thus, under very low crosswind conditions, the spray remains within the swath (higher release heights tend to move spray droplets downwind). Figure 10 (with the high crosswind of 4.47 m/s) shows that application properties are similar to the upper plot until the DP-12 speed reaches 6.5 m/s. This speed is near a critical speed which was discovered common to this problem, which is discussed below. Higher flight speeds increase airborne drift by capturing spray droplets and keeping them aloft.

A top view of the DP-12 wake is shown in Figure 11, illustrating the wide spread of the vortical wake and the deposition position of the droplet centroids. Note that the deposition is NOT centered directly behind the aircraft, even though spray droplets were released symmetrically from the boom, because of the effect of the crosswind. The ellipse in black (lower left in the figure, behind the UAS) denotes the locations of the centroids of the spray material released from each of the 9 nozzles. The location of the UAS is identified by the blue rotor blades near the front of the vortical wake (somewhat obscured by the wake they generated). The flight direction of the DP-12 is shown by the blue arrow; the wind direction is shown by the green arrow. Only the blade tip vortices are shown (in red).

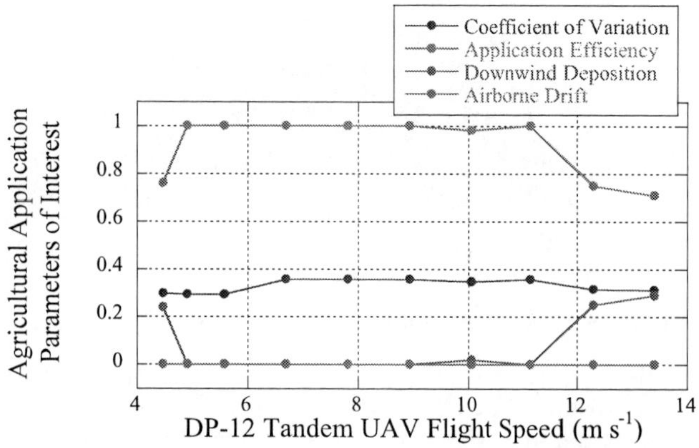

Figure 9. AGDISP-CHARM predictions of agricultural application parameters of interest for the DP-12 for a _low speed crosswind_ of 0.22 m/s. Airborne drift is measured at a maximum downwind distance of 6 m.

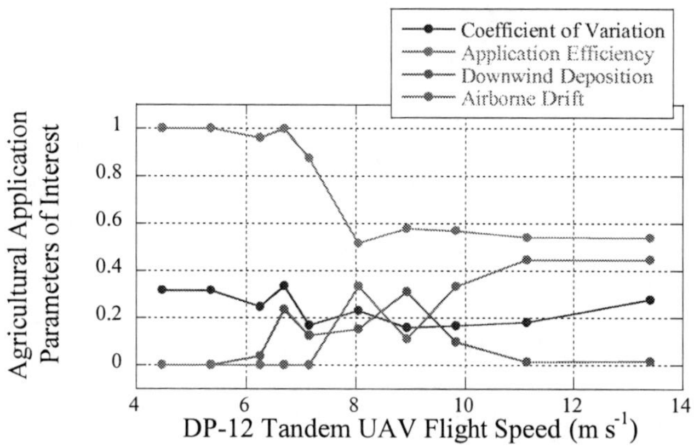

Figure 10. AGDISP-CHARM predictions of agricultural application parameters of interest for the DP-12 for a _high speed crosswind_ of 4.47 m/s. Airborne drift is measured at a maximum downwind distance of 6 m.

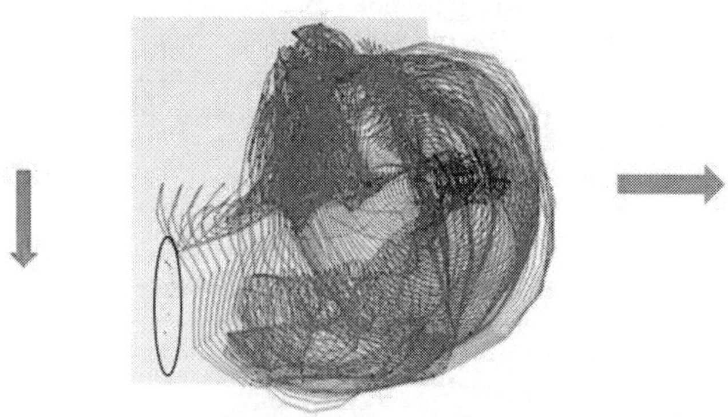

Figure 11. AGDISP-CHARM predictions of the DP-12 wake for an aircraft velocity of 8.94 m/s in a crosswind of 4.47 m/s.

Any pilot or autonomous controller operating the UAS should understand the effect of crosswind and make an appropriate adjustment to the flight path. This corrective action would be to improve the application efficiency and reduce the downwind drift (because of the repositioning of the swath relative to the UAS flight path), but may not necessarily improve COV or reduce airborne drift.

The critical aircraft + wind speed identified in this work is approximately equal to the speed of the induced airflow through the rotor blades in hover. The critical speed for the DP-12 is 8.7 m/s. Figure 10 shows DP-12 results for a high crosswind of 4.47 m/s. In this case the application efficiency is poor at DP-12 speeds higher than 7 m/s, which is slightly lower than its critical speed. The deterioration of the application efficiency is quite pronounced, emphasizing the importance of always operating below the critical speed in order to use the rotor downwash to assist in pushing spray material toward the ground. Otherwise, rotor outwash may entrain the spray upward, resulting in excessive airborne drift.

It must be emphasized that the predictions shown here are for two specific crosswinds, one release height, one UAS weight and one droplet size. Generalizations based on these results should be made with care. The application characteristics depend on many parameters, including release height, crosswind speed, flight speed, aircraft weight and nozzle droplet distribution. Different release heights alter the transition from downwash through the rotor blades to outwash away from the vehicle when the ground is reached. Different crosswind speeds alter the effect of the outwash, especially if the crosswind is sufficiently strong to impede the velocity of the outwash in a direction opposite to the crosswind direction. Increased flight speeds overtake the velocity of the outwash from the vehicle, creating additional problems. Also, aircraft weight – especially for light vehicles – varies as the spray mission continues because the ratio of spray weight to aircraft weight is higher for lighter vehicles. As the weight of the aircraft decreases, the downwash decreases, reducing its ability to push droplets to the ground. This results in increasing downwind drift. Finally, droplet distributions that contain fine droplet sizes will result in far greater spray drift than coarse droplet sizes.

Still, an overall pattern does emerge. Improved application characteristics are uniformly achieved when flying at lower altitudes, lower forward speeds, and in lower crosswinds. This result is not a surprise but can now be quantified. At low speeds the strong downwash pushes the spray quickly down toward the ground, and may potentially provide better distribution over individual plants, as opposed to merely coating their upper surfaces. As aircraft speed increases, however, a critical speed is reached where the downwash transitions to outwash well before the released droplets reach the ground. As these vortices form, the AGDISP-CHARM solution broadens the vortical field behind the aircraft and expands upward (because the ground prevents expansion downward). If spray material becomes trapped within this developing wake, that material could travel to unanticipated distances. It is seen from many simulations not discussed here that at speeds higher than the critical speed, if there is a crosswind, the spray drift may become considerable.

Release height plays a significant role in spray deposition as well. Higher release heights impact the deposition location, in-swath or downwind. The ground impact of released spray as a function of height for UAS's appears to follow the same patterns as for full-sized aircraft. The potential complexity of the UAS wake – the impact of multiple rotor blades – is the randomizer in this behavior. What is especially critical is to understand the eventual geometry of the multiple rotor wake and its possible ability to loft released spray droplets, an effect not present with full-sized helicopters because of their higher flying heights, speeds, and spray boom positions. Also, flying low creates greater safety concerns.

More details on the optimum spraying strategies work can be found in the NASA Contractor Report [16].

AUTONOMOUS CONTROL

There are two critical pieces of software associated with this project. The first is the AGDISP-CHARM software described previously that is used to model the UAS spray deposition and build the database of optimal trajectories for given vehicles, flight plans and wind conditions. The second is the software installed on the ground control computer in the field that performs the autonomous control of the UAS during the spraying process.

The first piece of software, AGDISP-CHARM, was completed and utilized to map out optimal spray strategies for given wind conditions for the DPI DP-12 tandem UAS and a Yamaha RMAX main rotor / tail rotor UAS. The optimal spray strategies correspond to the altitude, speed, and flight path required of the aircraft to provide uniform chemical deposition with minimal spray drift for a given wind condition and spray boom configuration. This information is required by the autonomous control software developed, which is called MakeFly.

A comprehensive series of AGDISP-CHARM calculations were performed to create the look-up tables required by the MakeFly software so that it can instantly determine the optimal flight path for the DP-12 in the field for changing wind conditions. Figure 12 shows example deposition and drift characteristics for a crosswind of 6 mph as a function of DP-12 speed and altitude. The aircraft speed can be found in the third column: ACvel(mph). The spray boom height above the ground can be found in the last column: ReleaseHeight(ft). The columns of greatest interest are the Offset(ft) column and the Spread column. The Spread is a normalized standard deviation that represents the ratio of the amount of spreading of chemical relative to the swath width. Values of less than 1 are acceptable. As seen in **Error! Reference source not found.**, when the release height is 2 ft, aircraft speeds up to 26.84 mph are acceptable, and when the release height is 10 ft, aircraft speeds up to 4.47 mph are acceptable. The Offset is the average offset of particles from the position of release due to the wind. The Offset is used to offset the flight path upwind an appropriate amount based on the crosswind, aircraft height and AGDISP-CHARM predictions.

Error! Reference source not found. is for one crosswind condition and one nozzle configuration. Similar tables were generated for crosswind speeds from 0 to 10 mph. Because the DP-12 is a tandem helicopter there was a difference in spray deposition with direction of crosswind, so calculations needed to be performed for both directions of crosswind.

Wind(mph)	WindDir	ACvel(mph)	COV	MeanDep	AppEff	DownDrft	Airborne	Evap	Swath(ft)	Offset(ft)	Spread	ReleaseHeight(ft)
6.00	-90.00	15.66	0.25	0.93	1.00	0.00	0.00	0.00	8.00	-0.04	0.24	2.00
6.00	-90.00	17.90	0.25	0.93	1.00	0.00	0.00	0.00	8.00	0.13	0.20	2.00
6.00	-90.00	20.13	0.25	0.93	1.00	0.00	0.00	0.00	8.00	0.04	0.22	2.00
6.00	-90.00	22.37	0.24	0.94	1.00	0.00	0.00	0.00	8.00	0.01	0.21	2.00
6.00	-90.00	24.61	0.18	0.97	0.30	0.70	0.00	0.00	8.00	0.23	0.29	2.00
6.00	-90.00	26.84	0.26	0.92	1.00	0.00	0.00	0.00	8.00	0.29	0.21	2.00
6.00	-90.00	29.08	0.23	0.88	0.83	0.00	0.17	0.01	8.00	-8.72	3.07	2.00
6.00	-90.00	31.32	0.23	0.90	0.77	0.06	0.17	0.01	8.00	-8.69	2.76	2.00
6.00	-90.00	33.56	0.20	0.90	0.83	0.00	0.17	0.01	8.00	-8.73	2.70	2.00
6.00	-90.00	15.66	0.20	0.96	1.00	0.00	0.00	0.00	8.00	-0.31	0.24	4.00
6.00	-90.00	17.90	0.23	0.94	1.00	0.00	0.00	0.00	8.00	-0.29	0.18	4.00
6.00	-90.00	20.13	0.25	0.93	1.00	0.00	0.00	0.00	8.00	-0.11	0.20	4.00
6.00	-90.00	22.37	0.18	0.97	0.30	0.70	0.00	0.00	8.00	-0.54	0.20	4.00
6.00	-90.00	24.61	0.18	0.93	0.50	0.16	0.33	0.01	8.00	-11.41	2.32	4.00
6.00	-90.00	26.84	0.28	0.91	0.22	0.62	0.17	0.01	8.00	-8.67	2.18	4.00
6.00	-90.00	29.08	0.26	0.91	0.51	0.15	0.33	0.01	8.00	-9.65	2.38	4.00
6.00	-90.00	15.66	0.20	0.96	1.00	0.00	0.00	0.00	8.00	-0.12	0.22	5.00
6.00	-90.00	17.90	0.18	0.97	0.31	0.69	0.00	0.00	8.00	-0.24	0.20	5.00
6.00	-90.00	20.13	0.12	0.98	0.54	0.46	0.00	0.00	8.00	-0.86	0.24	5.00
6.00	-90.00	22.37	0.12	0.98	0.64	0.20	0.17	0.01	8.00	-8.83	2.37	5.00
6.00	-90.00	24.61	0.19	0.94	0.32	0.52	0.17	0.01	8.00	-5.83	2.18	5.00
6.00	-90.00	26.84	0.22	0.89	0.36	0.30	0.33	0.01	8.00	-17.20	3.74	5.00
6.00	-90.00	6.71	0.20	0.96	1.00	0.00	0.00	0.00	8.00	0.46	0.36	6.00
6.00	-90.00	8.95	0.22	0.94	1.00	0.00	0.00	0.00	8.00	0.38	0.32	6.00
6.00	-90.00	11.18	0.20	0.96	1.00	0.00	0.00	0.00	8.00	-0.18	0.32	6.00
6.00	-90.00	13.42	0.24	0.94	1.00	0.00	0.00	0.00	8.00	-0.79	0.16	6.00
6.00	-90.00	15.66	0.12	0.98	0.81	0.19	0.00	0.01	8.00	-0.01	2.00	6.00
6.00	-90.00	17.90	0.12	0.98	0.64	0.19	0.17	0.01	8.00	-7.47	1.73	6.00
6.00	-90.00	20.13	0.18	0.97	0.97	0.03	0.00	0.00	8.00	-1.01	0.42	6.00
6.00	-90.00	22.37	0.14	0.94	0.67	0.16	0.17	0.01	8.00	-15.75	3.71	6.00
6.00	-90.00	24.61	0.16	0.92	0.61	0.22	0.17	0.01	8.00	-13.21	3.75	6.00
6.00	-90.00	26.84	0.12	0.97	0.46	0.21	0.33	0.01	8.00	-16.51	2.97	6.00
6.00	-90.00	2.24	0.20	0.96	1.00	0.00	0.00	0.00	8.00	-0.65	0.43	10.00
6.00	-90.00	4.47	0.23	0.94	1.00	0.00	0.00	0.00	8.00	-0.87	0.22	10.00
6.00	-90.00	6.71	0.18	0.97	0.32	0.68	0.00	0.00	8.00	3.32	1.26	10.00
6.00	-90.00	8.95	0.12	0.98	0.81	0.19	0.00	0.00	8.00	6.03	2.05	10.00
6.00	-90.00	11.18	0.24	0.94	1.00	0.00	0.00	0.00	8.00	-1.33	0.30	10.00
6.00	-90.00	13.42	0.18	0.97	0.14	0.69	0.17	0.01	8.00	-7.57	1.96	10.00
6.00	-90.00	15.66	0.20	0.92	0.44	0.40	0.17	0.01	8.00	-3.61	1.77	10.00
6.00	-90.00	17.90	0.20	0.95	0.00	0.78	0.33	0.01	8.00	-14.88	2.55	10.00

Figure 12. AGDISP-CHARM prediction of DP-12 spray characteristics – 6mph crosswind.

The information from the tables generated with AGDISP-CHARM is incorporated into the MakeFly software so that in the field the user can enter the desired area to be sprayed and current wind condition and the MakeFly software can design an optimal flight plan. If the wind changes during spraying, MakeFly can change the flight plan, adjusting to account for the new conditions.

The farmer wants to spray in the most economical manner. It is found that the greatest precision is achieved when flying low and slow. This reduces waste of costly chemicals and minimizes spray drift. However, this increases time and hazard. AGDISP-CHARM determines the maximum speed allowable for a given crosswind and release height for a given vehicle. Figure 13 shows example results for the Yamaha RMAX. This illustrates the fact that the maximum speed decreases with increasing crosswind and increasing release height. This information is condensed into tabular form and incorporated into the MakeFly software.

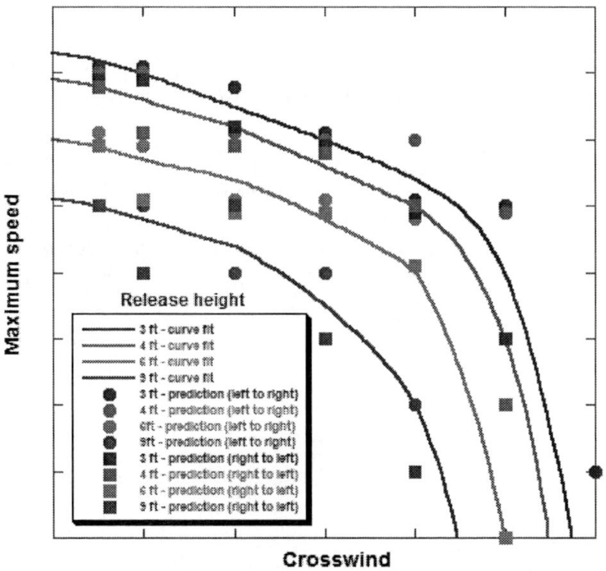

Figure 13. RMAX maximum speed vs crosswind and release height for good ground application characteristics with acceptable spray drift as predicted by AGDISP-CHARM.

MakeFly Software

A method was developed for translating the information from the AGDISP-CHARM generated database to a text file which can be uploaded to the aircraft and control the aircraft heading and direction in real-time in the field. These algorithms were developed with inputs from the autopilot manufacturer's documentation. The result was incorporated into the MakeFly software for auto-generating these text files, (.fly files), and tested in software simulations and flight test for a 10 lb UAS (DP-RH1 shown in Figure 2). The overall approach was as follows.

- Build a detailed model of the UAS within the AGDISP-CHARM software.
- Use AGDISP-CHARM to determine "optimal" spraying strategies. The definition of "optimal" will vary. In this effort, "optimal" was defined as minimum cost while satisfying two constraints: (1) application efficiency of .9 or above and (2) the Coefficient of Variation (COV) of the ground deposition within the designated area was below a specified value, (e.g., 0.3 which is typical of current requirements). There were hard constraints on the UAS system in terms of maximum speed, minimum height and maximum flow rate.
- Incorporate the information on optimal spraying strategies into the MakeFly software that will generate the .fly files that direct the UAS autopilot in the field.
- Demonstrate and validate the effectiveness of the fielded system through comparisons of AGDISP-CHARM predictions with actual ground deposition through testing. Adjust the AGDISP-CHARM model as necessary to match the test data.

The MakeFly software was developed to generate a .fly file for the MicroPilot autopilot used on the DP-12 to guide the flight and spraying process to match the optimum spraying strategy as determined by the AGDISP-CHARM analysis. Lessons learned in the process include:

- It is necessary to define aircraft speed as well as waypoints since speed is crucial for optimum spraying strategy.
- The .fly file command to commence spraying is performed while the aircraft is moving at optimal spraying speed and has just reached the front of the leg to be sprayed. Similarly it is turned off while the aircraft is moving at optimal spraying speed as well.
- In order to get up to speed the aircraft will have to begin each leg a distance away from the start of the leg. The same is true when finishing the leg and slowing down.

AUTONOMOUS FLIGHT TEST

An autonomous flight test was performed to demonstrate the developed technology including the sensing and data transmission capabilities. The types of data required can be grouped into three categories:

- Fixed information – known beforehand, does not need to be sensed by the UAS, e.g. aircraft characteristics, spray rig characteristics.
- Changing information - preflight – known beforehand, does not need to be sensed by the UAS, e.g. application area, application specs, temperature, humidity.
- Changing information - inflight – unknown beforehand, does need to be sensed by the UAS or control station, e.g. wind speed and direction, aircraft state, spray rig state.

The test vehicle was DPI's DP-RH1, a 10 lb main rotor/tail rotor configuration with a 5.2 ft diameter main rotor (see Figure 2). Aircraft position and velocity were determined on the UAS through an integrated inertial and GPS navigation system. The fuel burn rate is known, which provides an approximate aircraft weight given the initial vehicle weight. The DPI UAS stores information on roughly 70 variables of data on board during flight. This information is stored at 33Hz. A subset of this information is transmitted through radio telemetry to a ground controller at 5-10 Hz, more than adequate to support real-time calculations. The flight plan includes aircraft position, orientation, velocity and nozzle setting (open/closed).

Providing an ability for the aircraft to precisely determine the wind speed and direction in hover and at low speed is one of the key technological challenges. Pitot tube anemometers were present on board the UAS to determine aircraft + wind speed. The prevailing wind can be determined from this measurement given knowledge of the aircraft velocity determined through GPS. However, this method is not accurate in hover and very low speeds since the anemometer measurement loses accuracy due to the influence of the rotorwash. At low speed, the rotorwash creates large magnitude air flow relative to the wind speed, compromising the accuracy of the sensor reading. This issue is true for any type of anemometer on rotary-wing aircraft. The proposed solution to this problem is implementing an accurate, physics-based aerodynamic model of the aircraft within the sensor system based on CDI's CHARM rotorcraft analysis. This model could determine the prevailing wind directly from the aircraft control settings and state (position, heading, velocity, weight). In the current work, ground measurements of the prevailing wind were obtained by setting up directional anemometers around the field. This is a far less desirable solution than having an on board ability to measure the wind accurately. There is a significant cost in dollars and time associated with purchasing, maintaining and setting up ground anemometers around a farmer's field. Furthermore, each field is different, and calibrating the system would take time. Also, the anemometers would represent a flight hazard. Achieving the capability to measure wind speed and direction accurately on the aircraft at any flight speed would greatly reduce the time, cost, complexity and risk of the operation while improving the input into the controlling analysis. Furthermore, this capability would also benefit a multitude of other helicopter and UAS applications that require precise knowledge of prevailing wind conditions during low speed maneuvering flight. This is a prime topic for future research.

The MicroPilot Inc. MP2128HELI/HELI2 autopilot used supports a waypoint-based method for providing flight instructions to a UAS system. The flight instructions can be provided in the form of a series of single line commands in a text file. These text files are identified by the extension .fly and are referred to as "FLY" files. The FLY file used in the demonstration flight contained representative commands of those generated by the MakeFly software.

The flight demonstration was performed at the Rutgers Cooperative Research & Extension (RCRE), a rural, agricultural experimental facility run by Rutgers University in southern NJ. The test site was a cranberry bog operated by the RCRE. Figure 14 shows a screen capture of the display of the laptop computer that controlled the DP-RH-1 in the flight demonstration. The FLY file commands are seen on the left while the center of the display shows the waypoints in the FLY file super-imposed on a Google Maps aerial image of the test location. The flight consisted of 3 legs: (1) fly to Waypoint 1 (WPT1) and hover while spraying, (2) spray while flying from WPT2 to WPT3, (3) fly from WPT4 to WPT7 with spray on between WPT4 and WPT5, off between WPT5 and WPT6, and on again between WPT6 and WPT7.

On the day of the test, the wind was gusting to 17 mph, which made safety a challenge. Because of the need for a government representative to be present to operate under the FAA Certificate of Authorization, there was no flexibility to wait for a less windy day without incurring additional costs and schedule slippage. For the safety of personnel and the UAS in the strong winds, manual control of the aircraft was maintained during takeoff and landing and the flight plan was executed at a high altitude The strong winds affected the ability of the light UAS (10 lbs) to execute the flight plan accurately, though the autopilot was able to match the flight plan to within 3 meters over most of the demonstration. Four flight tests were performed with data recorded successfully for each. A panel of green LED lights was attached to the UAS (Figure 15). When the controller signaled the UAS to commence spraying, the panel of green LED lights turned on. The illuminated lights are easily seen in the video playback, though are not as evident in the series of screen captures from the video shown in Figure 16. A video of one of the flight tests can be found at the following URL address (https://www.youtube.com/watch?v=7AzLOyf0hMg).

Figure 14. Screen capture of the display of the laptop computer controlling the UAS during the flight demonstration.

Figure 15. DP-RH-1 outfitted with a panel of green LED lights to indicate spraying command is "on" during the flight demonstration.

Several lessons were learned from the flight demonstration. First, the autopilot commands need to be very specific to implement the level of precision required. It is necessary to command not just the aircraft position but also the aircraft orientation and velocity for each spraying leg. Second, low altitude flight is a key capability. The lower the altitude, the less chemical required and the less spray drift. GPS and/or barometric altimeters are not accurate enough to ensure the UAS maintains a uniform height close to the tops of the plants being sprayed. For some plants, like the cranberry bogs, a laser altimeter will be fine, because the tops of the plants cover almost all the exposed area. For more sparse plants, like blueberries, the laser altimeter may not work – or the output may have to be processed and averaged to maintain the proper height above the top of the plants. Third, the demonstration UAS was a small aircraft, so the strength of the rotorwash was not an issue. A larger aircraft will create a strong outwash that will impact application and spray drift in complex positive and negative ways depending on external conditions. This points to the importance of having a high fidelity, physics-based analysis like AGDISP-CHARM incorporated into the application process.

The aircraft state was successfully recorded at the anticipated frame rates for each of the four test flight demonstrations. The data necessary was successfully transmitted to the ground control station at between 5-10 Hz. Figure 17 overlays the actual flight trajectory on top of the commanded flight plan for one of the test flights. It was observed that despite the strong winds and the light weight of the aircraft, the autopilot was able to maintain the flight path to within 3 meters during spraying as determined by the GPS navigational system.

While the GPS navigational system used turned out to be quite accurate, officially it is only accurate to about 15 meters. Greater precision is required, and can be achieved at reasonable cost using more sophisticated GPS systems currently available. Second, roughly a five second phase lag was seen between when commands were uploaded to the aircraft and when they were observed in flight. In future work, this can be accounted for through either a hardware upgrade or through a software modification.

Figure 16. A series of screen captures from a video of the flight demonstration that can be found at (https://www.youtube.com/watch?v=7AzLOyf0hMg).

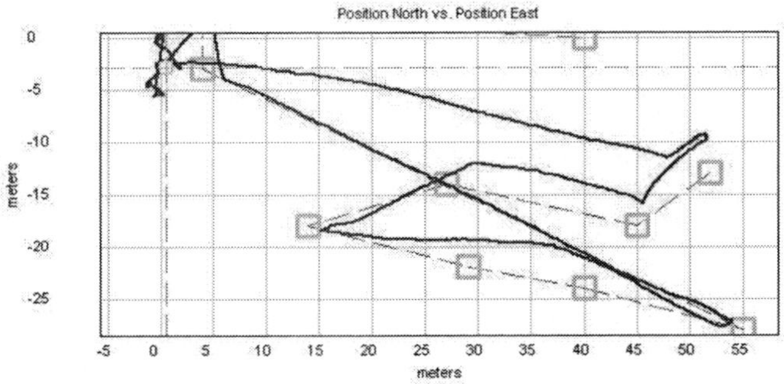

Figure 17. Overlay comparison of commanded and actual flight trajectories for one of the test flights.

CONCLUSIONS

This effort saw the development and demonstration of a prototype Unmanned Aerial System (UAS) designed to perform agricultural application economically and autonomously with an eye toward meeting FAA and EPA guidelines. The effort saw the design and development of a prototype system that can perform high-level decision-making during flight and satisfy NASA's draft certification basis for UAS performing precision agricultural spraying. The prototype system incorporates new software, (MakeFly), that can provide intelligent real-time flight control commands to a UAS autopilot system during agricultural application. The MakeFly software is "trained" for a particular UAS system using an upgraded and enhanced version of CDI's AGDISP agricultural application deposition and drift analysis. AGDISP was enhanced during this effort for modeling UAS, primarily by coupling AGDISP with CDI's CHARM rotorcraft aerodynamics module to support time-accurate modeling of rotary-wing UAS wakes operating at low speeds near the ground. The new software is called AGDISP-CHARM. A prototype system was developed for DPI's DP-12 mid-size (500 lb) tandem UAS which can be controlled autonomously by the MakeFly software by sending control commands to DP-12's MicroPilot autopilot system during spraying operations. A demonstration was performed using DPI's DP-RH1 10 lb main rotor / tail rotor UAS validating the ability of the system to autonomously command the UAS using information determined from the high fidelity AGDISP-CHARM software designed to maximize efficiency/economy while maintaining spray drift levels to within user-designated margins.

Suggested future work includes ongoing measurement of ground deposition for small UAS spraying obtained with cooperation from the U.S. Forest Service and EPA so that the data can be used to certify the accuracy of the AGDISP-CHARM software in this application as a step toward commercialization of the autonomous UAS agricultural spraying system in a manner certified for both airworthiness safety by the FAA and spray drift tolerance by the EPA. Subsequent to the completion of this effort, DPI completed construction of a DP-14 (1000 lb) UAS. Desirable future work would also include testing the system using the new, larger DP-14 UAS. The greater size of the DP-14 would make it more attractive to farmers in terms of economics, and the DP-14 was the subject of NASA's mock certification document [2]. Finally, the AGDISP-CHARM software should be improved to better model the outlet of the nozzles including the conical spread and outlet velocity.

Author contact: Daniel Wachspress dan@continuum-dynamics.com

ACKNOWLEDGMENTS

This work was sponsored by the NASA Langley Research Center through the Small Business Innovation Research (SBIR) Program. The authors wish to acknowledge and thank the NASA CORs for this effort, Kelly Hayhurst and Jeffrey Maddalon, for their assistance, guidance and support. The authors also wish to acknowledge Peter Oudemans who was instrumental in the success of the demonstration flight tests at the RCRE facility in NJ.

REFERENCES

1. Dragonfly Pictures Inc., "NASA and DPI Partner in Landmark Study," July 28, 2014, last accessed Jan 16, 2025: https://www.dragonflypictures.com/wp-content/uploads/2014/05/DPI-Press-Release_NASA-DPI_2014-07-28.pdf.

2. Hayhurst, K. J., Maddalon, J. M., Neogi, N. A., Verstynen, H. A., Buelow, B. and McCormick, G. F., "Mock Certification Basis for an Unmanned Rotorcraft for Precision Agricultural Spraying," NASA/TM-2015-218979, 2015.

3. Teske, M. E., Thistle, H. W. and Ice, G. G., "Technical Advances in Modeling Aerially Applied Sprays," *Transactions of the ASAE*, Vol. 46, (4), 2003, pp. 985-996.

4. Teske, M. E., Bowers, J. F., Rafferty, J. E. and Barry, J. W., "FSCBG: An Aerial Spray Dispersion Model for Predicting the Fate of Released Material behind Aircraft," *Environmental Toxicology and Chemistry*, Vol. 12, (3), 1993, pp. 453-464.

5. Teske, M. E. and Kaufman, A. E., "Field Measurements of Helicopter Rotor Wash in Hover and Forward Flight," *Journal of Aircraft*, Vol. 34, (6), Nov.-Dec. 1997, pp. 813,814.

6. Teske, M. E., Hermansky, C. G. and Riley, C. M., "Evaporation Rates of Agricultural Spray Material at Low Relative Wind Speeds," *Atomization and Sprays*, Vol. 8, 1998, pp. 471-478.

7. Teske, M. E., Thistle, H. W. and Eav, B., "New Ways to Predict Aerial Spray Deposition and Drift," *Journal of Forestry*, Vol. 96, 1998, pp. 25-31.

8. Teske, M. E., Bird, S. L., Esterly, D. M., Curbishley, T. B., Ray, S. L. and Perry, S. G., "AGDRIFT: A Model for Estimating Near-Field Spray Drift from Aerial

Applications," *Environmental Toxicology and Chemistry*, Vol. 21, 2002, pp. 659-671.

9. Teske, M. E., Wachspress, D. A., Quackenbush, T. R., McKillip, R. M. J. and Curbishley, T. B., "CADARS Model Verification and Validation Methodology Report," CDI/TN-99-04, 2002.

10. Teske, M. E., Kaufman, A. E., Thistle, H. W., Hewitt, A. J. and Valcore, D. L., "Droplet Evaporation Corrections for Aerial Spray Drift Modeling I: Theoretical Considerations," *Atomization and Sprays*, Vol. 13, 2003, pp. 243-250.

11. Wachspress, D. A., Whitehouse, G., R. and Keller, J. D., "Rotorcraft Brownout Aerodynamic Analysis and Flight Simulation," CDI/TR-08-30, 2008.

12. Wachspress, D. A., Whitehouse, G. R., Keller, J. D., McClure, K., Gilmore, P. and Dorsett, M., "Physics Based Modeling of Helicopter Brownout for Piloted Simulation Applications," Interservice/Industry Training, Simulation, and Education Conference (I/ITSEC) Proceedings, Orlando, FL, Dec. 2008.

13. Wachspress, D. A., Whitehouse, G. R., Keller, J. D., Yu, K., Gilmore, P., Dorsett, M. and McClure, K., "A High Fidelity Brownout Model for Flight Simulations and Trainers," American Helicopter Society 65th Annual Forum Proceedings, Grapevine, TX, May 2009.

14. Bird, S. L., Perry, S. G., Ray, S. L. and Teske, M. E., "Evaluation of the AgDISP Aerial Algorithms in the AgDRIFT Model," *Environmental Toxicology and Chemistry*, Vol. 21, (3), 2002, pp. 672-681.

15. Parkin, C. S. and Wyatt, J. C., "The determination of flight-line separations for the aerial application of herbicides," *Crop Protection,* Vol. 1, (3), 1982, pp. 309-321.

16. Wachspress, D., "Autonomous Agricultural Application Using Unmanned Aircraft," CDI/TR 19-10, 2019.

THE POTENTIAL ROLE OF DIGITAL TWINS FOR UAS / SHIP INTERFACE ARTIFICIAL INTELLIGENCE (AI) SYSTEMS

Dr. B. Ferrier, FRINA, FRAeS
Head, Adv Naval Systems Dept
Fincantieri Marinette Marine
Washington, D.C. USA

Professor Stefano Brizzolara
Dept of Aerospace & Ocean Eng.
Virginia Polytechnic Institute
Blacksburg, Va USA

Professor M. R. Belmont
School of Engineering
University of Exeter
Exeter, Devon UK

Professor J. T. Christmas
Department of Computer Science
University of Exeter
Exeter, Devon UK

ABSTRACT

The traditional perception of a digital twin in shipbuilding is as an initial CAD design to support the physical build process. More recently it has become a system in which data from the operational ship is maintained in approximately real time. One of the most challenging tasks in the deployment of basic ship-based VTOL Unmanned Air System (UAS) is the launch and recovery and deck handling operational performance on smaller ship platforms like Corvettes, Frigates and Destroyers. This is owing to the random nature of seaway created deck motion coupled with ship structure disturbed air wake patterns. For most of the traditional VTOL UASs, the precise measurement of these environmental data is essential for air vehicle safe launch and recovery events. Placing this real-time digital "shadow" within a realistic UAS x ship environmental simulation enables its use for planning and scenario testing, and for predictive operational mission planning. As Artificial Intelligence (AI) systems begin to be incorporated into ships, digital twins provide a new opportunity. A high-quality digital twin embedded in a realistic environmental simulation presents the possibility of generating large volumes of UAS x ship interface training data for the AI systems, without the need for expensive sea trials. An example is presented on of how this might work and discuss the benefits and challenges of such a system.

The class of technology represented by this AI and QPP approach is to measure remote sea surface profiles to predict the future wave forces acting upon a vessel. An example being closely followed achieves this by using a neural network (artificial intelligent) to predict deck motions based on photogrammetric remote sea surface measurements. The photogrammetric sea surface measurement system calculates the sea surface several hundred meters in advance of the ship. The Arial University/IAI computational methodology shows good correlation between observed forecasted and real ship motion recordings measuring up to 4 seconds for roll and pitch alone. QPP's approach functions in all-weather conditions. The photogrammetric method is not considered operationally all-weather. Nonetheless, the Arial University promises to be functional if it can select a more robust sea surface measurement system coupled with a neural network matrix composed of all 6-degrees-of-freedom.

NOMENCLATURE

DI	Dynamic Interface Study
DOF	Degree-of-freedom
EI	Energy Index (calculated by LPD)
HLA	High Level Architecture
LPD	Landing Period Designator
NATOPS	Naval Air Training and Operating Procedures Standardization
RAO	Response Amplitude Operator
RN	Royal Navy (UK)
RNAS	Royal Navy Air Station (UK)
SHOL	Ship-Helicopter Operational Limit
SMP	Ship Motion Program
SMS	Ship Motion Simulation (NAV routines)
TCS	Tactical Control Station
UAV/S	Uncrewed Aerial Vehicle/System
USV	Uncrewed Surface Vessel/System
WOD	Wind-over-deck

1. INTRODUCTION

The traditional perception of a digital twin in shipbuilding is as an initial computer-aided design (CAD) model to support the physical build process. While CAD is sparse and static, a digital twin is rich in detail and can be incorporated into a simulation where forces (or inputs in the case of non-physical systems) are applied to it over time and their effects actioned by the model. With the development of technologies such as Internet of Things (IoT), the digital twin has become a way of maintaining a digital representation of the real ship, with real-time, or near real-time, feeds of data that capture all relevant operational data at a suitable temporal frequency (figure 1).

We make the distinction between a digital twin and a simulation, where the digital twin is a static representation, but with the potential to be dynamic, while the simulation provides the temporal dynamism **[1]**.

An example of the use of a full digital twin is the system operated by OSC at the Norwegian Maritime Competence Centre in Aalesund, Norway **[2]**. An operator can submit the digital twin of their ship to be included in a maritime

Presented at the Vertical Flight Society's 11th Biennial Autonomous VTOL Technical Meeting, Phoenix, AZ, USA, Feb. 4-6, 2025.

simulator where they can perform scenario testing and training. The digital twin is then able to receive real-time data feeds from the real ship. The effects of this are displayed within the simulator, allowing the operator to observe the ship and its operations as they take place.

Figure 1 – Typical Use of Digital Twin Technology

The Digital twin model contains all the detail components of the original static 3-D CAD model, but with finite element definitions, physical and non-physical alike. The twin can be integrated into a maritime simulator with many scenarios providing the platform to conduct simulated operations dynamically. The Twin forms a model defined dynamically within a High-Level Architecture (HLA) program application. From the measured sea surface, a short-term deterministic wave model can be constructed allowing the wave system to be propagated to the ship's location. The ship's response to the wave spectrum is calculated to determine the level of quiescence at the time of encounter. The results are mapped dynamically on speed-polar visual graphics providing the ship's best course to steer to ensure encountering the quiescent zone in a timely manner. This procedure defines the ship motion aspect of the deck environment. A user advice system displays the predicted quiescent periods as the suggested best periods in which to execute an operation.

To understand how digital twins may play a role in the development and operation of artificial intelligence (AI) systems, we need to introduce some key concepts and how such systems are taught to perform specific tasks.

There are many types of AI systems, two major classifications of which are *supervised* and *unsupervised*, both requiring data from which to learn; often they require very large sets of data. Supervised learning systems need labelled data, i.e. sets of input data for which the desired outputs are known. Unsupervised learning systems use unlabeled data and learn to find patterns in the input data without the expectation of specific outputs, but with a defined overall objective such as dividing the data into a specified set of classes. Both classes of AI system undergo a period of *training*, where they learn to achieve the stated objectives given the inputs. This can be a time-consuming and computationally expensive task. Once trained, however, *testing* the trained model, by presenting it with previously unseen inputs to obtain its prediction of the outputs, is usually very quick. Creating or curating the training data, especially in the volumes many of these systems need, can be time-consuming and expensive, especially since in many cases the labelling for supervised learning must be performed by humans.

The development of the full digital twin described above, presents an opportunity for generating training data for AI systems. To facilitate the discussion of this, we start by introducing a specific case study where two different aspects of the system could benefit from this approach in different ways.

2. Case Study: Quiescent Period Prediction

Quiescent Period Prediction (QPP) is a class of technologies that use remote sea surface profile measurements to predict the future wave forces acting on a vessel in space and time. Such technologies have prediction time horizons measured in minutes and offer predictions of the actual motions of the vessel (figure 2). The purpose of this type of technology is to identify so called "Quiescent Periods" (the safest periods for launch and recovery) in the next few minutes. Knowing by how long a low motion period will begin and how long that period will last, makes it possible to safely execute a motion sensitive operation. These launch and recovery event may be possible even when the mean prevailing conditions exceed the traditionally considered safe deck limits.

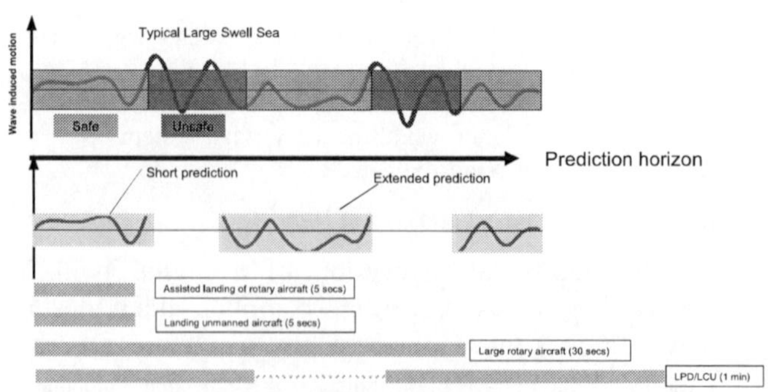

Figure 2 – Relationship between Wave induced motion and prediction.

A QPP system achieves this by using a remote wave sensor system to measure the sea surface several hundred meters in around the ship[3]. From the measured sea surface, a short-term deterministic wave model is constructed allowing the wave system to be propagated to the ship's location [4]. The ship's response to the wave spectrum is calculated to determine the level of quiescence at the time of encounter (figure 2).

The QPP purpose is to provide ship motion conditions and motion predictions with sufficient time to launch, recover and complete other motion sensitive tasks regardless of the seaway. Based on lessons learned from the use of a pre-prototype QPP on trials in the North Atlantic and Irish Sea, it is conceivable, that once integrated, shipboard communities could operate in much higher sea conditions recovering the air/sea vehicles with elevated confidence and reliability. The caveat is that SHOL can be constructed meeting basic deck limit x vehicle interface criteria [5]. As stated, the system is designed to safely expand ship operating and deck recovery envelopes. The computational flow diagram corresponding to QPP system components is displayed in figure 3a.

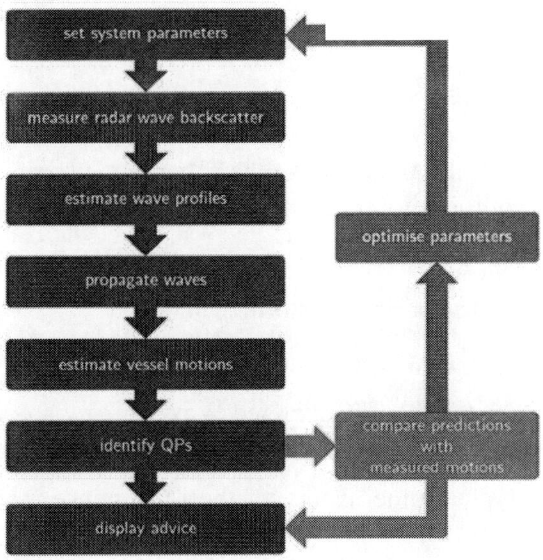

Figure 3a – QPP Computational Flow Diagram

A key part of the system is the wave sensor device, which can be composed of Light Detection and Ranging (LIDAR) or Radio Detection and Ranging (RADAR). A LIDAR system will require a mechanism to scan a laser beam over the sea surface, whereas a radar system can cover a larger area with a single radar pulse. The performance of a candidate sensor will need to be evaluated in a trial, and if proven, the data collected can then be used to validate a simulation of a complete QPP system.

The QPP software package is designed to be also capable of being integrated into a full motion simulator (in the crewed case) and a TCS (in the UAV case). This is a means to estimate system effectiveness as a function of simulated ship motion, visual environment, and synthetic operational systems, and to compare the results to related recorded at-sea data [5]. Such a simulator will also help ship operators gain confidence in the new technology (figure 3b).

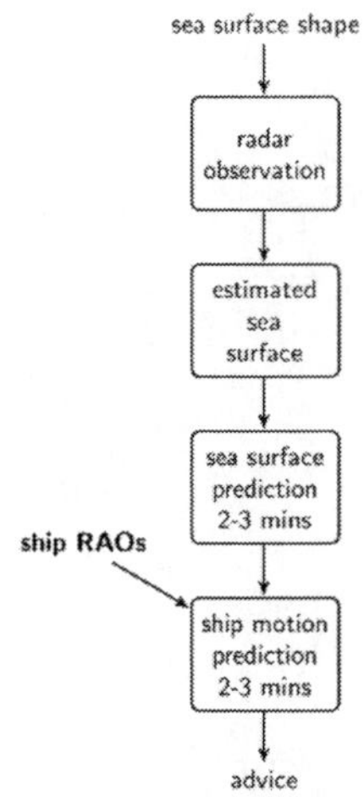

Figure 3b – QPP Digital Twin-based calculation

The QPP method of measurement uses a typical shipboard wave RADAR and fixed time or a shipboard LIDAR. The RADAR methodology is summarized in Figure 4.

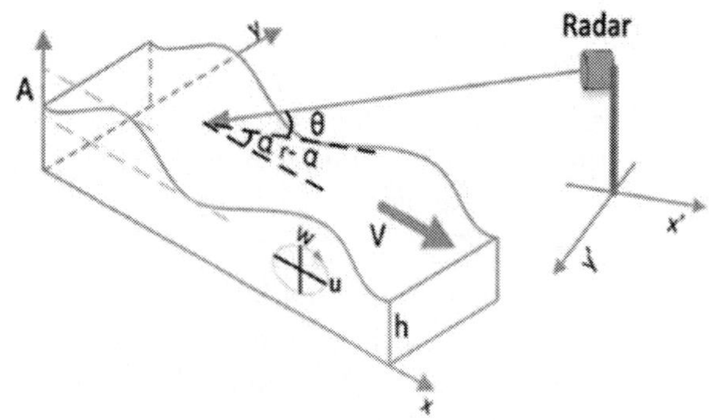

Figure 4 – Radar Wave Measurement

The Twin would attach a radar emulator as a model federate. Traditionally, marine radar is used to measure sea surface currents, and directional wave power density spectra. This only delivers statistical wave information. The novel aspect of RADAR use in QPP is to measure phase resolved directional wave spectra allowing actual wave shape to be determined. As described in [6], ship born wave radars typically operate at acute angled incidence to achieve the desired mapping range of coverage. This introduces gaps in the data due to so called wave shadowing as illustrated in Fig 5b. The development of methods to directly invert marine radar images that can cope with such errors to return accurate sea surface elevations is the subject of current research.

The wave data is then re-mapped onto a uniformly distributed equivalent. Typically, the backscatter image using algorithmic filters resembles real-world wave forms as shown in figure 5 [7]. QPP analysis uses a frequency domain computational methodology. The image is reduced to a sea spectrum containing all the frequencies of the converted wave form.

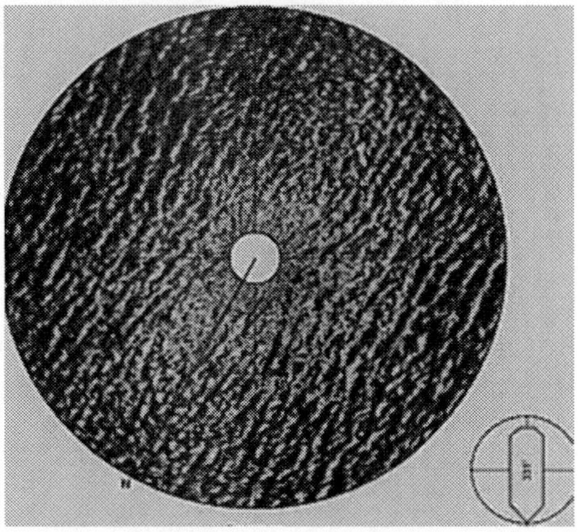

Figure 5: Marine radar Backscatter Image [7]

2.1. BACKSCATTER CONVERSION

The Twin accesses a backscatter convertor federate. The concept is to measure and then form the "backscatter" data into wave structures. Detailed explanation on the computation of wave forms is given in reference [8]. The definition furnishes wave radial velocity, amplitude, direction, and spatial speed. With wave definitions various computational methodologies may be applied, including both frequency domain and deterministic analyses.

In irregular wave fronts, the radar backscatter may miss wave frequency, period, and entire waves hidden behind a closer larger wave [9]. As might be anticipated typically, the backscatter image resembles real-world wave forms.

2.2 WAVE and VESSEL PREDICTOR

At this point the Twin has the information about the environment to compute ship dynamics and the deck motion characteristics. This part of the QPP process takes the backscatter image and converts them into measured wave components. Once the measured wave elevations are known and the wave frequency model adjusted for the point of the ship encounter, a virtual motion of the ocean ship model is computed [10].

As shown in figure 6, the seaway frequency distribution is crossed with the ship's transfer functions (response amplitude operator) producing a distribution of harmonic components. These a summed over a short period of time to create 6 degrees of freedom time histories. The deck limits of the motion sensitive task are compared to determine if a particular heading is appropriate.

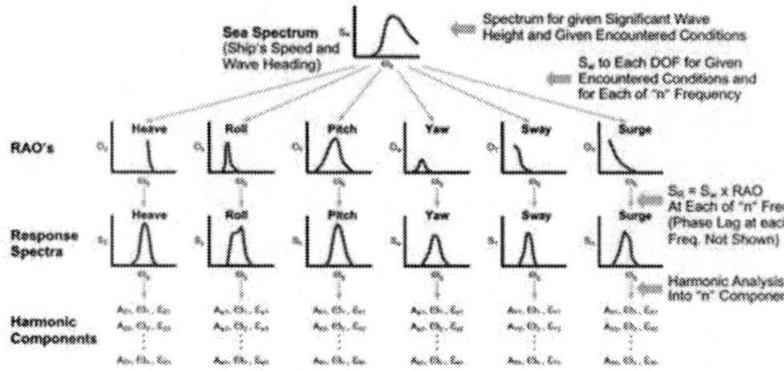

Figure 6 – Conversion of Sea to ship motion

The ship response spectrum is created as the product of the RAO and the driving sea spectrum over the entire range of frequencies. In QPP the conversion to phase-preserved real-time ship responses is given by the Fourier transform in Equation 1:

$$F_z(\omega_j,\theta) = W_{enc}(\omega_j,\theta,V,\phi)\ H_z(\omega_j,\theta) \qquad (1)$$

where:

ω_j : Circular frequency of the jth harmonic component of the natural sea directional (θ) spectrum

θ : Wave propagation angle

$F_z(\omega_j,\theta)$: Fourier transform z[th] mode

$H_z(\omega_j,\theta)$: RAO for z[th] mode

$W_{enc}(\omega_j,\theta,V,\phi_j)$: Encountered Wave Energy spectrum corrected ship heading, ϕ and Ship's speed, V

A typical time history of the ith response mode is obtained by Equation 2:

$$A_z = \sum_{j=1}^{N}\left[A_{zj}\cos\left(\omega_j - e_{zj}\right)\right] \qquad (2)$$

where:

A_{Zj}: Amplitude of the jth harmonic component of the zth response mode

ω_j: frequency of the jth harmonic component

e_{zj}: phase angle of the jth harmonic component of the zth response mode

This federate compute the simulated ship-oriented time histories as shown on figure 7.

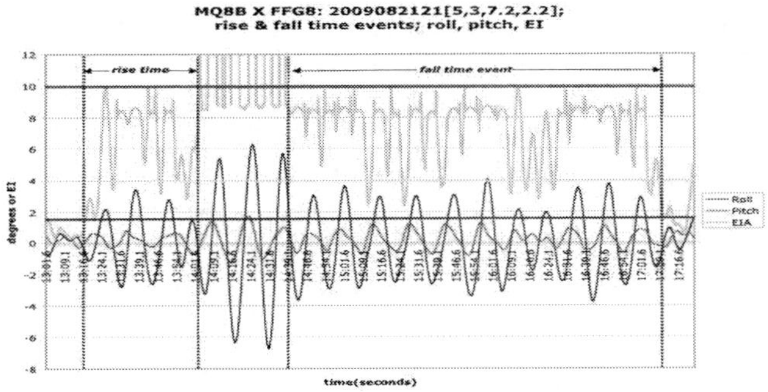

Figure 7 – Synthetic Ship Time Histories

2.3 APPLICATION PROGRAM

The long horizon forecast program objective is to produce a design for a sea going QPP system based around an intelligent decision-making wrapper (the AI shell) that exploits data from existing proven hardware and can learn from its previous activity. It applies verified short horizon prediction (Landing Period Designator or LPD) indices in mapping operationally acceptable seaway regions.

2.3.1. AIR VEHICLE ON DECK DEFINITION

To recap the purpose of employing a Digital Twin model to the complex problem of UAS automated launch and recovery is to learn the conditions to safely complete this task in real-time. QPP is to forecast (**more than a minute(s) depending on the ship and recovery platform**) acceptable deck motion periods for **safe and predictable** launch and recovery of air and sea vehicles (figure 8).

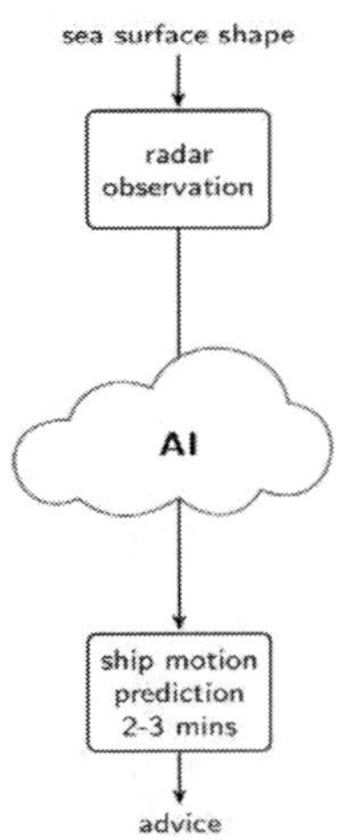

Figure 8 – AI Machine Learning in Digital Twin

Using the wave predictions as inputs to a vessel model makes it possible to predict the actual (deterministic as opposed to statistical) motions of a vessel. This means, there is enough advanced notice of the onset of a quiescent event to setup and recover, for example, a UAS vehicle. This impacts directly or indirectly mission effectiveness and deck tempo.

Motion of an aircraft on the flight deck is calculated in terms of ship motion as a function of the aircraft model. The aircraft model is considered an extension of the ship. The model is defined by aircraft landing gear footprint; on-deck location and orientation. The main difference is the availability in advanced of known timed conditions acceptable for air operations.

2.3.2 SHORT HORIZON MOTION PREDICTION

In the Twin modelling, an important addition to the QPP HLA system is the Landing Period Designator (LPD) federate. This determines and displays real-time information about current and near- term future motion (1/4 cycle) of any vessel as a function of airframe operational limits. As it does not attempt long horizon time predictions, only rely on actual vessel motions, it can be used to provide an emergency last minute waive-off capability for QPP. Having the ability to measure real-time and near-future deck motions eliminates the need for excessively conservative roll and pitch deck limits [11]. These limits are expressed as the ship's Energy Index, which is an empirical derivation representing ship motion as a function of aircraft limits to help guide air vehicles to safe deck landings regardless of the actual seaway.

The Energy Index formulation identifies ship's quiescent periods. These periods are interpreted in terms of aircraft and deck limit definitions. This index provides information on the motion that the ship could exhibit in the near-term future using real-time motion measurements of angular displacements and velocities and linear velocities and accelerations. It capitalizes on the rate at which a vessel can displace owing to natural hydrodynamic forces.

The energy index equation is based on an empirical derivation. It measures lateral and vertical velocities and accelerations as well as roll and pitch angular displacements and velocities weighted by dynamic coefficients. The equation is the least squares of the various parameters and terms representing real-time ship - aircraft interface motion.

Eq.(3)

$$EI=\sqrt{[a_1\phi^2 + a_2\phi'^2 + a_3\theta^2 + a_4\theta'^2 + a_5z'^2 + a_6z''^2 + a_7y'^2 + a_8y''^2]}$$

where:

EI Energy Index value

ϕ Roll ship angle

θ Pitch ship angle

ψ Yaw ship angle

x Longitudinal displacement

y Lateral displacement

z Vertical displacement

a_1 to a_8 Weighted dynamic coefficients

As indicated in Equation 3, the index contains acceleration, velocity, and displacement terms giving indications of the inertial content of the ship. It is an indication of the amount of motion the ship is likely to travel in the near-term future. All 8 degrees-of-freedom, their rates, and accelerations, are monitored and recorded. This includes the ship's vertical displacement, rate, and acceleration. as shown in Fig. 9.

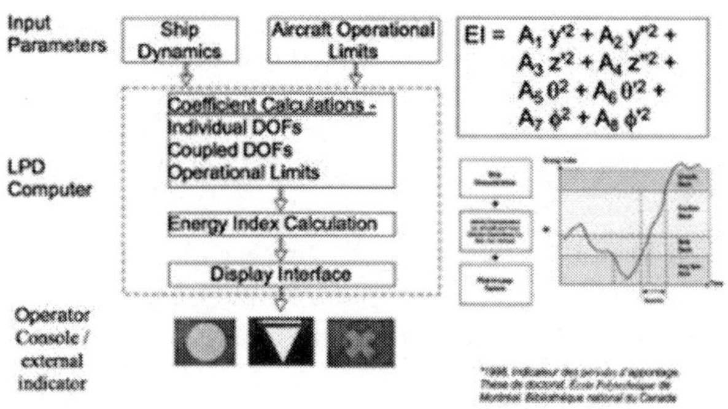

Fig. 9 - Deck Status and Risetime

The time required for the deck motion to rise from minimal motion (or very safe deck) to unacceptable motion is called the rise time. In terms of the EI scale, the rise time is defined as the period that is measured from the end of a green signal to the positive side of the red line. This is given as $(T_3 - T_1)$ as shown on Figure 11. The rise time is a thumb print characteristic of the ship's response and remains constant for each ship class.

The very safe deck is a special condition in which there is insufficient energy in the aircraft-ship system to raise the deck out of limit for some defined time-period or rise time, and it is this concept that was used in the simulators to indicate that the deck was safe to initiate landing [12].

Assuming the minimum average rise time for a given vessel remains constant, the aircraft descends within the rise time value and the aircraft is assured a recovery within the deck motion limits. Figure 10 displays graphically beyond static limit operations. The base envelope is taken from 10 knots of ship speed. Any points within the hour-glass structure are conditions within limits and contain no appreciable probability of out-of-limit deck motions. Outside of the structure contains motions which are considered by static reference as out-of-limit [13].

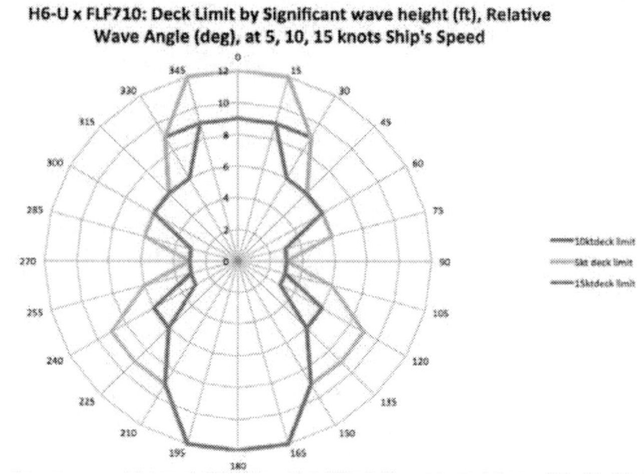

Figure 10: Safe Motion Operations

This is particularly important for robotic vehicles. A schematic representation of how the LPD is used for UAV auto recovery to signal the onset of the descent (and of deck quiescence) is displayed in Figure 11. Throughout the approach and initial hover, LPD is monitoring the deck. It is only at the low hover that the UAV autoland system would accept a Green Deck signal from the LPD. When the EI signals the onset of quiescence and through the uplink sets the air vehicle on its descent to the deck. Short horizon prediction is useful for predictions within 1/4 cycle of ship's motions. QPP is a dimension much longer in predictive horizons but uses the LPD least squares Energy Index to verify QPP's computation now of launch and recovery [14].

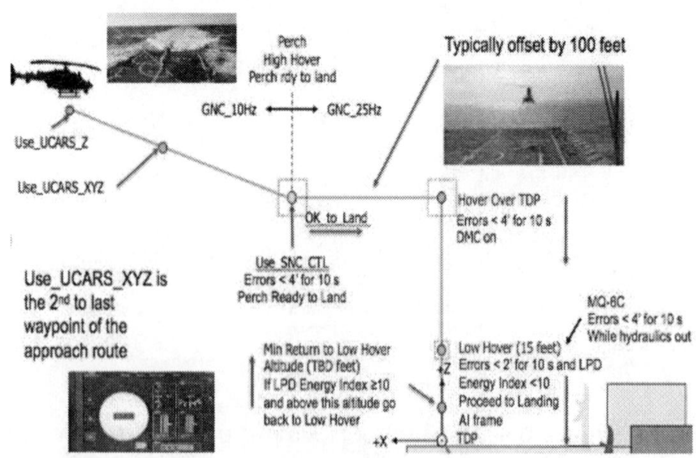

Figure 11: Notional Autoland System

In related trials it was possible to identify every period of low deck activity either by short or long horizon prediction. A simulated scenario is displayed in Figure 12. This compared to Figure 13 displaying actual launch and recovery opportunities.

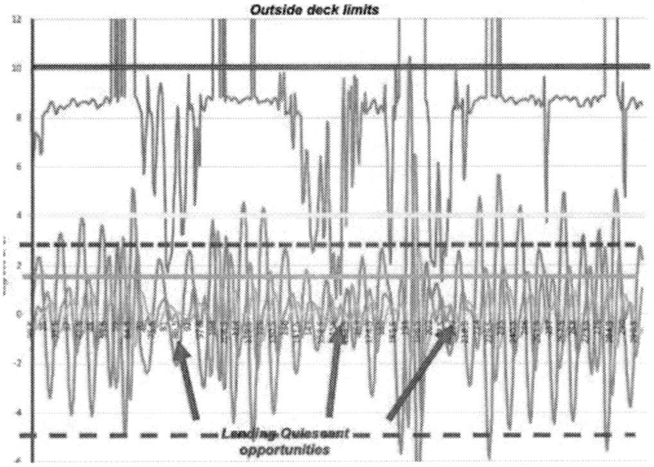

Figure 12 - Landing Opportunities, simulated

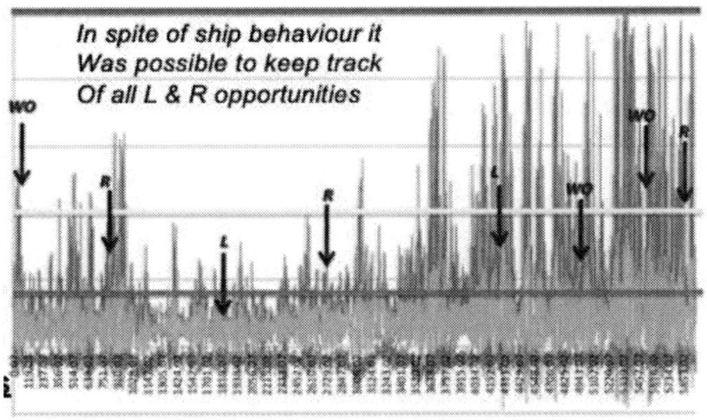

Figure 13 - Landing Opportunities, recorded

3.0 Proposed Digital Twin Method

The proposed solution is based on the digital twin of the ship, which is continuously updated with information from the real ship and its constituent systems, including its radar. In this way a library of real seas and real ship responses, gathered over extended periods of time and in a variety of sea and loading conditions, is created and augmented. This can happen for the whole life of the ship, for the benefit of other ships of the same class, capturing changes that occur, for example during refits. Although the trained model would be tailored to a specific ship, or class of ships, the form of the AI model would be the same for all ships, and so the development of that form of model would benefit from training data from a single ship in the first instance.

It is probably infeasible for the required volumes of data to be beamed back to shore in real time, either because of transmission bandwidth limitations or security concerns, so it would be stored locally on the ship and offloaded when convenient. For QPP and RAOs, the data would include the following: ship location, heading, speed, trim, loading, and motion (heave, pitch, roll, etc.); and radar returns. All data will need to be time-stamped from a single clock (perhaps that provided by the Global Positioning System) to ensure that data from different sources can be correctly aligned with each other and recorded at a temporal frequency that is appropriate for the data item. For example, the radar returns would have to be recorded at high speed, according to the rotation rate of the antenna (our radar records about 8 megabytes every 2.3 seconds), but it would be overkilled to record a container ship's speed at a rate of more than 8 bytes every 1 second.

In the first instance, this library can be used to tune the RAOs established during the design phase. Rather than establishing different RAOs for each rigid motion degree of freedom, and for each relative wave direction and sea state, the data could be used to learn a mathematical function along the lines of:

ship motion = RAO(ship trim, load, speed, multi-directional sea spectrum) (4)

The learning of this function could be achieved by training an AI model, as for instance successfully achieved on naval combatant hull forms [15] as well as small fast crafts [16], in an implicit way

This RAO function could be integrated into the existing design for the QPP system or as described in the previous section, QPP could be liberated from the constraints of a physical model (radar return – sea surface – ship motion – prediction) and become a purely AI model (radar return - prediction), trained using the digital twin data library.

4.0 CONCLUSIONS

At present the form of an AI model would be developed, and its training performed, on real data from a very small subset of environmental conditions. Such a system might be overfitted to the training data, and hence loose its prediction capability in new conditions.

Simulation can provide a means to develop the form of the AI model, based on as wide a range of conditions as can be conveniently generated in a timely fashion. If the simulation is based on the very thing the AI model is trying to learn, then this cannot be a reliable way of training the model, because if you can simulate something to a high degree of fidelity, then you most probably do not need an AI model to learn it. There is one exception to this, which is when something can be very accurately simulated, but it is very computationally expensive and time-consuming to generate the simulations. Then there may be real benefit to creating an AI model as a surrogate.

A full digital twin, with real-time data being continuously recorded, bridges the gap between simulation and reality. This requires the continuous real-time recording of, potentially, very large volumes of data, but this can be gathered during any time the ship spends at sea, rather than having sea trials devoted specifically to the task and can continue to be gathered during the whole lifetime of the ship. It may not require huge computation power, but it will require on-ship storage and, most probably, a technical expert in the crew who can support and maintain the systems making those data recordings.

Insofar as the UAS case study is concerned, a significant motivation for the development and deployment of VTOL UAS, is their exploitation in situations or conditions in which human exposure would by highly risky. Awareness and prediction of the ship's position and movement in the ocean is key to air operations by targeting the right time to deploy and recover. Developing both analysis and training by simulation is a method to avoid excessive and expensive sea trials. The Digital Twin method provides a means to observe the launch and recovery responses to a large matrix of scenarios and sea conditions. Finally, this may provide the means by which the AI model underlying the Digital twin may be trained for at-sea application.

ACKNOWLEDGEMENTS

This report is dedicated to the memory of NAWCAD's late Chief Engineer Robert Ernst, and to the late Dr. Peter J F O'Reilly and Dean Carico, fathers of the Dynamic Interface discipline.

REFERENCES

[1] CHRISTMAS, J., RIGBY, J., 'Changing Definitions of Digital Twins In Ship Design', *Conference Proceedings of EAAW*, 2023

[2] A-OSC, https://osc.no/, last accessed 4 July 2024.

[3] BELMONT M.R., CHRISTMAS J., DANNENBERG J., DUNCAN J., DUNCAN J.M and FERRIER B. An Examination of the Feasibility of Linear Deterministic Sea Wave Prediction in Multi-Directional Seas Using Wave Profiling Radar: Theory, Simulation and Sea Trials. *Journal of Atmospheric and Oceanic Technology*, Vol. 31, **No. 7**, pp1601-1614, 2014.

[4] CROSSLAND, P., FERRIER, B., APPLEBEE, TR. (2004). Providing Decision Making Support to Reduce Operator Workload for Shipboard Air Operations. Royal Institute of Naval Architects. London.

[5] HOENCAMP, A (2016). Helicopter- Ship Qualification Testing. Ipskamp Printing. Enschede.

[6] STRESSER, M. et al.(2022). On the Interpretation of Coherent Marine Radar Backscatter From Surf Zone Waves. *IEEE Transactions on Geoscience and Remote Sensing.* **Vol 60**. Toms River.

[7] SIMPSON, A., et al. (2019). Wave-by-Wave Forecasting via Assimilation of Marine Radar Data. *Oregon State University Press*. Corvallis.

[8] *POULTER, E.*M., SMITH, M.J., MCGREGOR, J.A. (2008). S-Band FMCS Radar Measurements of Ocean Surface Dynamics". *Journal of Atmospheric and Ocean Technology.* ***Vol. 10,*** *No 4.*

[9] CHEN, Z., WANT, Z., ZHAO, C., XIE, F. AND HE, C. (2017), S-Band Doppler Wave Radar System ". *Remote Sensing Journal. MDPI.* ***Vol. 12,*** *No.1POULTER, E.*M., SMITH, M.J., MCGREGOR, J.A. (2008). S-Band FMCS Radar Measurements of Ocean Surface Dynamics". *Journal of Atmospheric and Ocean Technology.* ***Vol. 10,*** *No 4*

[10] APPLEBEE, TR., BAITIS, AE., MEYERS, W.G. (1981) "User's Manual for the Standard Ship Motion Program, SMP" David W. Taylor Naval Ship R&D Centre Report DTNSRDC/SPD-0936-01.

[11] FERRIER, B. (2008). MoD Quiescent Period Prediction LPD Software Testing Programme. UK080229-1. Bristol.

[12] MORRIS, E.L., ZIENKIEWICZ, H.K., BELMONT, M. R (1998). Short term Forecasting of the Sea Surface Shape. *International Journal of Shipbuilding Prog.* **Vol. 45**, *No. 444*. Pp 381-400. Amsterdam

[13] RINA (2022). Ongoing relationship enhances sub rescue capability. *Warship Technology*. March 2015. London.

[14] FERRIER, B (2013). Deck Motion Characterisation Report. *LPD NSRS/QPP Operations*. UKSEA130219. Bristol.

[15] DEL ÁGUILA FERRANDIS, J., TRIANTAFYLLOU, MS., CHRYSSOSTOMIDIS, C., KARNIADAKIS, G.E. (2021). Learning functionals viaLSTM neural networks for predicting vesseldynamics in extreme sea states. Proc. R. Soc. A477: 20190897.https://doi.org/10.1098/rspa.2019.0897

[16] DEL ÁGUILA FERRANDIS, J., BRIZZOLARA, S. CHRYSSOSTOMIDIS, C. (2018). Ocean Engineering, 163,1: 207-222.

BIOGRAPHIES

Michael Belmont, BSc, PhD, hold a chair at the University of Exeter UK, where he heads the Marine Dynamics group. He pioneered the new discipline of Deterministic Sea Wave Prediction which allows the short-term estimation of the details of wave driven vessel motion. This makes possible the prediction of quiescent intervals in the motion of naval platforms allowing the safe extension of the range of sea states for the launch and recovery of air, sea surface and sub-surface craft. Recently the same principals have been applied to the prediction of air-wakes over vessels. The goal of his research is the ability to allow the real time deterministic prediction of all the major environmental factors affecting maritime launch and recovery operations.

Jacqueline Christmas is an Associate Professor of Machine Learning in the Computer Science Department at the University of Exeter, UK. After many years working in industry, she obtained a PhD in the field of statistical pattern recognition in 2011. The main focus of her work is the development of machine learning models for remote sensing, with an emphasis on explain ability and measures of uncertainty, especially for potentially infinite sets of real-time observations, such as are collected by long-term, continuously operating sensors. Her particular interest is in the use of inexpensive, off-the-shelf sensors, such as video, radar and still cameras, to remotely sense difficult dynamic environments in real time. Since 2012 this work has been directed towards the measurement and prediction of sea waves, based on physical oceanographic theory.

Bernard Ferrier, PhD, FRINA, FRAeS is a Systems Engineering and Technology Assistant (SETA) in NAVAIR PMA-266 with the primary task of integrating and deployment of Fire Scout MQ-8C and its technologies and defining the operational limitations of the deck environment. Ferrier holds an honourary Professor position at the University of Exeter's College of Engineering. He is the Engineering Head of Syntek Technology's Aircraft-Ship Dynamic Interface Lab in Fairfax, VA. Ferrier led the CL227 UAV dynamic interface program at Bombardier in Montréal. Ferrier was a rotor dynamist and project leader of the dynamic interface program of the marinised AH-64 (Apache) and LHX programs at the McDonnell Douglas Helicopter Company. He received his last doctorate in helicopter/ship interface engineering at the École Polytechnique de Montréal (Québec) Canada.

Dr. Stefano Brizzolara, PhD in Naval Architecture and Marine Engineering, is tenured Professor in the Kevin Crofton Department of Aerospace and Ocean Engineering. He founded and managed the Innovative Ship Design Lab (VT-iShip), co-director of the Center for Marine Autonomy and Robotics, directly advising a research group of more than 15 graduate students. In 2011 he debuted in the US academic world as Peabody Visiting Associate Professor in the MIT department of Mechanical Engineering. Research interests include numerical hydrodynamics (CFD) for ship design and its application to the design of advanced marine craft and high performance propulsors. The 5+ M$ research portfolio managed by Dr. Brizzolara is sponsored by the ONR, DARPA, DoE, ARPA-e and the shipbuilding industry. Author of more than 250 peer reviewed papers, designated inventor of eleven patents (two pending) for innovative marine crafts and ship systems. Awards and recognition include Crofton Faculty Fellow in 2020, the RINA 2018 Calder Prize for best paper about high-speed crafts, 2016 IEEE Excellence in Review for the Journal of Oceanic Engineering, the 2015 Mandel's Prize for Excellence in Hydrofoil Research advising. Dr. Brizzolara is chairing the newly formed SNAME H-10 'Marine CFD' technical panel, since 2021.

Real-Time Control Power Margin and Gust Estimation for Distributed Electric Propulsion (DEP) Aircraft

Robert McKillip, Jr.
Daniel Wachspress
Jeffrey Keller
Jean-Pierre Theron
Continuum Dynamics, Inc.

R. John Hansman
Professor
M.I.T. Aero-Astro Dept.

David Sizoo
Traci Stadtmueller
Dan Delmyer
FAA

ABSTRACT

A real-time dual-estimator of both remaining control power (RCP) margin and wind gust disturbance statistics for distributed electric propulsion (DEP) aircraft is described, for potential use as both a safety assessment system and a means for sharing observed wind disturbances while in flight. The algorithm is scalable across both piloted and uncrewed aircraft systems (UAS), and requires measurement of vehicle inertial response and actuator inputs. Analysis and simulation modeling for the DEP vehicle used CDI's CHARM (Comprehensive Heirarchical Aeromechanics Rotorcraft Model) code for representing the highly interactional aerodynamic environment for these aircraft, with the RCP and gust estimation algorithm development tested in both simulation and scaled flight experiments of representative DEP configurations.

INTRODUCTION

As autonomous systems encompass more capabilities of DEP aircraft, they will require additional monitoring to ensure their operations remain save, robust and reliable. A key component in that monitoring function is the assurance that the platform does not experience a loss of control (LOC) event, through inadvertent operation into a particular flight environment, or from encountering a limitation of the vehicle's own control capability. These two situations could be experienced in tandem, where a particularly strong gust disturbance could overpower the available moment generation capability of the aircraft, leading to LOC and possible catastrophic results.

The research work to be described in this paper is from an FAA-sponsored investigation on certification considerations for DEP aircraft, where having an on-board capability to monitor remaining control power (as in Ref. 1) would provide a safety assessment of the vehicle's operation in a turbulent environment. The monitoring function could function either during certification testing, or act as an onboard system for providing safety "credits" for use of the aircraft in its regular operation. Although such aircraft-centric safety information is very useful, a desire was expressed to include an assessment of the external flow disturbance environment itself, to be able to share that information with other aircraft and air traffic control operations in the vicinity.

While direct measurement of the applied control command to the available actuators can be, and is, monitored to determine its relative position to the available actuation limits, some predictive capability was desired to provide advance warning of the likelihood of encountering physical limitations and thus a potential LOC situation. In addition, a design requirement of any postulated algorithm for assessing both RCP and wind disturbance effects on the vehicle be *flight control system agnostic*, such that detailed representation of the internal functioning of the flight control system not be required in the implementation of the algorithm. Thus, the algorithm is structured on the assumption that any aircraft response that is not due to the current application of the vehicle actuators represents a disturbance due to atmospheric effects, as shown in block diagram form in Figure 1. This then allows one to represent a predictive model of the vehicle response as a dynamic system responding to either measured control displacements, or unmeasured gust inputs. It has the advantage that maximum performance commands, that may well drive the actuators to their limit values, would not represent LOC events if the aircraft responds appropriately to those command inputs.

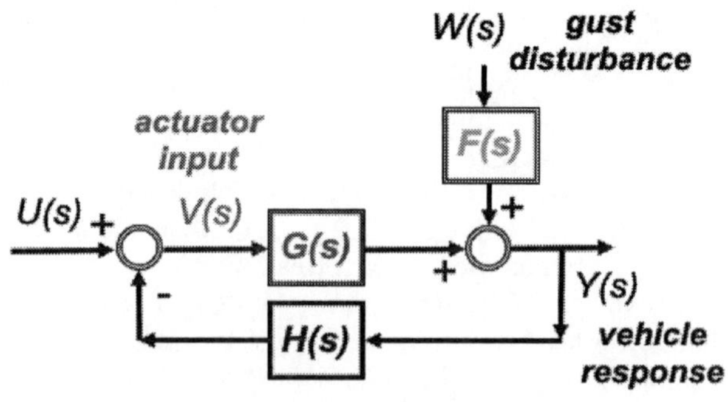

Figure 1. Vehicle response block diagram.

Presented at the Vertical Flight Society's 11th Biennial Autonomous VTOL Technical Meeting, Phoenix, AZ, USA, Feb. 4-6, 2025.

Creation of an algorithm for tracking differences in measured vehicle response to that predicted using a dynamic model of that response with measured control displacements is simplified by using a recursive estimation scheme such as an Extended Kalman Filter. The design aspects of the filter creation are in the representation of the level of dynamics included in the predictive response model, and the representation of the disturbance environment. For gust estimation, this latter design choice simplifies to either direct gust velocity representation, or additionally gust gradients, as additional "states" to estimate within the filter structure, but for predictive RCP use, one may instead treat the disturbance environment as a time-domain perturbation of equivalent control displacements that generate the discrepancy in response, as represented in Figure 2. This is exactly what has been done in the past, in a frequency-domain context, in estimation of a Control Equivalent Turbulence Input (CETI) (Ref. 2). When the disturbance is treated as additional control perturbations, one may then infer that the negation of this perturbation would represent the ideal additional control required to quell the disturbance effect on the vehicle. The RCP estimation can then assess both the current control displacement and the control displacement plus this additional perturbation offset as two separate values for comparison with the known control displacement limits to quantify the likelihood of a LOC event.

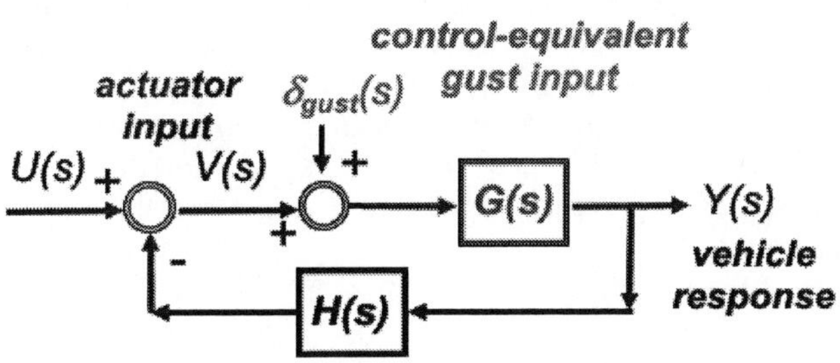

Figure 2. Block diagram of vehicle response showing control-equivalent perturbation model for gusts.

REMAINING CONTROL POWER

Air vehicle controllability is fundamental to both safety and utility, as the aircraft controls must be capable of: (1) trimming the vehicle in all phases of flight; (2) providing added forces and moment for maneuvers, and (3) compensating for disturbances (and possible inherent instability) encountered throughout the flight profile. Fundamental to the use of remaining control power (RCP) as a potential metric for safety monitoring is the assumption that situations that lead to control saturation constitute loss-of-control (LOC) events. While such a characterization is generally true, there exist situations when maximum performance is being requested from a flight vehicle, and thus some controls may well be at their limit values – such as throttle/power settings on takeoff, for example. Thus, the mere approach of vehicle/system control limits alone are not themselves representative of safety margins – instead, it is the sum of the current control power *and the remaining control power required to compensate for vehicle response being other than that commanded by the current control inputs* that is of interest. Figure 3 shows a conceptual picture of the contributions of ratioed trim inputs, maneuver commands, and disturbance rejection inputs on remaining available control power (RCP), shown shaded in green.

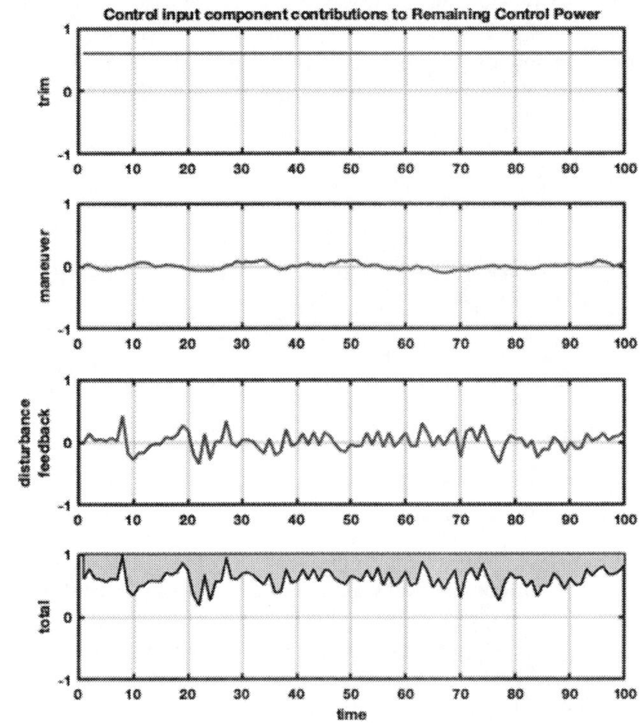

Figure 3. Contributions of control input sources to remaining control power (highlighted in green).

Prior work in this area (Ref. 3, 4) examined the application of this RCP estimation algorithm on multirotor configurations, both in simulation studies and on a commercial quadcopter with added instrumentation and telemetry capability. More recently, two approaches for RCP estimation were investigated, called input-oriented and output-oriented, each characterized by where the primary information source on the vehicle is located. Input-oriented RCP algorithms fundamentally ratio the present inputs to each force/moment generators with their total input range, solely monitoring actuator input ($V(t)$ in Figure 1); whereas the output-oriented algorithm compares the predicted vehicle response to that control input ($V(t)$) with the actual vehicle response ($Y(t)$) due to combined control ($V(t)$) and disturbance ($W(t)$) inputs.

Ratioing the current control input to an actuator (force/moment generator) on a flight vehicle to its input range would give a fraction of control expenditure; subtracting that value from its limits gives a resulting ratio of remaining control power (authority) for that input. Thus, an easily computed metric for Remaining Control Power (RCP) on actuator input "i" is:

$$RCP_i = min\left(2\frac{u_{max}-u_i}{u_{max}-u_{min}}, 2\frac{u_i-u_{min}}{u_{max}-u_{min}}\right) \quad (1)$$

This value generates a value of 1 when the control input is at the center of its range of travel, and 0 when it reaches an upper or lower control limit bound (Figure 4). This metric has the distinct advantage that only the control limits available on the prime mover are necessary for its computation, as *it is agnostic of vehicle type, configuration, or operating state.* And perhaps even more important, *this metric does not depend upon any particulars of the aircraft flight control system design.*

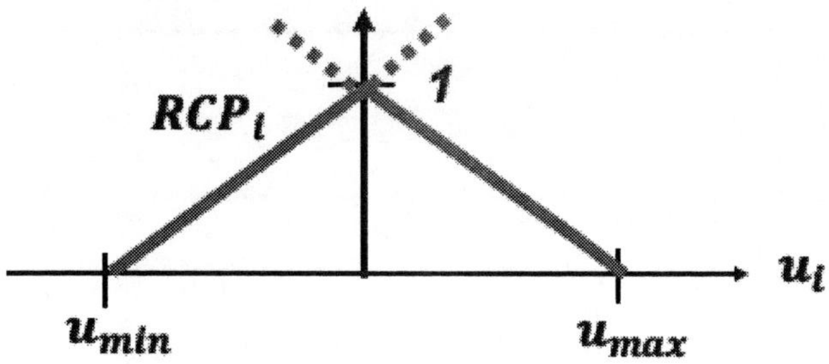

Figure 4. Input-oriented RCP algorithm for control "*i*".

While direct monitoring of RCP is of value to assess the current flight control system's closeness to a LOC event, this metric can be recomputed by adding in the equivalent feedback input required to mitigate/cancel that disturbance to the current control value. The output of this estimator of the control equivalent inputs is combined with the current inputs to the actuators on the vehicle to generate a total input that represents the current inputs plus those needed to exactly cancel the present disturbances (Equation 2), and this total input is then compared to the actual control limits, as done in Equation 1 above:

$$u_{total} = u - u_{equiv} \quad (2)$$

RCP computed using this assumption of instantaneous disturbance suppression will most likely be overly conservative, particularly if the bandwidth of the control actuator is significantly below that of the actual gust disturbance effect on the vehicle. But it is nonetheless an important metric as it indicates an ideal mitigation strategy and thus is indicative of the vehicle's capability to mitigate the disturbance through generation of countering forces and moments.

Dual Estimator Approach

Assignment of the discrepancy, or residual, between the measured and predicted response to two disparate sources, wind gusts or equivalent control perturbations, would normally require two separate filter formulations to generate those outputs. However, the algorithm that resulted from the work included both outputs as an algebraic mapping of estimate acceleration/rate perturbations from a dynamic model, where both the control sensitivity and gust sensitivity matrices of a linearized model are converted, via a pseudo-inverse operation, to generate equivalent perturbations of each quantity for further assessment. Each then becomes a higher-frequency representation of the quantities that would have resulted from individual filter use for these outputs, but thereby generates a conservative estimate for each quantity. Bandwidth control of the estimates is controlled through both the process model selection and the gain structure of the recursive estimator. A block diagram of this approach is shown in Figure 5, where the state rate perturbation in the estimator is represented as a signal in the red lines in the diagram.

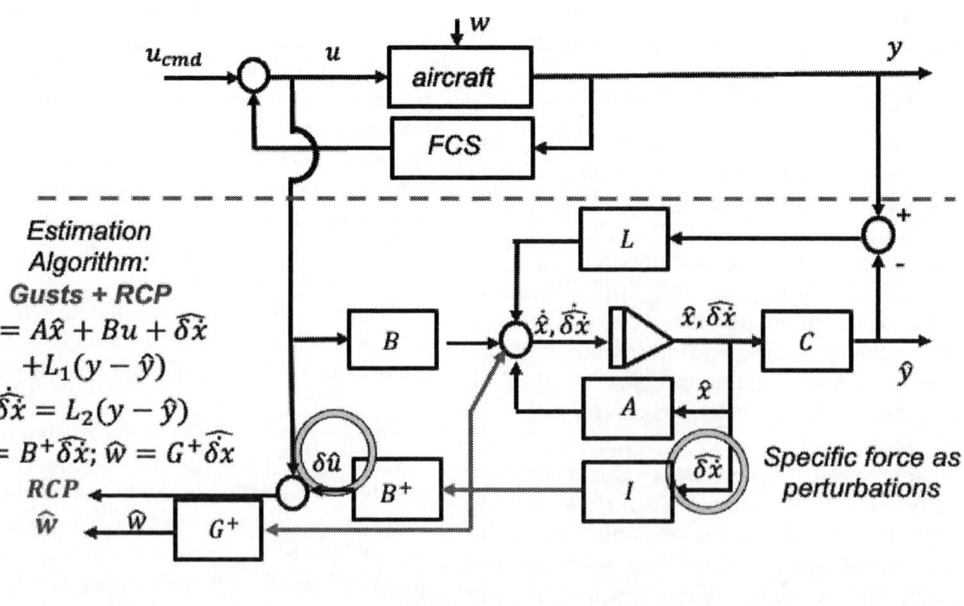

Figure 5. Combined RCP and gust estimates from estimated state rate perturbation vector.

SIMULATION ASSESSMENT

While the limited degree-of-freedom stability derivative models have suggested this dual estimation approach is a viable technique for online estimation of both remaining control power in the presence of disturbances and the gust disturbance magnitudes themselves, it is not sufficient for algorithm performance testing. Also, simplified models do not capture the potentially complex interactional aerodynamic effects that gusts can have on eVTOL/DEP aircraft having closely spaced lift and propulsion rotors. Thus, more detailed models of eVTOL/DEP aircraft for several base configuration types were studied to assess their gust response sensitivity and thereby develop appropriate dynamic models for use within the estimation algorithm. Two sources of models were used for assessment of this dual estimation approach: a "stitched" locally linearized flight dynamics model produced from the NDARC / FlightCODE software combination, and a fully nonlinear model incorporating CDI's CHARM Module software in its CHARM Toolbox for MATLAB.

NASA has created representative eVTOL/DEP vehicle models using its NDARC (for NASA Design and Analysis of RotorCraft) software (Ref. 5), and these vehicle definitions are inputs to the recently developed FlightCODE analysis tool for creation of locally linearized dynamic flight models of these aircraft (Ref. 6). FlightCODE generates extended stability derivative models of aircraft that have been sized within NDARC that represent small perturbation dynamic models at specified trim points identified within the NDARC input data. The linearized aerodynamic models are indexed with flight condition and vehicle configuration parameters and combined with trim control and body forces to couple with a nonlinear kinematic model that can create a continuous flight dynamics model using a "stitching" process described in Ref. 7.

The more extensive models developed for this use incorporate features of the CHARM Toolbox for MATLAB, a developmental environment that incorporates the CDI CHARM Module Wake/Panel model within a MATLAB-based component formulation of an aircraft model for fully nonlinear simulation of vehicle response to both control inputs and external disturbances (Ref. 8). The CHARM Module provides rapid vortex-based wake modeling for a wide assortment of flight vehicles, has been extensively correlated with wind tunnel and flight test data, and most recently is used within a fully nonlinear flight simulation tool for eVTOL/DEP aircraft called DEPSim (Ref. 9). Locally linearized models from DEPSim were also used in the assessment of modeling level of detail needed for acceptable disturbance effect estimation.

FlightCODE Models

Two flight dynamic models of the NASA single-place quadcopter, described in Ref. 5, were created using the FlightCODE tool for a hover condition, as takeoff and landing conditions in turbulence are likely to be defining metrics in control power requirements. The first was a simple quasi-static model with no rotor dynamics, and the second included first-order flapping along with dynamic inflow states for all four rotors. A normally distributed random vertical gust was imposed on each model to assess the difference in response between the two representations, as a way of assessing the difficulties in use of the quasistatic model to represent gust response behavior when additional dynamics are present.

Despite the difference in stability derivatives between the two models, their gust response to the imposed random vertical turbulence was almost identical. This is clearly due to the similarity in the frequency response of the two models over the low-frequency range, seen in Figure 6, where the vehicle is dominated by a heave damping response. This preliminary result suggests that, at this level of modeling fidelity, a low-order approximate model for vehicle response may be sufficient for estimating low to moderate frequency gust magnitudes. This intermediate fidelity model serves a useful role in assisting the understanding of the appropriate level of modeling detail needed to estimate the statistics of the equivalent control required for disturbance.

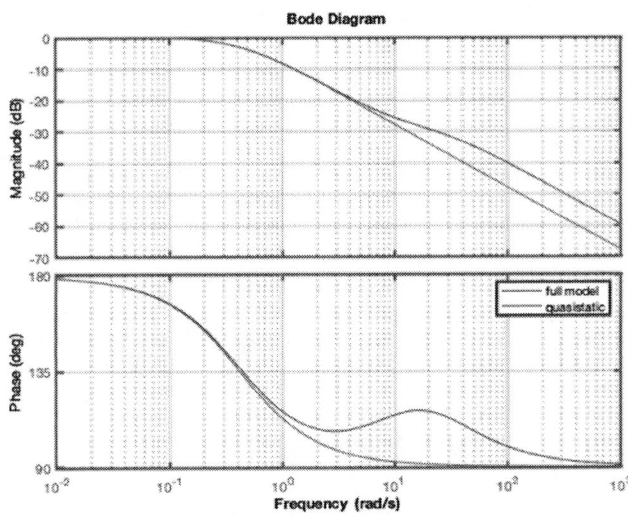

Figure 6. Quadcopter vertical gust to heave velocity frequency response, various model DOFs

CHARM Aerodynamic Models

Parameters from the same NDARC reference model used in the FlightCODE model generation were used to create a nonlinear CHARM Toolbox model in MATLAB of the quadrotor single-place concept vehicle. This model includes a full-span wake model coupled to a blade element rotor model, currently only using rigid flap response for the blade dynamics representation. Gust modeling uses uniform variation of the local flow field, thereby impacting local flows and wake filaments similarly, and represents a gust wavelength several times the fundament dimension of the quadrotor vehicle. A representation of the wake structure for this aircraft in a 40kt trimmed flight condition can be seen in Figure 7, where the interaction of the aft rotor wakes with the forward rotor wakes is evident, despite the vertical separation of the two lift systems.

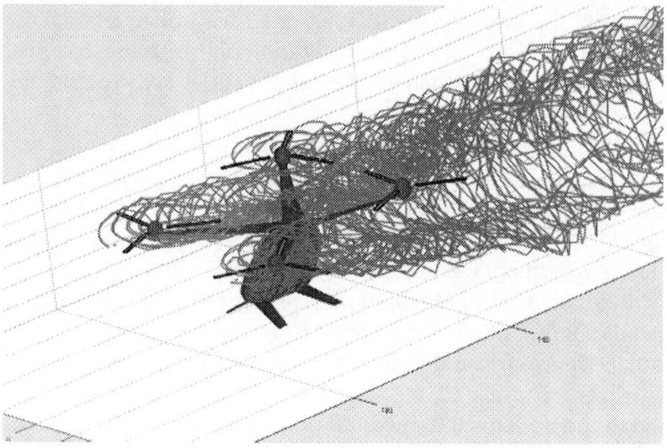

Figure 7. CHARM wake structure for NASA quadcopter at 40kts cruise.

This model was extended to include mechanical time constants for supporting control designs that modulate thrust on rotor/propellers via rpm variations. Use of rpm control instead of collective pitch control requires accelerating torques to change rotor speeds, and thus generates a transient (dynamic) effect that is compounded with the additional change in the wake characteristics with thrust changes. To appreciate the time lags associated with wake dynamics effects, a simulation of a 10% increase in rpm for an isolated rotor shows (in Figure 8) an oscillatory transient that settles into a new steady state condition after a quarter second following this step change. This response does not include rotor inertia in the applied torque requirement.

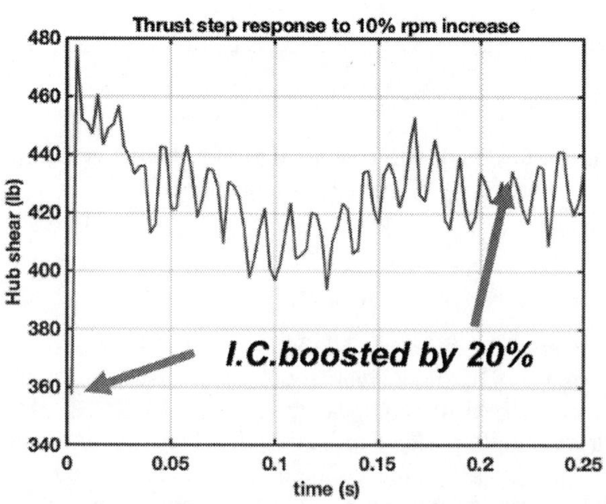

Figure 8. Hub shear transient following 10% step increase in rotor rpm.

As mentioned above, CDI and Penn State have developed a simulation tool dubbed DEPSim, for DEP aircraft Simulation, that provides detailed component models of representative distributed electric propulsion aircraft with included motor dynamics, rotor dynamics with multiblade coordinates, variable rpm and full rotor collective and cyclic controls, and an aerodynamic model that uses either dynamic inflow or the CDI CHARM module for rotor and fixed-surface airload calculations. While the full real-time executable form of this model was not yet available for manned simulation evaluation of the RCP and gust estimation algorithm, this same model has been used to support algorithm development in two ways. The first of these was to generate locally linearized flight dynamic models that include aircraft rigid body states as well as component dynamics, along with trim values for those state variables and control inputs. The linear models have been generated at 20 knot increments, for both rpm-controlled and full cyclic-controlled rotor configurations, from hover to 140 knots for the "generic" lift plus cruise (LpC) configuration shown in Figure 9. This aircraft represents a four-passenger eVTOL/DEP concept configuration that flies like a multicopter in hover and transitions to wing-borne airplane-like flight in cruise with the lift rotors stopped and indexed fore-aft on their respective support booms.

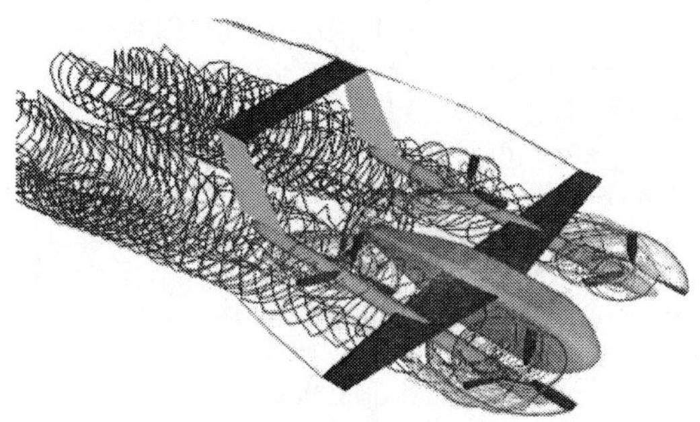

Figure 9. CDI/PSU lift plus cruise configuration with CHARM wake elements.

The second use of the model has included response simulation to gusts that incorporates the full CHARM wake model but running off-line in a time-marching simulation. The DEPSim model represented the configuration of Figure 9 in hover hold operation on the lee side of a hangar building, in the presence of a CFD-calculated turbulent air wake shed from the building edges, that represents the flight test conditions used in Ref. 2 for determination of CETI frequency response curves in that reference for modeling gust response for a UH-60 helicopter.

Evaluation Goals

Simulation studies were designed to check the assumptions associated with the algorithms for both RCP estimation and gust estimation, which include:

- The algorithms are applicable to the broad class of eVTOL/AAM configurations that include multicopters, lift plus cruise arrangements, and vectored thrust vehicles;

- Simplified dynamics of rigid body response that are applicable to handling qualities frequencies (e.g., short period dynamics) are appropriate for both equivalent disturbance control estimation and gust estimation;
- Estimation results are sufficiently agnostic to types and configurations of flight control systems and autopilots, as the algorithms use only vehicle-centric data;
- Algorithms can be adjusted for changing flight conditions and operating configurations through velocity and configuration scheduled linearized math models.

Simulation evaluations conducted using the above lift plus cruise used locally linearized dynamic response models that include motor dynamics, multiblade dynamics, dynamic inflow, and rpm variations along with rigid body dynamics. The range of underlying dynamic frequencies suggests a high complexity model that is useful for examination of the algorithm performance using lower order models for the vehicle response to control inputs.

The linear model was first reduced in detail to capture the vehicle rigid body states, roll and pitch attitude states, the first three multiblade rotor modes for all 5 rotors (these are collective and disk tilt modes), and the first three dynamic inflow modes for all 5 rotors, leaving a 38-state dynamic model reduced from an original 62-state representation. Since the linearized model of this vehicle is open-loop unstable, a low-level stabilization feedback was designed based on feeding back the rigid body states using standard linear quadratic regulator methods. Once the model was stabilized with these gains, as determined from an eigenvalue analysis, a simulated gust was applied in the vertical axis following a raised-cosine shape, with the gust duration of 5 seconds, followed by no gust excitation for a remaining 5 seconds giving a 10 second response record. The raised cosine gust was used instead of a filtered random white noise excitation on all axes, as a means of assessing both transient nature of the output-oriented estimation algorithms as well as the allocation of excitation energy within the algorithm to the estimated disturbance states (both gusts and control-equivalent disturbances).

The sequence just described resulted in simulation data for a high-order (38 state) dynamic model of a DEP aircraft that includes: the actual gust excitation time history, the aircraft state time history, and the control input time history for each actuator based on the rigid body state feedback gains operating on the vehicle response time history. These data were used in all algorithm evaluations as the truth conditions to which the algorithm estimates would be compared.

State Rate Estimation with Pseudo-Inverse

Estimation of vehicle state rate perturbations and subsequent pseudo-inverse mapping of those perturbations back to control equivalent inputs or gust excitation velocities, as diagrammed in Figure 5, was considered for low-order model assessment. A simple 8-state model was augmented with 6 perturbation states representing body axis linear and rotational accelerations, represented in the filter structure as random walk processes with white Gaussian inputs. These augmented states closely align with the differences between the model and the higher order simulation output and thus combine the effects of both disturbance effects and model mismatch into the perturbation estimates.

Shown in Figure 10 are the perturbations in state rates for this model, effectively generating acceleration perturbation states for the simulated gust response. While some effect can be seen for the longitudinal acceleration estimate, the responses are dominated by the vertical heave acceleration, directly driven by the raised cosine gust excitation of the simulation. Phasing of the vertical acceleration is slightly delayed from that of the actual gust velocity, reflecting the finite bandwidth of the recursive filter. Estimates of body axis velocities were found to track those from the simulation very well.

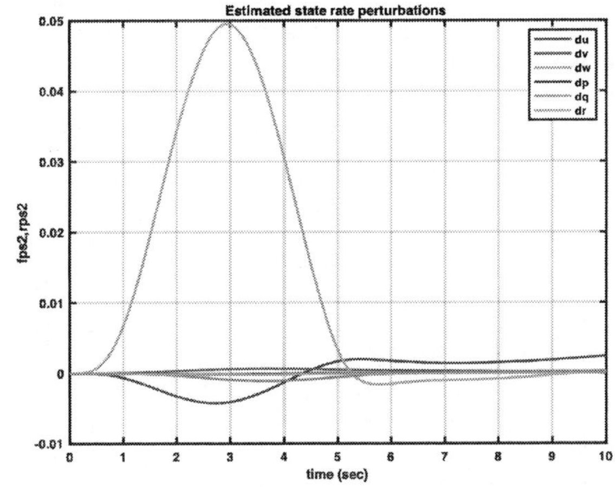

Figure 10. Estimated perturbation state rates from filtered simulated gust response.

When the control effectiveness matrix and the gust effectiveness matrix are applied as pseudo-inverse operators on these accelerations, the results are shown in Figure 11 and Figure 12. Since these are algebraic operations, there is no phasing difference from the generation of the perturbation accelerations, only a recombination to reflect the levels of sensitivity of each to their respective excitations.

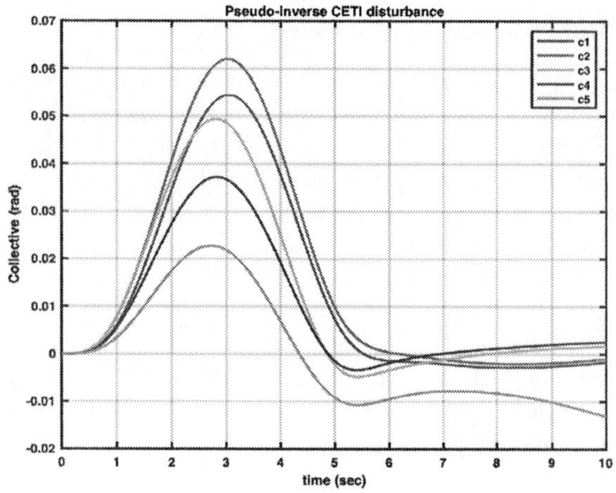

Figure 11. Control equivalent disturbance from pseudo-inverse operation on estimated perturbation accelerations.

While the estimated gust time histories from the pseudo-inverse application to the acceleration perturbations reconstructs the raised cosine input on the vertical axis, a more significant amount of gust energy is allocated to the horizontal gust estimate, as seen in Figure 12. But both the estimated control equivalent inputs and the estimated gusts using these applications of pseudo-inverse matrices will reconstruct the estimated acceleration perturbations exactly, due to the algebraic nature of their generation from those estimates.

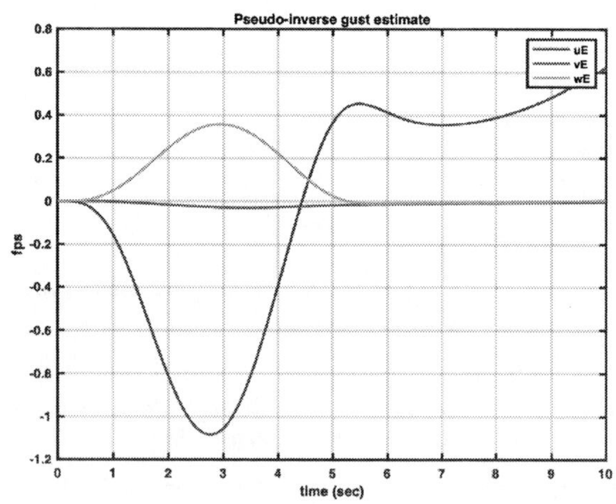

Figure 12. Estimated gust disturbances from pseudo-inverse application to acceleration perturbations.

Algorithm Application to Infinite-Dimensional Systems

As described above, the DEPSim LpC aircraft model of the configuration in Figure 9 is available in various formats that include a locally-linearized model with dynamic inflow for rotor wake effects, a locally-linearized model using the full free wake CHARM aerodynamics representation, an offline time-marching nonlinear model using dynamic inflow, an offline time-marching nonlinear model using the CHARM wake analysis, and a real-time nonlinear model with dynamic inflow for manned simulation applications. The real-time nonlinear model with the CHARM wake analysis has been demonstrated on graphics processing unit (GPU) machines but has not yet completed its port to a manned simulation facility at this writing. While the previous section explored algorithm application to a lower order model of a high order linearized model of the LpC configuration, this section applies estimation approaches on data generated from an offline time-marching CHARM-coupled simulation of the LpC configuration operating in the turbulent wake behind a hangar superstructure that represents the test conditions used in Ref. 2 for collecting data to model the spectral characteristics of control equivalent turbulence in simulation applications. The turbulent flow field is from Ref. 10 where a full boundary layer model was propagated over the local building structures that were present at the test site, and flow velocity data was stored within a box (shown in yellow in13) on the lee side of the hangar where the UH-60 was flown.

Figure 13. UH-60 CETI test condition and CFD model with box indicating flow used for wake disturbance.

Simulation data for the LpC operating in a hover-hold condition at the same location as the UH-60 in the flight test has been collected and includes the CFD effect on the CHARM vortex wake structure, along with the wake on wake effects seen between the different lift rotors on the vehicle. The complexity of the wake structure and the full CFD resolution of the turbulence environment make the simulation model effectively infinite-dimensional and provide an opportunity for algorithm evaluation to assess its effectiveness in estimating the gust magnitudes and statistics in the simulation from the collected response time histories. Hover station-keeping for this simulated case showed tight control of vehicle position, following an initial vertical transient of 1.5 feet from simulation initiation because of control transients.

Application of the state rate perturbation algorithm with pseudo-inverse equivalent control extraction and gust estimation resulted in the data shown in Figure 14 through Figure 16, where the initial transient in the state rate perturbations had the pseudo-inverses of both the control effectiveness matrix and the gust effectiveness matrix to generate the equivalent disturbance control of and the applied gust estimate of . Both estimates are shown following the transient response of the filter and aircraft after 15 seconds of simulated station-keeping. While the mean of the gust estimate is close to the simulated value of 25 fps at that condition behind the hangar, the fluctuations in the estimate are almost a factor of two greater than that seen in the CFD data itself, suggesting that the interactional aerodynamic effects of wake on wake load changes generate larger force and moment perturbations on the flight vehicle than modeled with the reduced order 8 DOF representation used in the filter formulation. Note that the estimated equivalent control disturbance is relatively small, indicative of the precision hold capability of the simulated flight control system.

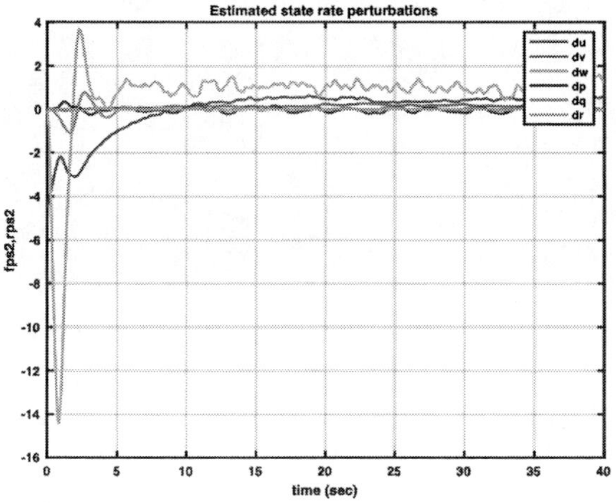

Figure 14. Estimates of state rate perturbations from the DEPSim-CHARM-CFD simulation

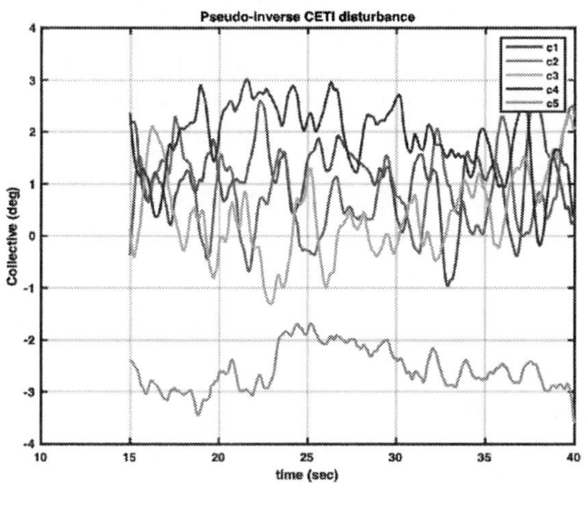

Figure 15. Estimated equivalent disturbance control inputs from coupled simulation

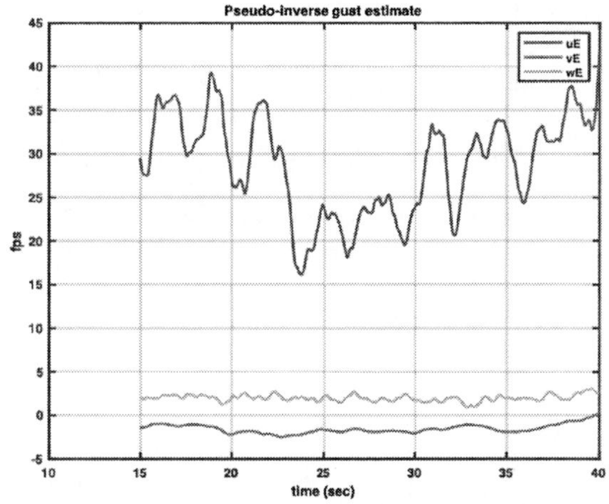

Figure 16. Estimated gust time history for DEPSim LpC in position hold simulation

MANNED SIMULATION TRIALS

The nonlinear DEPSim LpC model described above having dynamic inflow wake representations for each rotor was coupled to CDI's in-house fixed base simulation facility, as a means of evaluating both the RCP algorithm and the generation of cockpit displays of RCP during flight operations. Limited investigations were undertaken during this effort to study how best to communicate RCP information to both aircraft evaluators and pilots to provide actionable information that could aid assessment of vehicle safety for certification evaluation or suggest pilot behaviors to avoid LOC-I events from saturated control channels.

A notional display of RCP for each actuator input was constructed and simulated using MATLAB graphics primitives, shown in Figure 17, where the RCP limits represent ratios as defined above, and a bar extending from a neutral or center position to the current RCP value for each actuator is color coded to aid interpretation, with green indicating RCP limits below 80% of that direction's maximum input, yellow for values from 80% to 90%, and red for values 90% and over. An additional indicator is a pair of pointers in a bow-tie shape that indicate the sum of the current RCP based on the physical actuator displacement and the output-oriented RCP calculated value needed to completely null the current system disturbance. That bow-tie symbology has the same color-coding representation as the bar elements.

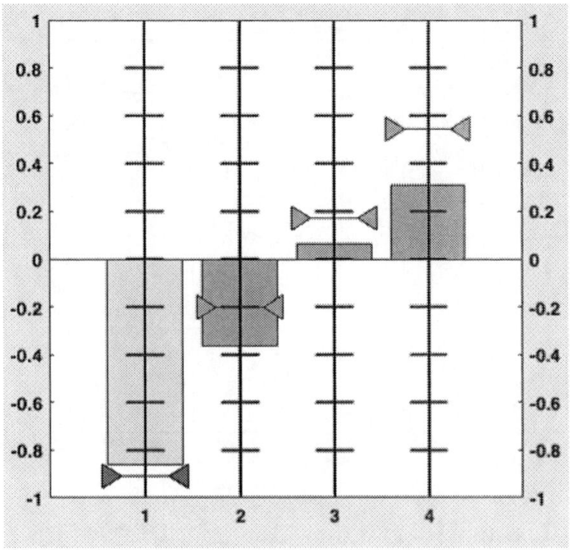

Figure 17. Notional RCP display showing input-oriented and output-oriented values.

A desktop simulation of the FlightCODE model of the single-place quadrotor NASA concept vehicle was manually flown with this display providing readings of input-oriented RCP to understand how such information transfer might benefit pilot monitoring of limiting flight conditions. Figure 18 shows the quadrotor in flight over a cityscape with the MATLAB graphic added as an overlay window onto the 3rd-person view of the flight simulation. This limited experience guided further development of a similar display used in manned simulation trials in CDI's fixed-base simulation facility, shown as an integrated display window on two touchscreen monitors located in the pilot station. That latter display uses Java Swing graphics primitives running on a Raspberry Pi computer and interpreting networked data on actuator displacements from the DEPSim simulation model.

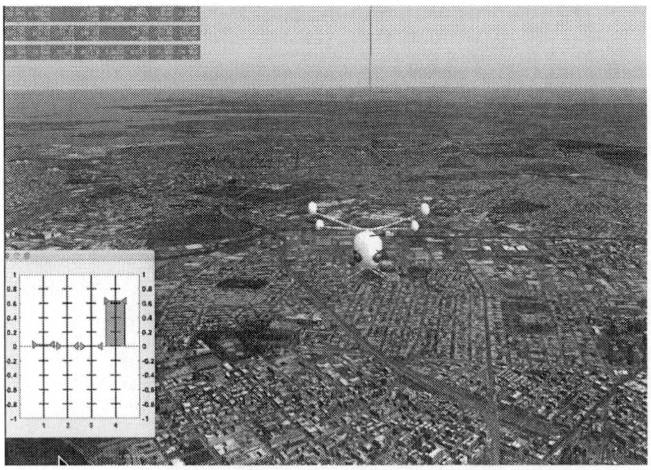

Figure 18. RCP display superimposed on NASA quadcopter flight simulation environment

Figure 19. RCP display on left cockpit monitor in CDI's fixed base simulation lab

The CDI DEPSim model was manually flown through a set of aggressive maneuvers to both evaluate the algorithm performance and to assess the utility of the linear representation of RCP values in this display format. The lateral response rates ares shown in Figure 20 from a test case to assess the control limits of the nonlinear simulation. For this simulation study, the commanded torques on the motors (Figure 21) were used in the RCP calculation process, as the LpC model itself was a representation of the rpm-controlled (versus collective-controlled) variant of that aircraft, and thus current limiting due to excessive rpm perturbation commands tend to limit the control capability of those DEP aircraft.

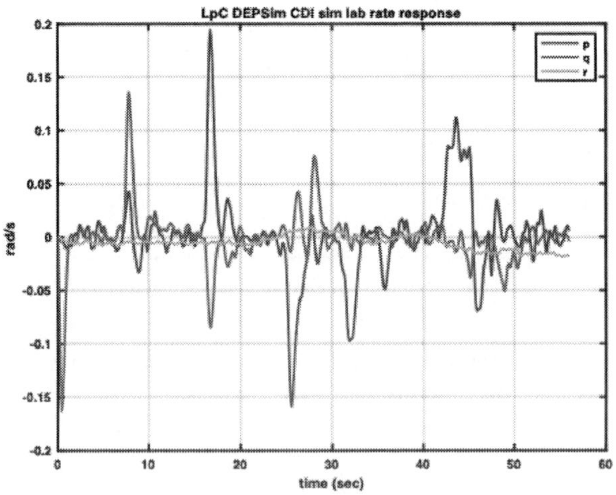

Figure 20. Angular rate time history for manned simulation trial event

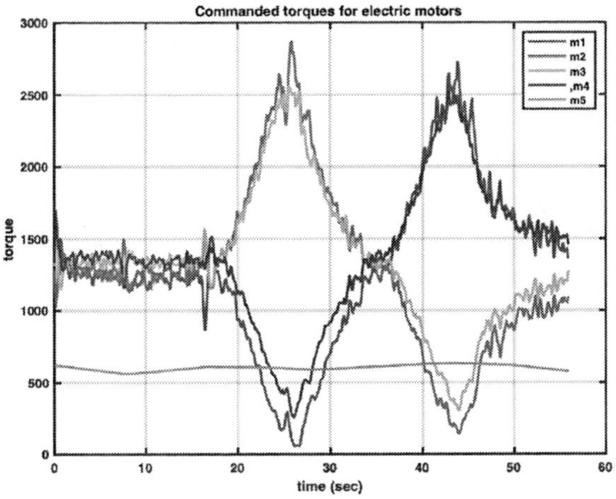

Figure 21. Torque commands for LpC lateral motion maneuver

The display of the RCP simply ratioed the difference of these torque commands with a nominal value of 1500 by half the maximum extent of available torque, to produce a similar time history. Pilot comments on the display were that the bars could be viewed using peripheral vision, but the indicators of output-oriented RCP were somewhat of a distraction. A suggested alternative was to eliminate them and either incorporate the output-oriented RCP directly into the bar chart display or generate an equivalent using an estimated RCP rate and future time prediction based on that rate. The determination of how to compute the averaged rate and what prediction interval is most useful remains as a future investigation in follow-on research.

CONCLUSIONS

This research has shown that application of algorithms for estimating remaining available vehicle control power, and local gust disturbance magnitudes, appear to provide a usable safety assessment for avoiding loss of control (LOC) events for eVTOL/DEP aircraft. Such real-time safety metrics may aid the challenge of providing safety assurance in certification of these aircraft for operations within the national airspace system. The algorithms have several notable features which ease their implementation for use in aircraft certification testing, or on-board monitoring for safety assurance that include:

- The input-oriented RCP estimation and the output-oriented gust disturbance and RCP estimation schemes are completely flight control system agnostic, as they only incorporate measurements associated with the base functionality of the aircraft.

- Methods for combining recursive estimation of both gust disturbances and equivalent control disturbances have shown to be compatible through pseudo-inverse mapping of estimated perturbations in vehicle state rates from those commanded from control inputs.

- Low-order models of aircraft response are appropriate for estimating RCP as they fall within the frequency range of both atmospheric disturbances and handling qualities for these eVTOL/DEP vehicles.

- Recursive estimation of both gust disturbances and RCP include Gaussian statistics that can further impact the assessment of possible vehicle LOC-I events during flight, providing an additional margin of safety.

Future work using these algorithms could investigate their applicability to other configurations of eVTOL/DEP aircraft, as well as conventional rotorcraft and fixed-wing aircraft, both piloted and uncrewed.

Author contact: Robert McKillip, Jr. bob@continuum-dynamics.com

ACKNOWLEDGMENTS

This work was supported under FAA contract 692M15-22-T-00019. The interactions with and support from the FAA monitors and the late pilot/builder M. Zeller during this effort are greatly appreciated.

REFERENCES

1. List, A. H., and R. J. Hansman, "Assessing Multi-rotor UAV Controllability in Low Altitude Fine-Scale Wind Fields," M.I.T. International Center for Air Transportation, Cambridge, MA, August 2019.

2. Lusardi, J. A., Blanken, C., and Tischler, M., "Piloted Evaluation of a UH-60 Mixer Equivalent Turbulence Simulation Model," Proc. 59th AHS Annual Forum, Phoenix, AZ, May 6-8, 2003.

3. McKillip, R., Jr., Wachspress, D., Hansman, R. J., Sizoo, D., and Dellmyer, D., "Investigation of Certification Considerations for Distributed Electric Propulsion (DEP) Aircraft," VFS Aeromechanics for Advanced Vertical Flight Technical Meeting, San Jose, CA, Jan. 25-27, 2022.

4. McKillip, R., Jr., Wachspress, D., Hansman, R. J., Sizoo, D., Stadtmueller, T., and Dellmyer, D., "Control Power Margin as a Certification Consideration for Distributed Electric Propulsion (DEP) Aircraft," Proc. VFS 79th Annual Forum, West Palm Beach, FL, May 16-18, 2023.

5. Johnson, W., Silva, C., & Solis, E. (2018). Concept Vehicles for VTOL Air Taxi Operations. *Aeromechanics Design for Transformative Vertical Flight.* San Francisco, CA: AHS.

6. McKillip, J. R., & Keller, J. (2021). *Development of a Rotorcraft Flight Dynamics and Control Modeling and*

Analysis Tool for Conceptual Design: FlightCODE Theory Manual. Ewing, NJ: CDI Technical Note 2108

7. Tobias, E., & Tischler, M. (2016). *A Model Stitching Architecture for Continuous Full Flight-Envelope Simulation of Fixed-Wing Aircraft and Rotorcraft from Discrete-Point Linear Models.* U.S. Army RDECOM Special Report RDMR-AF-16-01.
8. McKillip, J. R., Keller, J., Wachspress, D., Whitehouse, G., & Quackenbush, T. (2010). Simulation of Dynamic Interface Flight Control Concepts Using the CHARM Toolbox for MATLAB. *Specialists Conf. on Aeromechanics.* San Francisco, CA: AHS.
9. Theron, J., Horn, J., & Wachspress, D. (2020). An Integrated Simulation Tool for eVTOL Aeromechanics and Flight Control. *Aeromechanics for Advanced Vertical Flight Technical Meeting.* San Jose, CA: VFS.
10. Christoffel, T., Hendrick, C., Thedin, R., Schmitz, S., & Horn, J. (Oct. 2020). Towards Validation of a Dynamic Interface Simulation using Flight Test Data. *VFS 76th Annual Forum.* Virtual.

Novel Human Factors Challenges, Unexpected Events, and Safety Hazards Associated with Operation of Battery-powered eVTOL Aircraft

Kendall Carmody
Doctoral Graduate Researcher
Florida Institute of Technology
Melbourne, FL, US

Dr. Bhoomin Chauhan
Post-Doctorate Researcher
Florida Institute of Technology
Melbourne, FL, US

Maureen Namukasa
Doctoral Graduate Researcher
Florida Institute of Technology
Melbourne, FL, US

Yarisse Adorno
Flight Instructor
Florida Institute of Technology
Flight Line
Melbourne, FL, US

Dr. Meredith Caroll
Professor of Aviation Human Factors
Florida Institute of Technology
Melbourne, FL, US

ABSTRACT

As the AAM industry approaches its respective takeoff, novel dimensions of aircraft design, including propulsion systems and pilot interfaces, have emerged. These new and often complex dimensions result in operations differing from traditional aviation contexts, with the potential to influence pilot behavior and decision-making (Mendolia et al., 2022). There are also novel safety-critical events that pilots must be prepared to cope with on battery-powered aircraft. To examine these issues, we conducted an interview study with 11 pilots and engineers who have either flown, worked on, or conducted research pertaining to electrically powered aircraft. During the interview process, the participants were asked to describe and identify novel events pilots would face on electric aircraft, factors that would influence their decisions regarding power management, and challenges associated with transitioning to electric aircraft. We then conducted a qualitative analysis in which we identified themes that emerged across the interview responses related to the previous categories.

INTRODUCTION

Air transportation is considered one of the safest modes of transportation. This high level of safety is a result of a system that is based on rigorous procedures and specifications. Due to the inherent nature of aviation, which often includes encounter with off-nominal events that challenge the pilot's ability to operate the aircraft safely, the Federal Aviation Administration (FAA) has developed very stringent rules that govern flight operations such that the occurrence of unanticipated events is mitigated by pilot training and procedures that allow pilots to respond effectively. However, there persist "unknown unknowns" that result in unexpected events to which pilots must respond (Hancock et al., 2022).

The aviation industry is witnessing the introduction of novel aircraft designs, for example, electric vertical take-off and landing (eVTOL) aircraft characterized by vertical takeoff and landing capabilities, and distributed electric propulsion systems for enhanced efficiency. Electric propulsion and eVTOL aircraft have garnered considerable attention in recent years, especially within the Urban Air Mobility (UAM) sector. UAM represents a novel air transportation concept aiming to facilitate the swift movement of passengers and cargo in urban areas using electric-powered aircraft, thereby circumventing ground traffic congestion and enabling on-demand air travel (FAA, 2023). Significant investment has been directed towards developing eVTOL aircraft, which utilize electric propulsion systems for vertical take-off and landing, negating the need for traditional runways (Jiecho et al., 2024). As the aviation industry transitions towards eVTOL aircraft, associated technological advancements has introduced new complexities and unpredictability, leading to novel human factors, challenges, unexpected events, and safety hazards (Franciscone & Fernandes, 2023) for which pilots need to be prepared.

The FAA defines a challenge as a situation or a condition that requires adaption, problem-solving, or overcoming obstacles. This specifically relates to understanding the context of human interaction with systems, environments, and other elements that an operator might need to use to complete a task (FAA, 2015). According to the FAA, accidents frequently result from pilots being unprepared for challenges, especially those associated with novel aircraft (FAA, 2022). When confronted with such challenges, pilots are required to assess the operational situation and react appropriately, regardless of the underlying cause (e.g., mechanical issues, environmental

Presented at the Vertical Flight Society's 11th Biennial Autonomous VTOL Technical Meeting, Phoenix, AZ, USA, Feb 4-6, 2025.

factors, or security concerns). As outlined by the FAA UAM concept of operations (ConOps), the initial eVTOL operations will be crewed, meaning that a pilot would be responsible for safely flying the aircraft (2023). As the current cohort of commercial pilots will likley transition to flying this new type of aircraft, it will introduce significant human factors challenges, particularly regarding the utilization of the electric propulsion systems. Pilots will need to adapt to the unique characteristics of the electric propulsion system, including its impact on aircraft performance and energy management for their flight.

According to the FAA, an unexpected event is defined as any occurrence that catches individuals off guard, disrupting their anticipated course of action and affecting their cognitive processes for response (2015). In a simulated study conducted by Kinney and O'Hare (2020), general aviation (GA) pilots flew a series of flights in a flying simulator which included two experimental flights, either including an expected or unexpected engine failure. Results showed that pilots experienced heightened physiological arousal, reduced scanning behavior, and performance decrements following unexpected engine failures compared to anticipated events. To cope with unexpected events, the FAA provides checklists and procedures for a pilot to utilize when faced with an unexpected event. For example, if a pilot experiences an in-flight loss of electrical power, the FAA procedures outline that the pilot should first turn off all essential electrical equipment that can help them continue flying (FAA, n.d.).

Both the FAA and the International Civil Aviation Organization (ICAO) define a hazard as a condition or an object with the potential to cause damage to the equipment or reduce the ability of a system to perform a function, which can lead to or contribute to an unplanned or an unexpected event (FAA, 2022; ICAO, 2017). There are also novel hazards associated with operating an aircraft with an electric propulsion system. The integration of an electric propulsion system in an aircraft introduces hazards such as limited range due to battery capacity constraints, potential electrical system failures, sensitivity to adverse weather conditions, and electromagnetic interference (Thompson et al., 2022).

Several attributes of eVTOL aircraft contribute to the introduction of these novel challenges, unexpected events, and safety hazards. Lithium-ion batteries play a pivotal role in enabling electric propulsion, yet their specific energy remains considerably lower than that of jet fuel, significantly restricting both payload capacity and range for electric aircraft (Carmody et al., 2023; Fard et al., 2022). In contrast to most U.S.-certified aircraft relying on fossil fuels, eVTOLs typically depend solely on electricity for propulsion, with rechargeable batteries powering electric motors driving the propellers. One of the primary differentiators between an eVTOL aircraft and a conventional fossil-fuel-powered aircraft is the fuel or energy management characteristics. In a fossil-fuel-powered aircraft, fuel management necessitates meticulous planning and ongoing monitoring of onboard fuel (FAA, 2021). For example, as the flight progresses, the gross weight of an internal combustion engine aircraft decreases by 5 – 40 percent, leading to improved performance at the end of the mission (Viswanathan et al., 2022). The opposite is true for an eVTOL aircraft, in which, as the flight progresses, not only does the eVTOL aircraft not become lighter, but the power available is lower at the end of the mission when the state of charge is low (Adu-Gyamfi & Good, 2022). This may prove challenging to pilots who are trained in dealing with unexpected events and challenges associated with fossil-fuel-powered aircraft systems (Bauranov & Rakas, 2019).

Current pilot training is focused on familiarizing pilots with established procedures for known and unexpected events for a traditional, fossil-fuel-powered aircraft (Hancock et al., 2022). Regulatory and international aviation organizations, including the FAA, ICAO, and the European Union Aviation Safety Agency (EASA), advocate for comprehensive training programs to prepare flight crews for unexpected events, challenges of transitioning to a new aircraft and dealing with novel safety hazards. These programs usually include developing an understanding of the relationship between airspeed, altitude, and available power, along with identifying key parameters for appropriate responses. Additionally, pilots should possess a deep understanding of normal aircraft operations, potential unexpected events, and system integration effects related to flightpath and energy management systems. Usually, these training requirements are developed for events and challenges that are deemed most probable based on historical data and recommendations provided by authorities such as the FAA. However, considering the novelty of an electric propulsion system, the operational and flight data of an eVTOL that can be leveraged to understand the challenges, unexpected events, and safety hazards a pilot could face is fairly limited and often inaccessible due to the proprietary nature of the vehicles under development. As such, the purpose of this study was to identify potential novel human factors challenges, unexpected events, and safety hazards that pilots may face while operating an eVTOL aircraft.

LITERATURE

The human factors implications of challenges, unexpected events, and safety hazards in traditional aviation contexts are widely studied and understood. However, research examining these in an eVTOL context and from a human factor's perspective is limited. This section will provide an overview of previous research, and how the results can be leveraged.

While there is a growing body of academic research in terms of technological development and the potential hazards of electric batteries for an eVTOL aircraft, research from the perspective of how a pilot can manage hazards, challenges, and unexpected events related to electric propulsion is sparse. Additionally, numerous studies have shown the implications of unexpected events and challenges on the pilot's workload and performance in a traditional aviation context (Casner et al., 2012; Landman et al., 2017; Kinney and O'Hare, 2020;

Rivera et al., 2014). However, research from the perspective of eVTOL aircraft is limited. Thompson et al. (2022) conducted a survey in the form of a literature review of existing academic research, industry reports, and aviation safety data on the potential hazards related to AAM operations and their effect on individual aircraft, flight missions, and airspace. Through the literature review, Thompson et al. identified three areas of potential hazards afflicting eVTOL aircraft and unmanned aerial systems (UAS): (1) hazards caused by winds, (2) hazards from component level faults and degradation, and (3) hazards that arise from flight plan nonconformance and non-cooperatives. The results of the study revealed that unexpected events, such as a headwind, can result in the pilot having to use additional power to overcome the headwind. Using more than the anticipated power can result in additional flight time and can reduce available range, resulting in delays and the eVTOL aircraft not being able to complete the mission. In terms of the component level faults and degradations, the results of the study showed that hazards such as comprehending the nature of the state of charge (SOC), motor faults and degradation, and degrading state of health (SOH) of the battery (i.e., the overall condition of the batteries capabilities based on the number of flight cycles it has undergone) can be a challenge for pilots as it may affect the operation of the aircraft power train system. This can lead to reduced maneuverability, reduced available energy, and decreased energy output. In terms of flight plan conformance and non-cooperatives, loss or degradation of electric battery circuitry can lead to calibration loss and system overload, which can cause a power surge or extreme vibrations that can damage critical electrical connections, which in turn can cause the aircraft to deviate from the intended flight plan.

Hendricks et al. (2015) conducted research that studied and developed failure modes, mechanisms, and effects analysis (FMMEA) for lithium-ion batteries (Li-on). The study used the FMMEA methodology to identify failure mechanisms and models for all potential failure modes of complex applications. Specifically, the FMMEA provided a comprehensive list of the parts within a lithium-ion battery that can fail or degrade, the mode by which the failure is observed, the potential causes of the failure, whether the failure is brought on by progressive degradation or abrupt overstress, the frequency of occurrence, the severity of the failure, and the ease of detection of the failure mechanism. Using the FMMEA, Hendricks et al. identified that external factors such as extreme temperature and degradation resulting from prolonged use can lead to short circuits and thermal runaways. Overall, the study found that failure of battery pack components, such as the anode, cathode, lithium ions, and separator, can lead to reduced capacity, thermal runaway, and increased internal temperature and that the failures of these components can be difficult to detect. Similarly, issues arising from battery output from battery component failures can result in reduced aircraft capabilities and airspace safety concerns. Battery faults, including those originating from loading profiles and high-power consumption, pose risks of trajectory deviations and decreased aircraft maneuverability, affecting both individual flights and airspace safety. Additionally, faults in electronic speed controllers can lead to degraded control capability, hindering the aircraft's ability to perform critical maneuvers within prescribed timeframes, thus posing risks to both individual flights and airspace integrity. Other system hazards include faults affecting navigation and autopilot systems, including global positioning system (GPS) and inertial measurement units (IMU) loss, as well as communication failures, further compound safety risks, potentially leading to flight plan nonconformance, loss of separation, or even collisions with obstacles or other aircraft. The results from the study identified several hazards, challenges, and unexpected events pertaining to electric propulsion systems, highlighting that there are many novel challenges that eVTOL pilots could face.

Sripad et al., (2021), in their review of the safety considerations for using electric batteries in an aircraft, identified that the most pertinent hazards for an electric propulsion system are the issues related to exothermic heat-related events (i.e., thermal issues) and partial or complete loss of critical power supply. The analysis showed that one aspect of electric propulsion that the pilot might find challenging is the ability to estimate the battery's SOH in terms of capacity and power capability. Additionally, the review also determined that while safety mitigation strategies, such as the use of a distributed electric propulsion (DEP) system, can help alleviate safety issues, one of the biggest challenges a pilot will face is dealing with thermal runaway, where excessive heat generation within battery cells can escalate temperature, triggering cascading failures across multiple aircraft systems.

Karthik et al. (2021) conducted a meta-analysis of the current state of the electric propulsion system to establish the potential benefits, limitations, and opportunities of electric propulsion systems in the context of aircraft applications. The study utilized a systematic literature review using the Preferred Reporting Items for Systematic Reviews and Meta-Analysis (PRISMA) guidelines to identify potentially relevant studies related to electric propulsion systems for an aircraft. Karthik et al. selected 57 relevant articles for the review based on the inclusion criteria specified in the PRISMA guidelines. The review of the articles identified several hazards inherent in electric propulsion systems for aircraft. Among the key hazards delineated in the paper is thermal runaway, a prominent concern stemming from temperature and pressure fluctuations within batteries, which can result in fires and explosions. Additionally, the results also highlighted the uncertainty surrounding the non-linear burn discharge rate of the electric battery, which can be challenging for a pilot to comprehend, particularly during unexpected fluctuations or power supply failures, potentially compromising the reliability and performance of the electric propulsion system.

The General Aviation Management Association (GAMA) produced a report in 2023 that sought to identify requirements for managing the range and endurance of battery-powered aircraft. The report highlights the challenges pilots face in managing the range and endurance of electric aircraft. Particularly, in comparison to fossil-fuel-powered aircraft, accurate models of battery performance are essential for pilots to gauge their aircraft's capabilities safely. The report highlights that pilots might face challenges in developing accurate mental models of battery performance necessary to gauge the SOH and associated battery parameters. Without this information, pilots may not be able to interpret their aircraft's capabilities safely. Additionally, pilots face challenges in terms of understanding the critical, nominal, and energy depletion limits, knowledge required to effectively plan flights and ensure adequate endurance for unforeseen circumstances.

METHODS

The goal of the research described herein was to identify potential new human factors challenges, unexpected events, and safety hazards that pilots might face with eVTOL aircraft. It should be noted that this was part of a larger study, another aspect of which is published in Namukasa et al. (2024).

Study Design & Interview Questions

This research is focused on identifying the novel human factors challenges, unexpected events, and safety hazards that could stem from using an electric propulsion system, and how that could influence a pilot's ability to safely complete their mission. This research is not focused on the algorithms necessary to determine the information pilots need to successfully complete their mission. Semi-structured interviews were conducted with eVTOL test pilots and electrical aircraft engineers to identify human factors challenges associated with UAM operations, novel, unexpected events, and unique safety hazards. Two sets of questions were developed, reviewed, and pilot-tested by a team of six human factors researchers, including two commercial pilots and one Airline Transport Pilot (ATP), to validate their effectiveness in retrieving relevant information. Participants who were pilots received questions focused on their experiences with electric aircraft. Engineers were asked similar questions regarding challenges and novel events and hazards pilots may encounter. This approach ensured that we obtained diverse perspectives on challenges, unexpected events and hazards associated with eVTOL aircraft operations. The questions asked of each group are detailed in Table 1.

Table 1.
Interview Questions

Questions	Pilots	Engineers
1. What are the biggest challenges of an electric powered-flight compared to internal combustion engine powered-flight?	X	X
2. What do you think conventionally trained pilots will struggle with the most with managing electrically-powered aircraft?	X	X
3. What unexpected events could change power availability to the aircraft?	X	X
4. What safety hazards could a pilot encounter related to the battery?	X	X
5. In your experience flying electrically powered aircraft, what are some of the unanticipated events or surprises you have experienced?	X	
6. What strategies helped you to overcome these events?	X	
7. What are the factors (e.g., temperature, wind) that impact battery power availability? How do these factors influence the decisions you make regarding the mission during the flight?	X	X

Participants

A total of 11 participants (1 female) volunteered to be interviewed during this study. This number includes six electric-aircraft pilots, four electric-aircraft engineers, and one electric-automotive engineer with experience in the aviation industry. Most of the participants were employed by relevant eVTOL manufacturers and regulatory agencies at the time of their interviews. To participate in the study, they had to have had direct experience with, or knowledge of, electric propulsion systems and electric aircraft. The pilots had various aircraft experience, comprised of fixed-wing, rotary-wing, and both civilian and military experience; four participants reported flight time in electrical aircraft (M = 16.58 hours, SD = 25.86), and one pilot had only eVTOL flight simulator experience (50 hours). The engineers had direct work experience with electric aircraft propulsion systems or electric cars. Participants were obtained via convenience sampling and were not compensated.

Procedures

Interviews took place either in-person or over Zoom and lasted between 1-2 hours. Specific interview protocols and procedures were followed with a pre-determined script. Prior to the interview questions, the interviewer presented the participant with an overview of the study, the objectives and rules, and the reasoning for conducting the study. Upon completion, participants were thanked for their time and asked if they could be contacted for any follow-up questions if need be. During each interview, data was recorded and scribed. Because the interviews were semi-structured, the script guided the protocol but did not limit the participant's answers to those questions nor the interviewer's questions to those on the script. For example, the interviewer asked additional or follow-up questions where applicable or probed the participant if required.

Data Analysis

The data was analyzed using a thematic analysis approach (Braun & Clarke, 2006) and included five steps. First, the data was thoroughly cleaned and reviewed by researchers to ensure all information was captured and correct. This was carried out by reviewing the videos and scribe notes from the interviews. At this time, the data was deidentified and then merged into two spreadsheets, which reflected either pilot responses or engineer responses. Then we followed a structured plan to group similar questions. For instance, questions that sought challenges were grouped and organized in a cell, whereas questions that sought to understand the safety hazards were also grouped. Third, using an inductive thematic analysis (Braun & Clarke, 2006), codes were generated from participants' responses, independently identified by three researchers with regular consensus meetings. During the consensus meetings, a fresh spreadsheet was used to add the identified codes, and they were tallied with frequency counts. To maintain traceability, participant IDs and code locations were added to the spreadsheet. During this process, the raw data was reviewed to ensure the context in which a response was captured was accounted for. Fourth, the codes were then transformed into overarching themes enriched with descriptive layers. For instance, an interpretation of each theme was added using quotes from participants or extant literature. Themes were specific to unexpected events, hazards, and potential human factors challenges and could be related to various attributes, including: (1) what are the biggest challenges of an electric-powered flight, (2) how conventionally trained pilots will struggle when managing electrically powered aircraft, (3) what unexpected events could occur, (4) what safety hazards pilots could encounter, and (5) the strategies that help overcome these events. Additionally, cell locations of the raw data were mapped to the themes to provide easy access to supporting information for the specific theme. Fifth, four researchers met and comprehensively reviewed the themes for consistency. The process involved reviewing each theme and its description and consolidating any themes with similar meanings. We then examined frequency counts for challenges, unexpected events, and safety hazard themes and identified a threshold. To be considered for inclusion as a theme, the theme had to have at least two frequency counts/data points (i.e., resulting from data collected from at least 2/10 participants).

FINDINGS

The following section provides the thematic findings from the qualitative approach specifically related to (1) Human Factors Challenges that eVTOL pilots may face, (2) Novel Unexpected Events that eVTOL pilots may face, and (3) Novel Safety Hazards associated with battery-operated aircraft.

Human Factors Challenges

Participants highlighted 10 themes related to novel human factors challenges associated with electric aircraft. These include novel flight profiles, inconsistent energy burn rate, display differences, the potential impact of internal battery temperatures, novel impacts of environmental factors, novel flight profiles, non-linear power draw, limited operational range, novel control schemes, loss of power due to aircraft characteristics and lack of pilot battery knowledge. These human factors challenges associated with eVTOL operations are expanded in the following section and highlighted in Table 2

Table 2.
Human Factors Challenges

Human Factors Challenges					
Name	*Description*	*Freq.*	*Name*	*Description*	*Freq.*
Non-linear Energy Consumption	Inconsistent energy burn rate and array of factors that impact it makes it a challenge to estimate range	6	**Non-Linear Power Draw Impacts**	Variation in power draw across flight phases can significantly impact energy available and range.	4
Novel eVTOL Pilot Interfaces	Novel displays, including novel information and presentation formats.	6	**Limited Operational Range**	The shorter range and flight times associated with eVTOL operations limits reserves and challenges fight planning.	3
Internal Battery Temp. Impact on Performance	Challenges associated with changes in internal battery temperature impacts to aircraft performance and health.	5	**Novel Control Schemes**	Control set up is new, and unfamiliar with potentially different applications	3
Novel Impacts of Environmental Factors	Outside pressures, temperatures, wind conditions and other environmental factors has novel impacts on aircraft performance and range.	4	**Novel Impacts of Loss of Power**	Aircraft mechanism design (Multirotor, tilt, winged)	3
Novel Flight Profile	eVTOL phases of flight time have different characteristics and power demands.	4	**Lack of Pilot Battery Operational Knowledge**	Different designs for electric vs traditional	2

Inconsistent Energy Burn Rate. Inconsistent energy burn rate, a challenge identified by six of the 10 participants, is the non-linear rate at which electric energy from batteries is depleted as the mission progresses (Sarkar et al., 2022). For instance, during critical stages of the flight (e.g., vertical flight and hover), energy is expended faster than during the cruise stage. Additionally, the rate at which energy from an older battery is depleted is significantly higher than that of a newer battery because battery energy is also dependent on many other variables, such as the battery's SOH, power draw, and temperature variations (EUROCAE, 2021). In contrast, fossil-fuel-powered aircraft experience a more consistent and predictable fuel burn rate, making it easier for pilots to estimate. For instance, even when fuel is burned at a faster rate, the same amount of energy can be obtained out of the fuel tank. The SOC of the batteries is an indicator that reflects the amount of energy available for use at any given time. As the mission progresses, the SOC gradually decreases due to energy depletion. However, the rate of depletion is non-linear, influenced by factors such as power draw and environmental conditions. The SOC is similar to the idea of remaining fuel in fossil-fuel-powered aircraft; however, not synonymous. Fossil fuel-powered aircraft typically exhibit more predictable fuel burn rates, enabling pilots to estimate remaining flight time or range with greater accuracy. In contrast, electric aircraft face challenges in accurately estimating SOC due to the non-linear discharge characteristics of batteries and the influence of various factors like flight conditions, flight phases, temperature, and battery age (Rivera-Barrera et al., 2017). Pilots must contend with the challenge of accurately estimating the performance of the batteries to predict remaining flight time or range based on the current SOC, which can be confounded by inconsistent energy burn rates. As identified by Thompson et al. (2022) and commented by one engineer, "With SOC, it is tricky. Even if we have the same value of SOC [for different phases of flight], the performance of the battery is different." The SOH of the batteries adds another layer of complexity. Battery degradation over time, influenced by factors like usage patterns and environmental stressors, affects the overall performance and reliability of the energy storage system. Older batteries with reduced SOH may exhibit accelerated energy depletion rates compared to newer ones, exacerbating the challenge for pilots of managing energy reserves during flight (Mitici et al., 2023). To support pilots in addressing challenges related to inconsistent energy burn rates, training programs can be developed to familiarize pilots with the unique characteristics of electric propulsion systems, emphasizing torque management techniques and energy-efficient flying practices. Additionally, implementing intuitive cockpit displays with real-time data on SOC levels, and battery health can enhance situational awareness and aid in making informed decisions during flight. As such, the main human factors challenge associated with the inconsistent burn rate of batteries is that it can lead to pilots making inaccurate assumptions or estimations about the available energy and associated range.

Novel eVTOL Pilot Interfaces. An additional challenge is that eVTOL pilot interfaces have different characteristics than those of conventional aircraft, including what information is presented, how it is presented, and how the pilot controls the aircraft. For instance, the changes in control schemes of eVTOL aircraft such as SVO consequently alter the pilot interface by implementing a single control inceptor, removing unnecessary controls. Further, the introduction of electric propulsion system information changes the information displayed on the panels to support effective pilot decision-making. Six of the ten participants highlighted adapting to the new display changes as a challenge. One pilot commented, "...the interface for the power system is complex..." Further, another participant echoed that the non-linearity of the battery energy burn rate exacerbates the challenge of making intuitive and meaningful displays for pilots. For electric parameters, some participants suggested using the current conventional fuel-based design strategies to simplify the complexity of electric displays, adding necessary alterations to ensure the pilots do not misinterpret battery energy behavior as synonymous with fuel behavior. This way, pilots' current mental models could be leveraged through consistency with conventional fuel displays. The lack of academic research makes it difficult to determine which display designs would be most effective in supporting pilot decision-making. Mitigating this issue starts with conducting empirical human factors research to determine what information is needed and how best to display it. Few studies have conceptualized best-practice methods for presenting battery information (e.g., Namukasa, et al., 2024), and further research is needed to ensure effective and efficient displays for fostering a safe environment during eVTOL operations.

Internal Battery Temperature Impact on Performance. Battery internal temperature is described as the flow of the thermal energy in the battery (Hatherall et al., 2024) and is considered a good indicator of SOH (Ma et al., 2018). The battery's internal temperature was identified as a critical concern by five of the 10 participants. The main challenge associated with battery temperature is the inconsistencies related to fluctuations in temperature and the corresponding power draw. For instance, at low temperatures, electric batteries are not able to perform as efficiently due to reduced ionic conductivity and an increase in charge-transfer resistance, leading to a loss of capacity (Ma et al., 2018). Similarly, higher temperatures can lead to heat generation within the batteries, which can accelerate thermal aging and loss of capacity (Handel, et al., 2014; Ma et al., 2018; Maher & Boumaiza, 2025). In layman's terms, as the battery fluctuates to the lower and higher ends of the temperature scale, the battery will provide less energy, which impacts the available range. This relationship is not a concern for conventional aircraft operations, and therefore, pilots will need to adjust their mental model to accommodate and understand the influence the battery temperature has on other critical components within the aircraft. One pilot commented, "...certainly the battery temperature is of concern because as you continue on your mission...higher power draws, it is going to heat up...". From a human factor's perspective, one mitigation strategy includes educating pilots on the impacts of the variations in battery temperature and how that affects the other variables that determine the safety and efficiency of a flight. Simulation scenarios can help create mental models among pilots, so they develop an understanding of the impacts of the variations of battery temperature on the overall mission success.

Novel Impacts of Environmental Factors. Environmental factors include variables that originat outside of the battery system and impact power availability and energy. Environmental factors such as wind, outside temperature, and outside pressure pose significant challenges to pilots in managing power availability and energy usage. In line with the findings from Thompson et al. (2022), wind was identified by four out of 10 participants as an external factor that can influence power availability and energy management during a flight. Variations in wind speed and direction introduce higher than normal energy consumption, requiring constant adjustments to maintain desired flight paths and conserve battery power. For example, as mentioned by one engineer, "If you're flying into a heavy headwind, you'll have to use higher power to keep the same flight profile that will discharge the battery faster." Headwinds increase power demand during forward flight, potentially leading to faster energy depletion and reduced range; while tailwinds can complicate descent and hover phases, inducing a challenge for the pilot to avoid unexpected drops in altitude and complicating predicting the available energy for the remainder of the flight (Pradeep et al., 2020). The outside temperature was identified by two of the 10 participants as an environmental condition that could affect the performance of the electric battery of an eVTOL aircraft. Fluctuations in the outside temperature affect battery efficiency and internal resistance, leading to non-linear energy depletion rates and preventing pilots from accurately estimating remaining energy reserves. Further, colder temperatures can increase the internal resistance, leading to heat generation which can reduce battery efficiency, leading to decreased energy storage and shorter range (Ma, 2018). These characteristics present a challenge for the pilot, as pilots must account for the effect of outside temperature on the battery's performance capability. As commented by one pilot, "...if it's hotter externally [outside], then the batteries get hot. They [the batteries] can't perform as well." This is a challenge because, for traditional fossil-fuel-powered aircraft, performance is not significantly affected by changes in the outside temperature. Finally, changes in outside pressure were identified by six out of 10 participants as a significant challenge that could alter aerodynamic forces, necessitating adjustments in power output to maintain stable flight characteristics, adding another challenge related to energy management for eVTOL pilots. Changes in pressure can also significantly affect the battery's health. For example, as the altitude increases, the outside pressure decreases, which can decrease the thermal runway starting time, leading to a higher risk of occurrence (Xie et al., 2020). To address the challenges posed by outside temperature, outside pressure, and winds in eVTOL operations, pilots can employ a combination of pre-flight planning, real-time monitoring, flight profile optimization, energy management strategies, weather avoidance, battery thermal management, and emergency contingencies. Wind and external temperature information can be integrated into the eVTOL pilot interface design to support real-time monitoring. In terms of training, existing procedures and checklists can be leveraged to train the eVTOL pilots to account for environmental factors during their pre-flight planning.

Novel Flight Profiles. Four out of 10 participants identified the novel flight profiles of eVTOL aircraft as a challenge, highlighting that eVTOL aircraft require different power demands from the battery and different handling characteristics for each phase of the flight. The unique flight profiles of eVTOL aircraft will result in phases of flight that are similar to rotorcraft, but significantly different from conventional aircraft. For instance, an eVTOL aircraft flight profile includes vertical take-off, transition, climb, cruise, descent, transition, and vertical landing (Melo et al., 2020; Sarkar et al., 2022). In contrast, wing-borne aircraft's flight profiles include taxi, take-off, cruise, descent, landing, and taxi, without flight mode transitions. Rotorcrafts are most similar to eVTOL aircraft as they experience identical flight profiles, including taxi, hover, transition, take-off, cruise, transition, and landing. One pilot commented, "...when you hover, power changes...during distributed propulsion where there are changes or differential between energy, that is something you would not see in a conventional aircraft..." Further, an electric engineer explained that the transition phase is one of the most critical phases of the flight, one of the most demanding, and the most prone to any failures. From an eVTOL perspective, the critical or complex phases are the most power-demanding stages exacerbating the situation. This creates a challenge for pilots who will have to manage the unique power demands of these novel phases of eVTOL flight. One way to support pilots in coping with this challenge is to develop technology that can predict accurate energy variations during the various flight profiles to alleviate the need for pilots to estimate remaining energy. Another solution is to develop and test standardized procedures for pilots' responses to various events throughout each phase of flight. Finally, training should be developed that highlights the novel flight profiles of eVTOL aircraft, and the implications for energy management, to increase awareness among pilots.

Non-Linear Power Draw. Power draw is described as the amount of energy being consumed or expended from a battery at a given point in time (Donateo & Cinar, 2022). Four participants identified the non-linearity of power draw to be a challenge to eVTOL pilots. Although it is similar to the concept of fuel flow, power draw is not synonymous with fuel availability, because it has a non-linear burn rate, and is inconsistent as it varies with a range of factors. Like the previously mentioned inconsistent energy burn rate, this concept poses a challenge to pilots who are trying to make decisions that impact their mission. It will be necessary for pilots to understand how drawing power at certain rates may have long-term effects on their mission performance. Because of this nonlinearity, it makes displaying meaningful information to the pilot difficult. Battery power draw significantly differs from fuel flow such that high fuel flow does not impact the usable quantity of the remaining fuel in a fuel tank. One engineer commented, "...batteries have a non-linear burn rate, if you use them at a high power, they will give you less total energy than when you use them lower…that drives the challenge of making displays usable…" One mitigation strategy is designing power draw cues onto the display that facilitate effective monitoring and accurate interpretation of power draw effects to the entire mission. Such cues include estimated or projected power draw for the mission profile or various stages of the flight (EASA, 2022 & FAA, 2023). The participants also echoed that as conventional pilots' experience with eVTOL aircraft increases, so will their knowledge of the systems.

Limited Operational Range. Another challenge, articulated by four of the 10 participants, was the eVTOL aircrafts' limited range, specifically how far (range) or how long (endurance) eVTOL aircraft can travel based on the limited energy available in current eVTOL batteries (Staack et al., 2021). When compared to fossil-fuel-powered aircraft, current eVTOL aircraft are significantly limited regarding how far they can travel or how long they can endure before being required to recharge the batteries. This is because fuel has a higher energy density than electric batteries. Increasing the energy provided by onboard batteries would require increasing the size of the batteries, which would increase their weight and consequently limit the payload capabilities of the aircraft (Viswanathan et al., 2022). A particular challenge associated with the limited range of eVTOL aircraft is that it limits the amount of reserve available and, therefore, the time pilots have to respond to an unexpected event or emergency. This can create what is referred to as range anxiety among pilots, which is a stressful psychological experience that emerges from the presence of limited range and time available to mitigate a prevailing or future range-related off-nominal event (Rauh et al., 2015). Another challenge is that limited range further limits the type of maneuvers that pilots can perform, for instance, a vertical landing versus a conventional landing, and consequently limits the options of where the aircraft can land. One pilot commented, "…we're severely energy limited…that's my biggest concern…" Another engineer echoed that designing and certifying aircraft with a meaningful range that is commercially viable was the biggest challenge. One way to mitigate range anxiety is to develop range awareness among pilots such that they precisely plan for energy ahead of time to mitigate the limited time in case of an emergency. Range awareness can be developed by designing intuitive displays that not only indicate to pilots how many nautical miles they can fly, but also what airports that can reach, what maneuvers they can perform (e.g., conventional vs. vertical landing), and accurate predictions of future changes in range based on the current energy consumption or power draw. The theme of limited operational range is in line with the findings from the GAMA report that states that without appropriate knowledge of the battery's performance, a pilot will struggle to understand the aircraft's capabilities, particularly in estimating the range and endurance. Additionally, range awareness can be developed through training and experience with eVTOL aircraft over time. Such training can target improving pilots' high-level cognitive performance (e.g., projection and decision-making) by exposing them to range-focused simulated scenarios to which they can devise practical solutions.

Novel Control Schemes. Three out of 10 participants identified novel control schemes as a challenge, specifically, the difficulty of flying eVTOL aircraft due to the complexity of the flight control systems that maintain the aerodynamic axes. Such flight control systems are especially critical when eVTOL aircraft transition between various flight regimes such as hover to forward flight or vice versa. For instance, tilt-rotor and vectored thrust eVTOL aircraft have been cited as very complex to fly as evidenced by the numerous accidents of the V-22 Osprey combat aircraft and the AV-8B Harrier (O'Brien, 1992). One engineer commented, "…to transition the hover to vertical flight, to fixed wing flight is very difficult…" One way that manufacturers are mitigating this challenge is through the implementation of automation capabilities in the design of eVTOL aircraft, such as simplified vehicle operations (SVO) and unified controls. The idea behind SVO and unified controls is that they decoupled the axis such that only a single control inceptor is needed to make the transition between different flight regimes, drastically reducing the task complexity. These automated control schemes are similar to fly-by-wire systems and manage the aerodynamic control of the eVTOL aircraft by blending all the axes during flight maneuvers making the transition between different flight regimes easier (Goyal et al., 2021; Osman et al., 2022). Such automated control schemes have been cited to reduce pilots' workload pilots' training requirements and streamline flight operations (GAMA, 2023; Kish et al., 2021; Wing et al., 2020). Moreover, due to the simplification of the flying task by automation, pilots of eVTOL aircraft may not be required to be experienced in operating conventional fixed-wing aircraft or rotary-wing aircraft prior to beginning their training to operate an eVTOL aircraft (Emerson et al., 2023; Lombaerts et al., 2022).

Novel Impacts of Loss of Power. This challenge was identified by three out of 10 participants and is described as a potential loss of aerodynamic lift due to loss of supply of power from the electric propulsion systems in various eVTOL aircraft configurations. The human factors challenge associated with such designs is that in case of unexpected events such as a complete power failure in non-winged eVTOL aircraft, the aircraft could stall, causing it to effectively fall from the sky, and lack of power to the hydraulics could make the eVTOL uncontrollable. One engineer commented, "…a winged aircraft can glide... and helicopters can autorotate... but some eVTOL aircraft do not have big enough rotors to autorotate…" Further, a pilot reiterated that the availability of redundant systems (e.g., two batteries on the Pipistrel) allows the aircraft to glide in case of failure of the high-voltage battery responsible for thrust and propulsion. Manufacturers are currently mitigating this challenge by incorporating redundant systems similar to those in fixed-wing aircraft, such that eVTOL aircraft would remain controllable and not just "fall from the sky" during the loss of propulsion power. Examples of redundant systems cited by participants include separate batteries or power sources for controls and incorporating mechanical controls that do not require power. Hence, in the event of a total loss of propulsion power, pilots can still have mechanical control of the aircraft to guide it to touchdown. Additionally, from a human factor's perspective, pilots should be exposed to training associated with handling critical emergencies such as total loss of electric propulsion power to remove the element of surprise and reduce the newness of the event.

Lack of Pilot Operational Battery Knowledge. One challenge associated with the novelty of electric propulsion in eVTOL aircraft lies in the fact that existing pilot training programs need to cover the requisite level of battery knowledge necessary for operating eVTOL aircraft. This is compounded by the scarcity of pilots experienced in electric aircraft, leading to a lack of operational knowledge about batteries in the current pilot pool. This challenge was identified by two out of 11 participants. One participant commented, "...pilots will struggle with the new technology in general if they have been working with oil and hydraulics and fuel for their career, and now they have to talk about electricity...". In conventional aviation, pilots have training regarding their electrical systems, but it is mostly related to their cockpit components and navigational equipment. The electrical components are not directly related to their range or related aircraft capabilities. EVTOL pilots must fully understand the non-linear nature of battery energy depletion, including how variations in battery charge levels impact flight performance, range, and operational limitations. Pilots will need to become familiar with many relationships between the different variables that impact energy availability. For instance, how thermals and power draw impact energy, and how the battery performs in the lower SOCs. These are issues that pilots did not have to consider in operating fossil-fuel-powered aircraft. One way to mitigate this challenge is to update the training requirements for pilots and require a more thorough understanding of the electric propulsion system, and the subcomponents within that system.

Novel Unexpected Events

Participants highlighted 2 types of unexpected events that are novel to eVTOL aircraft and have human factors implications. These include novel electrical system failure modes and impacts, and battery thermal runaway are expanded in the following section.

Novel Electrical System Failure Modes and Impacts. In an eVTOL aircraft context, a failure mode refers to the loss of performance, functionality, utility, or malfunction of an essential component of the aircraft for operations, resulting from the failure of one or more components. Three different novel electrical system failure modes were identified by participants as novel unexpected events that could take place during eVTOL flight operations. Six out of 10 participants identified electrical/power train system failure; five of 10 participants identified battery or battery module failure; and three of the 10 participants identified motor failure. In the event one of these failures occurs the most immediate consequences could include loss of power, which directly impacts the propulsion system, flight control systems, communication systems, movement capabilities, and avionics (Ramoul, 2022). For example, if a battery failure occurs during critical phases of flight, it may require pilots to make an emergency or forced landing. In fossil-fuel-powered aircraft, these failures are similar to, but not equivalent to, an engine failure or loss of fuel cells. However, if you lose an engine in fixed wing flight, there are procedures and checklists to guide you, and usually, the aircraft still maintains flight capabilities, such as gliding. In the case of some eVTOL aircraft configurations, failures of critical components that influence the power, propulsion, or electrical capabilities, create a more immediate and dangerous event, which can result in total loss of aircraft mobility. EVTOL aircraft utilize electric propulsion systems, which rely heavily on the distribution of electrical energy from the batteries, and there is an array of interconnected electrical components. Electrical system failures may have more severe outcomes in an eVTOL context due to the highly integrated electrical systems. One participant stated the importance of decision-making in the event of battery failure, stating "So, one thing to consider is [what does] the pilot need to know if there's battery pack failure? Will it impact their choices and their decision-making?". Another participant emphasizes the need to understand the aircraft's capabilities in the event of thrust loss, proclaiming "If a battery event causes loss of thrust capability, that obviously must be known immediately because it fundamentally changes what the aircraft can and cannot do". Due to the complex, novel, and proprietary nature of these propulsion systems, there are no publicly available protocols for what to do when faced with one of these unexpected events. Participants identified designing in redundancy to mitigate the risk associated with these events. One participant stated, "Batteries are capable of having catastrophic or complete battery failures [therefore, you are] going to have multiple battery packs for that purpose". Another participant stated, "So, these types of aircraft with distributed electric propulsion should have... more power installed at each rotor system than they actually require... to account for failure modes with any other adjacent rotors, such that it's able to actually compensate". From a human factor's perspective, including training such as simulator-based training will allow pilots to practice responding to these events in a realistic yet controlled environment. This type of training provides pilots with the muscle memory to carry out critical procedures and should be utilized heavily in preparation of novel eVTOL unexpected events, as is the case with current commercial pilot training. Further, adopting stress inoculation and situational awareness training into the eVTOL pilot training space could help prepare pilots for an event like this. Finally, implementing incremental, salient, transparent alerts as a supplement to the redundant systems can help keep the pilots aware of the aircraft's current state, while lowering the workload needed to identify issues within the components.

Battery Thermal Runaway. In an eVTOL aircraft context, thermal runaway refers to a potentially hazardous scenario in which the temperature of an element, typically a battery or electrical system component, increases rapidly and uncontrollably due to a self-sustaining chemical reaction or other thermal process (Borger, et al., 2019). In an eVTOL, this can lead to a cascade of events, including further temperature increases, release of gases, and possibly ignition, explosion, or fire (Harrison, et al., 2019). Seven out of 11 participants identified either thermal runaway or a fire as a notable battery-specific safety concern. A thermal runaway event poses a unique hazard when compared to fossil-fuel-powered aircraft (Sripad, et al, 2021). In eVTOL aircraft, the dense battery systems and enclosed battery compartments lead to higher risks of thermal runaway occurring, and because the propulsion and power systems rely on battery power and electrical distribution, thermal runaway in an eVTOL aircraft can lead to dire outcomes in a short amount of time (Harrison, et al., 2019). One participant stated, "Thermal runaway is probably like the biggest [battery safety concern] I could think of in terms of a hazardous, pretty violent, event". One participant even identified the risk to structural integrity if thermal runaway gets out of control within an eVTOL aircraft, stating, "Thermal runaway [can lead to] a fire on some of these aircraft because they're carbon... So, if the heat gets excessive, and it will, it will soften the structure". Participants stated that the best way to mitigate the possibility of a thermal runaway becoming out of control is to include salient alerts of the changes in temperature of the battery, and alerts if there are failures in designed mitigations, such as cooling suppressants. Training and educating pilots on how to mitigate the risk of a thermal runaway event during flight is also critical.

Novel Safety Hazards

Participants highlighted two safety hazards that are novel to eVTOL aircraft and have human factors implications. These include electrical shock and poisonous fumes.

Electric Shock. In an eVTOL aircraft context, shock hazards refer to the potential risks of electrical injury posed by the aircraft's electrical systems and components. Three out of 11 participants indicated that the risk of shock was a notable unique battery safety hazard for eVTOL aircraft. Shock hazards can stem from various components in the aircraft, such as exposed wiring, malfunctioning equipment, inadequate maintenance keeping, and high voltage power systems (Lobitz et al., 2024). Due to the nature of eVTOL and their housing of multiple electrical distribution systems, the risk of electrical shock occurring is increased in comparison to conventional aircraft, making it a hazard that eVTOL pilots must be prepared for, or trained to mitigate. One participant stated, "There are, for example, operational hazards on the ground pre and post-flight [that are unique battery-related safety concerns]... if somebody were to come up and touch [the aircraft]... while the aircraft is on, if there are defects in the wiring or other things, you could have fatal electrocution...[especially] in a post-crash situation where the airplane is damaged. Also, there is a risk of, you know, electrocution if certain things are damaged, and the airplane doesn't have features to protect against it". From a human factor's perspective, implementing comprehensive training programs for pilots and ground personnel that emphasize electrical safety procedures, electrical hazard recognition and protection, and emergency response protocols would help mitigate the risk of electrical shock or electrocution.

Poisonous fumes. In the context of eVTOL operation, poisonous fumes refer to dangerous gases, vapors, or particulate matter that may be dispersed during an event related to the aircraft, such as a thermal runaway event, battery fire, or malfunctions within the aircraft's electrical systems (Chaturvedi, 2010). Three out of 11 participants indicated that some sort of fumes or toxic gas would be a notable and unique battery-related safety hazards. These fumes can be toxic and therefore pose a significant health risk if inhaled or exposed for an extended length of time. For pilots, even short-term exposure to substantial levels of hazardous or poisonous gases can cause symptoms that may impair the pilot's ability to make decisions, such as dizziness, nausea, headache, and impaired cognitive function (FAA, 2023). During eVTOL operations, there is a greater threat to pilots and occupants than compared to fossil-fuel-powered aircraft operations. This is due to proximity to onboard electrical equipment and components, making exposure to gasses more frequent in the event of a hazardous malfunction with the electrical systems. One participant stated "[a battery-related safety concern could be] off-gassing (toxic fumes) that can occur, right? And so, in some cases, the off gassing is...definitely going to be destructive to the battery system...[and] can be flammable as well". Participants suggested that to minimize the risk of exposure to poisonous fumes, eVTOL aircraft manufacturers need to design the electrical components with recurrent safeguards to prevent scenarios in which poisonous fumes result. From a human factors perspective, proactively implementing early warning systems and automatic shutdown mechanisms to detect and respond to specific critical events before they escalate is essential. It may also be important to train for specific fume detection and response, including identification of symptoms associated with inhaling poisonous fumes before they impair pilot decision-making.

Table 3.
Novel Unexpected Events and Safety Hazards

Novel Safety Events and Hazards					
Novel Unexpected Events			*Novel Safety Hazards*		
Name	*Description*	*Freq.*	*Name*	*Description*	*Freq.*
Electrical System Failures	E.g., energy disruption due to wiring issues	6	**Shock Hazards**	Can be fatal events that can result from getting in contact with live current from batteries or current/wires	3
Battery Failure & Battery/Module Pack Failure	Malfunctions or degradation in the battery systems	5	**Poisonous Fumes**	Poisonous which develops from overcharging, or overuse	3
Motor & Thrust Capabilities Failure	Failure of the motor/engine, loss of thrust capabilities	3			
Thermal Runaway/Fire	Incident during which there is an uncontrollable increase in the battery temperature that can lead to the destruction of the battery, or even fire hazards	7			

CONCLUSION AND FUTURE RESEARCH

This paper describes a research effort that employed qualitative approaches to identify novel human factors, challenges, unexpected events, and safety hazards associated with the operation of eVTOL aircraft. Eleven novel human factors challenges, 2 novel unexpected events, and 2 novel safety hazards were identified, each potentially requiring either design mitigations, specialized training, targeted procedural development, or a combination of these to ensure safe and effective eVTOL operation. These results should be interpreted with caution as there were several limitations associated with this research. First, there were limited participants, with only 10 subject matter experts (SMEs) interviewed. With such a small sample size, there may have been additional challenges, unexpected events, and safety hazards that were not identified by this group of SMEs. Further, the interviews were limited in time, limiting the depth in which these different items could be discussed. Third, eVTOL vehicles and associated procedures and regulations are rapidly evolving and some of these challenges, events and hazards may soon be mitigated, making the discussion herein obsolete. However, this paper provides a stepping stone to start the discussion regarding the importance of addressing the distinct human factors, challenges, and associated safety considerations of eVTOL aircraft to facilitate their safe and effective integration into the aviation landscape. Future research should assess best-practice methods for training practices and operational procedures designed to support pilots in coping with these novel challenges, events, and hazards that they may encounter in future UAM operations.

REFERENCES

1. Mendolia, A. S., Lascara, B., & Sanders, T. (2022). Exploring General Aviation Pilot Perspectives on Proximate Urban Air Mobility Operations. Journal of Air Transportation, 30(3), 102–112. https://doi.org/10.2514/1.D0290
2. Hancock, P.A., Cruit, J., Kochan, J.A., Gruber, M., Diaz, Y., Kaplan, A., Pruchnicki, S., *et al.* (2022), *Relevant Research Assessment Concerning Pilot Response to Unexpected Events. Task 2: Relevant Research Assessment*, doi: 10.21949/1528625
3. Federal Aviation Administration (2023) *Urban Air Mobility (UAM) Version 2.0 Concept of Operations.* Report. https://www.faa.gov/sites/faa.gov/files/Urban%20Air%20Mobility%20%28UAM%29%20Concept%20of%20Operations%202.0_1.pdf
4. Jiechao, Z., Yaolong, L., & Yao, Z. (2024). Overall eVTOL Aircraft Design for Advanced Air Mobility. Green Energy and Intelligent Transportation, 100150. https://doi.org/10.1016/j.geits.2024.100150
5. Franciscone, B.G. and Fernandes, E. (2023), "Challenges to the Operational Safety and Security of eVTOL Aircraft in Metropolitan Regions: A Literature Review", *Journal of Airline Operations and Aviation Management*, Vol. 2 No. 1, pp. 45–56, doi: 10.56801/jaoam.v2i1.2.
6. Federal Aviation Administration (2015) *Transition to unfamiliar aircraft*. 90-109A. https://www.faa.gov/documentlibrary/media/advisory_circular/ac_90-109a.pdf.
7. Federal Aviation Administration (2022) 'Emergency procedures,' in *Airplane Flying Handbook*. Federal Aviation Administration. https://www.faa.gov/sites/faa.gov/files/regulations_policies/handbooks_manuals/aviation/airplane_handbook/19_afh_ch18.pdf
8. Kinney, L. and O'Hare, D. (2020), "Responding to an Unexpected In-Flight Event: Physiological Arousal, Information Processing, and Performance", *Human Factors*, SAGE Publications Inc, Vol. 62 No. 5, pp. 737–750, doi: 10.1177/0018720819854830.
9. Federal Aviation Administration (2021) *Pilot's guide to a preflight briefing*. https://www.faa.gov/documentLibrary/media/Advisory_Circular/AC_91-92.pdf.
10. International Civil Aviation Organization (2014) *Hazard Definition and Usage, International Civil Aviation Organization.* https://www.icao.int/SAM/Documents/2017-SSP-BOL/CICTT%20Hazard%20Taxonomy.pdf
11. Thompson, E., Taye, A., Guo, W., Wei, P., Quinones-Grueiro, M., Ahmed, I., Biswas, G., *et al.* (2022). "A Survey of eVTOL Aircraft and AAM Operation Hazards".
12. Carmody, K., Chauhan, B., Namukasa, M., Sharma, V., Berkel, G., Adorno, Y., & Carroll, M. (n.d.). "Human Factors Training Implications for Urban Air Mobility Operations". *Interservice/Industry Training, Simulation, and Education Conference.*
13. Fard, M.T., He, J., Huang, H. and Cao, Y. (2022), "Aircraft Distributed Electric Propulsion Technologies—A Review", *IEEE Transactions on Transportation Electrification*, presented at the IEEE Transactions on Transportation Electrification, Vol. 8 No. 4, pp. 4067–4090, doi: 10.1109/TTE.2022.3197332.

14. Viswanathan, V. *et al.* (2022) 'The challenges and opportunities of battery-powered flight,' *Nature*, 601(7894), pp. 519–525. https://doi.org/10.1038/s41586-021-04139-1.
15. Adu-Gyamfi, B.A. and Good, C. (2022) 'Electric aviation: A review of concepts and enabling technologies,' *Transportation Engineering*, 9, p. 100134. https://doi.org/10.1016/j.treng.2022.100134.
16. Bauranov, A. and Rakas, J. (2019) *Urban air mobility and manned eVTOLs: safety implications*. 38th edn. IEEE. https://doi.org/10.1109/DASC43569.2019.9081685
17. Casner, S.M., Geven, R.W. and Williams, K.T. (2013), "The Effectiveness of Airline Pilot Training for Abnormal Events", *Human Factors*, SAGE Publications Inc, Vol. 55 No. 3, pp. 477–485, doi: 10.1177/0018720812466893.
18. Landman, A., Groen, E.L., van Paassen, M.M. (René), Bronkhorst, A.W. and Mulder, M. (2017), "Dealing With Unexpected Events on the Flight Deck: A Conceptual Model of Startle and Surprise", *Human Factors*, SAGE Publications Inc, Vol. 59 No. 8, pp. 1161–1172, doi: 10.1177/0018720817723428
19. Rivera, J., Talone, A.B., Boesser, C.T., Jentsch, F. and Yeh, M. (2014), "Startle and Surprise on the Flight Deck: Similarities, Differences, and Prevalence", *Proceedings of the Human Factors and Ergonomics Society Annual Meeting*, Vol. 58 No. 1, pp. 1047–1051, doi: 10.1177/1541931214581219.
20. Hendricks, C., Williard, N., Mathew, S. and Pecht, M. (2015), "A failure modes, mechanisms, and effects analysis (FMMEA) of lithium-ion batteries", *Journal of Power Sources*, Vol. 297, pp. 113–120, doi: 10.1016/j.jpowsour.2015.07.100.
21. Sripad, S., Bills, A. and Viswanathan, V. (2021), "A review of safety considerations for batteries in aircraft with electric propulsion", *MRS Bulletin*, Vol. 46 No. 5, pp. 435–442, doi: 10.1557/s43577-021-00097-1.
22. Karthik, A., Chiniwar, D.S., Das, M., M, P.P., Prabhu, P., Mulimani, P.A., Samanth, K., *et al.* (2021), "Electric Propulsion for Fixed Wing Aircrafts – A Review on Classifications, Designs, and Challenges", *Engineered Science*, Engineered Science Publisher, Vol. Volume 16 (December 2021) No. 0, pp. 129–145
23. GAMA (2023) *Managing range and endurance of Battery-Electric aircraft, General Aviation Manufacturer Association*. General Aviation Manufacturers Association. https://gama.aero/wp-content/uploads/Managing-Range-and-Endurance-of-Battery-Electric-Aircraft_v1-1.pdf
24. Namukasa, M., Chauhan, B., Oyola, Y. A., Carmody, K., & Carroll, D. M. (2024). Identifying the Requirements of eVTOL Battery Information Displays. In AIAA AVIATION FORUM AND ASCEND 2024 (p. 4213).
25. Braun, V. and Clarke, V. (2006) 'Using thematic analysis in psychology,' *Qualitative Research in Psychology*, 3(2), pp. 77–101. https://doi.org/10.1191/1478088706qp063oa.
26. Sarkar, M. *et al.* (2022) *A comprehensive EVTOL Performance Evaluation Framework in urban air mobility*. Intelligent Systems Conference (IntelliSys) 2022. https://ntrs.nasa.gov/api/citations/20220013717/downloads/intellisys2022_conf.pdf.
27. EUROCAE (2021) *Guidance on Determination of Accessible Energy in Battery Systems for EVTOL Applications - EUROCAE*. https://www.eurocae.net/news/posts/2021/may/ed-289-guidance-on-determination-of-accessible-energy-in-battery-systems-for-evtol-applications/
28. Rivera-Barrera, J.P., Muñoz-Galeano, N. and Sarmiento-Maldonado, H.O. (2017), "SoC Estimation for Lithium-ion Batteries: Review and Future Challenges", *Electronics*, Multidisciplinary Digital Publishing Institute, Vol. 6 No. 4, p. 102, doi: 10.3390/electronics6040102.
29. Mitici, M., Hennink, B., Pavel, M. and Dong, J. (2023), "Prognostics for Lithium-ion batteries for electric Vertical Take-off and Landing aircraft using data-driven machine learning", *Energy and AI*, Vol. 12, p. 100233, doi: 10.1016/j.egyai.2023.100233.
30. Hatherall, O., Barai, A., Niri, M. F., Wang, Z., & Marco, J. (2024). Novel battery power capability assessment for improved eVTOL aircraft landing. Applied Energy, 361, 122848. https://doi.org/10.1016/j.apenergy.2024.122848
31. Ma, S., Jiang, M., Tao, P., Song, C., Wu, J., Wang, J., Deng, T., *et al.* (2018), "Temperature effect and thermal impact in lithium-ion batteries: A review", *Progress in Natural Science: Materials International*, Vol. 28 No. 6, pp. 653–666, doi: 10.1016/j.pnsc.2018.11.002.
32. Handel, P., Fauler, G., Kapper, K., Schmuck, M., Stangl, C., Fischer, R., Uhlig, F., *et al.* (2014), "Thermal aging of electrolytes used in lithium-ion batteries – An investigation of the impact of protic impurities and different housing materials", *Journal of Power Sources*, Vol. 267, pp. 255–259, doi: 10.1016/j.jpowsour.2014.05.080.
33. Maher, K., & Boumaiza, A. (2025). Thermal challenges in lithium-ion battery technology: Investigating performance and thermal stability. Journal of Energy Storage, 111, 115396. https://doi.org/10.1016/j.est.2025.115396
34. Pradeep, P., Chatterji, G.B., Sridhar, B., Edholm, K.-M., Lauderdale, T.A., Sheth, K., Lai, C.F., *et al.* (n.d.). "Wind-Optimal Trajectories for Multirotor eVTOL Aircraft on UAM Missions".
35. Xie, S. *et al.* (2020) 'Influence of cycling aging and ambient pressure on the thermal safety features of lithium-ion battery,' *Journal of Power Sources*, 448, p. 227425. https://doi.org/10.1016/j.jpowsour.2019.227425.
36. Melo, S.P. *et al.* (2020) 'Life Cycle Engineering of future aircraft systems: the case of eVTOL vehicles,' *Procedia CIRP*, 90, pp. 297–302. https://doi.org/10.1016/j.procir.2020.01.060.
37. Donateo, T. and Çinar, H. (2022) 'Conceptual design and sizing optimization based on minimum energy consumption of lift-cruise type eVTOL aircraft powered by battery and fuel cell for urban air mobility,' *Journal of Physics. Conference Series*, 2385(1), p. 012072. https://doi.org/10.1088/1742-6596/2385/1/012072.
38. EASA (2023). *Special condition for VTOL and means of compliance.* https://www.easa.europa.eu/en/document-library/product-certification-consultations/special-condition-vtol
39. Staack, I., Sobron, A. and Krus, P. (2021) 'The potential of full-electric aircraft for civil transportation: from the Breguet range equation to operational aspects,' *CEAS Aeronautical Journal*, 12(4), pp. 803–819. https://doi.org/10.1007/s13272-021-00530-w.
40. Rauh, N., Franke, T. and Krems, J.F. (2015), "Understanding the impact of electric vehicle driving experience on range anxiety", *Human Factors*, Vol. 57 No. 1, pp. 177–187, doi: 10.1177/0018720814546372.
41. O'Brien, M.A., n.d. (1992). *The V-22 Osprey: A Case Analysis.* Thesis. Naval Postgraduate School, CA.
42. Goyal, R. *et al.* (2021) 'Advanced Air Mobility: Demand analysis and market potential of the airport shuttle and air taxi markets,' *Sustainability*, 13(13), p. 7421. https://doi.org/10.3390/su13137421.
43. Osman, M.C. *et al.* (2022) *Perspectives for the Future: Considerations for UAM Aircraft information requirements.* Interservice/Industry Training, Simulation, & Education Conference (I/ITSEC).
44. Kish, B. *et al.* (eds) (2022) *Flight Test Comparison of three air vehicles flying urban air mobility mission trajectories* [2022 IEEE Aerospace Conference (AERO)]. https://10.1109/AERO53065.2022.9843555
45. Wing, D. J., Chancey, E. T., Politowicz, M. S., & Ballin, M. G. (2020, June 15). Achieving Resilient In-Flight Performance for Advanced Air Mobility through Simplified Vehicle Operations. AIAA AVIATION 2020 FORUM. AIAA AVIATION 2020 FORUM, Virtual event. https://doi.org/10.2514/6.2020-2915
46. Emerson, S.N., Rizzardo, C., Waggenspack, L., Osman, M.C., Halverson, K.C., Ellis, S. and Haley, D. (2023), *eVTOL Flight: Forecasting Future Training Requirements in an Emerging Aviation Market*, SAE Technical Paper No. 2023-01–1009, SAE International, Warrendale, PA, doi: 10.4271/2023-01-1009.
47. Lombaerts, T., Shish, K. and Kaneshige, J. (2022) 'Trim envelope calculations for a tiltrotor in forward flight, hover and transitions,' *IFAC-PapersOnLine*, 55(6), pp. 611–616. https://doi.org/10.1016/j.ifacol.2022.07.195.
48. Ramoul, J. (2022). A Systematic Approach to Critical Electrical Fault Mitigation Strategies in an Electric Vertical Take-Off and Landing (EVTOL) Electrical Propulsion Unit [Thesis]. https://macsphere.mcmaster.ca/handle/11375/28198
49. Börger, A., Mertens, J., & Wenzl, H. (2019). Thermal runaway and thermal runaway propagation in batteries: What do we talk about? Journal of Energy Storage, 24, 100649. https://doi.org/10.1016/j.est.2019.01.012

50. Harrison, J., Charles, D., Zenker, J., & Frank, E. (2019). Using Multi-physics System Simulation to Predict Battery Pack Thermal Performance and Risk of Thermal Runaway During eVTOL Aircraft Operations. 2019 AIAA/IEEE Electric Aircraft Technologies Symposium (EATS), 1–13. https://doi.org/10.2514/6.2019-4406
51. Lobitz, L., Hahn, A., Vogt, D., Luplow, T., Michalowski, P., Heimbs, S., & Garnweitner, G. (n.d.). Conceptual Challenges for Crashworthy Battery-Electric Commercial Aircraft – A Review. In AIAA SCITECH 2024 Forum. American Institute of Aeronautics and Astronautics. https://doi.org/10.2514/6.2024-2156
52. Chaturvedi, A. K. (2010). Aviation Combustion Toxicology: An Overview*. Journal of Analytical Toxicology, 34(1), 1–16. https://doi.org/10.1093/jat/34.1.1
53. FAA. (2023). Pilot's Handbook of Aeronautical Knowledge | Federal Aviation Administration.

AUTHOR INDEX